Studies in Computational I

Volume 934

Series Editor

Janusz Kacprzyk, Polish Academy of Sciences, Warsaw, Poland

The series "Studies in Computational Intelligence" (SCI) publishes new developments and advances in the various areas of computational intelligence—quickly and with a high quality. The intent is to cover the theory, applications, and design methods of computational intelligence, as embedded in the fields of engineering, computer science, physics and life sciences, as well as the methodologies behind them. The series contains monographs, lecture notes and edited volumes in computational intelligence spanning the areas of neural networks, connectionist systems, genetic algorithms, evolutionary computation, artificial intelligence, cellular automata, self-organizing systems, soft computing, fuzzy systems, and hybrid intelligent systems. Of particular value to both the contributors and the readership are the short publication timeframe and the world-wide distribution, which enable both wide and rapid dissemination of research output.

Indexed by SCOPUS, DBLP, WTI Frankfurt eG, zbMATH, SCImago.

All books published in the series are submitted for consideration in Web of Science.

More information about this series at http://www.springer.com/series/7092

Krassimir T. Atanassov

Editor

Research in Computer Science in the Bulgarian Academy of Sciences

 Springer

Editor
Krassimir T. Atanassov
Department of Bioinformatics and Mathematical
Modelling, Institute of Biophysics and
Biomedical Engineering
Bulgarian Academy of Sciences
Sofia, Bulgaria

ISSN 1860-949X   ISSN 1860-9503 (electronic)
Studies in Computational Intelligence
ISBN 978-3-030-72286-9   ISBN 978-3-030-72284-5 (eBook)
https://doi.org/10.1007/978-3-030-72284-5

This Springer imprint is published by the registered company Springer Nature Switzerland AG
The registered company address is: Gewerbestrasse 11, 6330 Cham, Switzerland

The book is dedicated to the 150th anniversary of the establishment of the Bulgarian Academy of Sciences—the oldest institution in Bulgaria.

Preface

The present collection of papers is devoted to the emergence and development in Bulgaria of some of the areas of Informatics, including the Artificial Intelligence (AI) as a whole and as separate AI areas and tools as pattern recognition, acoustic analysis of voices, metaheuristic algorithms, artificial neural networks, multi-criteria decision making models, generalized nets, intuitionistic fuzziness, intercriteria analysis; the Computing and applications in different areas as computer algebra, coding theory, methods for modelling of telecommunication systems, time series analysis and machine learning, research for defence and security, in silico studies of biologically active molecules, approaches for forest fire spread prediction.

The papers are prepared by specialists from the Academy, some of whom are among the founders of these scientific and application areas in Bulgaria.

In general, the papers are ordered chronologically according to the timeline of events, described in them.

Sofia, Bulgaria   Krassimir T. Atanassov

Contents

In Short About Bulgarian Academy of Sciences

Julian Revalski and Krassimir Atanassov

Abstract Short notes concerning the history of computer science and informatics in Bulgarian Academy of Sciences are given.

Keywords Bulgarian Academy of Sciences · Computer Science · Informatics · History

Bulgarian Academy of Sciences is the oldest national institution in contemporary Bulgaria, being in existence for already 150 years now, despite the historical challenges facing the country.

The Academy emerged in 1869 in the Romanian town of Braila as Bulgarian Learned Society, about a decade before the Bulgarian state itself took its independence as a result of long battles of liberation.

After the liberation was fact in 1878, the headquarters of the Society was moved to the capital Sofia, where it remains to the present day (Figs. 1, 2, 3, 4, 5 and 6).

In 1911 the Bulgarian Learned Society became Bulgarian Academy of Sciences after the adoption of a special law.

In its 150 years history the Bulgarian Academy of Sciences has achieved numerous contributions to the development of European and global science and culture.

Bulgarian Academy of Sciences was the birthplace of the molecular-kinetic theory of equilibrium and crystal growth, proposed by Rostislaw Kaischew (1908–2002) and Ivan Stranski (1897–1979). Other achievements were the counter-pressure casting method of Angel Balevski (1910–1997) and Ivan Dimov (1926–1984), as well as the photo-electret state of matter discovered by Georgi Nadzhakov (1896–1981) that enabled the process of photocopying. A new scientific field—thracology—that studies the history, culture and religion of the ancient inhabitants of Bulgaria and the Balkan Peninsula was developed in the Academy by Alexander Fol (1933–2006). The botanists Nikolai Stojanov (1883–1968) and Boris Stefanoff (1894–1979) were

J. Revalski · K. Atanassov (✉)
Bulgarian Academy of Sciences, Sofia, Bulgaria
e-mail: krat@bas.bg; k.t.atanassov@gmail.com

K. T. Atanassov (ed.), *Research in Computer Science in the Bulgarian Academy of Sciences*, Studies in Computational Intelligence 934,
https://doi.org/10.1007/978-3-030-72284-5_1

1

Fig. 1 The building of Bulgarian Academy of Sciences at the end of 19th century

Fig. 2 The building of Bulgarian Academy of Sciences at the beginning of 20th century

Fig. 3 The building of Bulgarian Academy of Sciences at 1930s

Fig. 4 The building of Bulgarian Academy of Sciences today

Fig. 5 The building of Bulgarian Academy of Sciences—central entrance

Fig. 6 The building of Bulgarian Academy of Sciences—eastern entrance

among the pioneers in studying the flora of Balkan Peninsula. Blagovest Sendov (1932–2020) and Roumen Tsanev (1922–2007) developed mathematical models of cellular proliferation, differentiation and carcinogenesis in multicellular organisms.

The first nuclear reactor and the largest telescope of South-Eastern Europe, cosmic laboratory and geodesic observatory were put in operation in the Bulgarian Academy of Sciences. Many research units in the Academy have been working actively on space programs, before 1990 related mostly to the programs of the USSR. Thus, Bulgaria was the sixth country in the world to send a cosmonaut to space in 1979.

One of the first electronic calculators in the world was constructed in the Academy followed by a series of personal computers in the 1980s. Due to the studies on philosophy of physics, the 10th volume of the American Encyclopedia of Philosophy was dedicated to a European Scientist—and that was the Bulgarian Academician Azarya Polikarov (1921–2000).

In 1987, in Bulgaria was organized the First International Olympiad in Informatics for school students by Acad. Petar Kenderov.

Foreign members of the Bulgarian Academy of Science were the Nobel Laureates of Physics Frédéric Joliot-Curie (1900–1958, France), Nikolay Basov (1922–2001, Russia), the physicist Alexander Prokhorov (1916–2002, Russia) and Zhores Alferov (1930–2019, Russia), the biochemist John Kendrew (1917–1997, UK).

Among the foreign members of the Academy were also some of the world-famous pioneering specialists in the field of mathematics, informatics and computer technology as John Vincent Atanasoff (1903–1995, USA), Lotfi Zadeh (1921–2017, USA) and Ludvig Faddeev (1934–2020, Russia).

Actually foreign members of the Bulgarian Academy of Science are also a number of world famous scientists including Nobel prize laureates.

Since its very creation, Bulgarian Academy of Sciences has served to the society and has actively worked to ensure the prosperity of Bulgaria.

Computer Science Research—Unity in Variety

Avram Eskenazi⬤, Neli Maneva⬤, and Krassimira Ivanova⬤

Abstract The history of the computer science department at the Institute of Mathematics and Informatics of Bulgarian Academy of Sciences since 1963 is briefly described. Starting with small groups of enthusiastic people initiating the new for Bulgaria research in the field of Informatics, nowadays the department can report some significant achievements. The paper presents some of them in three areas—software quality evaluation and management, data mining and semantic analysis, information systems and communities—which mark some key areas of research during the years.

Keywords Quality models · Quality assessment · Software quality assurance · Data mining · Semantic analysis · Emotion recognition · Information systems · Context-awareness · Crowdsourcing

1 Introduction

The "Software Engineering and Information Systems" Department of the Institute of Mathematics and Informatics at the Bulgarian Academy of Sciences (IMI-BAS) was established in October 2014 as a union between two departments of IMI, which are the basis of the development of informatics in Bulgaria—"Software Engineering" (SE) and "Information Systems" (IS).

A. Eskenazi · N. Maneva · K. Ivanova (✉)
Institute of Mathematics and Informatics at the Bulgarian Academy of Sciences, Block 8, Acad. G. Bonchev Str., 1113 Sofia, Bulgaria
e-mail: kivanova@math.bas.bg

A. Eskenazi
e-mail: eskenazi@math.bas.bg

N. Maneva
e-mail: neman@math.bas.bg

K. T. Atanassov (ed.), *Research in Computer Science in the Bulgarian Academy of Sciences*, Studies in Computational Intelligence 934, https://doi.org/10.1007/978-3-030-72284-5_2

The "Software Engineering" department originates from the "Theory of Finite Automata Group", established in 1963 as part of the then called Institute of Mathematics with Computing Center. In 1966, the group increased and was renamed to "Theoretical Problems of Cybernetics". In 1970, after the Institute itself was renamed to Institute of Mathematics and Mechanics and was included in a larger research and educational unit, a new department was created with the name "Foundations of Cybernetics and Control Theory". Fifteen years later, the department "Automation of Software Production" emerged, which later received the name "Software Engineering". For years, the department has been headed by Prof. Dr. Avram Eskenazi, and since 2013 by Prof. Dr. Neli Maneva.

The "Information Systems" department originates from the Computer-Aided Programming group formed in July 1963 within the Numerical Methods Department at the Institute of Mathematics of BAS. In November, 1968 the Department "Computer-Aided Programming" was established from this group. Later, in February 1971, it was renamed to "Computer Science", in 1995 was renamed to "Information Research", and in 2009 was renamed as "Information Systems" Department. Since its establishing for many years the Department was leaded by Prof. Dr. Peter Barnev, and from 2006 by Prof. DSc Peter Stanchev. Since March 2011 the Laboratory for Digitization of Scientific and Cultural Heritage (former Digital Humanities Department) became part of the Department. Starting in 1976, for 30 years, the Department had been the main organizer of the annual International Programming School which during the years has grown up in the International Conference on Information and Communication Technologies and Programming—a key informatics event held in Bulgaria during these years.

When in 2014 the scientific council of IMI-BAS decided to merge the SE and IS departments into one, our community was skeptical about the possibility that two groups of specialists in different fields could be effective in their research work. Currently, almost 5 years after the merge, we can say that it was successful because everyone embraced the idea of "unity in variety". The strength of the department comes from the joint work of people with different knowledge, research interests, experience, skills and attitudes. In a creative collaborative environment the members of the department manage to find topics for investigation, which are of common interest. They started to share methods and tools providing non-trivial problem solutions.

The department's research is currently centered on the following main topics: evaluation of software quality, quality assurance and benchmarking, context-aware systems, data mining, knowledge discovery, semantic analysis, digital libraries, interoperability, digital humanities, citizen science, crowdsourcing, virtual communities.

In the next sections we will present some results of the research, which prove the validity of the principle "unity in variety" in the field of computer science.

2 Software Quality Evaluation and Management

At the Institute of Mathematics and Informatics at the Bulgarian Academy of Sciences and especially in the software engineering department, the questions of evaluation and quality management have been developed over the last 35 years. Software technologies are a constructive discipline and, ultimately, the goal is to satisfy the end user. In fact, this is the definition of quality according to today's understanding—"the degree to which a set of inherent characteristics fulfills a set of requirements" [1]. This definition can also be applied to the quality of the software and serve as a good basis for constructing a clear and concrete view, primarily with regard to the pragmatic tasks of measuring, evaluating and managing it. From a methodological point of view, the natural step leading up to this goal is the creation of a software quality model and a relevant assessment method. This is actually one of the directions in which positive research results have been achieved.

2.1 Quality Assessment by Means of Classification Methods

Quality assessment is based on models. The first one appeared in 1973, it is not very well structured (it resembles a hierarchy without being exactly such a structure) but played a very positive role. Practically applicable models are published in the mid-1980s [2, 3]. Some members of the Software engineering department took part in the creation of [4]. Later on some standards appeared [5]. These models and the standards are strictly hierarchical, where quality is considered as a tree structure. At the first level are the characteristics or factors. Each of them is defined as user-oriented property that represent a certain aspect of the quality of the software from the user's perspective. Most often, they include flexibility, correctness, reliability, usability, (performance) efficiency, functionality, maintainability, portability. On the second level are the sub-characteristics or criteria, which are software-oriented properties that represent features of the software product. Each characteristic/factor is determined by several sub-characteristics/criteria. For example, the (performance) efficiency is determined by time behavior, resource utilization, and capacity. The last third level contains the attributes or assessment elements, which can be verified or measured in the software product.

This structure follows immediately the evaluation procedure, i.e. providing for a quantitative assessment of quality. Only the attributes are directly evaluated. The values of all attributes are determined by experts on the base of some general guidelines that have been prepared in advance. The values of higher-level elements are calculated as weighted sums of the values of the lower-level elements. On every element at each level is assigned a fixed weight in the interval [0, 1]. Usually, the sum of the weights of the elements defining a characteristic from a higher level is 1 to achieve normalization. The final result is a number in the interval [0, 1], which is the quality measure of the particular software product obtained by this method.

The described quality model and the associated evaluation procedure have important advantages—the structural simplicity of the model, simple assessment procedure, which can be easily automated, uniqueness and clarity of the final result. Certainly, this hierarchical model has also drawbacks—the evaluation procedure contains too many elements of subjectivity (the guidelines for determining the values of the attributes, the evaluation by the experts itself, the determination of the weights of the elements for the different product types), a huge labor intensity due to the already mentioned need from the multilateral and varied work of experts on weighting, assessment guidelines, as well as individual work to determine the values of the attributes (they are approximately 250 in [4]) for each particular software product.

In [6] and [7], another model of the quality of software products is proposed. It tries to overcome some of the disadvantages of hierarchical models, above all subjectivity, and labor-intensiveness. Naturally, since every improvement has its price, in this case it is a certain loss of precision. The latter, however, as will be seen, is not always vital.

The basic idea of this method stems from the understanding that in evaluating the quality of a software product it is not so important to know an exact number reflecting this quality but rather to gain clarity about the position of the evaluated software product among several popular same type products of well-known quality. It is based on classification methods, using [8] as a starting point.

Without going into detail, the mathematical basis of the proposed model and method consists of the following. Let's look at some type of software (text editors, payroll system, etc.). For this type a set of n attributes are outlined. Each attribute can be assigned a value of 1 if the particular program product has this characteristic, 0 if it does not have it, and x if there is no information about whether it possesses it. From the other side, there are several products of this type with determined values of attributes, which on the base of the experience of experts and end-users are classified into several classes (two classes—{good, poor}; tree classes—{excellent, good, poor}, or more). These products are used as learning samples. Thus, each sample is represented by the vector $E_i = (a_{i1}, a_{i2}, ..., a_{in})$, where n is the number of attributes and a_{ik} can take values from the set $\{0, 1, x\}$, E_i belongs to exactly one class K_g, where $g = 1, 2, ..., s$. If m is the number of samples, the data can be presented as T_{mn} table, usually referred to as a learning table.

Created once, the table is relatively constant within a type of software. When a new software product E becomes, it is formed its vector $E = (a_1, a_2, ..., a_n)$. Using T_{mn} learning table and the descriptor of E with appropriate methods of classifying the new E product to one of the s predefined classes gives the relative information about its quality, which in many cases is quite sufficient.

The essence of the method consists of searching for subsets of attributes that reflect the differences between classes. There were proposed different algorithms to determine a measure to a cluster, based on the terms *Irreducible test* and *Irreducible representative set*. An *Irreducible test* is a minimal subset of variables, for which the measurements in vectors for different clusters are different. An *Irreducible representative set* for the cluster K_i is a minimal subset of variables, for which the values (measurements) in the vectors of K_i and all others vectors are different. Hence, the *tests* are dissimilarity units for all clusters, the *representative sets* are dissimilarity units for one cluster and all another clusters.

With respect of the membership of a given variable in such units, this variable receives certain weight to reflect its contribution to the diversity of the vectors in the learning table 3 types of weights were proposed: p_i (part of all tests whose member is the variable i), q_i (the same with respect of the length of the test), and r_i (the same applied to the representative sets).

By definition, the vectors in the cluster K of the learning table have 100% coincidence of values for all tests and representative sets in this cluster, and 0% for values in other clusters. If a new vector is given, according to the number of coincidence of values for all tests and representative sets, it can be established the similarity of this new vector (object) to each cluster, based on the dissimilarity units.

So, the algorithms for pattern recognition are three types:

1. Using the minimum distance of the objects' weight sum (resp. with p_i, q_i, r_i). The new object belongs to the cluster with the nearest object.
2. Using the city-block distance with weights (resp. with p_i, q_i, r_i). This metric is the total distance between all objects in a cluster and the new object.
3. By voting, taking into consideration the number of coincidences between the new object and the tests or the representative sets of a class.

The algorithm use hierarchical clustering—each vector is considered to be one cluster at onset. Iteratively, when two vectors are recognized as closest each to other by the used algorithm for pattern recognition, they will be associated. The first two groups of algorithms are convenient for variables with quantitative attributes, the last one—for not fully defined and for qualitative attributes.

The proposed method was implemented through a program that was last updated in 2011. With its help, the quality of various types of software products (authoring systems, salary programs, text editors, database management systems, etc.) was evaluated and the results analyzed.

2.2 Quality Management

As noted, significant interest in software production is the effective management of the quality. Despite the availability of earlier developments, the Software Process Maturity Model (CMM) [9, 10] and its refinements (CMMI) [11] are the founding and leading-edge approach today. This model is designed to improve software processes, to evaluate them, and to evaluate by qualified experts the ability of potential professionals to implement a software project.

The mentioned improvement is done by means of precise and detailed defined and structured actions in the planning, technology and management of software development and support.

Another basic instrument of quality assurance is the set of well-known ISO 9000 standards. However, from the point of view of software production, even ISO 9000-3 with its software-directed recommendations is not very constructive and is therefore not so useful for practical application.

CMM is indivisible from the so-called "classical" methodologies for software development. However, in the last two decades, the so called agile methodologies (in particular, extreme programming (XP) [12] and Scrum have emerged and have developed rapidly. Their specificity (especially of XP) does not allow direct and complete application of CMM. We tried quite early (2000–2002) to carry out an in-depth analysis [13]. We decided that XP has characteristics that are associated with the high levels of CMM, including the highest level 5. This gave us reason to argue that XP is a vertical intersection of CMM. But at that time there were no attempts to improve XP processes as is one of the main goals of CMM.

We suggested in [13, 14] that there was still place to improve some of the practices and their adjustment to specific environment and this should be accomplished by applying a methodology for XP process improvement that will propose required activities to achieve better quality of code and design.

The other possibility was to concentrate only on the quality assurance in agile methodologies. Such an approach has the drawback not to ensure the predictability of the organization's processes, but requires much less efforts and still brings a major result—a final software product of quality. Certainly, a third option was the further development CMM, what actually happened with the appearance of CMMI [11, 15]. However the problem still keeps attracting a substantial interest [16].

2.3 Software Quality Assurance Through Comparative Analysis

The goal of software quality assurance is to provide the adequate confidence that quality is being systematically built into the software. A method, facilitating the proper decision making during the software development process will give us possibility to control this process and its results.

Let us describe briefly a formal method for a reasonable choice, proposed in [17]. It supports the multi-criteria decision making and has been developed in accordance with the best achievements in this area.

Comparative Analysis (CA) is a study of the quality content of a set of homogeneous objects and their mutual comparison so as to select the best, to rank them or to classify each object to one of the predefined quality categories.

The compared objects can be products, processes or resources, identified as significant for the software engineering activity under consideration. When apply the CA method, we distinguish two main players: the **Analyst**, responsible for all methodological and technical details of CA implementation, and a **CA customer**—a single person or a group of persons, who should make a decision in a given situation and wants to use the CA as a supportive method. The context of the desired CA is specified through a **case**, described by the following six elements:

case = {**View**, **Goal**, **Object**, **Competitors**, **Task**, **Level**}

The **View** describes the customer's role and the perspective from which the CA will be performed.

The **Goal** expresses the main customer's intentions in CA use and can be to describe, analyze, estimate, improve, predict or any other, formulated by the Customer, defining the case.

The **Object** represents the item under consideration.

For each investigated object a hierarchical quality model has to be created—a set of characteristics, selected to represent the specific quality content, and the relationships among them. This modeling activity is the most difficult one due to its high cognitive complexity. A repository for already constructed object quality models is maintained thus facilitating their re-use in similar situations.

According to the goal, the set **Competitors**—the instances of the objects to be compared—should be chosen.

The **Task**, described as an element of a case can be Selection (finding the best), Ranking (producing an ordered list), Classification (splitting the objects to a few preliminary defined quality groups) or any combination of them.

The depth **Level** defines the overall complexity of the CA and depends on the importance of the problem under consideration and on the resources planned for CA implementation.

On the basis of the Comparative analysis a methodology ISPIRE for software engineering activities has been developed [18]. A step-wise procedure for its transfer to practice is described. It has been successfully applied for improving a number of software engineering activities: usability assurance, outsourcing software development, quality models evaluation, software quality assurance, teamwork training, etc.

2.4 Quality-Related Research in the Field of Programming Languages

Studies in programming languages began in the 1990s and were mostly centered on finding out properties of programming languages and programs, related to user perception and comprehension. A number of such properties were identified and defined, such as e.g. straightforwardness, adequacy, explicitness, apparency, conciseness, memory load, notationality, mnemonicity, and uniformity [19]. They can be considered as characteristics of a program quality model and the issues of their analysis, evaluation and improvement through some constructs of the programming languages are worth to be investigated.

Special attention was given to the notion of explicitness. Various manifestations of it or lack of it, in programming were studied in order to reveal potential deficiencies in programming languages. A number of programming constructs have been proposed to enhance expressiveness in programming through ensuring that a wider range of properties and relations can be stated explicitly.

For example, a model of repetitive computation based on consumable actions, and a number of specific constructs within that model were proposed. It has been shown that these means of program expression enable covering of a significantly wider class of repetitive patterns, thus enriching the programmer's repertoire while also preserving well-structuredness [20].

In programming, unlike traditional mathematics, often an expression does not only produce a value but carries a more complex meaning. The Icon programming language is a well-known example in this respect. In a similar fashion, drawing inspiration also from other programming languages, a number of enhancements at expression level have been considered, aiming at avoiding unnecessary repetition and other benefits in (mostly) imperative programming [21].

Also related to the syntax and semantics of expressions but in the functional programming paradigm, a novel, simple, uniform, and sufficiently universal pattern for constructing expressions have been devised and incorporated in a functional programming language. An early description of that language is given in [22].

The study of expressiveness has also lead to an improvement to the BNF syntax-description language (and any of its variants such as EBNF) for representing sequences of any kind.

At a more abstract level, some efforts were dedicated to developing a general concept of what constitutes structure in programs. Principles of structuredness of computing systems that give prominence to the dualism between their static and dynamic content were formulated in [19, 23]. A series of notions related to this were defined and put into an emerging taxonomy.

At present, work on studying the properties of programming and other computing-related languages continues with building experimental implementations of the said and other languages and constructs.

3   Data Mining and Semantic Analysis

Over the past few centuries, the quantity of accumulated information in digital form is exponentially growing. Because of the rapid development in all areas of human activity in modern society, the production, economic and social processes have become more complex. Most organizations using information technology resources collect and store huge amounts of data. The challenge that all those organizations face today is, not how to collect and store the data needed, but how to derive meaningful conclusions from this massive volume of information. The solution is in the technology of data mining as essential part of artificial intelligence.

The forerunners of the research in the field of artificial intelligence in our section were the research related to the study of the mechanisms of human thinking, especially the ways of solving problems, understanding and decision making by analogy. Understanding the mechanisms of memory activation, and, in particular, the constructive element that brings the context, are the basis of the research of Boicho Kokinov—the promotor of Cognitive Science in Bulgaria. On the other hand, an example of the application of classification methods for quality assessment of software products has already been discussed in the previous section. Below we will present some of the latest developments in the department related to this area.

3.1   Class-Association Rules Classifier PGN

Class-Association Rules (CAR) algorithms have a special place within the family of classification algorithms. This type of classifiers offers a number of advantages: efficiency of the training regardless of the training set; easy handling with high dimensionality; very fast classification; high accuracy; classification model comprehensible for humans. Usually CAR classifiers uses support of the rules as radical criterion in the process of choosing the rules to be observed. So, we elaborated a new CAR-classifier PGN that questions this common approach to prioritize the support over the confidence and focuses primarily on the confidence of the association rules and only in a later stage on the support of the rules. One of the main advantages of PGN is that it is a parameter-free classifier. The association rule mining is executed from the longest rules to the shorter ones until no intersections between patterns in the classes are possible. In the pruning phase the contradictions and inconsistencies of more general rules between classes are cleared, after that the pattern set is compacted excluding all more concrete rules within the classes. Later, we developed a method for effective building and storing of pattern set in multi-layer structure MPGN during the process of associative rule mining [24]. The software of proposed algorithms and structures has been implemented in the frame of data mining environment system PaGaNe [25]. The conducted experiments prove the vividness of proposed approaches showing the good performance of PGN and MPGN in comparison with other classifiers from CAR, rules and trees that build similar behavior of creating the task model, and especially in the case of multi-class datasets with uneven distribution of the class labels [26]. At the same time, PGN has an essential advantage over the other classifiers being parameter free. The results provide evidence that the confidence-first approach yields interesting opportunities for knowledge discovery.

Additionally, a simple approach for association rule mining, which uses the possibilities of the multidimensional numbered information spaces as a storage structures was proposed in so called ArmSquare. The main focus in that realization was to use the advantages of such spaces, i.e., the possibility to build growing space hierarchies of information elements, the great power for building interconnections between information elements stored in the information base, and the possibility to change searching with direct addressing in well-structured tasks [27].

3.2   Semantic Image Retrieval

The latest advances in the development of information and communication technologies, high performance computing and data storage methods have made it possible to widely use and manage multimedia data. The rich content of multimedia data built up by merging information stored in different modalities has prompted the search for new approaches to modeling, processing, retrieving, organizing and indexing that data. The ultimate goal is to give users the ability to extract a desired content from large amounts of visual data in a fast, efficient, semantically-relevant, user-friendly and site-independent environment. The systems for automatically retrieving metadata from images and video were limited by the fact that they can only operate at the level of primitive features while users operate at a highly semantic level. This discrepancy is usually referred to as the semantic gap. Particularly, in the field of art image retrieval this challenge is more visible. The digital technologies are limited within pixels capture, but the human expectation are connected with a spread palette of semantic, aesthetic and cultural messages which usually the artwork sends to the viewer. There is a hard task to develop appropriate machine algorithms to analyze the image. These algorithms are based on completely different logic and "instruments" compared to the human process of perception, but would give similar results in interpreting the input image. The challenges are even bigger when we focus the efforts on image analysis of the aesthetic and semantic content of art images. Naturally, the interpretation of what we—humans—see is hard to characterize, and even harder to teach to a machine. Yet, over the past decade, considerable progress has been made to make computers learn to understand, index and annotate pictures representing a wide range of concepts.

In general, the categories describing the content of art images can be assigned to one of the following spaces—image space, semantic space and abstract space. This taxonomy was suggested in [28] as a refinement of proposition of [29, 30] with more accurate dividing between semantic and abstract space (Fig. 1):

- Image space contains visual features, needed to record an image through visual perception;
- Semantic space is related to the meaning of the elements, their potential for semantic interpretation;
- Abstract space is connected with the aspects that are specific to art images and reflect cultural influences, specific techniques as well as emotional responses evoked by an image.

Using as a starting point Itten's color theory, a color model suitable for color characteristics that define harmonics and contrasts is proposed. On the basis of this model we elaborate a formal description of harmonies and contrasts from the point of view of three main characteristics of the color—hue, saturation and luminance [31]. The investigation on the possibilities to integrate such characteristics on a local level within specialized resource discovery [32] were connected with using

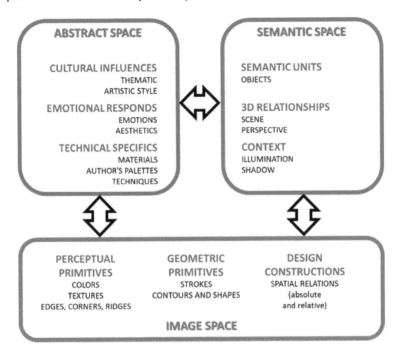

Fig. 1 A taxonomy of art image content

already discussed CAR classifier PGN, ArmSquare association rule miner and implemented statistical analyzing tools for extracting regularities for artists' and movements' styles. Proposed and developed types of harmonies and contrasts features strive to narrow the semantic and abstraction gap between low-level automatic visual extraction and high-level human expression.

3.3 Emotion Recognition from Speech

Another track that has undergone a booming success in recent years is smart system development with an emphasis on sentiment recognition of text written communication and emotion recognition from speech signals using digital signal processing. It extends the applied research area in digital media entertainment through emotions in speech signals. The later research had a direct application in recommendation and personalization of digital media content including movies, TV shows, music, paintings, etc. Smart systems are also applicable in numerous important fields such as: healthcare, education, banking, security, etc.

The main findings suggest a novel method for recognizing the emotional state of a person based on speech. More specifically it was shown how extracting the glottal signal from real-life speech could be used in smart system development. Furthermore, it was proven through extensive robustness testing that, the glottal signal is one of the best informational channels for determining the emotional state of a person, which survives a wide number of signal intrusions [33].

Recent research aimed to fuse areas of Natural Language Processing (NLP) and speech emotion recognition was posed in [34]. From one hand, inverse filtering methods applied to speech signal processing is merged with Natural Language Processing methods, applied to five volumes of The Game of Thrones. Every sentence from each of the five books was tokenized in order to obtain polarity count that determined the emotional content of words in a succeeding step. The results from the first step were applied to the second step in order to create a recommendation of a book based on emotion detected from speech.

The research is extended to study glottal signal extraction using gender separation in four emotional states (Happy, Angry, Sad and Neutral) as applied to Human–Computer Interaction systems in order to widen the scope of application in real-life scenario [35]. It was shown that gender could improve accuracy in a real-life scenario when it comes to multiple users of a personalization and recommendation digital media systems.

4 Information Systems and Communities

From the point of view of the development of the information systems and the main means for their effective construction in terms of collecting, storing, processing and disseminating information over the years, a number of up-to-date themes with original developments have been created in the department. Let's only mention the few of them. At first place, the merit of Peter Burnev for the development of office automation in Bulgaria. An original development of a context-free access method ArM, based on a multi-dimensional coordinate space, was proposed by Krassimir Markov, who over the years proved its applicability in many different tasks (for example, on its base is built the above-mentioned ArmSquare). Peter Stanchev participated in several revolutionary medical imaging studies, such as mathematical methods for correcting images obtained from a magnetic resonance scanner by detecting movements and their correction in the K-space and in the real space, which is implemented in modern scanners; segmentation of the vessel's central lines based on ridge descriptors, implemented in real database with retinal images DRIVE; identifying statistically significant differences in brain structures of healthy and sick Alzheimer's disease proposing original algorithms for segmentation, boundary definition, and three dimensional image reconstruction, etc. At a later stage, Milena Dobreva became the pioneer of a Digital Humanities in Bulgaria and contributed to the development of digital libraries worldwide, participating in the creation of the DELOS Digital Library Reference Model. Below we will stop more attention on two of the newest research field in the Department, connected with context-aware ICT applications and virtual communities.

4.1   Context-Aware ICT Applications

In many cases, in order to be effective, software applications need to allow sensitivity to user context state changes. Different approaches have been used to achieve this goal. Considering the SDBC (Software Derived from Business Components) approach that brings together enterprise modeling and software specification, in [36] a base context-awareness conceptualization is made. This allows to extend the research and to identify three categories of context-aware systems and to propose a corresponding modeling approach [37]. One of the perspectives in this regard is featuring a context-awareness driven by the goal of optimizing system-internal processes: this is a matter of maintaining the system "performance" within some "boundaries" and adapting the system behavior accordingly. Another perspective concerns a context-awareness driven by the goal of maximizing the user-perceived effectiveness: this is a matter of adapting the system behavior to the situation of the user. Finally, value-sensitive context-aware systems have been addressed, considering possibilities to adapt the system behavior to relevant public values (if any).

Context data has been especially considered, mainly in two perspectives: (i) sensor-based data capturing vs prediction-based data capturing; (ii) rule-based processing of context data vs training-data-based processing of context data. Two corresponding relevant methods were identified, namely the Norm Analysis Method (part of Organizational Semiotics) and the Naïve Bayesian Classification Approach. This was reflected in a proposed approach featuring the consideration of context data [38].

These conceptual results have been justified by applications, such as land border security and operation drone in a context-aware and value-sensitive way.

The recent research and achievements are presented in [39], where context-awareness was considered in the broader context of enterprise information systems, touching upon the challenge of bringing together enterprise modeling and software specification.

4.2   Virtual Community as a Production Environment

The creative potential of the virtual community phenomenon is an object of numerous theoretical and empirical studies since 1990s'. The efforts to provide a mechanism to effectively employ the resources of the members of the community in the process of creation of virtual or even physical goods lead to various solutions with the open-source model [40] and the crowdsourcing [41] as two of the foremost.

The general objective of the research presented in [42] is to outline an alternative approach that facilitates participation of virtual community members in goods development process. The research effort is motivated by the notion to find a middle ground between less structured and liberal open source model and firmly structured but more restrictive crowdsourcing model. In some sense it is a continuation of D.

Brabham's studies [43] and is inspired by it. As a result a systematic community-sourcing approach is proposed [44]. Community-sourcing implies development of virtual community that provide managed environment for production of goods by teams of community members that use both own resources and services provided by the community government. On that foundation, a conceptual model of resource management system that use some IT industry current best practices in service and project management is proposed. The results are validated in a virtual learning community that implements the model. That community (VIVA Cognita, https://vivacognita.org) has more than 12,000 registered users in 2019. On the basis of the community-sourcing approach several potential applications scenarios of the approach are described, with some of the most promising in the linguistics and language studies [45]. In 2017, in pursuit of better differentiation of the tens of existing variations of the crowdsourcing, the research topic forked towards development of tuples of lexical (extracted from definitions of the researchers in that field) and visual descriptions (BPMN diagrams) of the variations.

5 Conclusion

The "Software Engineering and Information Systems" department has demonstrated its capacity for successful research. A wide spectrum of problems in different areas of Informatics has been covered.

All researchers in the department have a clear vision how to proceed. Among their future intentions are to share the achievements with a broader community of both scientists and practitioners so as to be able not only to have new scientific results, but also to validate the existing ones by applications.

References

1. ISO 9000 Quality Management Systems—Fundamentals and Vocabulary. International Organization for Standardization (2005)
2. Boehm, B.: Software and its impact: a quantitative assessment. Datamation **19**(5), 49–59 (1973)
3. Bowen, T., Wigle, G., Tsai, J.: Specification of software quality attributes. In: Software Quality Evaluation Guidebook, RADC-TR-85-37, vol. III (1985)
4. General methodology for software quality evaluation. The intergovernmental committee for the cooperation of the socialist countries in the field of computing. Bulletin **1**(37), Moscow (in Bulgarian) (1988)
5. ISO/IEC 9126: Information Technology—Software Product Evaluation—Quality Characteristics and Guidelines for Their Use (1991)
6. Angelova, V.: Evaluation of the quality of software by classification methods. PhD thesis—IMI-BAS, Sofia (in Bulgarian) (1987)
7. Eskenasi, A.: Evaluation of software quality by means of classification methods. J. Syst. Softw. **10**(3), 213–216 (1989)
8. Dmitriev, A., Zhuravlev, J., Krendelev, P.: About mathematical principles of object and phenomena classification. Discreet Anal. **7**, 3–11 (in Russian) (1986)

9. Humphrey, W.: Characterizing the software process: a maturity framework. IEEE Softw. **3**, 73–79 (1988)
10. Paulk, M., et al.: Key practices of the capability maturity model for software, version 1.1. TR CMU/SEI-93-TR-025, SEI (USA). Carnegie Mellon University, Pittsburgh (1993)
11. CMMI Maturity Profile Report, 30 June 2018 https://cmmiinstitute.com/resource-files/public/quality/maturity-profiles/maturity-profile
12. Beck, K.: Embracing change with extreme programming. Computer **32**(10), 70–77 (1999)
13. Eskenazi, A., Balkanski, P.: Quality improvement in XP development. In: Proceedings of the International Conference CompSysTech, pp. II.9.1–II.9.6 (2002)
14. Balkanski, P.: Quality assurance in extreme programming. Int. J. Inf. Theor. Appl. **10**, 113–117 (2003)
15. A Guide to Scrum and CMMI: Improving Agile Performance with CMMI. CMMI Institute (2016)
16. Silva, F., et al.: Using CMMI together with agile software development: a systematic review. Inf. Softw. Technol. **58**, 20–43 (2015)
17. Maneva, N.: Comparative analysis: a feasible software engineering method. Serdica J. Comput. **1**(1), 1–12 (2007)
18. Maneva, N.: INSPIRE: a software engineering methodology. J. Inf. Technol. Control **1**, 22–26 (2005)
19. Bantchev, B.: Issues of explicit expression in program texts. PhD thesis—IMI-BAS, Sofia, 127 p. (2014)
20. Bantchev, B.: Terminable statements and destructive computation. ACM SIGPLAN Not. **29**(2), 33–38 (1994)
21. Bantchev, B.: Putting more meaning in expressions. ACM SIGPLAN Not. **33**(9), 77–83 (1998)
22. Bantchev, B.: A language for compositional programming: a rationale and design. In: Proceedings of 40th Spring Conference of the Union of Bulgarian Mathematicians, pp. 236–243 (2011)
23. Bantchev, B.: Understanding computing through defining its lexicon. In: Proceedings of 39th Spring Conference of the Union of Bulgarian Mathematicians, pp. 171–177 (2010)
24. Mitov, I.: Class association rule mining using multi-dimensional numbered information spaces. PhD thesis—Hasselt, Belgium, 200 p. (2011)
25. Mitov, I., Ivanova, K., Markov, K., Velychko, V., Vanhoof. K., Stanchev, P.: PaGaNe—a classification machine learning system based on the multidimensional numbered information spaces. In: World Scientific Proceedings Series on Computer Engineering and Information Science, vol. 2, pp. 279–286 (2009)
26. Mitov, I., Depaire, B., Ivanova, K., Vanhoof, K.: Classifier PGN: classification with high confidence rules. Serdica J. Comput. **6**(2), 143–164 (2013)
27. Mitov, I., Ivanova, K., Depaire, B., Vanhoof, K.: ArmSquare: an association rule miner based on multidimensional numbered information spaces. In: Proceedings of the 1st International Conference on Advances in Information Mining and Management (IMMM 2011), pp. 143–148, Barcelona, Spain (2011)
28. Ivanova, K., Stanchev, P., Vanhoof, K., Ein-Dor, P.: Semantic and abstraction content of Art images. In: Proceedings of the Fifth Mediterranean Conference on Information Systems, Tel Aviv, Israel, AIS e-Library, paper 42 (2010)
29. Burford, B., Briggs, P., Eakins, J.: A taxonomy of the image: on the classification of content for image retrieval. Vis. Commun. **2**(2), 123–161 (2003)
30. Hurtut, T.: 2D Artistic Images Analysis, A Content-Based Survey. https://hal.archives-ouvertes.fr/hal-00459401_v1/ (2010)
31. Ivanova, K., Stanchev, P., Vanhoof, K.: Automatic tagging of art images with color harmonies and contrasts characteristics in art image collections. Int. J. Adv. Softw. **3**(3 & 4), 474–484 (2010)
32. Ivanova, K., Stanchev, P., Velikova, E., Vanhoof, K., Depaire, B., Kannan, R., Mitov, I., Markov, K.: Features for art painting classification based on vector quantization of MPEG-7 descriptors. In: 2nd International Conference on Data Engineering and Management (ICDEM 2010), India, LNCS, vol. 6411, pp. 146–153 (2012)

33. Iliev, A.I.: Emotion Recognition from Speech, 168 p. Lambert Academic Publishing (2012)
34. Iliev, A.I.: Content discovery using perceptual automation. In: Proceedings of the 10th International Conference on Management of Digital Ecosystems (MEDES'18), pp. 233–238, Tokyo, Japan (2018)
35. Iliev, A.I., Stanchev, P.L.: Glottal attributes extracted from speech with application to emotion driven smart systems. In: Proceedings of the 10th International Joint Conferences on Knowledge Discovery, Knowledge Engineering and Knowledge Management (IC3K 2018), vol. 1: KDIR, pp. 297–302 (2018)
36. Shishkov, B., van Sinderen, M.: From user context states to context-aware applications. In: Filipe, J., et al. (eds.) Enterprise Information Systems, ICEIS 2007. Lecture Notes in Business Information Processing, vol. 12, pp. 225–239 (2008)
37. Shishkov, B., Larsen, J.B., Warnier, M., Janssen, M.: Three categories of context-aware systems. In: Shishkov, B. (ed.) Business Modeling and Software Design, BMSD 2018. Lecture Notes in Business Information Processing, vol. 319, pp. 185–202 (2018).
38. Shishkov, B.: Tuning the behavior of context-aware applications, using semiotic norms and Bayesian modeling to establish the user situation. In: Shishkov, B. (ed.) Business Modeling and Software Design, BMSD 2019. Lecture Notes in Business Information Processing, vol. 356 (2019)
39. Shishkov, B.: Designing Enterprise Information Systems, Merging Enterprise Modeling and Software Specification, 246 p. Springer Nature Switzerland AG (2019)
40. Lerner, J., Tirole, J.: The simple economics of open source. J. Ind. Econ. **52**, 197–234 (2002)
41. Prpić, J., Shukla, P.P., Kietzmann, J.H., mccarthy, I.P.: How to work a crowd: developing crowd capital through crowdsourcing. Bus. Horiz. **58**(1), 77–85 (2015)
42. Branzov, T.: Managing the production of goods in a virtual community. PhD thesis—IMI-BAS, 140 p., Sofia (In Bulgarian) (2017)
43. Brabham, D.: Crowdsourcing as a model for problem solving: an introduction and cases. Convergence **14**(1), 75–90 (2008)
44. Branzov, T.: Community-sourcing in virtual societies. Serdica J. Comput. **10**(3–4), 263–284 (2016)
45. Branzov, T.: Approaches for development of online teaching materials employing resources from virtual communities. Bul. Lang. Lit. **60**(2), 191–200 (2018)

The Achievements of the Technical Sciences in the Bulgarian Academy of Sciences (BAS) in Automation, Robotics and Computers During the Past Century

Vassil Sgurev

Abstract Technical sciences come and develop in BAS relevantly late: at the end of the nineteenth century and at the beginning of the twentieth century. Then for BAS members are elected the first representatives of these sciences. Only just in 1959 a state act forms the Department of Technical Sciences (DTS) as a self-dependent structural division within the range of the academy. Regardless of this delay the department succeeds very fast to realize significant achievements in the realized in the country industrialization and in particular—in the high-tech production. In TSD-BAS appear the first for our country scientific teams in the field of automation and cybernetics. There have been started and implemented the first automated systems sand also the first Bulgarian system with a specialized computer and periphery for a real-time control of an industrial transport. Noteworthy are the successes of technical sciences in the field of initiation, development and the rapid and broad development of the serial production and distribution of the personal computers as a new highly efficient trend in the information technologies. Following an analogical scenario the technical sciences in Bulgaria initiated the research in the field of robotics the result of which was the industrial production and the implementation of industrial robots. All this to a considerable extent helped the creation and the development of high-tech production in Bulgaria during the past century.

Keywords Technical sciences · Academy · Robotics · Computers · Achievements

During 1959, 90 years after the foundation of the Bulgarian Literary Society (BLS) the Council of Ministers took the decision published in Decree No. 236 from November 3, 1959 about the creation in the Bulgarian Academy of Sciences (BAS) of a Department of Technical Sciences (DTS) aiming at 'the further development of the Bulgarian science and the increase of its role in the socialist construction'. The first Academic secretary of the DTS at the Bulgarian Academy of Sciences has been elected Correspondent Member Angel Balevski [1].

V. Sgurev (✉)
Institute of Information and Communication Technologies at BAS, Sofia, Bulgaria

© The Author(s), under exclusive license to Springer Nature Switzerland AG 2021   23
K. T. Atanassov (ed.), *Research in Computer Science in the Bulgarian Academy of Sciences*, Studies in Computational Intelligence 934,
https://doi.org/10.1007/978-3-030-72284-5_3

The establishment of the Department of Technical Sciences at the Bulgarian Academy of Sciences is not an accidental idea of a high-ranking state actor. It is based on the assumed state rate—after the completion of the post-war reconstruction period, to the accelerated industrialization and modernization of the individual sectors of the economy. The goal is to catch up with the countries ahead, the task—long-lasting and difficult to solve.

The Presidium of the Bulgarian Academy of Sciences reacted quickly and at its meeting on November 20, 1959, it determined the structure of the new ward with the main directions in which it will develop and the necessary means for this development. On the basis of the taken decisions, in the following years the following fields of technical sciences will be found at the DTS-BAS: automation and cybernetics; metal casting and metalworking; building mechanics; research and management of water resources.

In the years after 1959, various scientific units, laboratories and institutes will be set up and reorganized, in which activities will be carried out on the above mentioned four strands.

This paper describes the achievements of DTS in one of four directions: scientific and applied results on metal casting and metalworking, structural mechanics and water resource management research and management are the subject of other investigations. Achievements in computer technology are related to other departments of the academy.

The section 'Relay Protection and Automation of Energy Systems', headed by Assoc. Prof. Bojidar Popov, will be set up in the first direction to the existing Institute of Power Engineering (IPE) at the Bulgarian Academy of Sciences directed by Assoc. Eng. Georgi Rasheev. A separate section will be formed at DTS-BAS 'Automation and Telemechanics', headed by Assoc. Eng. Denyo Belchev. It follows from these facts that in 1959 there was foresight in the academy and the development of the high-tech direction 'automation' was planned. Institutionally, this outpaces the establishment of the Automation Central Laboratory (ACL) and the Central Research Institute for Complex Automation (CRICA).

In 1962, the two automation units were reorganized in the unified section 'Automation and Telemechanics' with the Compatibility Director Assoc. Eng. Georgi Rasheev, while retaining the same subject. In August 1963 it was transformed into the Central Laboratory of Automation and Telemechanics (CLAT) at the Bulgarian Academy of Sciences, headed by Assoc. Prof. Eng. Nikolay Naplatanov, who is the Compatibility Head of the Department of Industrial Automation at the Mechanical–Electrical Engineering Institute (MEEI) Sofia. It is comprised of 40 associates and with the following two sections: 'Technical Instruments for Automated Control, Regulation and Management', headed by Assoc. Eng. Deniu Belchev and 'Systems for Centralized and Telemechanical Control and Management', headed by Assoc. Eng. Bojidar Popov.

There is an active preparatory activity for the establishment of the Institute of Technical Cybernetics (ITC) at the Bulgarian Academy of Sciences, based on the ACL-BAS. The consultant in the establishment of the institute is Prof. B. Sotskov—Corresponding Member of the Academy of Sciences (AS) of the USSR.

On April 1, 1964, the Presidium of the Bulgarian Academy of Sciences approved the establishment of ITC at the Bulgarian Academy of Sciences with the following sections: Applied Mathematical Problems of Technical Cybernetics (AMPTC), headed by Assoc. Dr. N. Naplatanov, Elements and Tools of Technical Cybernetics (ETTS), led by Assoc. Eng. D. Belchev, System for Central and Telemechanical Control and Management (SCTCM), headed by Eng. F. Velkov, Automated Control and Regulation Systems (ACRS) with Assoc. Eng. B. Popov and Bionics, headed by Compatibility Assoc. Dr. A. Gidikov.

Assoc. Dr. N. Naplatanov was elected as the Director of ITC-BAS, and as Scientific Secretary—Eng. P. Petrov. The last change in the structure of this institute was in 1978 when its leadership consisted of the director Corresponding Member Prof. N. Naplatanov, the deputy directors Assoc. Dr. I. Popchev and Prof. DSc. D. Mladenov, and the scientific secretary Assoc. Dr. V. Sgurev. The sections are now eight and they have the following names: Applied Mathematical Problems of Technical Cybernetics (AMPTC), headed by Assoc. Dr. I. Popchev, Ergonomic Management Systems, headed by Correspondent Member Prof. N. Naplatanov, Management of Industrial Systems, headed by Assoc. P. Petrov, Informatics, headed by Assoc. Z. Nikolov, Bionics, headed by Correspondent Member Prof. N. Naplatanov, Automation of Intellectual Activity, headed by Assoc. Dr. V. Sgurev, Information Transmission Systems, headed by Assoc. F. Velkov and Hybrid Modeling, headed by Prof. DSc. M. Zlatev. The Institute has two subsidiary units: Hybrid Computing Complex (HCC), headed by Assoc. N. Milev and Development and Implementation Base, led by Eng. K. Kurshumov.

For almost the 20th anniversary of the establishment of the Department of Technical Sciences until 1978, when a large-scale reorganization of ITC-BAS was carried out, not a small number of theoretical investigations were carried out and a number of systems and devices were created, some of which were implemented in practice. They enable the Department of Technical Sciences to take a prominent place in the academy and beyond in automation, cybernetics, robotics and computer technology. It is not possible to describe these achievements in detail in one article, but the more important of them are worth mentioning.

Such are: The System for Automated Measurement and Regulation of Molding Mist Humidity (1961), The Complex Tele-mechanical System TP-62 for Constructing Automation Technical Devices (1965), The Electrocardiogram Rhythm Analysis Method and Device ANDROS-66 (1966), The Electronic Diagnostic Machine EDM-4 (1967), the production of ANDROS-R by the Electronics Plant (1967), the Technical Diagnostics Fluid-Elements System (1967), the PH-Meter (1967), The Liquid Electrical Conductivity Meter CONDUCTOMETER (1968), the Fluid Jet Elements System (1969), the System of Software and Technical Devices for the Management of Freight Transport in Mining and Quarrying Plant 'Medet' Astra 71 (1970), The Transmission System for Operative Statistical Information CENTRONIC-1001-TC (1970), the Managing Computing Machine with Integrated Circuits Astra 71–25 (1972), the implementation of the Astra 71 Industrial Transport Operative Management System in Mining and Quarrying Plant 'Medet' (1973), the range of Hydraulic Jet Elements FALOMA (1973), the Screening Diagnostic

Information System of the Ministry of Public Health (1973 г.), the INTERTEST Automaton for Functional Control and Diagnostics of Devices with Integrated Circuits (1973), the Algorithms and Programming Systems for Optimal Distribution of Water Resources (1975), the Design Automation System SAPR (1976), the Multi-Dimensional Adaptive Systems (1977), the Complex of Systems for Managing a Class of Mobile Objects VISEUR (VIEWFINDER, 1977), The Adaptation of Early Diagnostics of Cancer Systems (1977), The Development of a Class of Multi-Microprocessor Management Systems (1978), the Human–Machine System EXATEST (1978), The Automated Operative Management System (AOMS) of the Technological Transport in Copper Extraction Plant 'Elatzite' (1978).

Some of these developments have not been implemented, but—with all their criticism as a whole—it must be noticed that, first, their number impresses, and secondly, that they are the first and very often successful attempts to entry into important areas of information technologies that are actively developing today.

The Astra 71 system for operational management of industrial road transport in Mining and Quarrying Plant (MQP) 'Medet' deserves special attention. For it, the first Bulgarian specialized computing machine for real-time operation was created with incredible difficulties. The development and deployment of this computer controlling configuration was entirely done by a team of ITC-BAS in the period 1970–1974. For those years, at that time, the development, training of operating personnel and the deployment of such system required too much time, patience, persistence and skills to overcome enormous difficulties. Researchers and developers had these qualities and this enabled the Astra 71 system to work successfully for 12 years—until the resources of the mine and its closure were exhausted. With numerous patents and efficient solutions, as well as the successful operation of the system, Astra 71 has taken a lead in similar systems outside of our country.

Mining and quarrying specialists closely followed the periphetics of the implementation of the system and this helped accelerate the use of electronic computing devices in industry.

These developments are based on relevant theoretical research, the most significant of which are the following: Methods and Algorithms for Synthesis of Recognizing Diagnostic Systems (1968), Research of the Hierarchical Structure of Managing Large Systems (1968); Methods and Algorithms for Operative Management of Industrial Transport Systems in Real Time (1971); Methods and Technical Tools for Presenting Navigational Information for Piloting Unmanned Controlled Flying Objects (1973); Approach for Optimal Synthesis of a Specialized Digital Computing Machine (DCM) (1974); Algorithms and Programs for Optimal Distribution of Water Resources (1976); Closed Transport Systems Description Method (1976); Synthesis of Automata in a Homogeneous Environment (1976); Method-Oriented Programming (1976); Multidimensional Adaptive Systems (1977); Research and Synthesis of Multi-Microprocessor Management Systems (1978).

ITC at BAS organized a number of international conferences: Applied Aspects of Automata Theory (Varna, 1970); International Conference on Fluidics (Varna, 1972); Optimization of Systems for Collection, Transmission and Processing of Analogous and Discrete Information in Local Information Networks (Minsk, 1976); Information

Processing and Transmission Systems (1977); Design Automation Systems (DAS) (VGS-2, Sofia, 1977).

In 1963, the academic journal 'ACL-BAS News' began to go out, which in 1964 was renamed "The Cybernetics Institute's News". In 1964, the Scientific Council of ITC at the Bulgarian Academy of Sciences received the right to award the scientific title and the grade 'senior research associate (s.r.a.) 2nd grade' and 'candidate in the technical sciences (c.t.s.)'.

At the end of 1978 a decision was made at the highest state level on a radical, profound reorganization of ITC at BAS with the institute focusing on solving important national tasks with a practical focus. The leadership of BAS has not been informed of this decision, which has been dropped on top of it 'for execution'. Later, such practice will be applied to other BAS institutes.

On November 30, 1978 was published Order No. 73 of the Council of Ministers which established the Institute of Technical Cybernetics and Robotics (ITCR) at BAS on 1 December 1978 on the basis of ITC at BAS and three scientific-research laboratories of the Center for Accelerated Deployment (CAD) 'Progress' of the State Committee for Science and Technical Progress (SCSTP), namely, 'Unique Electronics', 'Specialized Robots' and 'Mechanization and Automation of Welding Processes'. The purpose of this new academic institute is defined as 'experimental and applied research in areas related to cybernetic systems for managing technical processes and objects, research and implementation in the field of robots and their accelerated application in the national economy, experimental production as well as training, preparation and post-graduate qualification of specialists in these fields'.

The text follows mainly in two directions: 'cybernetic systems for management of technical objects and systems' and 'research and development activities in the field of robotics'. For the new institute, there are resources of 600 thousand lv. for the second direction. The assets of ITC-BAS, including the newly built block 2 of BAS pass through to it.

It is acted decisively—all employees and associates of the old institute are released, but they can participate in the competition for 360 new posts at the new academic institute. Since 1980, the former 'Sasho Kofardzhiev' Scales Factory has been given as the Experimental Base by the industry to the ITCR at BAS. The list reached a total of 870 people.

The director was appointed Assoc. Dr. Angel Angelov until then Deputy-Chairperson of the State Committee for Science and Technical Progress (SCSTP), Deputy-Director—Assoc. Prof. Petar Petrov, and Scientific Secretary—Dr. Nikolay Iliev. There are two directions: 'Technical Cybernetics' led by Dr. Nikolay Iliev and 'Industrial Robots', led by Nedko Shivarov. The six sections are divided between these directions. Eng. Bogdan Stoyanov was elected head of the development and deployment units.

With the resources and effort made available, the Experimental Base and Development Units have been transformed into a truly experienced plant (a kind of 'techport', as they would now say). It includes: an electric-constructive assembly; machine-building department; technological documentation and archives; breeding base; electro-assembly workshop for rapid implementation of prototype models,

models and experimental installations; unit for the automated design of the experimental plants. Several associates of departmental science and development have joined the ITCR and this has established this academic institute to get the same opportunities as the leading industrial institutes. This has allowed ITCR-BAS to send in series of production plants verified technical documentation for the manufactured products, which can be immediately used in production.

The activity of development and construction of industrial works, the electromechanics of which was produced by the Beroe plant in Stara Zagora, and the control devices from the State Storage Devices (SSD)—Stara Zagora, started very quickly. The specialists from the Institute participated in the selection and purchase of machinery and equipment, as well as in the selection and training of staff for the Beroe plant and the auxiliary workshops working in cooperation with him. The ITCR at BAS initiated the establishment of a departmental robotics institute in the town of Stara Zagora and provided significant assistance in its consolidation.

The first task was to design a microprocessor control unit for the hydraulic robot of the Versatran-AMF series, for which the hydro-mechanical construction was purchased by the Beroe plant. The controller was embargoed, and its design and prototyping was carried out in ITCR. The times were short and the difficulties were considerable. Everything was completed successfully; the plant and ITCR have gained confidence that they can produce even more sophisticated products.

The next much more difficult task was to design a robotic complex for arc welding (РБ-251). This development is entirely Bulgarian and is protected by several patents. The development was supported by the Paton Institute of the Ukrainian Academy of Sciences. The Paton Institute and ITCR-BAS established in Sofia a joint international laboratory 'Interrobosvarka', which helped greatly to raise the level of robotic welding in both countries.

The RB-251 complex consisted of a portal-type robot for arc welding, a welding current source, a water-cooled burner and a number of technological devices. Modern solutions with multiple feedbacks were used. Until now, this complex remains one of the most complicated and expensive products ever made in Bulgaria and produced by Bulgarian machine building! At the Beroe plant in Stara Zagora, 15 RB-251 complexes were produced for the needs of the USSR's Military-Industrial Complex (MIC). The reviews of their work were very good.

Another successful direction was the initiation and creation under the leadership of Prof. Nedko Shivarov of the 'Robco' robot family. They have a relatively simple mechanical design, but with their multifunctionality they are a good platform for training in schools and universities. And so far, these robots can be seen in some study halls in our country.

On the base of the used base structures were developed and manufactured 'takeplace' robots for the needs of machine building as well as robots for painting. The latter have more sophisticated control devices and software because the robot had to be trained to repeat the painting operations properly.

By the time of the changes in Bulgaria since 1990, the Stara Zagora plants had enough orders from Bulgarian and foreign clients for the production of various types of industrial robots. All this gives serious grounds to BAS and its ITCR Institute to acknowledge the decisive contribution to the initiation and creation of the industrial robots industry in our country. In those years this fact was not denied by the professional circles in our country and in the Eastern European countries. Following the changes, due to the inappropriate privatization, the two well-equipped Stara Zagora high-tech factories—State Storage Devices (SSD) and Beroe—were brought to a destitute position. This is the end of serial production of industrial robots in Bulgaria! New small businesses were born, which in their own way continue the Bulgarian robotics traditions.

The next major breakthrough in ITCR-BAS was related to the initiation and production of personal computers in our country [2–5]. The well-known eminent professional engineer Ivan Marangozov, the first in Bulgaria, has anticipated the prospects of the emerging class of personal computers designed for personal use by the particular user. Through them it is possible to communicate in networks and with other computers. Eng. I. Marangozov grouped around him a small team of specialists in the computing equipment for creating a Bulgarian personal computer: engineers Georgi Zhelyazkov, Petar Petrov, Boris Vachkov and others. Such a computer was developed for a short time in 1980. It is called IMCO-2, 8-bit and fully compatible with the best Apple II Plus, DOS 3.2 and Pascal compiler. Success is complete; the leadership of ITCR and especially its director, Assoc. Prof. Angel Angelov, actively support this direction. The Ministry of Education has commissioned 200 copies of IMCO-2 for schools; they are produced in the Expert Base of ITCR. At the end of 1981, hundreds of IMCO-2 units created in the same Testing Base were distributed to many institutions and organizations in our country and made this computer very popular.

The ITCR leadership is looking for a production site of IMCO-2, focusing on the State Economic Holding (SEH) 'Instrumentation and Automation (IA)'. A branch for ITCR is established on the territory of town of Pravets.

SEH 'Izot', though late, develops an 8-bit personal computer. In 1982, the State Committee for Scientific and Technical Progress (SCSTP) and two other interested ministries developed a Program for the development of personal computers in the People's Republic of Bulgaria for the period 1982–1985. Resources were provided for the equipment of the factories and the production of 100 000 pieces of personal computers. A precise expertise has been carried out on which of the two types of computers—IMCO-2 or that of SEH 'Izot'—to be produced in series. For all indicators, a higher score is given to IMCO-2, which is widely distributed under the name 'Pravets-82'. The Head of the Experimental Base of ITCR, Assoc. Prof. Plamen Vachkov, is appointed Director of the 'Pravetz' Microprocessor Engineering Complex (MEC) and is also Deputy Director of ITCR-BAS. SEH 'IA' was transformed into SEH 'Microprocessor Engineering', and Prof. A. Angelov was appointed Chairman of its Management Board. The whole further work of MEC 'Pravetz' takes place in the close interaction and coordination of its activity with ITCR at BAS.

In the middle of 1984, the group of Eng. I. Marangozov developed the 16-bit personal computer IMCO-4, which is fully compatible with the IBM PC-XT computer. SHE 'Izot' has developed competing variants, but in the end the computer of ITCR's Eng. I. Marangozov has better performance, and under the name 'Pravetz-16' it is produced in series at the 'Pravetz' Microprocessor Engineering Complex (MEC). In general, all serially produced personal computers in Bulgaria were made by the group of I. Marangozov, the ITCR and the Bulgarian Academy of Sciences in general; this is quite clear even in those years [6]. The production of Pravets is constantly increasing not only because of the demand in Bulgaria, but also because of the export of these computers abroad. The popularity of personal computers 'Pravets' is constantly growing and the annual commodity production increased from 8 million lv. in 1983 to 400 million lv. in 1987.

In general, the initiation and creation of the PC manufacturing industry is the third major achievement of BAS in serially manufactured industrial products, counter-pressure casting machines and industrial robots [3, 6].

The ITCR-BAS staff also carries out other important theoretical and applied research in the field of automation, robotics, personal computers and information technology. The most important of these are: Planetary wired devices Isolan, Isomodul and Pollinaplan (1978); the Universal Casting Control System (1979); the Microprocessor Industrial Robot Control Device PB-230 (1979); the Electronic System for Sorting Tobacco Sheets 'Deltachrom-01' (1979); the Isomatik-TM Transmanipulator Management System (1979); the Automatic Fire Extinguishing System SAP (1979); the Robotic System for Arc Welding in the Plant for Electric Trucks 'September 6-th' PC-501 (1981); the Functional Modules for Building a System and Tools for Information Processing and Automation MIK-68 (1981); the Microprocessor System for Automated Control of the Technological Process in the Concrete Block 'Betoncontrol' (1981); the Automated Non-Destructive Magnetic Control System for Physico-Mechanical Indicators of Rolled Ferrous Metals Using a Microprocessor 'Microsom' (1981); the Microprocessor 'Icar-10' 'Flying Scissors' Control System (1981); the Functional Modules for Construction of Tools for Management of Industrial Robots and Aggregate Machines 'Isomatik' (1981); the Basic Microprocessor Regulator 'MIK-09' (1984); the Uuniversal Ccurrent Source for Welding Current (1984); the MIC-2000-C Continuous-Process Distributed-Management System (1984); Managing a Small Training Robot 'Robco' with a Personal Computer (1984); the Standard Robotized Workstation with Industrial Robot PB-232 for Palletizing in the Ceramic Industry (1984); The Robco-01 Training Anthropomorphic Robot (1985); the Automated Welding Process Control with High Frequency Current Sources for Robot Welding (1985); the Portable Personal Computer for Management of Technological Processes (1985); the Portable Info Terminal PIT-01 (1986); the 32-bit Interlab Microcomputer System (1986); the Intel Microprocessor Functional Modules for Building Real-Time Management Systems MC-1000 (1986); the Microprocessor System for Processing and Interpretation of Data from Experimental Research 'Compex' (1986); the Managing Device for Robot RB-241 and RB-242 (1987); the Tracy-30 Operative Management System of Industrial Traffic for Mining and Quarrying Plant 'Assarel' (1987); The Family of Microprocessor Modules for Computing Devices and Automation Systems for the Design of 32-bit VME-Highways (1987); the Professional PC-MIC-16-Turbo and Real-Time Operating System (1987); the Laboratory Data Collection and Processing System

CSY-10 (1987); The ESIAA Expert Technical Assessment System (1987); The Universal Controller 'Isomatic-1001-UK' (1987); The Personal Computer 'Pravets-16A' and 'Pravetz-8A' (1987); the Intelligent Controller for Control and Process Management, CSY-16 (1987); The Local Area Network for 16-bit PCs (1987); the Local Network Concentrator Linking Single-Circuit Regulators to a Personal Computer (1987); the RISK Emulators (1988), the Family of Power Supply Units for Electric-Truck Robots (1988) The C-16 Compiler (1988); The MIC-2000 System, Implemented in Neftochim, Burgas (1988); The MIC-2000 Management System for Biotechnological Processes (1989); the Integrated Data Collection and Management System for Industrial Processes 'Mikroscan-16' (1989); the Power PC Modules (1989); the Application Package 'Multipack-II' (1989); the Distributed Mmicroprocessor Ssystems for Ppersonal Aaccess (1990); the NLP-16 Nonlinear Optimization Software Product (1990); the Adaptive Fault-Resistant Distributed Computing Systems (1990); the Industrial PCs for Industrial Applications (1990).

In spite of this far incomplete list of research and development, there are significant efforts of the numerous ITCR staff that have led to a number of economic developments in our country and to significant theoretical results.

During this period, significant work in the field of Artificial Intelligence (AI) has also taken place in ITCR [7]. In this field, the AIMSA international conference, which has so far produced 17 editions of the same, has been initiated.

The institute is also the initiator of the international conferences 'PersComp' and 'RobCon', which were held in 1985 and 1987, and also the 'Yablonka' Fluidic Conference. There is also considerable international cooperation in volume and scope.

ITCR has established a number of branches of the institute in 'Pravetz' Microprocessor Engineering Complex (MEC), Stara Zagora, Veliko Tarnovo and Plovdiv. It, together with the Bulgarian Standardization Institute (BSI) and the National Production Holding 'Metal Constructions', participated in the creation of a research and production base ROSSA, as well as in the Bulgarian-Japanese association MEDICOM SYSTEMS.

With the help of BAS, ITCR bought two apartments in Tokyo, Japan, where a joint Bulgarian-Japanese Robot Laboratory and Computer-Based Laboratory began. Work stopped after the 1990 changes.

ITCR independently financed and built a residential building and by 1990 there were no employees registered as needing housing in the Institute. Also, ITCR has built a creative home in the mountain above Vladaya. The construction of such a home and the Veleka river on the Black Sea coast began.

During the period from October 1982 to the beginning of 1987, the Director, Assoc. Prof. Angel Angelov, was appointed ambassador of the People's Republic of Bulgaria in Japan and during that period the position of Director of ITCR-BAS was executed by Prof. Dr. Vasil Sgurev.

The Institute has two deputy directors, two secretaries of science and six scientific fields of application in each of which are positioned groups of six sections. At the head of these relatively independent directions and units are experienced and authoritative developers, researchers and scientists in the ITCR: Assoc. Prof. Dr.

Nikolay Iliev, Eng. Ivan Marangozov, Assoc. Prof. Georgy Nachev, Eng. Bogdan Stoyanov, Correspondent Member Angel Angelov, Prof. DTSs. Vasil Sgurev, Assoc. Prof. Nedko Shivarov. ITCR has a specific status in BAS, which gives it considerable flexibility and effectively stimulates the scientific and applied activity. All this allows ITCR to take a special place in the professional backgrounds of focused, fundamental and applied research.

At the end of the 1980s, the authority of ITCR at BAS in the professional and academic community was incredibly large. A significant number of young professionals and researchers aspire to work in this prestigious academic institute.

After the 1990 changes, the Department of Technical Sciences (DTS) was joined to other departments in another structure, and ITCR at BAS was divided into five independent academic institutes and a technology center with an experienced base. Over the next few years, the new ITCR institutes have been reorganized a few more times, and their composition and material abilities have sharply decreased. But this is a completely different story that deserves self-study.

ITCR at BAS can be considered as a specific unique project of the Bulgarian Academy of Sciences—as it would be said today, which in general proved to be very successful and left a lasting trace in the technical sciences and computer technologies in Bulgaria. In many ways, it was ahead of its time, for which it not only gathered many approvals, but not a little envy and malice. Its unique and successful experience can help in our country's ongoing efforts to develop high-tech industries.

Since 2000, the Department of Engineering Sciences (DES) has been established at the Bulgarian Academy of Sciences, which should be regarded as the successor of DTS at BAS.

The discussion about the role of the technical sciences in Bulgaria during the past century in the field of automation, robotics and computers must include the firm remark that this role is multidirectional and rather effective. This is related most of all to the initiation, the development and the serial production of the personal computers and the industrial robots. It also concerns the creation of computer cybernetic systems for a real-time control of industrial objects.

This activity helps to a great extent the development of high-tech production in Bulgaria during the past century.

References

1. Petrov, P.: The contributions of BAS institutes for research and production of computer and communication systems. J. BAS **2**, 53–62 (2016). (in Bulgarian)
2. Angelov, A., Sgurev, V., Petrov, P.: Some aspects of the automation and informatics development at the Bulgarian Academy of Sciences, Technicheska Myssal, XXXI, Anniversary Issue, pp. 14–19 (1994) (in Bulgarian)
3. Marangozov, I., Petrov, P., Hristov, M.: Personal computer 'Pravets-8A'. In: Proceedings of PERSCOMP'87, vol. 1, pp. 167–171. BAS (1987) (in Bulgarian)
4. Sgurev, V.: State, developed and perspective of personal computers of the 'Pravets' family. In: Proceedings of PERSCOMP'85, vol. 1. BAS (1985) (in Bulgarian)

5. Angelov, A., Petrov, P., Zhelyazkov, G.: In: Marangozov, I. (ed.) 'Technicheska Myssal' **1**, 9–31 (2004) (in Bulgarian)
6. Vachkov, P., Boyanov, K., Yanev, K.: Prospects for the development of personal computers in the 'Microprocessor Systems' economic holding. In: Proceedings of PERSCOMP'87, vol. 1, pp. 51–63. BAS (1987) (in Bulgarian)
7. Sgurev, V.: State of research on artificial intelligence in Bulgaria. J. BAS **3** (1986) (in Bulgarian)

Brief History of the Bulgarian Computing

Kiril Boyanov

Abstract The development of the computing technology in Bulgaria is examined for the period from 1961 till 1990 and insight is presented of the research and development of central processing unit and external storage devices. Data is provided for the manufacturing process and export from Bulgaria of computing devices.

Keywords Computing technology · Hardware · Software

The history of the Bulgarian Computing starts in the mid 1930-ies when Bulgaria imports "Powers" punching-card machines for statistical purposes. The first IBMpunching-card machines are hired for the needs of the railways, mining, insurance, and statistics in 1937. During those years, Bulgaria starts importing various mechanical and electromechanical machines and devices for accounting, to be used in different organizations and enterprises. Most of these machines were widely used until the 1960-ies. During the 1950-ies, the banking institutions and clerical departments of various organizations use punching-card and electromechanical machines not only for accounting but also for solving banking and insurance tasks.

In the 1950-ies, Bulgarian scientists were acquainted with the major tendencies in Cybernetics, and the abilities of the computers at the time. Bulgaria imported the French computer "Gamma 10" in 1962, and the Bulgarian company "Balkancar" acquired IBM1460 in 1963. The establishment of the first Computing Centre at the Institute of Mathematics at the Bulgarian Academy of Sciences (BAS) in 1th of June 1961 marks the birth of the Bulgarian electronic computing. In the Centre, under the leadership of Prof. L. Iliev, the first Bulgarian computer—"Vitosha" was designed, built and was fully functioning at the end of 1963. "Vitosha" is a computer with about 1500 tubes, and set of 32 instruction using index registers. Its main memory was based on magnetic drum and had 4096 words. The programming was in machine code. In the team which designed and developed the first machine were G. Alipiev,

K. Boyanov (✉)
Institute of Information and Communication Technologies, 25A, Ac. G. Bonchev Str., 1113 Sofia, Bulgaria
e-mail: boyanov@acad.bg

© The Author(s), under exclusive license to Springer Nature Switzerland AG 2021
K. T. Atanassov (ed.), *Research in Computer Science in the Bulgarian Academy of Sciences*, Studies in Computational Intelligence 934,
https://doi.org/10.1007/978-3-030-72284-5_4

Fig. 1 View of first Bulgarian computer Vitosha

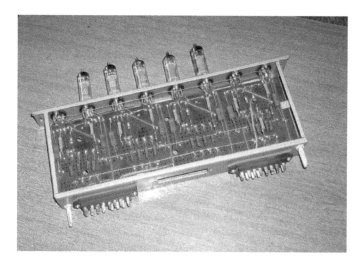

Fig. 2 Changeable module of Vitosha computer

R. Aslanyan, K. Boyanov, M. Dimitrova, D. Bogdanov, E. Kurmakov, S. Pashev, D. Rachev, Bl. Sendov, I. Stanchev, I. Yulzari [1].

The word size of the machine was of 40 bits and represented: fixed point integer in two's complement representation, two one-address commands (instructions). They are represented in octal numeral system. The main blocks of the machine are: Arithmetic unit, Control unit, Main memory, Input device, Output device (Fig. 1). The main registers of the machines are latches, built on the double triode ECC 862. The electronic circuits are built on changeable modules (Fig. 2).

The construction of the machine is about 4 m long and 2 m high, where some 200 modules and the magnetic drum are situated. The average time for access to data is 10 ms, the recording system is two levels with return to zero. The input device is punched tape operating at a speed of 7 lines per second. The output device uses electrical typewriter with wide roller and it reaches a speed of 15 characters per second, converting the numbers from binary into decimal by a program. The cooling of the machine is mandatory. The total power consummation is 12 kW. The power supply is 220/380 V and is stabilized by motor-generator group. The control of the machine is carried by a control panel at the front.

The Russian machine Minsk 2 was installed in the Institute of Mathematics with Computing Centre (IMCC) in 1964. A team of mathematicians and engineers enhance the software for tasks from scientific and research institutes and design organizations.

The first Bulgarian electronic calculator—ELKA was designed in 1965 at IMCC by St. Angelov, L. Antonov and P. Popov. This was a significant achievement of its time, having in mind that there were only three known devices in the world of this kind (products of "IME-84," "FRIDEN" and "ANITA"). The calculator used transistors, it had 16 bit registers with a number of advantages and improvements compared to the existing at the market at that time: square root, integer division, average numbers, fixing of decimal point with rounding up, etc. ELKA found soon a wide number of applications and became rapidly employed. Following it was the design of a series of calculators ELKA 22, ELKA 25 (with printer). The Scientific and research institute on electronic calculators (NIPKIEK) was found in 1969. The first calculators on Integrated Circuits—ELKA 42 were developed. This was the model, Bulgaria presented at the World's Fair EXPO 1970 in Osaka and this was the only calculator built using IC—at the top level at the time [2].

The design of ELKA 42 was with MOST IC—the UNIMOST series, developed in the Institute of microelectronics and it consisted of standard set of counters, latches, and logical elements.

The logical circuits of the calculator were realized in "logical matrixes". According the modern terminology—micro-programmed ROM. The unification improved significantly the technical and economical features of the products. The next series are ELKA 50, ELKA 55 where new Bulgarian electronic circuits are embedded with some 1000 elements in a 24 pin chip. This allowed the number of IC to be up to ten. Due to the use of structures of the type "logical matrixes", a new calculator with new design, smaller dimensions and weight was created—ELKA 40. Four bit microprocessor configuration CM-500 was developed. These were the first in the country and in the socialist block 4-bit microprocessor series. Using them, ELKA 51 was produces, where the number of MOS IC was reduces from 10 to 4.

Pocket calculators ELKA 130 and ELKA 135 were produced following an order of a Swiss company and 50 000 have been exported to Switzerland and also some quantity for Italy.

The period from 1964, when the first calculator was developed until 1974, when the last calculators using domestic IC, was a very successful one both in techno-logical and production terms. The design and production of calculators in Bulgaria stimulated the development of many technologies in Bulgaria—for example—the printed circuits board technology.

The microelectronic was developed because it had the specific task to created and produce calculators. Since the foundation of the Plant for microelements in Botevgrad, the main user of those elements, starting with germanium transistors was the calculator production, and then followed the integrated circuits—the UNIMOS series, which had small and medium level of integration and last came the microprocessor systems for calculators and other automats.

The most original solution of the UNIMOS series were the universal base IC: counters, registers, logical blocks, etc.

The Institute of microelectronics was one of the institutes that started production of IC for calculators and reached modern IC on MOST technology. Its first director—corresponding member of BAS Yordan Kasabov, a prominent scientist and constructor left us rather early. His followers—I. Zarkov and K. Filiov did a lot for its growth.

For the period 1971–1985 Bulgaria exports calculators at a volume of more than 480 million leva, including for countries in the West—for 4 million leva. In their design a number of prominent experts take part: L. Antonov, Y. Kasabov, S. Hritova, V. Chilov, D. Shishkov, I. Stanchev, V. Elenkov, Z. Aleksandrova, E. Parvanova, I. Staneva, Y. Petkov, G. Ganchev, G. Kazandjiev, E. Pandov, M. Zaharieva, G. Chakyrov, M. Medarov, I. Minchev, P. Gerginov, D. Borshukov, S. Nachev, I. Zarkov, Ch. Bogoev, D. Vateva, D. Peshin, S. Hovanesyan, etc.

In 1965, under the directions of Prof. I. Popov—Head of the State committee for scientific and technical progress, a decision of the Council of Ministers (CM) on the development of the computing and organizational technique has been prepared. Experts from IMCC—E. Karmakov, S. Srebrev, P. Popov, B. Bonchev, B. Hristova traveled to Japan. Upon their return in Bulgaria, under the leadership of I. Marangozov, they start the design of the ZIT 151 machine, following a license of the Japanese company FACOM. In the Plant for Computing technique in Sofia, 20 machines have been produced. Following a decision of the CM № 25 from March 1st 1966, a group of experts from IMCC, working in the field of computing and software is transferred to a newly established Central Institute for Computing Technique (CICT) with 233 employees out of which 101 experts with university education and 24 research associates. In 1980 IMCC provides the fertile environment that allows CICT to have some 2700–3000 staff out of which 1/3rd is research associates and experts.

At the beginning of 1967, DSO "IZOT" was established, which in a very short period of time created a base for modern industrial production of computing systems and devices. According to data from V. Nedev, for just 17 years, DSO "IZOT" provided income for Bulgaria for 11,230 billion currency leva and a profit of 6.480 billion leva, while the expenses were only 633,350 million leva. Here the proper credit should be given to the professionals, who under the leadership of Prof. Ivan Popov formed the national policy in the area of Computer science. The general directors of DSO "IZOT" that actively participated were S. Chavdarov, V. Nedev, I. Tenev, A. Shopov, as well as other economic managers: S. Markov, P. Kisyov,

J. Mladenov, A. Stamenov, L. Vitanov, L. Kozlev, V. Hubchev, L. Guturanov, V. Tzarevski, P. Vachkov, M. Marinov, D. Dimov, A. Trifonov, etc. A long term strategy with clear objectives was prepared, with specific intentions in which areas Bulgaria should produce what its specialization should be and in what directions the efforts must be aimed. The development of the first products, that were compatible with Western products started soon. For example—IBM for mainframes and DEC for mini machines [2, 3].

The design and construction of the Plant for mechanical constructions in Blago-evgrad, the Plant for magnetic disks in Stara Zagora, the Plant for printed circuit boards in Ruse, the Plant for magnetic tapes in Plovdiv, the Plant for magnetic-disk packets in Pazardjik and the Plant for instruments and non-standard equipment in Shumen started in 1970.

In the DSO "IZOT" system a number of organizations worked very successfully—CICT, NIPKIEK, the foreign trade organization IZOTSERVIZ and one for the design of systems—SISTEMIZOT.

The Eastern block countries (COMECON) established an Intergovernmental Commission for Cooperation in the areas of Computing in 1969. One of the founders and Bulgarian member of the commission was Prof. Ivan Popov. He is regarded to be the founder of the Bulgarian Computing industry. The main task of the Commission was coordination of the promotion and manufacturing of the computers within COMECON. The Commission took a decision to develop and start production of a Unified Series (US) of computers and software for them. This concept was adopted because the import of such products from the developed countries required hard currency. USSR, Bulgaria, Hungary, Poland, East Germany, Czechoslovakia were the first participants in the design and production of the products from the US. Cuba and Romania joined the Commission in 1973 [2].

The Executive Boards of the Commission were a Coordination Centre based in Moscow and the Council of the Chief Constructors. Several Councils were established—for mainframes, minicomputers, personal computers, etc. Each Council consisted of representatives from each member country—in most cases the director of a leading institute in a country. In the Council of the Chief Constructors Bulgarian representatives were: for mainframes—A. Angelov and Zh. Zhelezov, for mini-computers—S. Srebrev and V. Elenkov, for personal computers—K. Boyanov.

At the councils of the Chief Constructors have been organized councils of experts on the main computer system blocks: architecture, Input/Output devices, storage devices, channels and interfaces, software, standards and documentation, etc. In each council several experts from each country have been nominated. The councils gather on regular basis 2–3 times per year and the meetings are held mainly in Moscow in one of the buildings of NICEVT (scientific and research centre on electronic and computing technique), known as "Detskij sad" (Kindergarden). The member countries have also hosted meetings on a rotating principle [2, 3].

The Unified series of computers (US C) was also known as "RYAD". This name meant machine family with growing performance and program/software compatibility—the software working on the models with lower performance should also work on the models with higher. The technological and constructive bases were unified in term of use of standard integrated circuits (IC) and common constructive parts. The main elements of the construction were standardized, unified codes and security access levels were introduced, unified connection of the devices (standard interface) was used. The adopted standardization led to favorable environment for international labor division during the design and production of the systems and the various devices. This allowed to great extend to facilitate the use and service of the products. A naming scheme for the devices of the Unified Series was adopted—code—the letters EC (for Unified Series) and four digit number [4, 5].

The first digit of the number defines the class of the device, following the classification: 0—nodes and components, 1—computing systems, 2—CPU, 3—main memory, 4—channels (interfaces), 5—external storage devices and their control, 6—input devices, 7—output devices (print, etc.), 8—teleprocessing devices, 9—preprocessing data devices.

The second digit carries the classification further and denotes a group of devices—for example—50 denotes external storage, 55—control for external storage, teleprocessing devices start with 8, 80 is the code for modem, 81—for error protection, 84—for multiplexors, 85—for terminals.

The third digit concerns the parameters of the devices and those from certain group can be further subdivided according to the most important parameters. For the computing machines this number shows the models with better performance: EC1010, EC1020, EC1030, etc.

The fourth digit is used to denote the various devices in a subgroup or their modification.

As an example, let us look at the EC1020 system. Central processor (CPU)—EC 2020, console with typewriter EC 7074, punch card input device EC 6012, printing device EC 7071, magnetic tape control device—EC 5512, magnetic tape storage EC 5012, disk storage control device EC 5552, magnetic disk storage device EC 5052.

A similar system was deployed for the design and development of minicomputers, where the family was known as (CM). During time, the coding system (the digit) was modified according to the new requirements and technologies. Integral circuits were built using TTL in addition with silicon semi-conductor elements. Gradually ECL, big IC and others were introduced. The unified construction nodes were four: printed circuit board (TEZ—type element for change), cassette, frame and cabinet. The printed circuit boards were standardized and with the change of technologies, new standards were introduced. The consolidation of the boards was in cassettes.

The cassettes were mounted in frames and the lower part of the frame accommodated the power supply and the ventilation blocks. The last level—the cabinet allowed placement of up to 3 frames. For the rest of the devices that were outside the CPU, unified constructive elements were used, according to their specific technical requirements.

The production of computing systems and devices had to be in line with the Unified system for constructor's documentation (ESKD). This system encompasses technical and organizational requirements that lead to documentation exchange between various enterprises. ESKD allowed enhancement and unification during the design and development of industrial goods, simplified the form of various documents and allowed for their automation. It was compliant with the adopted standardization documents in COMECON and was constantly improved in accordance to the introduction of new technologies. Several classification groups of standards were drafted and each one of them had the option of including of 99 standards [5].

Bulgarian computing advanced in three main directions: mainframes, mini and personal computers, disk and magnetic tapes systems.

The mainframe machines from the Unified System (US)—EC 1020, EC 1022, EC 1035, EC 1037 were sold primarily in the Eastern block countries. The very first mainframe computer produced from this series was EC 1020. It was a result of the joint efforts of a Bulgarian team from the "Central Institute for Computing Technique" in Sofia and a team from the "Scientific Research Institute for Computing" in Minsk (НИИЭВМ). The computer was compatible with IBM360/40. Amongst the Russian specialists were G. Lopato, V Prijalkovski, G. Smirnov, N. Maltzev, V Kachkov, R. Astzagurov, A. Fleorov, etc. On the Bulgarian side were S. Angelov, G. Alipiev, K. Boyanov, S. Srebrev, N. Sinyagina, V. Lazarov, T. Velichkov, I. Georgiev, N. Ganchev, B. Drumev, K. Batmazian, S. Namliev, P. Popov, V. Kisimov, R. Papazov, T. Kanchev, H. Turlakov, G. Hadjidimitrov, I. Dalbokov, K. Yanev, Y. Raikov, N. Sherev, G. Ganchev, G. Draganov, N. Ikonomov, Z. Zlatev, M. Ivanova, K. Kirov, I. Minev, D. Petrov, P. Daskalov, K. Stankov, A. Spasov, A. Takov, etc.

EC 1020 has micro programmed control with 142 instructions, 256 KB main memory with a cycle of 2 μsec and performance of 10 000 operations per second. It has 1 multiplex and 2 selector channels at speed of 200 KB/sec. Eight controllers for peripheral devices could be attached to each channel. The improved version of that machine—EC 1022 had performance up to 0.08 MIPS [2].

The capacity of the main memory was up to 512 KB. Both EC 1020 and EC 1022 were software compatible with the IBM 360 series machines. The next stage of the main frame development was the EC 1035 machine, a joint development with НИИЭВМ, from Minsk (Belorussia).

The production of this machine started in 1978 and it was capable of producing 200 thousand operations per second. The maximum volume of main memory was 1 MB, the number of channels was increased to 5, with improved throughput of 1 MB/sec. An important characteristic for EC 1035 was that it used not a disk based OS (like EC 1020 and EC 1022) but OS 6.1 and operational system OS 351, which is compatible with the IBM 370 architecture.

The development of EC 1035 is an important moment in the history of the creation of mainframes as a number of new devices with bigger capacity was designed for it—disk and tape subsystems. The joint teams that developed EC 1035 overcame the significant difficulties with the transition to the new technology.

The first independent design of electronic computer in Bulgaria was of EC 1037, the production of which in 1987 (Fig. 3). EC 1037 is mid class universal computer aimed at solving a wide number of technical and economical tasks and is compatible with the systems built on the IBM 370 architecture. It was alternative to the models IBM 4331 and IBM 4341. It had CPU, disks using the Winchester technology and tape device with improved characteristics. The system included character-digit printing device EC 7033 M, punched card input device EC 6019, multi-console

Fig. 3 View of Bulgarian super computer with matrix processor EC1037

station EC 8566, processor for video data processing EC 8371, matrix processors EC 2706 and EC 2707. The processor performance was about 2 million operations per second and the main memory varied between 2 and 16 MB. The data speed transfer reached 2 MB, while in the byte-multiplex channel it went up to 350 KB. The channel subsystem had up to 12 modified channels, each of which could work in each of the following modes: multiplex, block-multiplex, selector. A special service subsystem supported the link between the operator and the computer and additionally carried out control and diagnostic services for the processor. The diagnostic system was very well designed. Matrix processors produced in Bulgaria could have been attached to EC 1037. Using this machine, the Institute of space research in Moscow has been controlling the space station "Vega". The team that carried out the design included: V. Lazarov, D. Minev, E. Naumov, Z. Yancheva, Z. Zlatev, Y. Ivanova, K. Kirov, M. Tashev, M. Ivanova, D. Petrov, P. Popov, O. Kostadinov, P. Daskalov, S. Serbezov, T. Velichkov, N. Tashev, F. Filipov, H. Setyan, P. Kozhuharov.

The development of the first Bulgarian mini-computer started in 1974, using as a prototype PDP8L of the company DEC. The mini-computer was known as IZOT 310 and its word was 8 bits and time for execution of the instruction- 2–12 μsec, main memory—64 KB and UNIBUS interface.

With a decision of the Intergovernmental Commission for Cooperation a development of series of mini-computers started—CM2, CM3, CM4, which are compatible with DEC's PDP11/34, VAX11/730, and VAX11/750.

In organizational aspect, the policy remained the same—a Council of Chief Constructors on minicomputers and specialization of various countries. 16 and 32 bit mini-computers were designed. The former had a processor cycle of 0,2—0,4 μsec, main memory up to 124 KB, UNIBUS interface, while the later had main memory of 2 MB and UNIBUS interface.

Bulgaria has been designing the models CM 1426 (IZOT 1054 C), CM 1706 (IZOT 1055 C), CM 1504 (IZOT 1056). The team working on the series of mini-computers was: S. Srebrev, K. Boyadjiev, I. Aleksiev, P. Popov, N. Gelibolyan, A. Velichkov, G. Kukureshkov, A. Kamenov, L. Bonchev, T. Valchev, B. Simeonov, Y. Yankova, D. Micev, M. Angelova, K. Cocheva, etc.

The design and development of personal computers in Bulgaria started at the end of the 1970-ies, when in the Institute of technical cybernetics at BAS, Inko1 was presented. Its constructor was Ivan Marangozov, who, together with his team was successful in developing a working model of the famous at its time Apple computer. In the next few years, a powerful modern technical base was created, which allowed the production of these computers.

Following a reorganization of the Institute of technical cybernetics in 1978, work on the next generation personal computers, IBM compatible, started. In such way Pravetz 8 appeared (still Apple compatible), and soon after that—Pravetz 16, which was IBM XT/AT compatible. The first personal computer that was distributed in a wider scale was Pravetz 82, being in serial production since 1982. It had main memory of 48 KB and had options for graphical mode, which allowed use of color monitor.

The personal computer IZOT 1031, which has been designed in CICT, went into production in 1984. It was compatible with Atari 3. The construction was based on the Z80 microprocessor and had main memory of 64 KB. The chief constructor of this personal computer was Vladimir Chilov with a team of very capable experts.

The production of Pravetz 8A, which is software compatible with Apple 2E started in 1987. It uses new hardware solutions and the main hardware base was with elements of Bulgarian production. The main memory was enhanced to 1 MB.

In parallel with the perfection of the 8-bit personal computers, the teams of I. Marangozov and Vl. Chilov designed 16 bit personal computers. The production of Pravetz 16, which was build using the processors 8086 and 8088 and was software compatible with IBM PC/XT, started in 1985. Its basic configuration included 256 KB main memory, which could be enhanced to 640 KB, two floppy disk devices, each with capacity up to 500 KB, video controller and color monitor. Additional modules for main memory enhancement, for parallel interface, for hard disks had been created. The production of the improved model Pravetz 16A started in 1987. These models had been designed under the leadership of I. Marangozov and a team—G. Jeliaskov, K. Dosev, N. Popov, P. Petrov, K. Koruchev, G. Georgiev, St. Kuzarov, V. Chilov, J. Kisiov, N. Iliev, R. Raichev, Cr. Hristov, P. Somov, M. Israel, etc [3].

Since 1987 started the production of: Pravetz 16I, which is a laptop with embedded monitor and up to 3 floppy disk devices, Pravetz 16 V—with vertical construction and reduced size, with two 5.25 inch floppy disk devices, and Pravetz 16A, which is a desktop personal computer. For those personal computers, a number of additional modules on the basis of Intel 8087 and Intel 80,286 have been developed—such are the 2 MB main memory module, controllers for external storage, color video controllers with high resolution, LAN adapters, and interface adapters for the IEEE 488, RS 232 and RS 432 standards.

The cooperative "Microprocessor system" with general director Plamen Vachkov was founded in 1987. It developed in a very short period of time its base, including in it the newly created Institute of microprocessor technique in Sofia, the Plant for microprocessor system in Pravetz, the Plant for power supply devices "Analikik" in Mihailovgrad (today Montana), the Plant "Elektronika" in Gabrovo, the Plant for printed circuit boards in Pravetz, the Plant for production of instrumental equipment in Gorna Malina.

The personal computer IZOT 1036 (EC 1831) was created in CICT in 1985 and started being produced, being used mainly as intelligent terminal. The personal computers IZOT 1036 (EC 1831) and IZOT 1037 (EC 1832) were adopted for production in the factories of the newly created union "Inkoms" in Veliko Tarnovo and Silistra. These personal computers were developed by teams with leaders Hristo Turlakov and Todor Kanchev with the following experts: N. Vecev, V. Getov, A. Simeonov, O. Gorchakov, S. Machev, N. Petrov, V. Barbutov, S. Voinov, M. Simeonova, S. Stanchev, B. Filipov, N. Dabov, S. Dimitrov.

The personal computer Pravetz 286 (EC 1838), based on Intel's 20,286 microprocessors were created in the Institute for microprocessor technique (director K. Boyanov) at the end of 1986 and later adopted in production. It had main memory 3 MB, address space—16 MB, multilayer and color graphical monitor with options for use of arithmetic coprocessor, synchronous and asynchronous communication modules, and controllers for LAN with ring and bus topology. The developing team was lead by Zlatka Alexandrova and prominent experts: S. Pishtalov, I. Saraivanov, D. Lilov, Y. Visulchev, E. Aleksieva, B. Bachvarova, N. Germanova, H. Karagetliev, Sh. Koen, G. Marinov, L. Nedeva, S. Rashkova, M. Ribarska, K. Todorov, M. Treneva, I. Cankova, A. Aleksandrov.

The creation of external storage started in 1970 when the first storage device, magnetic tape—analogue of Fakom 603E was produced. On the basis of the experience gathered by the leadership of Ivan Arshinkov, in 1971, the first product of the Unified System—a magnetic tape storage device EC 5012/01 was designed. The mini-tape device IZOT 5003 was adopted in 1975. During the next couple of years, a series of storage devices on magnetic tape had been manufactured. At the end of the 1970-ies, a quality improvement of their parameters took place—phase modulation was introduced and also automatic load of the magnetic tape and a group-coded write method. Magnetic heads with hard chrome and ceramic coat were adopted. During the 1980-ies, a number of devices with dataflow mode of operation had been developed. Some modifications of magnetic tape storage devices and their parameters are given in Table 1. For the control of the devices—the appropriate controllers had been manufactured [2].

The team that has developed the magnetic tape storage devices and the controllers for them was: I. Arshinkov, D. Dyakov, H. Momerin, T. Popov, M. Kolarov, L. Markov, M. Strahilova, I. Dimitrov, T. Topalov, M. Tarpesheva, Y. Raikov, A. Spasov, P. Manolov, H. Rashev, K. Stankov, V. Tenev, L. Fenerdjiev, etc.

Table 1 List of tape drives production

Tape drives production				
EC 5012.03	Real tape unit	96 KB/s	32 bits/mm	1972
EC 5612	Real tape unit	190 KB/s	63 bits/mm	1977
EC 5026	Real tape unit	492 KB/s	246 bits/mm	1985
EC 5027	Real tape unit	738 kB/s	246 bits/mm	1986
EC 5028	Real tape unit	1230 KB/s	246 bits/mm	1988
EC 5710	Stream tape unit	160 KB/s	63 bits/mm	1987
	20 MB cartridge tape unit	90 KB/s	394 bits/mm	1988
	60 MB cartridge tape unit	55 KB/s	315 bits/mm	1989

Table 2 List of disk drives production

Disk drives production				
EC 5052	7,5 MB	Disk driver 14"	156 KB/s	1971
EC 5061	29 MB	Disk driver 14"	312 KB/s	1973
EC 5066.01	100 MB	Disk driver 14"	806 KB/s	1977
EC 5067	200 MB	Disk driver 14"	806 KB/s	1977
EC 5063	317 MB	Disk driver 14"	Winchester	1982
EC 5063	635 MB	Disk driver 14"	Winchester	1986
CM 5508	10 MB	Disk driver 5,25"	Winchester	1987
CM 5510	160 MB	Disk driver 3,5"	Winchester	1990

The design of an external storage device with magnetic disk was carried out under the leadership of Zhivko Paskalev, having disk packages developed for it. In the next couple of years, a series of magnetic disks had been adopted for both the Unified Series of machines and the mini-machines series.

Disk devices for personal computers have been developed in 1985. Table 2 lists some of the manufactured products. In parallel, controllers, disk packages and floppy disks were created.

The team that developed these products was: Zh. Paskalev, N. Botev, B. Conev, B. Hristova, B. Cenkulov, L. Yordanov, K. Mitev, L. Petrov, G. Mutafov, I. Kovachev, V. Denishev, R. Kadijska, A. Blagoeva, M. Avramova, D. Aleksandrov, D. Todorov, L. Mihov, N. Sinyagina, G. Malinovski, O. Carnorechki, etc.

In 1984 the design and development of teleprocessing systems begun. The ESTEL system had been assembled, including a machine from the Unified System, multiplexor, modems and terminal devices.

Software was also written, including base software and application programs for a wide range of problems. Different devices for computer networks had been developed: multiplexors, synchronous and asynchronous adapters, etc. Members of the team were: I. Yulzari, K. Vitanov, V. Altanov, V. Videv, A. Dochev, Z. Zlatev,

R. Iliev, E. Yonchev, S. Karagyozov, A. Matrazov, P. Pavlov, M. Grueva, R. Savov, T. Kanchev, G. Hadjidimitrov, K. Yanev, E. Dimitrov, V. Markov, V. Vladova, A. Pamukchiev, P. Chernokozhev, B. Raichev, M. Petrov, S. Basmadjieva, L. Zabunov, B. Iliev, S. Krastev, I. Vladikov, V. Damyanova, etc.

The industrial organization in Bulgaria could have been viewed in hierarchical structure. The plants and enterprises belonged to 3 groups.

The plants for main components and base clients belonged to the first group; for big components for the computers, which can be sold as separate devices—to the second group, the plants for computers, end devices and complex system—to the third group. Close links with mutual dependencies and coordination existed between the plants, which led to good management and effective production. During the development of the computing industry some important production areas were neglected, mainly in the field of microelectronics and passive elements. For them, the country relied on the cooperation with various socialist countries but later it became evident that this approach did not lead to very good results. In total, in the area of computing, more than 30 plants were functioning, some of which listed below [2].

The first group includes:

1. **The Plant for printed circuit boards and technological devices in Ruse**. It had a capacity of production of 30 000 to 50 000 square meters printed circuit boards per year. The plant had about 2500 employees.
2. **The Plant for mechanical constructions in Blagoevgrad**. It produced mechanical constructions for mainframes and mini-computers, cases for personal computers, constructions for magnetic disk and magnetic tape devices, etc. The plant was had about 1800 employees.
3. **The Plant for instrumental equipment and non-standard equipment in Shumen**. In that plant, instruments, plastic press forms, stations, matrices, and others were designed and manufactured. It had a well-developed constructor-design unit and more than 1500 workers and engineers that created more than 5000 instruments annually.
4. **The Plant for power supply devices in Harmanly**. It produced various power supply devices: traditional, high-frequency and no-transformer for mainframes and mini-computers, and also for disk and tape devices.
5. **The Plant for magnetic heads in Razlog**. The plant had 1500 employees and used to produce big quantities of magnet heads for tape and disk devices.
6. **The Plant on electronics in Gabrovo**. It produced keyboards for terminals and personal computers, digitizers, various types of plotters, etc. The plant had more than 800 employees.

The second group included:

1. **The Plant for magnetic disks (ZZU) in Stara Zagora**. This was the most advanced and modern plant in the Bulgarian machine building and electronics. It used to produce magnetic disks for the countries of the ex-socialist block. It had an export for other countries as well, and the annual export exceeded 1.2 billion rubles, which was more than 1 billion dollars at the exchange rate at the

time. The plant had more than 5000 employees and had excellent for its time equipment.

2. **The Plant for magnetic tapes (ZZU) in Plovdiv**. It was the second most important in the Bulgarian electronics, where magnetic tape devices for mainframes and mini-computers were produced, and also devices for data preprocessing.
3. **The Plant for magnetic disks in Pazardjik**. It manufactured changeable packets and after the introduction of new technologies, it started producing videocassettes, floppy-disk devices and test systems. It had more than 1500 highly trained employees.
4. **The Plant for typewriters in Plovdiv**. Its yearly production reached 200 000 portative mechanical typewriters. The plant used to assemble copying machines following according to a cooperation agreement with the Rank Xerox company.

The third group included the plants for final production:

1. **The Plant for computing technique in Sofia**. was one of the oldest in"IZOT". It manufactured predominantly mainframes.
2. **The "Elektronika" plant in Sofia**. It has a profile for the production of mini-computers, compatible with DEC machines, series PDP 11 and VAX 730/750. Its employees were almost 2000 and yearly turnover of 1700 million dollars.
3. **The Plant for systems and teleprocessing in Veliko Turnovo**. It manufactured products for teleprocessing, close range converters, magnetic tape devices, modems, etc. The plant had about 1500 employees.
4. **The Plant for personal computers production in Pravetz**. It united 3 plants: for printed circuit boards, for mechanical constructions and for personal computer assembly. The later was equipped with production pipelines, fully automated with a capacity for production of 100 000 computers per year. The plant had more than 2000 employees.

During the period 1975–1989, the Bulgarian share of export of computers, computing devices and software products reached 40% of the total volume of exports within COMECON. This industry employed more than 120,000 people in 9 companies (Tables 3 and 4).

The volume of export had been very big for a small country like Bulgaria and the profit from these products was very big, far more than 20–22% (for some products it exceeded 200–300%), i.e. the maximum for a market economy. One reason for this was the special position of the computer industry and electronics, as such products were still in high demand on the market of the ex-socialist countries and such demand dictated their higher prices. Another reason was the monopoly position of Bulgaria on the market because the country was able to modernize its production in a very short period of time, making it relatively efficient and the equipment in the Bulgarian plants were at world level. Almost all machines, technological pipelines, equipment were imported from the most prominent world companies from Europe, Japan and sometimes from the USA.

For some of them the existing embargo had been broken—the foreign-trade organizations managed to find a way for delivery of banned for Bulgaria goods, on higher

Table 3 IT volume of exports within COMECON (in rubles)

MEA countries

Export (mil rubles)	Total	Bulgaria	Hungary	DDR	Cuba	Poland	Romania	USSR	Czechoslovakia
	3174	1653	245	472	14.6	404	36	153	197
(%)	100	52	7.70	14.90	0.46	12.70	1.13	4.81	6.21
Import (mil. rubles)	3174	36	28.6	223	28	80.7	67.7	2390	321
(%)	100	1.14	0.90	7.02	0.88	2.54	2.13	75.30	10.10
Total turnover (mil. rubles)	6348	1689	273.6	695	42.6	484.7	103.7	2543	518
(%)	100	26.60	4.30	10.10	0.67	7.63	1.63	40	8.17

Table 4 Some economical parameters for IT sector in the period 1980–1990

Year	1980	1985	1988	1990
Total amount of enterprises	144	165	204	206
Total personal involved (thousands)	126	148	169	181
Percent of total Bulgarian labor force (%)	9.3	10.6	11.7	13.1
Assets (mil.USD)	1154	1935	3162	3949
Total production (mil.USD)	3861	4951	7387	5436
Percent from Bulgarian total Industrial production (%)	9.3	11	14.5	12

Table 5 Number of specialists and managers working in the IT sector

Fields of science	Total	In research and technological organizations	In Universities
Radio technique, electronics, communication technique	945	654	291
Device manufacturing, automation, telemechanics	2379	1949	430
Physics[a]	1351	839	512
Ph.D. students	615	–	–
Total (1, 2 and 3)	4675	3442	1233

[a]Taken as a representative of the natural sciences, serving to a great extend the field of electronics

prices, of course. As a final result—the Bulgarian factories and plants had a world and European class equipment and were lagging behind only in few areas.

In the field of technological equipment of the production of disk devices, compact discs, personal computers and many other products, our plants and enterprises were some of the best equipped in Europe and were only behind those from USA. We talk only about certain plants and factories but the rest were also on a very high technological level. This was due to the huge investment on behalf of the state and the high level of the personnel and the management staff, who were able to determine the technological process (Table 5).

The production of several devices typical for the Bulgarian export list is given in Table 6.

The economy managers, factory directors and their teams and the institutes that were designing and developing the products and the technologies for their production were at a very high professional level.

Not surprisingly after the 1990-ies most of the found jobs in the West and are still in demand. Very often it is said that the products have been copied from Western companies. This is not exactly so. Most of the world producers, starting with Japan, have copied their initial products or parts of them. And this happened in the environment of full market economy, where copying is far easier, as each company or trade organization can buy any element, detail or component of a product, to buy the product itself and to copy it one-to-one. For the Bulgarian conditions this was

Table 6 Export list for typical Bulgarian devices

Name	1989	1990	Till 06.1991
Minicomputers CM-ЕИМ	80	58	–
Micro calculators	32,806	32,346	20
Magnetic tapes for ES ЕИМ	4776	1087	638
-„- CM ЕИМ	52,861	12,885	848
-„- Magnetic drivers for ES ЕИМ	44,612	85,854	101,103
-„- CM ЕИМ	125,138	112,220	5831
Magnetic data inscriber	5878	1798	–
Devices for teleprocessing	57,586	–	–
Disk packets	166,897	156,630	16,156
Diskettes	15,185	6037	11,241

impossible as the new product felt under the embargo and their components could not be purchased, and just 3–5% from the embedded details could had been Western production. So—in no way one can speak on copying and prototyping.

Our task was to create products, which had to operate in the same functional way as those, produced by Western companies. This meant that it was very difficult to make a product that completely corresponds to the Western product and its features, including its look, construction, etc. However, in regard to the main features, technical parameters, applications, use of software, it had to have a similar functioning, so it is better to say that the work was on similar functionality products or **analogues**.

Sometimes the parameters of our similar products were better than the Western goods and this was due to two factors: first—the shortcomings of the products in production were already known and our constructors had been correcting them, and second—due to the weaker competitiveness, we had in many cases more time for design and detailed tests, experiments in extreme conditions of the products—all this allowed getting better results and having products with better qualities.

Let me make a note here. The opinion that our production were with lower quality, give more defects, fail more often, etc. was well known but was not always true and was to a great extend exaggerated. Most of the exported products, and even more—those with special purpose for the military industry were of remarkable quality and reliability during their use. If some of the initial series had some defects, during the production cycle in few months the complaints were dropping down significantly, and even more—their exploitation lasted long after the period of time, given in the documentation. Even nowadays, some countries continue to use our devices, manufactures in the 1990-ies, which is very rare for a computing product.

In order to develop new generation computers, with parameters close to the best available computer systems in the world, a Commission for International Cooperation for the Academies of Sciences of the Eastern Block countries was established. To implement the program of the Commission, a Centre for Informatics and Computing Technologies (CICT) at BAS was founded in 1985. The first director of CICT, and deputy president of the Commission, was Acad. Bl. Sendov. The research group of the Centre developed and built new generation parallel computers—matrix processors, a transputer based (T414 and T800) parallel computer, and a dataflow parallel machine. Highly parallel computer architectures were the solution for satisfying computationally intensive tasks.

However, the computing power attainable through parallel processing had to be combined with the ability to reconfigure the topology of the interprocessor structure in order to provide a smoothly expansible range of facilities. The Advanced Parallel Systems (APS) combined parallel processing with a reconfigurable topology to cover a wide spectrum of applications [3].

The transputer was used as a basic microprocessor element for building multiprocessor structures. With its local memory, built-in floating point unit, four communication channels and its parallel programming language Occam, the Transputer was one of the most eligible microprocessors for parallel systems. The transputer architecture allowed modular design of large networks with up to thousands of processor nodes.

Using transputer technology, the APS family offered a computing environment with a network structure, easily reconfigurable according to the characteristics of the application task. The computing environment was comprised of a set of computing nodes and parallelism could be easily scaled by adding additional high performance modules.

The APS family included personal computers, workstations, mini-supercomputers and supercomputers. All models were highly parallel systems, the personal computers having from one to 16 nodes, the workstations from 16 to 96 nodes, the mini-supercomputers from 64 to 256 and supercomputers from 512 to 1024.

The form of modularity used had allowed the development of a whole range of program-compatible machines, starting from a single board system. The performance ranges from several millions to several billions of operations per seconds.

Programmable reconfiguration of the network topology was provided for all models of the APS family. The processor nodes and the host computer communicated through programmable switches.

The APS 48 was designed to be supported by existing system software for parallel computers. The following software packages could be used: the Inmos Transputer Development System, the Helios operating system, and compilers for parallel languages.

The transputer development system was used for writing and debugging of parallel programs. It works under the control of MS DOS.

The distributed operating system Helios was multitasking and supported the execution of parallel user programs on one transputer or on a network of interconnected transputers.

FORTRAN, PASCAL, PROLOG and C compilers were available for the development system and for the operating system. Along with the standard features, these languages were supplemented with extensions for writing parallel programs.

In IMPS and CCIT, a family of transputer boards was designed and developed and they were embedded in personal computers, turning them into Parallel Computations

Workstation. For example, ten-transputer enhanced computing board for a PC based workstation for parallel computations had the following technical parameters:

- 10 IMST800 20 MHz transputers giving 80 MIPS or 15 MFLOP speak performance
- Link speed at 10 or 20 Mbits/s
- PC interface—IMSC012 link adapter
- 20 edge connectors providing 40 serial Imnos links.

Cresta Marketing Company (UK) reported "And yet the Sofia Academy of Sciences has produced transputer boards, which were tested by DTI at Strathelyde University where they were certified as excellent".

New industrial corporations were created in 1987, aimed to provide the industrial base for Bulgaria's Computing industry. They were: "Information and Communication Systems" (general director L. Guturanov), "Microelectronics" (general director M. Marinov), etc. Most of the research and design activities continued to be in the "Central Institute for Computing Technique", "Institute for Microprocessor Systems", "Institute for Microelectronics", etc. The system software support was provided by various research organizations, and also by the joint Bulgarian-Soviet institute "Interprograma". Those with the most significant contribution for the establishment of that institute on the Bulgarian side were R. Angelinov and V. Spiridonov.

Several types Local Area Networks (LAN) were developed during the period 1985—1990 and they were for export and for the internal market. The MicroLIM was a bus topology local area based on the ES 1838, ES 1839 and IBM PC, PC/XT PC/AT compatible personal computers. The MicroLIM was an open system. Every system configuration was built on the basis of specific user requirements. The MicroLIM might had been used in office automation and CAD systems. It was designed for administrative and office work automation and is a basis for the development and building of complex local area network configurations. Data transmission rate was 10 Mb/s, the max number of stations per segment were 100, and the max coaxial cable segment length was 300 m.

The MicroSTAR communications system was built around a star topology local area network. The network comprised of two types of stations: central and peripheral. Several peripheral stations might have been connected to a single central station. The stations were based on the personal computers EC 1838, EC 1839, and compatibles: IBM PC/XT/AT.

The central station had to be equipped with MicroSTAR adapter board. The specialized software might have defined one station either as a central, or as a peripheral, or both.

The peripheral stations were connected to the central station via serial channels. The channel data transmission rate depended on the distance between the central and the peripheral stations. The team that developed the transputer systems and LAN included: A. Ananiev, B. Anachkov, V. Sabev, V. Barbutov, V. Getov, V. Filipov, E. Elicina, I. Pavlov, I. Popov, I. Radev, I. Cikandelov, K. Boyanov, K. Yanev, K. Arabadjiiski, L. Manikov, L. Zekova, M. Simeonova, M. Iliev, N. Avramov, N. Vecev, N.

Vapcarov, N. Dabov, N. Petrov, O. Gorchakov, O. Chipev, P. Malinovski, P. Ruskov, P. Tomov, R. Salchev, S. Bezuhanova, S. Machev, S. Voinov, T. Kanchev, T. Kardjiev, H. Turlakov, etc.

The software for transputer systems was developed by R. Lazarov, S. Margenov, Kr. Georgiev, P. Marinov, A. Andreev, Ch. Djidjev etc. More than 1200 transputer systems were exported to Russia and APS station was used as development platform for diagnostic software for nuclear reactors.

The volume of the Bulgarian Computing production started to decrease rapidly after 1991. The total number of people, employed in this field, fell from about 120,000 to 30,000. With the collapse of the state controlled sector, a number of private companies were established in Bulgaria. They were engaged primary in trade, or assembling machines from imported modules.

At present, the research activities are carried out in the institutes of BAS, and Computer Science departments of the Bulgarian Technical Universities. About 400 hardware private companies are distributing and supporting machines and materials. The production in Bulgaria is limited—mainly based on assembly of computers and devices with imported printing boards and components. In the same time more than 300 software companies successfully work in local and international market. More than 5% of GDP is coming from IT sector. The existing infrastructure in Bulgaria can be used more effectively and a number of governmental programs consider options for the revival of the Bulgarian computer industry.

Bulgaria was one of the co-founders and amongst the first members of IFIP, which was founded in 1959. Bulgarian scientists held the posts of vice-president—Acad. L. Iliev (1974–1977), President—Acad. Bl. Sendov (1989–1992), member of board Acad. K. Boyanov (1998–2000) in this prestigious international organization. Acad. L. Iliev received I. Auerbah prize. Acad. A. Angelov and Acad. L. Iliev were elected as "Computer Pioneer" of IEEE.

Finally, it is worth mentioning the Bulgarian connection in the development of the world Computing. In 1973, a US court ruled in favour of the American citizen of Bulgarian origin John Atanasoff, in his case against Eckert and Mouchly, which were regarded until then as the pioneers of the first electronic computer—ENIAC. Atanasoff together with Cliff Berry designed the prototype of the first electronic computer ABC, using binary arithmetic, with capacity storage device, and is rightfully regarded as the "father of the modern computer".

References

1. Alipiev, G., et al.: Digital electronic computer, Technika **2** (1964). Sofia (in Bulgarian)
2. Boyanov, K.: Notes of the Development of Computing in Bulgaria. Marin Drinov Academic Publishing House, Sofia, Prof (2010).(in Bulgarian)

3. Boyanov, K.: Truth is a well… My life in the computer age. Prof. M. Drinov Publishing House of Bulgarian Acdademy of Sciences, Sofia (2018)
4. Градиль, В. П. и др. Справочник по Единой системо конструкторской документации, Харьков, изд. Прапор 1988 (in Russian)
5. Андреев А. и др., Система документации единой системы ЭБМ, под ред. А. М. Ларионова, изд. Статистика, Москва 1975 г. (in Russian)

Soft Computing: Three Decades Fuzzy Models and Applications

Ivan Popchev

Abstract The paper attempts to give protection of soft computing in the investigations of the scientist form the Institutes of Informatics, Information Technologies and Information and Communications Technologies of the Bulgarian Academy of Sciences in recent three decades. There are discussed researches and results on multicriteria fuzzy decision making, fuzzy decision making software in economic clustering, fuzzy inference system in support of financial investments, genetic fuzzy system for asset management and fuzzy methods and applications. It is determined that in the coming periods soft computing will expand and enrich with new subsuming principal partners. Presented is a short list of 60 publications on Soft Computing.

Keywords Berkeley initiative of soft computing · Soft computing · Fuzzy methods · Fuzzy models · Artificial intelligence · Software system · Fuzzy inference system · Genetic fuzzy system · Financial investment · Asset management · Economic clustering

1 Introduction

Lotfi A. Zadeh in many of his studies and writings (https://www.cs.barkely.edu/-zadeh) introduces the idea of **Soft computing as an example of a new kind of artificial intelligence**.

In a paper published in 1981, "Possibility Theory of Soft Data analysis" L. Zadeh employed the term "soft data analysis" to described data that are partly probabilistic and partly possibilistic. Ten years later, in 1991 the Berkeley Initiative in Soft Computing (BISC) was launched. This initiative was motivated by the fact that in science, as in other realms, there is a tendency to be nationalistic—to commit oneself to particular methodology and employ it as if it were a universal tool.

The essence of soft computing in that, unlike the traditional, hard computing, it is aimed at an accommodation with the pervasive imprecision of the real world.

I. Popchev (✉)
Bulgarian Academy of Sieneses, Akad. Georgi Bonchev Blvd., bl. 2, 1113 Sofia, Bulgaria
e-mail: popchev@iit.bas.bg

© The Author(s), under exclusive license to Springer Nature Switzerland AG 2021
K. T. Atanassov (ed.), *Research in Computer Science in the Bulgarian Academy of Sciences*, Studies in Computational Intelligence 934,
https://doi.org/10.1007/978-3-030-72284-5_5

Thus, the principle of soft computing is: "… exploit the tolerance for imprecision and partial truth to achieve tractability, robustness, low solution cost and letter rapport with reality". In the final analysis, the role model for soft computing is the human mind.

Lotfi A. Zadeh determined that soft computing is not a single methodology. Rather, it is a partnership. The principal partners at this juncture are fuzzy logic, neuro-computing and probabilistic reasoning, with the latter subsuming genetic algorithms, chaotic systems, belief networks and parts of learning theory.

Moreover, Lotfi A. Zadeh wrote in the following periods that in coming years, the ubiquity of intelligent systems is certain to have a profound impact on the ways in which man-made intelligent systems are conceived, designed, manufactured, employed and interacted with.

With the development since 1991, the **Berkeley Initiative of Soft Computing**, the guiding principle and the principal partners of soft computing have found evidence not only in academic research but also in a wide range of practically useful solutions for people and society in many countries.

Noting in these few lines only one aspect of Lotfi Zadeh's work it must be emphasized his enormous contribution to Bulgarian and world science.

Deservedly Prof. Lotfi Aliasker Zadeh since 2005 is foreign member of the Bulgarian Academy of Sciences.

2   Soft Computing in Development

In the departments of "Soft Computing" and "Intelligent Systems" of the Institutes of Informatics, Information Technologies and Information and Communication Technologies of the Bulgarian Academy of Sciences in the period 1991–2019 under European programs and contracts, between academic and national projects with the National Science Fund and Innovation Fund, in collaboration with teams of researchers from the Plovdiv University "Paisiy Hilendarski" and Burgas Free University, institutes and companies have designed new methods, models and software tools that have been developed and widely used. These results, a product of the efforts of many teams of scientists, have proved the claim of L. A. Zadeh for Soft computing as an example of a new kind of artificial intelligence.

The limited pages allow present in abstract format only several research directions: towards the multicriteria fuzzy decision making, fuzzy decision making software in economic clustering, fuzzy inference system in support of financial investments, genetic fuzzy system for asset management and fuzzy methods and applications.

3 Towards the Multicriteria Fuzzy Decision Making

The scientific investigations in the Institute of Information Technologies of Bulgarian Academy of Sciences (BAS) on Soft computing are directed toward multicriteria decision making and applications of fuzzy sets theory in Decision support systems (DSS). These systems are modelled on given predetermined system behavior, it has to be in condition to bear resemblance to its conduct and to match the following characteristics:

– The fuzzy modelling has to use as much as possible expert and decision-maker's knowledge in the investigated area and
– If the input and output data are known, the application of standard techniques for the system identification is possible.

The decision-making process for multi-criteria tasks goes through three phases [4, 15, 33, 37]: a uniform, aggregating and exploitation phase.

A. Uniform phase. If the criteria are in different scales, the information is required to be unified. One basic approach to achieving this is to use fuzzy relationships over multiple alternatives as a fundamental element of uniform representation. This requires the use of transformation functions to determine the relations between the pair of alternatives by each criterion. In this respect, fuzzy relations instead crisp are preferable because of their more convenient and adequate form of presenting the relationship between alternatives. Fuzzy relations can model situations where interactions between alternatives are not well defined [1–4]. They also reflect the interests of the experts or the decision maker. These transformation functions define relations with different properties, such as similar or preference relation. Such a transformation function has been proposed in [5–9], which as a result gives the degree of membership to several alternatives of a fuzzy preference relation. This relation has some properties that are useful for solving the problems of ordering alternatives. Membership degrees have been shown to vary slightly with slight changes in alternative estimations.

Qualitative criteria are represented by qualitative terms through linguistic variables [19], i.e. variables whose values are not numbers but words or sentences in natural or artificial language. In first possibility, the linguistic variables are represented as fuzzy numbers. Therefore, the problem is how to compare the fuzzy numbers to obtain the corresponding fuzzy relation. There are different methods for comparing or sorting fuzzy numbers—using a ranking function or computing a comparison index for each pair. In [7, 9] a new index for the comparison of fuzzy numbers is proposed. This index is based on the geometric properties of the fuzzy numbers. It is tested on a group of selected examples and compared with other well-known indexes. A method for comparing sequences of fuzzy numbers and an algorithm for comparing subsets (clusters) of similar, closed vectors of fuzzy numbers are presented as well [1, 9].

B. Phase of aggregation of the performance values with respect to all criteria for obtaining a union performance value for the alternatives. An appropriate approach to uniting individual evaluations corresponding to an alternative is to use the aggregation procedures that realize the idea of compensation and compromise between conflicting criteria when compensation is allowed. These procedures can be performed by aggregation operators [10, 12, 13, 16]. There are a large range of operators that can be used in the confluence of the criteria. The choice of operator for a specific application depends on various factors. Choices must be made depending on:

– the mathematical model of operators;
– the properties of the operators for deciding problems of ranking or choice or clustering the set of alternatives;
– the sensitivity of operators to small variations of their arguments.

The dependence between the properties of the aggregate relation and the properties of the individual relations by each criterion for some operators has been investigated in [5, 7, 8, 10–12, 18]. The dependencies of each operator are presented in a table that contains the implication of the kind: if the initial relations possess given properties, then the aggregated relation has proved properties. Some of the most often used operators are presented and their properties are proven and presented in this table. The properties of the aggregate fuzzy relation to assist the solution of problems of choice or ranking alternatives are shown. The results obtained by different aggregation operators are compared with other well-known multi-criteria decision-making methods. The sensitivity of operators to deviations in their arguments is defined and computed in [14]. In [13], the aggregation of the sequences of fuzzy numbers representing alternatives is done by aggregation operators as the aggregated estimates are also fuzzy numbers. The model is tested with a real example and compared with the results obtained from other well-known multi-criteria decision-making methods.

The weighted aggregation is other very important problem in the decision-making problems. Weighed transformations in aggregation operators are used for this purpose. The problems of preserving the properties of the fuzzy relations in the application of these transformations are considered. In [17, 20, 22], the use of criteria weights is investigated in cases where they are not presented in the aggregation operator formula. Problems when weights are presented as a fuzzy relation between the importance of criteria are considered in [21, 24, 26, 28] or fuzzy numbers in [27]. Models connected with weighting functions as a criteria importance, depending on the membership degrees of the fuzzy relations are proposed in [28–30, 34, 35]. The results of the weighted aggregations discussed above are summarized in [23, 25–27, 30, 32]. Illustrative examples are given for comparing the suggested models [36, 37].

C. The exploitation phase of the union performance value for obtaining a rank ordering, sorting or choosing the alternatives. This phase addresses the problems of selecting a subset of the "best" alternatives in a sense; ordering over the whole set of alternatives; parturition the set of alternatives to subsets of similar, close, i.e. partition from clusters. The results obtained in this area are:

The fuzzy cluster comparison method proposed in [9]—makes it possible to compare clusters from fuzzy numbers and order these subsets. The proposed algorithm is based on the results obtained in [3], where it is proved that the comparison between two fuzzy sets can be made on the base of the comparison of results only on the fuzzy sets without loss of information. The proposed algorithm is tested with an example. A decrease of computations is illustrated as well.

The applications of fuzzy logic in multicriteria problems for assessing the quality of an asset and making an investment decision are discussed in [31, 45, 48–50].

4 Fuzzy Decision Making Software in Economic Clustering

The choice under uncertainty is based on a system of a priori knowledge of decision maker (DM) about the behavior of environmental factors. In [38] describes a feasible application of multicriteria choice problems in structuring and analysis of economic clusters by MAP—CLUSTER software system as a decision-making support tool in economic clustering (EC). The system utilizes a specially designed approach to solve analysis problems that deal with planning, structuring and prediction variants of horizontal network integration of small business enterprises (SBE) in a technological network chosen by DM. The system allows the use of a modern approach to the management of integrated economical structures including balanced scorecard (BCS) based assessment. Integration of these tools allows finding solutions while simultaneously considering the state of different resources.

In essence, the decision is subjective, which increases the responsibility of DM. Formally, the methods of fuzzy sets theory can be applied [37]. These methods require introduction of fuzzy relations regarding qualitative values of the environmental factors and the target function (criterion of optimality). The fuzzy relation is characterized by a membership function, which expresses the subjective measure of the degree of fulfillment (credibility), for example, of the factor/criterion relation.

In [53] proposes a decision support approach for selection of small and medium enterprises (SMEs) in economic cluster. The suggested decision is based on descending ranking of their integral score of business performance within each technology network (TN).

The tests are performed by **system FuzzyPro**, designed for studying and solving problems of one and multi-criteria selection or ranking of alternatives of final set with a set of criteria with or without weighted coefficients. The alternatives are presented according to criteria and their respective weighted coefficients:

A. *Criteria are real numbers, fuzzy relations and/or fuzzy numbers*
1. Weighting coefficients are real numbers:

   - Algorithm with aggregation operators with weighting coefficients real numbers (ATOKRI1);
   - Algorithm with aggregation operators without weighting coefficients for criteria (ATOKRI2).

2. Weighting coefficients are fuzzy relations:

   - Algorithm with ranking function (ARAKRI1);
   - Mixed data algorithm (ARAKRI3).

3. Weighting coefficients are real functions:

    – Algorithm for crisp criteria with weighted coefficients—weighting functions
      (ATOKRIF).

B. *Criteria are only fuzzy numbers.*
4. No weighting coefficients—direct aggregation fuzzy numbers algorithm
    (ARAKRI2).

The economic clustering problem on a given TN with participation of SME is
defined as a multicriteria selection task.

In [53] are accepted following pre-determined assumptions:

1. TN of the cluster.
2. A mixed type integration of the cluster.

There are three tasks defined as:

Task 1: Making a list of SMEs—potential participants in the cluster with an
indication of their membership to a particular node of TN.

Task 2: Developing of a "passport" for each SME, according to a predetermined
system of criteria, that allows adequate assessment of their economic performance
and development.

Task 3: Assessing and ranking of SMEs scores in decreasing order on TN nodes.
Based on this decision, selection of lists of potential SME participants for each TN.

The diagram of proposed approach is shown in Fig. 1. Tasks 1 and 2 carry out the
preparatory phase, Task 3 implements the analytical stage and the decision-making
stage.

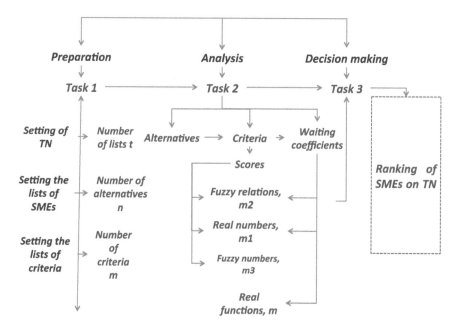

Fig. 1 General diagram of the decision-making stages

Task 1 **is considering following steps**:

Step 1: Description and setting the number of nodes t in TN, their branch affiliation, and relations in technological chain of production.

Step 2: Setting lists of SME_{tn}, where n is the number of potential participants. Such list is prepared for each node t.

Step 3: Developing a "passport" structure for each node t, according to economic specialization. Criteria included in the "passport" should allow performance level evaluation assessment of each SME_{tn}, taking under consideration the branch and the specific terms of the task. Assessments are performed for each node t.

Step 4: Collecting, selecting and processing "passport" data for SME_{tn}. (For the purposes of this study, a universal "passport" structure was adopted for all TN nodes.)

Task 2 **consists of the following steps**:

Step 1: Determination of ranking lists according to the number of nodes of TN.

Step 2: Determination of number of alternatives and criteria under consideration. The number of alternatives in ranking list t depends on the number of SMEs—candidates. A "passport" determines the number of criteria.

Step 3: Processing the primary information and selection of the type of input data.

The presented approach proposes transformation of data obtained as outputs from Tasks 1 and 2 and a solution to Task 3 to be through the tools of fuzzy set.

It is necessary to define a finite set of alternatives $A = \{a_1, \ldots, a_j, \ldots, a_n\}$, evaluated by number of criteria $K = \{k_1, \ldots, k_j, \ldots, k_m\}$. The weights of criteria are $W = \{w_1, \ldots, w_j, \ldots, w_m\}$. The final set A of alternatives is the list of SME_{tn} (the list of enterprises in a node n of TN). The final set K contains Passport's criteria for assessing the alternatives.

The input data are presented in Table 1. The number of tables equals the number of technological nodes in a cluster.

Table 1 The input data of alternatives and criteria

Alternatives	Criteria				
	k_1	...	k_j	...	k_m
a_1	x_{11}	...	x_{1j}	...	x_{1m}
$\vdots$	$\vdots$	$\vdots$	$\vdots$	$\vdots$	$\vdots$
a_i	x_{i1}	...	x_{ij}	...	x_{im}
$\vdots$	$\vdots$	$\vdots$	$\vdots$	$\vdots$	$\vdots$
a_n	x_{n1}	...	x_{nj}	...	x_{nm}
Weights of criteria	w_1	...	w_j	...	w_m

Table 2 Classification of algorithms by type of criteria

	Criteria data	No weighting coefficients	Weighting coefficients real numbers	Weighting coefficients fuzzy relations	Weighting coefficients real functions
One type criterion	*Either real numbers m1*	ATOKRI 2	ATOKRI 1 or ATOKRI 2	ATOKRI 1 -> ARAKRI	ATOKRI 1 -> ATOKRIF
	Either fuzzy relations m2		ATOKRI 1 or ATOKRI 2	ARAKRI	ATOKRIF
	Either fuzzy numbers m3		ARAKRI 1 or ARAKRI 2	ARAKRI 1 or ARAKRI 2	ARAKRI 1 or ARAKRI 2
Mixed type criteria	*Real numbers m1*		ARAKRI 1, ATOKRI 1 or ATOKRI 2	ATARR, ARAKRI 1	ARAKRI 1, MMTRR, ATOKRIF
	Fuzzy relations m2		MMTRR		
	Fuzzy numbers m3		ARAKRI 3		

The real number x_{ij}, $i = 1, \ldots, n; j = 1, \ldots, m$ in the table is the assessment of the alternative a_i by the criterion k_j. To each criterion k_j is assigned a real number w_j weighting coefficient. On the basis of these data, a ranking of alternatives in descending order is obtained. The total number of criteria which DM selects is equal to m, such as $m = m_1 + m_2 + m_3$, where:

m_1 is the number of criteria that are set as crisp numbers;

m_2 is the number of criteria that are set as fuzzy relations;

m_3, is the number of criteria that are given as fuzzy numbers.

The type and combinations of selected by DM data determine algorithms by which the final solution is obtained. Different combinations of data for criteria assessments and type of weighting coefficients that DM can work with, as well as the types of algorithms used are systematized in Table 2.

Case 1: Crisp criteria assessment and Crisp weights

All criteria are crisp ($m_1 = m$) and weighting coefficients are real numbers ($w_1, \ldots, w_m$). Here, the crisp criteria assessments are limited to fuzzy relations with certain properties. The calculations are using aggregation operator algorithm ATOKRI1 or the algorithm with aggregation operator without weighting coefficients ATOKRI2, where the transformations of ATOKRI1 fuzzy relations generated by ATOKRI1 algorithm are performed and then aggregation operators without weighting coefficients are used.

The algorithm ATOKRI1 [37] uses the data from Table 1. Since criteria assessments can be set in different measuring scales and unit, procedures for transformation of each column of Table 1 (criteria) into fuzzy preference relations are used for normalization. The following function is used:

$$\mu_k(a_i, a_j) = \begin{cases} 1 & \text{if } i = j \\ 0.5 + \dfrac{x_{ik} - x_{jk}}{2\left(\max\limits_i\{x_{ik}\} - \min\limits_i\{x_{ik}\}\right)} & \text{if } i \neq j \end{cases}$$

where $\mu_k(a_i, a_j), a_i, a_j \in A, k \in \{k_1, k_2, \ldots k_m\}$.

For each column of Table 1 the corresponding fuzzy relation R_k is calculated, i.e. for k-criteria, the corresponding fuzzy relation in the matrix is:

$$R_k = \begin{bmatrix} \mu_k(a_1, a_1) & \cdots & \mu_k(a_1, a_j) & \cdots & \mu_k(a_1, a_n) \\ \vdots & \vdots & \vdots & \vdots & \vdots \\ \mu_k(a_i, a_1) & \cdots & \mu_k(a_i, a_j) & \cdots & \mu_k(a_i, a_n) \\ \vdots & \vdots & \vdots & \vdots & \vdots \\ \mu_k(a_n, a_1) & \cdots & \mu_k(a_n, a_j) & \cdots & \mu_k(a_n, a_n) \end{bmatrix}, k = 1, \ldots, m \quad (1)$$

The minimizing criteria should be transformed into maximizing. They are calculated as additions to the relations corresponding to the minimizing criteria, i.e. the values of the respective matrices are subtracted from 1.

All relations $R_k, k = 1, \ldots, m$ are shrunk to obtain an aggregated relations R with the following matrix:

$$R = \begin{bmatrix} \mu(a_1, a_1) & \cdots & \mu(a_1, a_j) & \cdots & \mu(a_1, a_n) \\ \vdots & \vdots & \vdots & \vdots & \vdots \\ \mu(a_i, a_1) & \cdots & \mu(a_i, a_j) & \cdots & \mu(a_i, a_n) \\ \vdots & \vdots & \vdots & \vdots & \vdots \\ \mu(a_n, a_1) & \cdots & \mu(a_n, a_j) & \cdots & \mu(a_n, a_n) \end{bmatrix} \quad (2)$$

Each element of the matrix is computed by the formulas for aggregation operators with weighting coefficients $\{w_1, \ldots, w_j, \ldots, w_m\}$. The following operators are used and values for $\mu_k(a_i, a_j)$ are derived form (1):

$$\mu(a_i, a_j) = \sum_{k=1}^{m} w_k \mu_k(a_i, a_j), \text{ were } 0 \leq \mu_k \leq 1, \sum_{k=1}^{m} w_k = 1 \quad (3)$$

$$\mu(a_i, a_j) = \prod_{k=1}^{m} [\mu_k(a_i, a_j)]^{w_k}, \text{ were } 0 \leq w_{ki} \leq 1, \sum_{k=1}^{m} w_k = 1. \quad (4)$$

For two operators above weights of the criteria are normalized, i.e. recalculated to be in the range [0,1] and their sum is one.

$$\mu(a_i, a_j) = \overset{max}{\underset{k}{}} \{min(\mu(a_i, a_j), w_k)\}, \text{ were } 0 \le w_k \le 1 \overset{max}{\underset{ki}{}} \{w_k\} = 1, k, \ldots m.$$

$$(5)$$

$$\mu(a_i, a_j) = \overset{min}{\underset{k}{}} \{max(\mu(a_i, a_j), 1 - w_k)\}, \text{ were } 0 \le w_k \le 1 \overset{max}{\underset{ki}{}} \{w_k\} = 1, k, \ldots m.$$

$$(6)$$

For two operators above weights of the criteria are normalized, i.e. recalculated to be in the range [0,1] and their sum is one.

There are obtained 4 aggregated relations, i.e. 4 matrices of type R(2). Each of matrices is recalculated to obtain matrices R' in the following manner:

$$\text{if } \mu(a_i, a_j) \ge \mu(a_j, a_i), \text{ then } \mu'(a_i, a_j) = \mu(a_i, a_j) \text{ and } \mu'(a_i, a_j) = 0. \quad (7)$$

The matrices R' are rearranged to triangular matrices that show the order of alternatives from best to worst depending on selected aggregation operator.

The difference between ATOKRI2 and ATOKRI1 algorithms is in the choice of aggregation operators. ATOKRI2 [37] uses operators without weighing coefficients. If the criteria weights are not set, then after obtaining the fuzzy relations $R_k, k = 1, \ldots, m$ (1) and recalculation of the minimizing criteria from the corresponding steps of ATOKRI1 algorithm, aggregated relations are calculated by the following aggregating operators:

$$\mu(a_i, a_j) = \alpha \overset{max}{\underset{k}{}} \{\mu_k(a_i, a_j)\} + (1 - \alpha) \overset{min}{\underset{k}{}} \{\mu_k(a_i, a_j)\}, \alpha \in [0, 1], i,$$

$$j = 1, \ldots, n, k = 1, \ldots, m. \quad (8)$$

$$\mu(a_i, a_j) = \frac{\lambda}{m} \sum_{k=1}^{m} \mu(a_i, a_j) + (1 - \lambda) \min_{k} \{\mu(a_i, a_j)\}, \lambda \in [0, 1], i, j = 1, \ldots, n.$$

$$(9)$$

$$\mu(a_i, a_j) = \left[\prod_{k=1}^{m} \mu_k(a_i, a_j)\right]^{1-\gamma} \left[1 - \prod_{k=1}^{m}(1 - \mu_k(a_i, a_j))\right]^{\gamma}$$

$$\text{if } \mu_k(a_i, a_j) \ne 0 \mu(a_i, a_j) = 0 \text{ otherwise} \quad (10)$$

The values $\mu_k(a_i, a_j)$ are taken from (1), coefficients α, λ, γ are set by DM according to the particular task being solved. It can be experimented with different values.

Three aggregated relations are obtained (3 aggregated matrices of type R(2)) then the other steps of ATOKRI1 algorithm are followed. Each matrix is recalculated to matrices R' by (7). The matrices R' are ranked and produce triangular matrices that show the order of alternatives from best to worst depending on selected aggregation operator. In order to use the above aggregation operators, if criteria weights are set, the steps of the algorithm are as follows. The values of the matrices R_k, $k = 1, \ldots, m$ (1) are transformed by expressions:

$$\mu_k^1(a_i, a_j) = (1 - w_k) + \mu_k(a_i, a_j) - (1 - w_k)\mu_k(a_i, a_j), i, j = 1, \ldots, n \quad (11)$$

or

$$\mu_k^2(a_i, a_j) = w_k \mu_k(a_i, a_j), i, j = 1, \ldots, n, \quad (12)$$

Where $\mu_k(a_i, a_j)$ is a corresponding element of the matrix (1). For each relation R_k, $k = 1, \ldots, m$, two matrices $R_k^1, R_k^2, k = 1, \ldots, m$ are obtained:

$$R_k^1 = \begin{bmatrix} \mu_k^1(a_1, a_1) & \cdots & \mu_k^1(a_1, a_j) & \cdots & \mu_k^1(a_1, a_n) \\ \vdots & \vdots & \vdots & \vdots & \vdots \\ \mu_k^1(a_i, a_1) & \cdots & \mu_k^1(a_i, a_j) & \cdots & \mu_k^1(a_i, a_n) \\ \vdots & \vdots & \vdots & \vdots & \vdots \\ \mu_k^1(a_n, a_1) & \cdots & \mu_k^1(a_n, a_j) & \cdots & \mu_k^1(a_n, a_n) \end{bmatrix}$$

$$R_k^2 = \begin{bmatrix} \mu_k^2(a_1, a_1) & \cdots & \mu_k^2(a_1, a_j) & \cdots & \mu_k^2(a_1, a_n) \\ \vdots & \vdots & \vdots & \vdots & \vdots \\ \mu_k^2(a_i, a_1) & \cdots & \mu_k^2(a_i, a_j) & \cdots & \mu_k^2(a_i, a_n) \\ \vdots & \vdots & \vdots & \vdots & \vdots \\ \mu_k^2(a_n, a_1) & \cdots & \mu_k^2(a_n, a_j) & \cdots & \mu_k^2(a_n, a_n) \end{bmatrix}$$

Aggregating operators (8), (9), (10) are used to obtain R(2) and calculation are as follows:

$$\mu(a_i, a_j) = \alpha \max_k \{\mu_k^2(a_i, a_j)\} + (1 - \alpha)\min_k \{\mu_k^1(a_i, a_j)\}, \alpha \in [0, 1], k = 1, \ldots, m.$$

$$\mu(a_i, a_j) = \frac{\lambda}{m} \sum_{k=1}^{m} \mu_k^2(a_i, a_j) + (1 - \lambda)\min_k \{\mu_k^1(a_i, a_j)\}, \lambda \in [0, 1], k = 1, \ldots, m.$$

$$\mu(a_i, a_j) = \left[\prod_{k=1}^{m} \mu_k^1(a_i, a_j)\right]^{1-\gamma} \left[1 - \prod_{k=1}^{m}(1 - \mu_k^2(a_i, a_j))\right]^{\gamma} \text{ if } \mu_k(a_i, a_j) \neq 0$$

$$\mu(a_i, a_j) = 0 \text{ otherwise}$$

There are 3 aggregated relations obtained (3 matrices of R(2) type). Calculations proceeds following the other steps of ATOKRI1 algorithm.

Case 2: Crisp criteria assessments, without weighing coefficients.

All criteria are crisp, $m_1 = m$ and criteria weights are not specified - the crisp estimates of criteria are reduced to fuzzy relations with certain properties and calculations are performed with ATOKRI2 algorithm;

Case 3: Crisp criteria assessments and weighting coefficients—real functions.

All criteria are crisp, $m_1 = m$ and weighting coefficients are real functions $f_1(x), \ldots, f_m(x), x \in [0, 1]$—crisp criteria estimates are reduced to fuzzy relations using ATOKRI1 algorithm, then ATOKRIF is used for crisp criteria with weighting coefficients—weighting functions.

The difference between ATOKRIF and ATOKRI algorithms is the choice of weighting coefficients for criteria. In ATOKRIF [37] weighting coefficients are weighting functions $f_1(x), \ldots, f_m(x), x \in [0, 1]$ with arguments the elements of corresponding matrices (1). For example, for function $f_k(x), x$ the elements are of the matrix R_k.

The input data for this algorithm is fuzzy relations. If relations are derived from ATOKRI1 algorithm, they have the proper properties to obtain a ranking. If relations are set by the expert, they must be checked whether they meet the required properties. If not, they are transformed into respective relations with properties required. So, the set relations are processed with the MMTRR algorithm. Let the obtained from ATOKRI1 algorithm relations are $R_k, k = 1, \ldots, m$. Aggregate relations are calculated as follows: one of the following convenient weighing functions is selected by setting the relevant function parameters that must meet certain conditions:

- Linear Functions

$$f_k(x) = 1 + \beta_k x, 0 \leq \beta_k \leq 1, k = 1, \ldots, m, m \geq 2 \qquad (13)$$

- Parametric linear functions

$$f_k(x) = a_k \frac{1 + \beta_k x}{1 + \beta_k} = \gamma_k(1 - \beta_k x), 0 < a_k \leq 1, 0 \leq \beta_k \leq 1,$$

$$\gamma_k = \frac{a_k}{1 + \beta_k}, k = 1, \ldots, m \qquad (14)$$

- Quadratic functions

$$f_k(x) = 1 + \beta_k - \gamma_k x + \gamma_k x^2, \beta_k \geq 0, \gamma_k \geq 0, k = 1, \ldots, m \qquad (15)$$

Function's parameters are unrelated and different for different functions.

For each relation $R_k, k = 1, \ldots, m$ a new degree of preference for each criteria k are calculated considering corresponding weighting function:

$$\mu_k^w(a_i, a_j) = \begin{cases} 1 & ifa_i = a_j \\ \frac{f_k(\mu_k(a_i,a_j)\mu_k(a_i,a_j))}{S(a_i,a_j)} & ifa_i \neq a_j \end{cases} i,j = 1,2,3, k = 1,2$$

were $S(a_i, a_j) = \sum_{k=1}^{m} f_k(\mu_k(a_i, a_j))$, and $f_k(.)$ is one of weighting functions (13), (14), (15). The result is m matrices:

$$R_k^w = \begin{bmatrix} \mu_k^w(a_1, a_1) & \cdots & \mu_k^w(a_1, a_j) & \cdots & \mu_k^w(a_1, a_n) \\ \vdots & \vdots & \vdots & \vdots & \vdots \\ \mu_k^w(a_i, a_1) & \cdots & \mu_k^w(a_i, a_j) & \cdots & \mu_k^w(a_i, a_n) \\ \vdots & \vdots & \vdots & \vdots & \vdots \\ \mu_k^w(a_n, a_1) & \cdots & \mu_k^w(a_n, a_j) & \cdots & \mu_k^w(a_n, a_n) \end{bmatrix}, k = 1, \ldots, m$$

These matrices are combined into a matrix of the type R (2), the elements of which are obtained by sum of corresponding elements of matrices R_k^w, $k = 1, \ldots, m$. Based on the R-type matrix, other steps of ATOKRI1 algorithm are followed, i.e. recalculation to the matrices R' by (7). The resulting matrix R' is reordered to obtain a triangular matrix that shows descending ranking of alternatives.

Case 4: Crisp criteria assessments and fuzzy relation of preference for criteria weights.

All criteria are crisp $m = m_1$, and it is set a fuzzy relation of preference W for criteria weight—the crisp criteria estimates are reduced to fuzzy relations using ATOKRI1 algorithm, then ARAKRI algorithm is used for fuzzy relations of alternatives and weights of criteria [37].

In ARAKRI the input data are fuzzy relations of type (1) between alternatives by each criterion and a fuzzy relation between criteria. The input data for the criteria can be both crisp (as in Table 1) and fuzzy relations of type (1). For crisp criteria, fuzzy relations between alternatives are calculated by the first steps of ATOKRI1 algorithm. If the criteria assessments are set directly by fuzzy relations, then it is checked for additive transitivity properties of reciprocal relations. If not, the initial relations are transformed (using ATARR algorithm described in Case 7) into new fuzzy relations with the necessary properties. The weighting coefficients of criteria are given as a fuzzy relation with certain properties of the type:

$$W = \begin{bmatrix} w_{11} & \cdots & w_{1j} & \cdots & w_{1m} \\ \vdots & \vdots & \vdots & \vdots & \vdots \\ w_{i1} & \cdots & w_{ij} & \cdots & w_{im} \\ \vdots & \vdots & \vdots & \vdots & \vdots \\ w_{m1} & \cdots & w_{mj} & \cdots & w_{mm} \end{bmatrix}, w_{ii} = 0.5, w_{ij} = 1 - w_{ji}, i,j = 1 \ldots, m \quad (16)$$

After receiving the matrices by all criteria R_k, $k = 1, \ldots, m$, i.e. one obtained from ATOKRI1 algorithm and the other verified for the properties required, there are given two possibilities for joining each matrix pair to include the elements of the matrix W. If the fuzzy relations are consistent with criteria $R_1, R_2, \ldots, R_m$, then by merging R_i and R_j, taking into account respective elements of the matrix W it is obtained a new relation R_{ij} and since $R_{ij} = R_{ji}$ then the number p of the new relation will be equal to the combination of two elements on m, i.e. $p = \frac{m(m-1)}{1.2}$. For aggregation of these new pairs p, ATOKRI2 algorithm can be used, i.e. aggregation operators without weighting coefficients. Then the steps of this algorithm are followed.

Case 5: Criteria evaluations—fuzzy relations, weighting coefficients—real numbers.

All criteria set fuzzy relations of preference between alternatives, $m_2 = m$ and weighting coefficients are real numbers. Fuzzy relations are checked for max–min transitivity (using MMTRR algorithm) and those that do not possess these properties are transformed into new relations with the required properties. ATOKRI1 with aggregation operators with weighted coefficients is used and/or ATOKRI2 where weight transformations of fuzzy relations are performed and then aggregation operators are used without weighting coefficients. The algorithm MMTRR includes:

A. Verification algorithm for max–min transitivity of relations:
1. The relation $R = \|r_{ij}\|$, $r_{ij} = 1$, $i, j = 1, \ldots, n$ is given
2. If $r_{ij} \geq \min(r_{ik}, r_{kj})$, $\forall 1, j, k = 1, \ldots, n$ then relation is max–min transitive, otherwise it is transformed into max–min transitive relation using the following algorithm:
B. Algorithm for obtaining a max–min transitive relation:
1. The relation $R = \|r_{ij}\|$, $i, j = 1, \ldots, n$ is given
2. Calculate

$$R^2 = R \circ R = \|r_{ij}^2\|, \quad r_{ij}^2 = \max\{\min(r_{ik}, r_{kj})\}, \, i, j, k = 1, \ldots, n$$

$$R^3 = R^2 \circ R = \|r_{ij}^3\|, \quad r_{ij}^3 = \max\{\min(r_{ik}^2, r_{kj})\}, \, i, j, k = 1, \ldots, n$$

3. For a matrix of $n \times n$ dimensions, n is the number of matrices, where:
4. The transitive closure of R is a transitive fuzzy relation that contains R_T and

$$R_T = R \cup R^2 \cup \cdots \cup R^n.$$

Case 6: Criteria assessments—fuzzy relation, weighting coefficients—real functions.

All criteria are set as fuzzy relations of preference between alternatives, i.e. $m_2 = m$, and weighting coefficients are real functions $f_1(x), \ldots, f_m(x)$, $x \in [0, 1]$—by MMTRR algorithm fuzzy relations are checked for certain properties. Those that do not possess these properties are transformed into new relations with the required properties, then ATOKRIF algorithm is used.

Case 7: Criteria assessments and weighting coefficients—fuzzy preference relations.

All criteria are set as fuzzy relations of preference between alternatives, i.e. $m_2 = m$, and fuzzy preference relation W is given between the importance (weights) of criteria. Fuzzy relations are checked for the additive transitivity of reciprocal relations with ATARR algorithm, and those that do not possess these properties are transformed into new relations with the required properties, then using the ARAKRI algorithm.

A. ATARR algorithm is used for checking additive transitivity reciprocal relation:
1. The relation $R = \|r_{ij}\|$, $r_{ii} = 0.5$, $r_{ij} = 1 - r_{ij}$, $i, j = 1, \ldots, n$ is given.
2. If $r_{ij} + r_{jk} + r_{ik} = 1.5 \forall i, j, k = 1, \ldots, n$, then R is additive transitive, if not it is transformed by the following algorithm:
B. Algorithm for constructing additive transitive reciprocal relation:
1. The following $n - 1$ elements of the matrix R are taken $r_{12}, r_{23}, \ldots, r_{n-1,n}$;
2. These elements give new additive transitive reciprocal relation with elements:

$$r_{ii} = 0.5, i = 1, \ldots, n,$$

$$r_{21} = 1 - r_{12}, r_{32} = 1 - r_{23}, r_{43} = 1 - r_{34}, \ldots, r_{n,n-1} = 1 - r_{n-1,n}$$

$$r_{31} = 1.5 - r_{12} - r_{23}, r_{13} = 1 - r_{31},$$

$$r_{41} = 2 - r_{12} - r_{23} - r_{34}, r_{14} = 1 - r_{41}$$

$$r_{42} = 1.5 - r_{23} - r_{23} - r_{34}, r_{24} = 1 - r_{42},$$

$$r_{ij} = \frac{j - i + 1}{2} - r_{i(i+1)} - r_{(i+1)(i+2)} - \cdots - r_{(i+n-1)(i+n)} - r_{(i+n)i}, r_{ij} = 1 - r_{ji}$$

Case 8: Criteria assessments—fuzzy numbers and weighting coefficients—real numbers.

All criteria give fuzzy numbers for evaluations of alternatives, i.e. $m_3 = m$ and weighting coefficients are real numbers $w_1, \ldots, w_m$. One or more algorithms ARAKRI are used when evaluating the fuzzy numbers. For these algorithms, estimates of alternatives it is required for all criteria to be fuzzy numbers or real numbers (real numbers are a private case of fuzzy numbers). Two approaches are then proposed for decision-making. For one of these (ARAKRI1), the fuzzy numbers are replaced by a corresponding real index by a ranking function. The aggregation of these numbers is done by aggregating operators depending on set weights of criteria. In the other approach (ARAKRI2) no ranging function is used, but aggregated fuzzy numbers are directly generated using aggregating operators. The advantage of ARAKRI2 is the smaller amount of input data needed. As this method does not require weighting coefficients, memory complexity is reduced and expert's subjectivity is almost minimum. [41].

In ARAKRI1 [13] evaluations of the alternatives according to criteria are fuzzy numbers, given in a matrix form:

$$
\begin{bmatrix}
& k_1 \ldots k_j \ldots k_m \\
a_1 & \tilde{A}_{11} \ldots \tilde{A}_{1j} \ldots \tilde{A}_{1m} \\
& \vdots \\
a_i & \tilde{A}_{i1} \ldots \tilde{A}_{ij} \ldots \tilde{A}_{im} \\
& \vdots \\
a_n & \tilde{A}_{n1} \ldots \tilde{A}_{nj} \ldots \tilde{A}_{nm}
\end{bmatrix},
\tag{17}
$$

where the fuzzy numbers are of the type:
$\tilde{A}_{ij} = \left(a_{ij}^1, a_{ij}^2, a_{ij}^3, a_{ij}^4\right)$, $i = 1, \ldots, n, j = 1, \ldots m$, were $a_{ij}^1 \leq a_{ij}^2 \leq a_{ij}^3 \leq a_{ij}^4$ are real numbers and for corresponding criteria these assessments could be of different scales.

If estimates are of different scales and are not in the range [0,1], the procedure is performed for unification and normalization of fuzzy numbers. Suitable for this purpose is the following procedure which does not change the order of the numbers. Let:

$$
a_j^{max} = \max_i \left\{a_{ij}^4\right\}, a_j^{min} = \min_i \left\{a_{ij}^1\right\}, da = a_j^{max} - a_j^{min}
$$

Then the unified and normalized fuzzy number $\tilde{Z}_{ij} = \left(z_{ij}^1, z_{ij}^2, z_{ij}^3, z_{ij}^4\right)$, $i = 1 \ldots, n, j = 1, \ldots, m$ is estimated by formula $\tilde{Z}_{ij} = \frac{\left(\tilde{A}_{ij} - a_j^{min}\right)}{da}$, $i = 1, \ldots, b, j = 1, \ldots, m$.

If some of criteria are maximizing and some minimizing, then for the minimization of criteria the additions of the fuzzy numbers to fuzzy number $(1, 1, 1, 1)$ are calculated to obtain the same type criteria. The unified and normalized fuzzy indexes $F(\tilde{Z}_{ij})$ are calculated using the following ranging function:

$$
\tilde{Z}_{ij} = kF_1(\tilde{Z}_{ij}) + (1 - k)F_2(\tilde{Z}_{ij}), k \in [0, 1],
\tag{18}
$$

were:

$$F_1(\tilde{Z}_{ij}) = a_{ij}^1 + \frac{\left(a_{ij}^4 - a_{ij}^1\right) + \left(a_{ij}^3 - a_{ij}^2\right)}{2} \times \frac{1}{\sqrt{\left(a_{ij}^4 - a_{ij}^1\right)^2 + 1}},$$

$$F_2(\tilde{Z}_{ij}) = a_{ij}^4 + \frac{\left(a_{ij}^4 - a_{ij}^1\right) + \left(a_{ij}^3 - a_{ij}^2\right)}{2} \times \frac{1}{\sqrt{\left(a_{ij}^2 - a_{ij}^1\right)^2 + 1}}.$$

Since assessments are already real numbers, it comes to a classical decision-making problem.

- If criteria are equally important, aggregation operators without weighting coefficients (ATOKRI2 algorithm with matrices rows) are used.
- If weights of criteria are different real numbers, then aggregation operators are used in models with obvious weights (ATOKRI1) or aggregation operators in models where the weights are not obvious but weighting transformations of assessments are used (ATOKRI2 algorithm with matrix columns).
- If criteria weights are different real functions, ATOKRIF algorithm is used to aggregate the scores according to different criteria.

Then the steps of these algorithms are followed.

ARAKRI2 [9] does not calculate fuzzy number indices to obtain aggregate estimates, and aggregated fuzzy numbers are obtained using direct aggregators and operations between fuzzy numbers. The first steps are the same as in ARAKRI1, and the index matrix of unified and normalized fuzzy numbers of (17) is used to determine the maximum and minimum fuzzy number of a given series.

- If criteria weighting is not set, aggregation operators without weighting coefficients (ATOKRI2 algorithm) are used.
- If the weighting coefficients are real numbers, they are normalized and then either aggregation operators with weighting coefficients (ATOKRI1) or aggregation operators without weighting coefficients are used, but the fuzzy numbers are multiplied by the corresponding weights (ATOKRI2 algorithm).
- If weighting functions are specified, ATOKRIF algorithm with matrices rows is used.

As a result, for each alternative aggregated fuzzy number are obtained, which should be ranked by descending order. For this purpose, the indices (18) of aggregated estimates are calculated. The ranking of alternatives corresponds to the ranking of obtained indices.

Case 9: Criteria evaluations—fuzzy numbers and weighting coefficients—real functions.

All criteria set fuzzy numbers for alternative evaluations, i.e. $m_3 = m$ and criteria weights are different real functions $f_1(x), \ldots, f_m(x), x \in [0, 1]$—the algorithms ARAKRI1 and/or ARAKRI2 are used, then ATOKRIF algorithm is used.

Case 10: Criteria and weighting coefficients—different.

The criteria are different, i.e. $m = m_1 + m_2 + m_3$. ATOKRI1 algorithm is used for m_1 criteria, MMTRR for m_2 criteria, and ARAKRI3 for m_3 criteria [8]. Since these algorithms reduce initial information to fuzzy preference relations depending on weighted coefficients of the criteria, the task is reduced to one of the Cases 5, 6, 7.

With ARAKRI3 algorithm, the set of alternatives has been evaluated by various criteria, e.g. quantitative (utility functions), qualitative (crisp, nonfuzzy rankings), fuzzy relations, fuzzy numbers. It is necessary to bring this information to a common scale. A basic approach is to obtain fuzzy relations from available data by comparing the evaluations of alternatives pairwise for each criterion. Here the problem is addressed when some of criteria evaluate alternatives by fuzzy numbers (17). In order to unify information, fuzzy relations on these criteria must be obtained. Using the index (18) it is possible to compare each pair of fuzzy numbers to obtain a membership degree to the fuzzy relation. The ARAKRI1 algorithm for obtaining the indexes (18) of the fuzzy numbers from the matrix with unified and normalized fuzzy numbers is used. Then for each pair of alternatives a_i, a_j and criteria is calculated the following value (e.g. for the *kth* criteria):

$$\mu_k(a_i, a_j) = 0.5 + \frac{F(\tilde{Z}_{ik}) - F(\tilde{Z}_{jk})}{2(F_k^{max} - F_k^{min})},$$

where F_k^{max}, F_k^{min} are indices of the largest and smallest fuzzy number of the fuzzy number's series for *kth* criteria.

Thus, for each criteria whose scores are fuzzy numbers, a fuzzy relation of type (1) is obtained. This fuzzy relation is used for calculations in algorithms ATOKRI.

A block diagram of algorithms application is given in Fig. 2.

The solution of Task 3 is to analyze the rankings received in Task 2 and to select SMEs to be included in the cluster structure.

In [53] proposes an approach to solving a problem for the selection of SMEs—potential participants in integration of economic cluster using system FuzzyPro. The approach implementation is illustrated by an example performed by FuzzyPro software system with results compared to those of PROMETHEE II multicriteria analysis method. The test results confirm the imbedded in the approach concept that, when solving application problems, the use of fuzzy algorithms improves the quality of solution. It allows the DM to make decisions in situations of greater awareness about specific conditions compering to solutions provided by a classical approach.

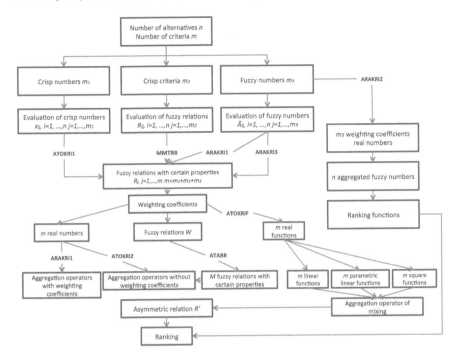

Fig. 2 Block diagram of algorithms application

From the example shown, fuzzy algorithms require significantly more complex input information, which can be interpreted as a difficulty in using them. In this regard, it is appropriate to clarify the specific conditions that would require DM to prefer fuzzy algorithms to classical ones [53].

The answer should be sought in the terms of problem formulation, which initially was defined as a "poorly structured problem under uncertainty", because the goal is to design a cluster structure from SMEs where it is difficult to set a clear target function. Presumably, the SME management models, including integrated into business clusters are burdened with poorly structured information that makes it difficult to implement traditional decision-support methods. This means that the development of such technologies requires this factor to be considered.

Unlike traditional economic problems, which can be reduced to a project of evaluation of investment, efficiency, increase of wealth, i.e. objectives with a single, finite, easily interpretable result, expressed as a value and directed to the achievement of maximum performance, in business—clustering the goal is difficult to assess unequivocally. This ambiguity predetermines "ill structured under conditions of uncertainty", both in terms of the target function, including the local criteria and the primary information. Given that SME business-clustering is a voluntary act of associating self-owned enterprises, it is of great importance that when developing strategies for the activities of these integrated groups, the DM has a toolbox that allows a correct solution to the above tasks.

As a result of the study, an approach is proposed that uses fuzzy algorithms united in the FuzzyPro system. The approach has been verified using a test example. As a result of the decision, SME rankings have been obtained, which can be included in horizontal integration at TN's nodes. The results obtained make it possible to consider fuzzy input data for criteria evaluation. They are compared with the results of the same task using the PROMETHEE II outranking method, without considering fuzzy data. The comparison shows that the rankings obtained using fuzzy logic tool allow refinement of the rankings and improve the awareness of DM in decision-making process [53].

5 Fuzzy Inference System in Support of Financial Investments

Fuzzy inference systems (FIS) are computational structures based on the theory of fuzzy sets, if–then rules and fuzzy logic. Since fuzzy inference systems vary in structure and purpose, different names such as *fuzzy expert system, fuzzy model, fuzzy associative memory, fuzzy logic controller, fuzzy system* and others are used. There are various algorithms for building a fuzzy system [29, 30, 35, 37].

The general structure of a FIS has three conceptual components (Fig. 3):

- database where all the membership functions, all terms used in fuzzy rules and linguistic variables of the fuzzy system are defined;
- rule base including all fuzzy rules for decision-making;
- inference machine performing the procedure for deriving conclusions using given rules and facts to get the correct output.

The rule base and the database form the knowledge base.

A FIS operates in the following sequence: first the input data are fuzzified in order to obtain membership degrees to each of the terms of the input fuzzy variables; then the inference machine applies the aggregation rules, using the knowledge base and thus membership degrees to the terms of output variables are calculated and finally, after defuzzification, the output result is obtained.

Mainly, there are three types of FIS: Mamdani-type, Sugeno-type and Tsukamoto-type Fuzzy Systems.

The core part of a FIS is the fuzzy inference machine (Fig. 4).

Let N be the number of the input fuzzy variables K_i, $i = 1, 2, 3, \ldots, N$, and n_i be the number of terms X_{ij} of K_i for each i with $j = 1, 2, 3, \ldots, n_i$.

Let S be the number of output fuzzy variables Q_s, $s = 1, 2, 3, \ldots, S$, and p_s be the number of terms Y_{sp} of Q_s for each s with $p = 1, 2, 3, \ldots, p_s$.

Let $\mu_{ij}(x)$ be the membership function of the term X_{ij} and $\mu_{sp}(y)$ be the membership function of Y_{sp}. The overall number of the membership functions in the knowledge base is

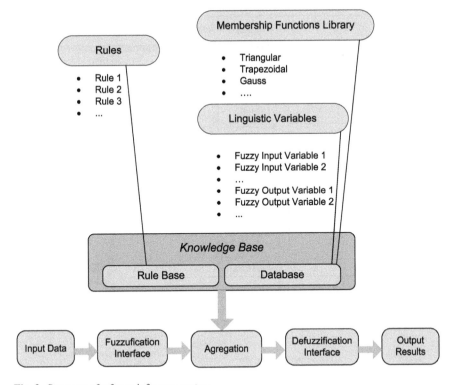

Fig. 3 Structure of a fuzzy inference system

$$N. \sum_{i=1}^{N} n_i + S. \sum_{s=1}^{S} p_s.$$

The crisp input values form a vector $x^* = \left(x_1^*, x_2^*, \ldots, x_N^*\right)$. This vector is fuzzified by calculating $\mu_{ij}\left(x_i^*\right)$ for each i and j. At this point there are

$$N. \sum_{i=1}^{N} n_i$$

membership values, stored in the database after that calculation.

The next step is to aggregate. For simplicity let min operator be used for the T-norm and T-norm be used for the AND operator. Let M be the number of rules and the m-th rule R_m has the form:

$$\textbf{if } \left\{K_{m_1} \text{ is } X_{m_1 j_{m_1}}\right\} \textbf{ and } \left\{K_{m_2} \text{ is } X_{m_2 j_{m_2}}\right\} \textbf{ and } \ldots \textbf{ and } \left\{K_{m_k} \text{ is } X_{m_k j_{m_k}}\right\}$$

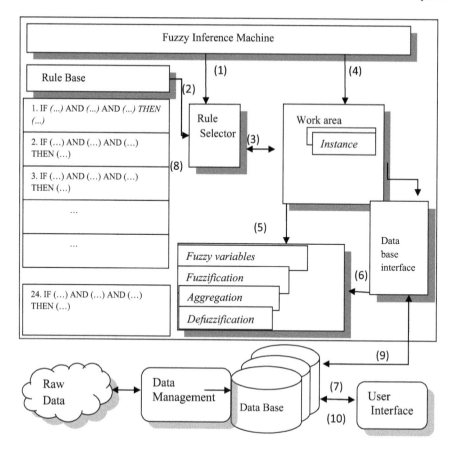

(1) selector activation;
(2) rule choice;
(3) template;
(4) rule activation;
(5) DO: fuzzification, fuzzy aggregation, defuzzification;

(6) interface connection;
(7) reading from the database;
(8) processing the next rule;
(9) writing the results in the database;
(10) display results

Fig. 4 Control of the fuzzy inference machine

$$\text{then } \left\{Q_{m_1} \text{ is } Y_{m_1j_{m_1}}\right\} \text{ and } \left\{Q_{m_2} \text{ is } Y_{m_2j_{m_2}}\right\} \text{ and } \ldots \text{ and } \left\{Q_{m_l} \text{ is } Y_{m_lj_{m_l}}\right\}$$

and each rule has its weight w_m, $m = 1, 2, 3, \ldots, M$.

Once the m-th rule is selected and put into the template (Fig. 4), two consecutive calculations are made:

1. $\Theta_m = min\left\{\mu_{m_1j_{m_1}}\left(x_1^*\right), \mu_{m_2j_{m_2}}\left(x_2^*\right), \ldots, \mu_{m_kj_{m_k}}\left(x_k^*\right)\right\}$ and then

2. $\Theta_m^o = \Theta_m . w_m$.

After firing all the rules, the corresponding values of the membership functions $\mu_{sp}{}^m = \Theta_m{}^o$ for each term Y_{sp} of the output variables are obtained. The number of these values depends on the number of rules in which they are used.

The aggregation applies after calculating

$$P_{sp} = max\{\mu_{sp}{}^1, \mu_{sp}{}^2, \ldots, \mu_{sp}{}^M\}$$

for each Y_{sp}, s $= 1, 2, 3, \ldots, S$ and $p = 1, 2, 3, \ldots, p_s$.

The last step is defuzzification. For implementing any of the methods for defuzzification a numerical integration could be applied.

Financial investments are subject to a simple concept: every investor aims at achieving maximum return while taking minimum risk. Therefore, the key point in the process of managing financial investments is finding a reliable estimator for changes in asset prices. Most financial models are built on the assumption that asset returns have some type of probability distribution. However, empirical tests conducted on real data (e.g. from Bulgarian Stock Exchange) prove the opposite.

Fuzzy logic provides adequate tools for dealing with large amount of data (such as time series of asset prices) and sometimes vague and imprecise information (as is economic information) [33]. What is more, in fuzzy modelling there are no require-ments for existence of probability distributions. Assessing the characteristics of potential assets is a key issue in managing individual finance investments as well as in managing finance portfolio investments. Since an investor seeks maximal possible return combined with minimal possible risk, in the decision-making process at least two criteria have to be taken into consideration. Return and risk are the most common measurements in different models. They are obtained either from historical data or by subjective probabilities. The precise determination of these characteristics is the basis for a good initial investment and proper investment management over time. Investors' decisions are based on the degree of satisfaction of certain criteria often in conditions of vague and imprecise available economic information and are an optimal solution in a set X of alternatives. This optimization problem could be solved with the tools of fuzzy logic if the degrees of satisfaction of the criteria are treated as membership functions of fuzzy variables $\mu_i : X \rightarrow [0, 1]$.

The goal of deciding is to find an optimal solution in a space of possible solutions, satisfying predefined constrains and goals.

Let C_i be a fuzzy set of constrains and G_j be a fuzzy set of goals. Then, the fuzzy decision D, taken under the constrains C_i and meeting the goals G_j, is the fuzzy set:

$$D(p) = \underset{i \in N_n, j \in N_m}{aggregation}\{C_i(p), G_j(p)\},$$

for each $p \in P$, where P is the universe of possibilities [37].

Following this idea, Fuzzy Logic Q-measure Model (FLQM) and Fuzzy Software System for Asset Management (FSSAM) have been published in [43, 45, 48, 50]. In both applications (FLQM and FSSAM) annual log return is used as an estimator of asset return, the standard deviation of log returns—as an estimator of investment risk, and their quotient q-ratio -as an indicator for stability in return changes. The econometric argumentation can be found in details in [31, 42, 44, 50].

FLQM. The calculations of the crisp values of the input variables: annual return, risk and q-ratio are derived in [44]. These crisp values are fuzzified with the predefined linguistic variables (LVs). The output variable is one: *Q-measure* of an asset. The names of LVs are $X_1 \triangleq return$, $X_2 \triangleq Risk$, $X_3 \triangleq q$-$ratio$, $Y \triangleq Q$-$measure$. The term-sets of LVs are $T(X_1) = \{X_{1j}\}$, $T(X_2) = \{X_{2j}\}$, $T(X_3) = \{X_{3k}\}$, $T(Y) = \{Y_j\}$ for $j = 1,..., 5; k = 1,2,3$ and

$$X_{ij} \triangleq \begin{pmatrix} Very\ low\ i = 1,2\ j = 1 \\ Low \qquad i = 1,2\ j = 2 \\ Neutral \quad i = 1,2\ j = 3 \\ High \qquad i = 1,2\ j = 4 \\ Very\ high\ i = 1,2\ j = 5 \\ Small \qquad i = 3 \quad j = 1 \\ Neutral\ i = 3\ j = 2 \\ Big \qquad i = 3\ j = 3 \end{pmatrix} ; Y_j \triangleq \begin{pmatrix} Bad \qquad j = 1 \\ Not\ bad\ j = 2 \\ Neutral\ j = 3 \\ Good \qquad j = 4 \\ Very\ good\ j = 5 \end{pmatrix}$$

The universes of discourse of LVs are $U_{X1} = U_{X2} = U_{X3} = U_Y = R$.

Three types of membership functions are used in FLQM: Gaussian membership function $\mu_G(x) = e^{-\frac{1}{2}\left(\frac{x-\beta}{\alpha}\right)^2}$; Bell membership function $\mu_B(x) = \frac{1}{1+\left|\frac{x-\gamma}{\alpha}\right|^{2\beta}}$ and Sigmoid membership function $\mu_S(x) = \frac{1}{1+e^{-\alpha(x-\beta)}}$. The corresponding type of the membership functions and the values of their parameters can be found in [31, 42, 50]. For each input variable a degree of membership to the corresponding term is calculated.

In FLQM a Mamdani-type fuzzy inference system is chosen. A fuzzy output is obtained as a result of a four-stage fuzzy inference process: (1) evaluation of the antecedent for each rule; (2) obtaining a conclusion for each rule, (3) aggregation of all conclusions and (4) defuzzifying. The AND and THEN operators are implemented by min fuzzy T-norm, whereas the aggregation is implemented by max fuzzy T-conorm. For defuzzification of the output is used center of gravity method.

As there are three input variables with 5, 5 and 3 terms accordingly, the universe of all possible rules consists of 75 rules. In the proposed model 24 of the rules are chosen due to the expert opinion of the authors. Although these rules adequately describe the most important possible situations that might arise in the process of investment decision-making, the list of fuzzy rules can be extended without changing the system's architecture. The fuzzy rules model intuitively the decision-making process and have IF–THEN form:

$$IF\left(r^* \text{ is } X_{1i}\right)AND\left(s^* \text{ is } X_{2j}\right)AND\left(q^* \text{ is } X_{3k}\right) THEN \left(Q - measure \text{ is } Y_p\right)$$

for $i = 1,...,5; j = 1,...,5; k = 1,...,3$ and $p = 1,...,5$.

In the next stage the fuzzy rules are fired. At this point additional expert knowledge is considered by assigning weights to rule in the structure. In this way for a crisp input (r^*, s^*, q^*) the obtained membership values are:

$$\theta^* = min\{\mu_i(r^*), \mu_i(s^*), \mu_i(q^*)\}$$

and then respectively

$$\theta^{**} = w.\theta^*,$$

where w is the corresponding weight of the rule (Fig. 5).

After applying all the rules, several values for each term of the output variable Q-measure are calculated. Aggregation is the process of bringing together the outcomes of all the fuzzy rules. Choosing a suitable aggregation operator is a key issue when a fuzzy system is designed. As an aggregation method the *max* fuzzy *T-conorm* is applied in the proposed model and thus the fuzzy output variable is obtained.

The center of gravity (CoG) has been chosen as a defuzzification method. According to it, the crisp output value is calculated as:

$$\widehat{Q} = C_o G(Q) = \frac{\int\limits_{-\infty}^{+\infty} x.Q(x).dx}{\int\limits_{-\infty}^{+\infty} Q(x).dx},$$

which is illustrated on Fig. 6.

FSSAM is an independent software system which consists of procedures for data collection and data storage, asset evaluation and investment portfolios construction.

The application software system consists of three modules:

Data managing module (DMM) with the following features: automatically submits queries to the Web server of a particular stock exchange; extracts data from the downloaded pages; writes data to the database; fills in the missing data; calculates *return*, *risk* and *q-ratio* for each asset in the database.

Q-measure fuzzy logic module (QFLM);

portfolio construction module (PCM), in which various portfolios are constructed.

The modules are described in detail in [44, 50].

Individual financial asset management

In investment management, the most important point is gaining high profit with lowest possible risk. However, for the non-speculative investor it is essential to what extent these two characteristics (return and risk) are stable over time. FSSAM is built on one additional characteristic: the *q-ratio*, which is the quotient of return and risk and reflects the degree to which the taken risk is justified by adequate returns.

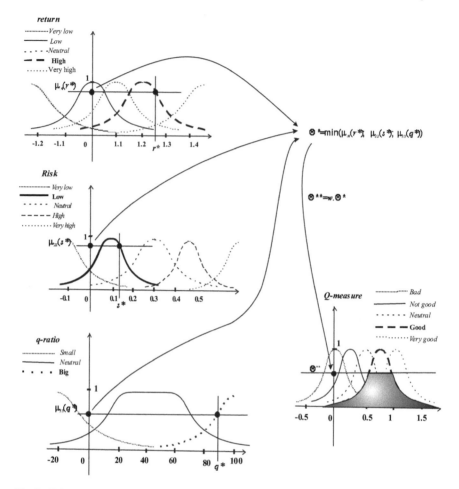

Fig. 5 Firing the rule: IF (return is high) AND (Risk is low) AND (q-ratio is big) THEN (Q-measure is good)

A series of empirical tests have been conducted for all assets listed on Bulgarian Stock Exchange for different periods of time, and they show that the *Q-measure* is a proper indicator of the quality of the asset over time. If the Q-measure is less than 0.4 (whatever return and risk) a dramatic decrease of price occurs in up to about 3 months. At the same horizon and a *Q-measure* between 0.4 and 0.6 the price of the asset does not change significantly and even if it increases, the transaction costs will exceed the potential benefits. When the *Q*-measure is greater than 0.6 the asset price increases steadily and such an asset is considered suitable for purchase. Detailed results for individual asset management are published in [31, 42, 50].

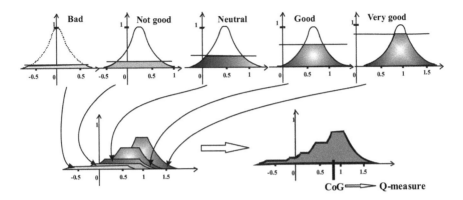

Fig. 6 Aggregation and defuzzification for obtaining the *Q-measure* of an asset

Currency rates

FSSAM is also applied on time series of currency rates. The used data are for currency pairs EUR/USD and AUD/USD. The goal is to assess the performance of FSSAM if applied for forecasting the rates. It is applied for different sequential periods— 5 days period, 6 days period, 7 days period, 8 days period and 9 days period. Thus, an array of values of *Q-measure* is obtained. The value of *Q-measure* obtained from FSSAM clearly shows the tendency to substantial decrease in the pair rates. There is, however, little or no evidence that the increase of the value of *Q-measure* indicates growth in currency rates. FSSAM is designed and originally used for assessing financial assets—individual as well as financial portfolio investments and it supports the process of investment decisions in economic environment with an enormous amount of data, which is often incomplete and imprecise. However, in order to use it for currency rates, it is necessary to tune the parameters of the system [47].

Portfolio management

When applying FSSAM for supporting the financial portfolio management process, the cardinality problem has to be very carefully considered [44, 45]. Then, all the assets in the database are sorted in descending order by their Q-measure. Following the allocation algorithm [49], if the investor would like to hold a portfolio with no more than k assets, the first k assets from the sorted list are taken. Let $A = \{A_1, A_2, A_3, \ldots, A_k\}$ be the set of the top k financial assets. Now all possible combinations with $1, 2, \ldots, k$ elements are constructed and recorded. The number of these combinations is $2^k - 1$, because the empty set is not taken into consideration. Next, for each combination of assets a portfolio is constructed in the following manner.

Let x_j be the share of asset A_j. Then according to x_j is calculated as:

$$x_j = \frac{Q_j}{\sum_{j=1}^{k_0} Q_j},$$

where k_0 is the number of financial assets in the particular portfolio, $k_0 = 1, 2, \ldots, k$ and Q_j is the Q-measure of A_j, obtained in QFLM.

Next, the three characteristics R_p, σ_p and q_p of the portfolio are calculated for all $2^k - 1$ portfolios according to the formulas:

$$R_p = \sum_{j=1}^{n} x_j . r_j, \ \sigma_p = \sum_{j=1}^{n} x_j . s_j, \ q_p = \frac{R_p}{\sigma_p}$$

Finally, each portfolio is put through the Q-measure Fuzzy Logic Module of the software application in order to obtain the portfolios' Q-measure. The portfolios with their characteristics are stored in the database and displayed for further use. Detailed results on real data for different time periods are published in [42–45, 48, 50].

6   Genetic Fuzzy System for Asset Management

Conceptually, fuzzy systems are considered to be able to solve non-linear problems in a variety of applications such as classification, modelling, management and others [3, 11, 32, 37]. Their key feature is the ability to model expert human knowledge [53, 60], but the main deficiency of such systems is the lack of ability to learn and adapt. The first attempts to add other techniques to fuzzy systems start around 1990. Two types of hybridization are considered to be the most successful and they are neural-fuzzy systems and genetic fuzzy systems.

Genetic processes are applied in fuzzy systems for solving problems of different degree of complexity: from the simplest case for optimization of the parameters of a fuzzy system to the most complex—for training its rule-base. Two basic approaches exist in creating hybrid systems based on fuzzy sets theory and genetic algorithms: (1) fuzzy sets theory is used for modelling the components and adjusting the parameters of genetic algorithms and these are called fuzzy genetic algorithms (GA_FIS = Genetic Algorithms with Fuzzy Inference System); (2) genetic algorithms are used to solve optimization or search problems related to fuzzy systems and thus genetic fuzzy systems (GFS) are obtained.

The process of designing a fuzzy rule-based system can be defined as an optimization problem for finding most suitable built-in variables, parameters and rules. Moreover, genetic algorithms are a widely used technique for global extrema search, as they show the ability to find nearly optimal solutions with the possibility of using a priori knowledge concerning the search space. For a fuzzy rule-based system a priori knowledge is the information about the type of membership functions, fuzzy rules and the architecture of the fuzzy system itself.

Fuzzy rule-based systems mimic the decision-making process by handling the available information in a human-like manner. The behavior of a fuzzy rule-based system depends on three sets of parameters: (1) fuzzy sets, associated with linguistic variables that define the semantics of the rules; (2) fuzzy rules, determining how the output variables are to be derived and (3) t-norms and s-norms used for aggregation and defuzzification.

The process of creating a fuzzy system starts with designing the system's architecture, which is a relatively easy task. The most difficult, and requiring significant amount of resources, is the stage of setting up the parameters of the system so that the obtained output results are feasible.

Following the general structure of a fuzzy system, each fuzzy software rule-based system consists of a knowledge base, incorporating the data base and the rule base, and an inference machine. A precise description of the inference machine of FSSAM is published in [50].

The explicit structure of a fuzzy rule-based system follows the logic of the fuzzy reasoning process. In the process of hybridization, it is required to determine the number of parameters to be optimized and this number depends on the total number of the membership functions of input and output linguistic variables, the number of parameters of each function, the number of rues in the rule-base and the weight of each rule.

The ultimate goal of every decision-making process is to find an optimal solution for a certain problem when a number of possible solutions exists. Historically, Bellman and Zadeh are the first to propose a fuzzy model for decision-making. They treat objectives and goals as fuzzy sets and the solution is calculated by aggregating these sets.

At the core of genetic fuzzy systems lies the idea that the advantages of evolutionary computation and fuzzy systems can be combined. There are different ways to apply evolutionary computing to fuzzy systems—genetic algorithms, genetic programming or evolutionary strategies can be used in order to obtain a genetic fuzzy system.

Genetic Fuzzy Software System for Asset Management (GFSSAM) aims at an automatic set up of the system's parameters. In this case the fuzzy system architecture, the domains of the linguistic variables and the used operators are known in advance. Three distinctive situations exist with regard to the fuzzy system knowledge base:

- the rule base is known, but the fuzzy variables are either unknown or only approximated and need optimization;
- the semantics of fuzzy rules is at least approximately known, but the rules themselves are not defined precisely;
- neither the fuzzy variables nor the fuzzy rules are known.

In each of these situations, the genetic algorithm is applied differently.

In the preliminary stages of designing a genetic fuzzy system the main goal is to automate the process of generating knowledge base, which is actually an optimization problem or a search problem for an optimal solution. This process consists of finding a suitable for the situation knowledge base, and after the parameterization of this

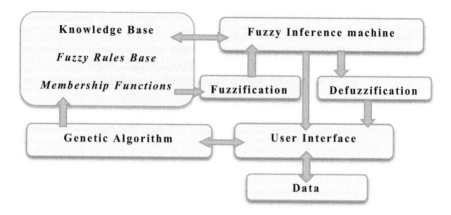

Fig. 7 Genetic fuzzy system for tuning the knowledge base

knowledge base—adjusting the values of the parameters so that they are optimal depending on the optimization criteria.

In case of unknown fuzzy variables and known fuzzy rules, the problem is reduced to an optimization problem with constrains on an unlimited domain. If the fitness function is smooth, the conventional methods for optimization are applicable and generally they are faster than the genetic algorithms. The reason for using genetic algorithms in this situation is that first—the fitness function is not smooth, and second—a genetic algorithm gives several optimal solutions as an output result.

The main objective in designing the model of GFSSAM is to use a genetic algorithm optimization on an already existing fuzzy system. The preliminary created system is FSSAM which econometric base, architecture and performance are described in details in [42, 45, 48, 50].

As can be seen from the general scheme of a genetic fuzzy system, presented on Fig. 7, the genetic algorithm is used for optimization of the fuzzy system's knowledge base. The knowledge base of Fuzzy Software System for Asset Allocation Management consists of the membership functions of the terms of input and output fuzzy variables, as well as the rule-base.

The design of the hybrid system has to provide the simultaneous tuning of all parameters of the knowledge-base by finding optimal values without changing the type of membership functions or the rule-base.

Genetic algorithm (GA) is an adaptive algorithm, defined by an ordered septenary of operators and parameters:

$$GA = (\mathcal{P}, \boldsymbol{PP}, \mathcal{F}, \mathcal{S}, \Omega, \Psi, \varsigma),$$

where $\boldsymbol{PP}$ is a population with size $\mathcal{P}$ consisting of chromosomes c^j:

$$c^j = \left(c_1^j, c_2^j, \ldots, c_l^j \right) \in \boldsymbol{PP}, j = 1, 2, \ldots, \mathcal{P},$$

which are l-dimentional binary vectors;

$\mathcal{F}$ is a function of l variables over $\mathbb{R}^+$, called fitness function:

$$\mathcal{F} : c^j \rightarrow \mathbb{R}^+, j = 1, 2, \ldots, \mathcal{P};$$

$\mathcal{S}$ is a selection operator for choosing u parents p^k from the population PP:

$$\mathcal{S} : PP \rightarrow \{p^1, p^2, \ldots, p^u\};$$

Ω is a set of genetic operators:

$$\Omega = \{\Omega_{Cross}; \Omega_{Mut}; \ldots\},$$

where Ω_{Cross} is a crossover operator, Ω_{Mut} is a mutation operator, generating v children q^m from u parents p^k:

$$\Omega : \{p^1, p^2, \ldots, p^u\} \rightarrow \{q^1, q^2, \ldots, q^v\};$$

Ψ is a removal operator for v chromosomes from the i-th population and then the $i + 1$ population is obtained according to the formula:

$$PP(i + 1) = PP(i) - \Psi(PP(i)) + \{q^1, q^2, \ldots, q^v\};$$

ς is a criterion for end.

The $\mathcal{S}$ and Ω operators are always probabilistic, while Ψ can be either probabilistic or deterministic.

Every GA is a consecutive computation of populations:

$$PP(0), PP(1), \ldots, PP(i), PP(i + 1), \ldots$$

with a randomly chosen initial population $PP(0)$ [54].

GA can also be described as a procedure for solving an optimization problem:

$$max\{F(c)|c \in \{0, 1\}^l\} \text{ or } min\{F(c)|c \in \{0, 1\}^l\}$$

where F is the objective function, $c \in PP$ is a binary vector of length l, PP is the space of possible solutions of size $\mathcal{P}$.

On the other hand, any optimization problem can be reduced to finding $x_0 \in X$, such that

$$f(x_0) = \sup_{x \in X} f(x),$$

where X is the search space and $f : X \rightarrow R$ is an objective function.

When a GA is used for optimization, encoding and decoding functions are needed. These two functions are defined as follows:

$$c : X \rightarrow S \text{ and } \bar{c} : S \rightarrow X, \text{ such that } c \circ \bar{c} \equiv id_S,$$

where S is the set of binary strings and thus the task of GA is to find $s_0 \in S$, such that

$$\bar{f}(s_0) = \sup_{x \in S} \bar{f}(s),$$

where $\bar{f} := f \circ \bar{c}$.

FSSAM is an independent software system which consists of procedures for data collection and data storage, asset evaluation and investment portfolios construction. The system consists of three modules—data managing module (DMM), Q-measure fuzzy logic module (QFLM), portfolio construction module (PCM), that are described in full detail in [42, 44, 50]. The software application is based on FLQM [31, 43, 49]. Input data are the crisp numerical values of asset characteristics—return, risk and q-ratio. These crisp values are fuzzified and after applying the aggregation rules, a fuzzy variable (Q-measure) for each of the assets is derived. The output is the defuzzified crisp value of Q-measure.

The linguistic variables are four: three input variables and one output variable. The input linguistic fuzzy variables K_i are three ($N = 3$) and their names correspond to the characteristics of an asset: $K_1 \triangleq return$, $K_2 \triangleq Risk$ and $K_3 \triangleq q - ratio$. The term-sets are $T(K_1) = \{X1j\}$, $T(K_2) = \{X2j\}$, $T(K_3) = \{X3k\}$ with $j = 1, ...,5$, k $= 1, 2, 3$ and thus $n_1 = n_2 = 5$ and $n_3 = 3$. The output linguistic fuzzy variable is one ($S = 1$) and it is $Y \triangleq$ Q-measure with a term-set $T(Y) = \{Yp\}$ for $p = 1, ...,5$ and that means that $p_1 = 5$ [50].

Thus, the total number of the membership functions in the knowledge base is

$$N. \sum_{i=1}^{N} n_i + S. \sum_{s=1}^{S} p_s = 3.(5 + 5 + 3) + 1.5 = 44.$$

The universes of discourse of the linguistic variables are $U_{K_1} = U_{K_2} = U_{K_3} = U_Y = R$ and the types of membership functions of the terms, used in the system, are Gaussian membership function $\mu_G(x) = e^{-\frac{1}{2}\left(\frac{x-\beta}{\alpha}\right)^2}$, Bell membership function $\mu_B(x) = \frac{1}{1+\left|\frac{x-\gamma}{\alpha}\right|^{2\beta}}$ and Sigmoid membership function $\mu_S(x) = \frac{1}{1+e^{-\alpha(x-\beta)}}$.

The overall number of parameters of the membership functions of the fuzzy functions terms is 37.

For each input variable the degree of membership to the corresponding term is calculated and the numerical values of the input data form a 3-dimensional vector

$x^* = (x_1^*, x_2^*, x_3^*)$. Each coordinate of x^* is then fuzzified by calculating the values of the membership functions $\mu_{ij}(x_i^*)$ for each i and j.

The t-norm, used in the aggregation procedure, is defined by a min operator while the s-norm is defined by a max operator. There are $M = 24$ fuzzy rules implemented in the system, for the t-norm is used for the logical AND operator. Under these considerations the m-th rule R_m in the rule base has the form:

$$\text{IF } \left\{ K_{m_1} \text{ is } X_{m_1 j_{m_1}} \right\} \text{ AND } \left\{ K_{m_2} \text{ is } X_{m_2 j_{m_2}} \right\} \text{ AND } \left\{ K_{m_k} \text{ is } X_{m_k j_{m_k}} \right\}$$

$$\text{THEN} \left\{ Q_{m_1} \text{ is } Y_{m_1 j_{m_1}} \right\}$$

where $m = 1, 2, \ldots, M$ and $M = 24$ is the number of the rules.

After the execution of the m-th rule the values of Θ_m and Θ_m^o are calculated sequentially according to (3) and (4) for $m = 1, 2, \ldots, 24$. The rule weights are $M = 24$ additional parameters in the fuzzy system.

Finally, FSSAM depends on $37 + 24 = 61$ parameters.

After the execution of all the rules in the rule-base the membership degrees are derived for each term Y_{1p} of the output variables. The aggregation is obtained by calculating the values of P_{sp} for each Y_{sp} for $s = 1$ and $p = 1, 2, 3, 4, 5$. In the last procedure, which is defuzzification, a crisp output value is obtained. In FSSAM the centre of gravity method is implemented by composite trapezoidal numerical method for approximating the integrals.

The system's parameters are stored in the knowledge base of the fuzzy system and form the population of the GA—each of them being a chromosome, stored as a vector.

As the genetic algorithm is used for optimizing the parameters of the fuzzy system, the objective function is connected with the defuzzified value of the output variable Q of FSSAM. The objective function F for GFSSAM is defined as a sum of the squared differences of k consecutive values of Q:

$$F = \sum_{i,j=1}^{k} (Q(i) - Q(j))^2.$$

The optimization is focused on finding the values of the parameters of FSSAM aiming at achieving stability of the derived estimation and so the optimization problem is in its essence searching a minimum value of F.

Coding fuzzy variables

All linguistic variables are defined on intervals and consist of a finite number of terms with corresponding membership functions. These variables are initially represented as binary strings.

Coding a finite closed interval $[a; b]$

The coding of the interval $[a; b]$ is realized with two calculations:

1. applying the coding function $c_{m,[a;b]}$ defined as follows:

$$c_{m,[a;b]} : [a; b]\{0, 1, 2, \ldots, 2^m - 1\}$$

$$c(x) = [(2^m - 1).\frac{x - a}{b - a}],$$

where m is the length of the binary string and [.] is the integer part of the number;

2. the integer $c(x)$ obtained in (1) is converted into a binary string.

Decoding a finite closed interval [a; b]

Decoding is also realized with two calculations:

1. the binary string is converted to a decimal number;
2. the corresponding decoding function is $\tilde{c}_{m,[a;b]}$ defined as follows:

$$\tilde{c}_{m,[a;b]} : \{0, 1, 2, \ldots, 2^m - 1\}[a; b]$$

$$\tilde{c}(x) = a + \frac{b - a}{2^m - 1}x.$$

Thus, the condition $co\ \tilde{c} \equiv id\ \overline{\{0, 1, 2, \ldots, 2^m - 1\}}$ is met.

Coding membership functions defined on finite closed interval [a; b]

One direct method for coding membership functions, defined on a finite closed interval of real numbers[a; b], is by applying a linear interpolation on equidistant nodes. The nodes for the interpolation are encoded by applying the function $c_{m,[a;b]}$ and for any membership function the functional values in the corresponding nodes and their encoding by $c_{m,[0;1]}$ are used.

The problem that occurs after such interpolation is related with the length of each of the binary strings. For this reason, it is appropriate to look for another approach. If the membership function is triangular, its encoding may be simplified by simply coding the three real numbers that define it. If the membership function is trapezoidal, four parameters are encoded.

Coding the linguistic variables of FSSAM

In order to apply a genetic algorithm to FSSAM, it is necessary to encode differentiable functions—*Gaussian, Bell* and *sigmoidal*. Instead of encoding functional values as in interpolation, only the parameters of these functions can be coded which means encoding only two parameters (α, β) for the Gaussian and Sigmoid functions and three parameters (α, β, γ) for the Bell function. The coding functions $c_{m,[a;b]}(\alpha)$ and $c_{m,[\varepsilon;\delta]}(\beta)$ are used for coding the parameters of Gaussian function; $c_{m,[a;b]}(\gamma)$, $c_{m,[\varepsilon;\delta]}(\alpha)$ and $c_{m,[\varepsilon';\delta']}(\beta)$ are used when coding the parameters of Bell function. Similarly, $c_{m,[a;b]}(\beta)$ and $c_{m,[\varepsilon;\delta]}(\alpha)$ are used to encode the sigmoidal function parameters. For each of the functions ε, δ, ε' and δ' are the lower and upper limits of the corresponding parameter.

The range, the coding functions and the binary representation of the terms $X_{11}, X_{12}, X_{13}, X_{14}$ and X_{15} of the input variables $T(X_1)$ and $T(X_2)$ of FSSAM are published in [54] and [55]. The terms of the other input variable $T(X_3)$ and the output variable $T(Y)$ are encoded analogously.

Implementation of GA

After finding the coding functions, but before the genetic algorithm is started, parameters that affect its performance are initially declared. These are number of generations, population size, number of variables, number of variable bits, probability of crossover p_c, probability of mutation p_m, threshold of elitism, upper and lower bounds. The upper and lower bounds define the intervals in which the variables of the objective function can be changed and thus define the search space.

Then, a population of random candidate solutions for the optimization task is created. Chromosomes need to be coded from binary to real values. Vector transformation takes place according to the population size, number of variables, number of bits, and lower and upper bounds. Once a population has already been formed, the fitness of each of its individuals for the given objective function is calculated.

The objective function is F:

$$F = \sum_{i,j=1}^{k} (Q(i) - Q(j))^2$$

where $Q(i)$ are the defuzzified output values of FSSAM and k is a parameter.

The system FSSAM is designed and implemented with one purpose: to assess the quality of a financial asset at a particular point of time. Because this assessment needs to be stable, the GA aims at finding a minimal value of the objective function F. The so constructed objective function of the optimization problem is actually the fitness function of the algorithm. In this research, the number of the variables of the fitness function F is equal to the number of the parameters of the linguistic variables of the system—37. The traditional optimization techniques are inapplicable in this case.

After calculating the values of the fitness function, the best most suitable individuals are selected from the selection operator. In this research, the roulette wheel method is used, so the probability for selection for each individual equals the ratio of its fitness and the sum of the values for all individuals. Once the parents have been selected, reproduction for a new generation begins. Crossover is performed and it is a crossover at one point is used—the parents exchange their genetic information at a point calculated with a probability p_c, chosen at the beginning of the algorithm. The next step is mutation—a mutation with a change of one bit has been selected and the probability of that change is p_m. For the newly created generation, the fitness function values are again calculated and the elitism check is performed for outlining the most fit individuals, because each new generation may lose same of the most fit individuals. The genetic material may be lost due to selection, recombination and mutation and this loss may lead to deterioration of results. In order to preserve the good genes, these individuals must be kept in the algorithm. This process is called elite.

After the successful implementation of hybrid system GFSSAM (in the software environment MATLAB, numerous experiments have been conducted to study the performance of the genetic algorithm with respect to changes of its basic parameters. The results about the tuned parameters of linguistic variables and observations about the convergence of GA can be found in [54, 55].

Let M be the number of rules and the m-th rule R_m has the form:

$$\textbf{if}\left\{K_{m_1}\text{is}X_{m_1j_{m_1}}\right\}\textbf{and}\left\{K_{m_2}\text{is}X_{m_2j_{m_2}}\right\}\textbf{and}\dots\textbf{and}\left\{K_{m_k}\text{is}X_{m_kj_{m_k}}\right\}$$

$$\textbf{then}\left\{Q_{m_1}\text{is}Y_{m_1j_{m_1}}\right\}\textbf{and}\left\{Q_{m_2}\text{is}Y_{m_2j_{m_2}}\right\}\textbf{and}\dots\textbf{and}\left\{Q_{m_l}\text{is}Y_{m_lj_{m_l}}\right\}$$

and each rule has its weight w_m, $m = 1, 2, 3, \dots, M$.

Once the m-th rule is selected and put into the template, two consecutive calculations are made:

1. $\Theta_m = min\left\{\mu_{m_1j_{m_1}}\left(x_1^*\right), \mu_{m_2j_{m_2}}\left(x_2^*\right), \dots, \mu_{m_kj_{m_k}}\left(x_k^*\right)\right\}$ and then
2. $\Theta_m{}^o = \Theta_m.w_m$.

After firing all the rules, the corresponding values of the membership functions $\mu_{sp}{}^m = \Theta_m{}^o$ for each term Y_{sp} of the output variables are obtained. The aggregation applies after calculating

$$P_{sp} = max\left\{\mu_{sp}{}^1, \mu_{sp}{}^2, \dots, \mu_{sp}{}^M\right\}$$

for each Y_{sp}, $s = 1, 2, 3, \dots, S$ and $p = 1, 2, 3, \dots, p_s$.

The last step is defuzzification of P_{sp} and thus obtaining the crisp value of the output Q.

In FSSAM are implemented 24 rules with corresponding 24 weights. For example, the sixth rule is:

Rule 6: If {*Return* is *Very High*} and {*Risk* is *Neutral*} and {*q* is *Big*} then {*Q* is *Good*} with a weight $w_6 = 0.8$.

The selection of the weights was made using expert opinion and they are set as initial values in the system.

The individuals of the population are 24-dimensional vectors:

$$c = (w_1, w_2, w_3, \ldots, w_{23}, w_{24})$$

and the GA aims at finding the minimum of the objective function:

$$F = \sum_{i,j=1}^{k} (Q(i) - Q(j))^2.$$

A number of tests have been performed under different values of the GA parameters $nGen$ (number of generations), $nPop$ (size of population), Pc (probability for crossover), Pm (probability for mutation) (Table 3).

The obtained results show that no rule in the rule-base of FSSAM is obligatory, indeed $w_m < 1$ for $m = 1, 2, 3, \ldots, 24$. For a substantial number of rules ($m = 1, 2, 3, 4, 5, 6, 8, 10, 11, 12, 13, 14, 15, 17, 18, 21, 22, 23, 24$) the optimized values of the weights w_m are approximately equal to the initial weights. But for some rules this is not the case. Such are the rules:

Rule 7: If {*Return* is *High*} and {*Risk* is *Low*} and {*q* is *Big*} then {*Q* is *Good*}.

Rule 16: If {*Return* is *Very low*} and {*Risk* is *High*} and {*qisSmall*} then {*Q* is *Bad*}.

Rule 20: If {*Return* is *Very low*} and {*Risk* is *Neutral*} and {*q* is *Small*} then {*Q* is *Not bad*}.

Table 3 Results from the GFSSAM for rule-base optimization

NGen	nPop	Pc	Pm	w_1	w_2	w_3	w_4	w_5	w_6
initial values of w_m				1	1	1	1	0.8	0.8
10	20	0.75	0.20	0.9767	0.8205	0.7666	0.8683	0.7173	0.6566
20	20	0.50	0.05	0.7458	0.9075	0.6729	0.8724	0.3594	0.7721
100	8	0.20	0.02	0.8179	0.7136	0.8024	0.7467	0.5547	0.5372
500	8	0.05	0.05	0.9615	0.8275	0.8546	0.8862	0.7374	0.4601

w_7	w_8	w_9	w_{10}	w_{11}	w_{12}	w_{13}	w_{14}	w_{15}	w_{16}
0.8	0.8	0.8	0.8	0.7	0.7	0.7	0.7	1	1
0.1303	0.3152	0.1141	0.8725	0.9636	0.8208	0.7697	0.8635	0.7732	0.7506
0.0741	0.7664	0.0648	0.8437	0.8427	0.9013	0.3778	0.8727	0.9754	0.4104
0.3249	0.9388	0.2843	0.7089	0.8967	0.7160	0.8007	0.7670	0.8543	0.2782
0.2446	0.8172	0.5641	0.9901	0.6730	0.9475	0.8463	0.8617	0.8338	0.3603

w_{17}	w_{18}	w_{19}	w_{20}	w_{21}	w_{22}	w_{23}	w_{24}	*minF*
1	1	0.8	0.8	0.8	0.8	0.8	0.8	
0.8266	0.6055	0.1141	0.2725	0.6304	0.7115	0.8630	0.8891	**0.00074098**
0.5534	0.9816	0.0649	0.1438	0.7741	0.8643	0.6927	0.9051	**0.00063489**
0.6930	0.8293	0.2843	0.2089	0.9249	0.8388	0.9062	0.7949	**0.00003211**
0.8285	0.9392	0.5641	0.1901	0.8447	0.7172	0.9052	0.7675	**0.00000069**

These three rules may be excluded from the rule-base and in this way, it will contain fewer rules, or their weights could be changed.

There are two rules for which the results are inconsistent:

Rule 9: If {*Return* is *Very high*} and {*Risk* is *Very low*} and {*q* is *Neutral*} then {*Q* is *Good*}.

Rule 19: If {*Return* is *Very low*} and {*Risk* is *Very High*} and {*q* is *Small*} then {*Q* is *Not bad*}.

The last two rules either contradict with or overlap other rules from the rule-base of FSSAM. In both cases additional tests are needed.

In the last two decades, technology changes have inspired numerous applications in an area, broadly called Soft computing. GFSSAM, presented in this paper, is a genetic fuzzy system, aiming at optimizing the knowledge base of a fuzzy inference system. As genetic algorithms have the potential to bring flexibility, they are suitable for optimizing fuzzy systems for decision-making in diagnostics, monitoring and management. The hybridization of the genetic algorithm and the fuzzy system is considered successful because the behavior of the system with the found optimal values is stable and reliable.

7   Fuzzy Methods and Applications

In [39] the author investigates two families of fuzzy multi-criteria algorithms for bidding strategy selection in continuous double auctions (CDA)—Fuzzy Techniques and Negotiable Attitudes (*FTNA*) and Aggregation of fuzzy Relations between Alternatives and a fuzzy relation between the weights of the CRIteria 1 (*ARAKRI1*). The described methods for decision making can be used by software trading agents in electronic auctions not only for preliminary selection of the most suitable strategy from a predefined set of bidding strategies, but also for changing the current strategy during auction.

To find the multi-criterion ordering in [41], author use a fuzzy algorithm *ARAKRI2* with direct aggregation operators *MaxMin* and *MinAvg*. The key difference between this new approach and known from the literature solution *FTNA* is in the lack of weighted coefficients [12, 36–38]:

A series of experiments has been conducted: with equal value coefficients α and γ, which change stepwise in the interval [0,1]; with coefficients α and γ, whose sum equals 1; with coefficient k, which varies in the interval [0,1], with minimizing and mixed criteria. The obtained results have confirmed the applicability of the *ARAKRI2* (*MaxMin* and *MinAvg*) method for solving the given problem.

In [40] the author suggests a bidding strategy selection method for online auctions via direct aggregation of fuzzy numbers *ARAKRI2* which uses the *Gamma* operator [12, 29, 30, 34, 35].

A series of experiments with different values for γ and k has been conducted with minimizing and mixed (minimizing and maximing) criteria. The obtained results confirmed the applicability of the described method for solving the strategy's selection task at hand.

The obtained experimental results in [39–41] indicate that the described method for decision making can be applied in electronic auctions for preliminary selection of the most appropriate strategy from a given set of available strategies as well as for changing a strategy in the course of an auction. The suggested method can become a part of a software decision making module in which various heuristics increasing reliability, sustainability and robustness when choosing a strategy can be implemented. Using fuzzy numbers and their direct aggregation will allow software agents to increase efficiency and adaptivity by selecting an optimal bidding strategy.

According to [46] the problem of the financial state comparison of construction companies can be solved by multiple-criteria decision making (MCDM) with fuzzy relations between important financial ratios. The multi-criteria ranking method called *ARAKRI* is in the basis of the proposed solution for comparing the competitiveness of nine leading Bulgarian construction companies by using financial assessment index for performance evaluation of construction companies.

The proposed method ensures an objective and unbiased multi-criterion ordering of construction companies. The described procedure can improve compliance with the principles of transparency, equal treatment and non-discrimination in Bulgarian public procurement and serve as an objective method of impartial winner determination. Although we are focused here on the *ARAKRI* algorithm, the proposed approach is flexible and can be easily extended to deal with a variety of decision-making problems.

In [51] the author proposes a new Technique for Order Preference by Similarity to Ideal Solution (TOPSIS) modification via type-2 fuzzy numbers.

The feasibility and effectiveness of the proposed TOPSIS modification are illustrated by a numerical example. The obtained rankings by using two different approaches are found as the same. The new method does not require a complicated computation procedure and it is beneficial to decision analysis. The disadvantage of the method is its dependence from fuzzy numbers' shapes.

The purpose of [52] is developing and applying an Interval Type—2 Fuzzy Numbers modification of the Decision Making Trial and Evaluation Laboratory (DEMATEL) and ViseKriterijumska Optimizacija I Kompromisno Resenje (VIKOR) in a hybrid group MCDA model.

The feasibility and effectiveness of the proposed DEMATEL and VIKOR combination are illustrated by a numerical example for business intelligence (BT) software selection. All BI platforms rankings in the example derived by three different approaches (the new modification, fuzzy TOPSIS extension and DtraT TOPSIS) are found to be very similar.

In [56] the author proposes two new **Evaluation based on Distance from Average Solution (EDAS)** modification via IT2FSs similarity measure and via **Interval type-2 fuzzy sets (IT2FSs)** graded mean integration representation.

Analysing the metrics the authors show that they are all suitable for measuring the distances between alternatives being compared and the ideal solution in classic and fuzzy EDAS modifications [57].

The advantages of the abovementioned EDAS algorithms are as follows:

1. expand the applicability of EDAS in more uncertainty environments;
2. enrich the tools for distance measure in EDAS MCDM.

Using a numerical example, it was proven that the new algorithms are suitable for MCDM in the absence of dependencies between assessment criteria. [57].

To select a big data platform from three cloud providers and two user's requirements in [58] author applies TOPSIS, VIKOR and EDAS methods via fuzzy triangular numbers. The produced results are consistent with the results mentioned by other researchers. The advantages of the proposed solution are in fact that it is effective, computationally simple and easy to use. The proposed fuzzy algorithms' combination could be useful to the different stakeholders in big data evaluation, for example, architects, analysts, developers, designers, testers, consultants and managers in big data providers.

In [59] author suggests, analyzes and assesses a new supervised machine learning algorithm for time series (TS) prediction with Fuzzy Transition Relationships Matrix (FTRM).

The new fuzzy-fluctuation time series (FFTS) prediction method consists of several main parts: time series data fuzzification, new FTRM model establishment, a model-based FFTS output prediction and finally, defuzzifications of predicted fuzzy time series values into crisp ones.

A detailed step-by-step description of new algorithms can be found below:

Step 1. Enter the time series elements $F(t), t = 2, 3, \ldots, T$.

Step 2. Calculate the differences between every two adjacent periods $G(t), t = 2, 3, \ldots, T$. After that, determine the number of intervals g and calculate the boundaries of each interval L. Let $g = 7$ and $L = \{l_1, l_2, \ldots, l_7\}$. Then define 7 levels of fuzziness as follows: $u_1 = (-\infty, l_2), u_2 = [l_1, l_3), u_3 = [l_2, l_4), u_4 = [l_3, l_5), u_5 = [l_4, l_6), u_6 = [l_5, l_7), u_7 = [l_6, +\infty)$.

Step 3. Convert the differences $G(t)$ into their corresponding triangular fuzzy numbers $S(t), t = 2, 3, \ldots, T$ in a given universe of discourse U. The membership degree μ_i of each index value is defined as follows:

$$\mu_1 = \begin{cases} 1, x < l_1 \\ \frac{l_2-x}{l_2-l_1}, l_1 \leq x < l_2 \\ 0, l_2 \leq x \end{cases}$$

$$\mu_k = \begin{cases} 0, x < l_{k-1} \\ \frac{x-l_{k-1}}{l_k-l_{k-1}}, l_{k-1} \leq x < l_k \\ \frac{l_{k+1}-x}{l_{k+1}-l_k}, l_k \leq x < l_{k+1} \\ 0, l_{k+1} \leq x \end{cases}, k = 2.$$

$$\mu_7 = \begin{cases} 0, x < l_6 \\ \frac{x-l_6}{l_7-l_6}, l_6 \leq x < l_7 \\ 1, l_7 \leq x \end{cases}$$

Step 4. Let *FTRM* be a fuzzy relationship matrix between each k successive FFTS elements. In case of $k = 2$, *FTRM* is a two-dimensional array of fuzzy numbers with l rows and l columns. Let the first dimension of the matrix correspond to the first factor $S(t-1)$, and the second dimension – to the second factor $S(t-2)$. Let denote $S(t-2)$, $S(t-1)$ and $S(t)$ from FFTS with fuzzy numbers $M_a = \{\mu_{a1}, \mu_{a2}, \ldots, \mu_{al}\}$, $M_b = \{\mu_{b1}, \mu_{b2}, \ldots, \mu_{bl}\}$, and $M_c = \{\mu_{c1}, \mu_{c2}, \ldots, \mu_{cl}\}$. To determine the output change over time periods $t-2$, $t-1$ and t, compute the relationship between M_a, M_b and M_c by the next formula:

$$\Delta_{ij} = M_c * \mu_{ai} * \mu_{bj}.$$

Add Δ_{ij} matrix to the current *FTRM* state for $t = 4, 5, \ldots, T$. Normalize the final *FTRM* $_{lxl}$.

Step 5. To forecast next value, multiply *FTRM* and k previous periods values $M_{S(t-k)}, \ldots, M_{S(t-2)}, M_{S(t-1)}$ and $M_{S(t)}$. In case of $k = 2$ the equation is:

$$FTRM = FTRM * M_a * M_b$$

By using contraction, we convert the result FTRM into a type-1 fuzzy number:

$$\hat{S}(t) = \sum_{i,\ldots,o,p=1}^{l} FTRM$$

Step 6. Defuzzify $\hat{S}(t)$ into $G'(t)$, the forecasting value of future fluctuation $G(t)$. Calculate the predicted $F'(t+1)$ value by using the following formulae:

$$F'(t+1) = F(t) + G'(t).$$

Repeat Step 5 and Step 6 for each element in the test dataset.

Step 7. In order to assess prediction performance, calculate the differences between forecasted actual values. Apply one of the widely used indicator in time series models evaluations—Root of the Mean Squared Error (RMSE), defined as follows:

$$RMSE = \sqrt{\frac{\sum_{t=1}^{n} (forecast(t) - actual(t))^2}{n}}$$

The robustness of the new algorithm is proven by univariate TAIEX prediction with window size equal to two. The experiments have shown that the new *FTRM* method achieves better performance than existing fuzzy prediction models with only one exception.

8 Presenting the Results to the Scientific Community

In recent three decades, selected publications have emerged from numerous contracts and projects and have found acceptance on the pages of:

A. Journals:

A1. Fuzzy Sets and Systems (1993, 1998, 2003).

A2. Fuzzy Optimization and Decision Making (2007).

A3. International Journal Cybernetics and Systems (1998, 1999, 2001).

A4. Advanced Studies in Contemporary Mathematics (2005, 2006, 2007).

A5. International Journal of Soft Computing (2012).

A6. Computational and Applied Mathematics (2018).

A7. International Journal of Engineering Science and Innovative Technologies (2015).

A8. Cybernetics and Information Technologies (2002, 2006, 2008, 2009/1/4, 2010, 2012, 2015, 2016, 2017/1/3, 2018/2/2, 2019).

A9. Problems of Engineering Cybernetics and Robotics (1991).

A10. Comptes Rendus de l'Academie bulgare des Sciences (1998, 2001, 2002, 2003/1/10, 2004, 2005, 2006, 2007, 2008/3/5, 2009, 2010, 2013).

A11. Engineering science (2016, 2019).

A12. Management and Sustainable Development (2012).

B. Proceedings (Transactions):

B1. International conference of Fuzzy Systems, Fuzzy IEE'96, New Orleans, LA, 1996.

B2. Thirteen European Meeting on Cybernetics and Systems Research, Vienna, 1996.

B3. XIV European Meeting on Cybernetics and Systems Research, EMCSR'98, Vienna, 1998.

B4. Xth International Conference on Multiple Criteria Decision Making, Taipei, 1992.

B5. International IEEE Conference "Intelligent Systems", Varna, 2004.

B6. Twenty-fifth Conference of Union on the Bulgarian mathematicians "Mathematics and Education", Kazanlak, 1996.

B7. Innovative Techniques in Instruction Technology, E-Learning, E-assessment and Education, Springer, 2008.

B8. Workshop on Applications of Software Agents, Aachen, 2011.

B9. 11th International Conference (ICANNGA 2013), Lausanne, 2013.

B10. Machine Learning and Artificial Intelligence, Manchester, 2016.

B11. 19th International Symposium on Electrical Apparatus and Technologies (SiELA), 2016.

B12. Seventh International Conference on Telecommunications and Remote Sensing (ICTR'18), Barcelona, 2018.

B13. 42nd International Convention on Information and Communication Technology, Electronics and Microelectronics (MIPRO 2019), 2019.

C. Handbook:

C1. Fuzzy Sets Comparison—Theory, Algorithms and Applications (Ed.: George A. Papakostas, Anestis G. Hatzimichailidis and Vassilis G. Kaburlasos), Gate to Computer Science and Research Vol. 7, 2016.

9 Instead of a Conclusion

These publications are recognized in Scopus, Web of Science, Google Scholar and other international scientific databases and refereed journals. Full texts of the publications can be found in https://www.researchgate.net and https://www.bas.academia.edu and in the websites of the respective institutes and university publications. A substantial part of the results in these publications are repeatedly cited and developed in the works of many academic publications, PhD thesis, monographs, handbooks, projects and find their place in bachelor's, master's and PhD programs in national and international educational and research institutes. It is clear that in the coming periods, soft computing will expand and enrich with new subsuming principal partners.

Many open issues in research and applications in new kind of artificial intelligence are awaiting their new methods, models, algorithms and programs as a product of the efforts of teams of researchers.

References

1. Popchev, I., Peneva, V.: Comparison of cluster partitions. Probl. Eng. Cybern. Robot. **33**, 33–43 (1991). ISSN: 0204-9848
2. Popchev, I., Peneva, V.: A fuzzy multicriteria decision making algorithm. In: Proceedings of the X-th International Conference on Multiple Criteria Decision Making, vol. II, pp. 11–16. Taipei (1992). Last accessed19–24 July 1992
3. Popchev, I., Peneva, V.: An algorithm for comparison of fuzzy sets. Fuzzy Sets Syst. Elsevier Science Publishers, North-Holland, Amsterdam, vol. 60, No 1, pp. 59–65 (1993). ISSN: 0195-0114
4. Peneva, I. Popchev, V.: Rating alternative using fuzzy numbers. In: Proceedings of the Thirteenth European Meeting on Cybernetics and Systems Research (Edited by Robert Trappl), Cybernetics and Systems '96, vol. 1, pp. 347–349. Vienna (1996). ISSN: 0196-9722. Last accessed 9–12 April 1996
5. Peneva, V., Popchev, I.: Fuzzy relations in decision making. Fuzzy—IEEE'96. In: Proceedings of International Conference on Fuzzy Systems, pp. 336–341. New Orleans, LA (1996). 0-7803-3645-3/96$5.00©1996IEEE. Last accessed 8–11 Sept 1996
6. Peneva, V., Popchev, I..: Fuzzy numbers in decision making systems. In: Proceedings of Twenty-Fifth Conference of Union on the Bulgarian mathematicians "Mathematics and Education", pp. 221–228. Kazanlak (1996). Last accessed 6–9 April 1996

7. Peneva, V., Popchev, I.: Decision making with fuzzy relations. In: Proceedings of the XIV European Meeting on Cybernetics and Systems Research (EMCSR'98), pp. 195–200. Vienna, Austria (1998). Last accessed 14–17 April 1998
8. Peneva, V., Popchev, I.: Fuzzy ordering on the basis of multicriteria aggregation. Int. J. Cybern. Syst., vol. 29, No 6, pp. 613–623. Taylor and Francis (Editor Robert Trappl) (1998). ISSN: 0196-9722
9. Peneva, V., Popchev, I.: Comparison of clusters from fuzzy numbers. Fuzzy Sets Syst. **97**(1), 75–81 (1998), 0165-0114/98/$19.00©1998 Elsevier Science B.V. ISSN: 0195-0114
10. Peneva, V., Popchev, I.: Aggregation of fuzzy relations. Comptes Rendus de l'Académie Bulgare de Sci. **51**(9–10), 41–44 (1998). ISSN: 1310-1331
11. Peneva, V., Popchev, I.: Fuzzy logic operators in decision—making. Int. J. Cybern. Syst. (Robert Trappl R (ed.), **30**(6), 725–745 (1999). ISSN: 0196-9722
12. Peneva, V., Popchev, I.: Aggregation of fuzzy relations in multicriteria decision making. Comptes Rendus de l'Académie Bulgare des Sci. **54**(4), 47–52 (2001). ISSN: 1310-1331
13. Peneva, V., Popchev, I.: Aggregation on fuzzy numbers in a decision making situation. Int. J. Cybern. Syst. **32**(8), 871–885 (2001). ISSN: 0196-9722
14. Peneva, V., Popchev, I.: Sensitivity of Fuzzy Logic Operators. Comptes rendue de l'Académie Bulgare des Sci. **55**(3), 45–50, (2002). ISSN: 1310-1331
15. Peneva, V., Popchev, I.: Fuzzy multicriteria decision-making. Cybern. Inf. Technol. **2**(1), 16–26 (2002). ISSN: 1311-9702
16. Peneva, V., Popchev, I.: Properties of fuzzy relations obtained by aggregation operators. Comptes Rendus de l'Académie Bulgarie des Sci. **56**(1), 9–16 (2003). ISSN: 1310-1331
17. Peneva, V., Popchev, I.: Weighted aggregation of fuzzy relations. Comptes Rendus de l'Académie Bulgarie des Sci. **56**(10), 29–35 (2003). ISSN: 1310-1331
18. Peneva, V., I. Popchev, I.: Properties of the aggregation operators related with fuzzy relations. Fuzzy Sets Syst. **139**(3), 615–633 (2003). ISSN: 0165-0114
19. Peneva, V., Popchev, I.: Fuzzy decisions in soft computing. In: Proceedings of 2nd International IEEE Conference "Intelligent Systems", vol. II, pp. 606–609. Varna, Bulgaria (2004). IEEECat.No. 04EX791. ISBN: 0-7803-8278-1. Library of Congress Control Number: 2003115853. Last accessed 22–24 June 2004
20. Peneva, V., Popchev, I.: Transformations by parameterized t-norms preserving the properties of fuzzy relations. Comptes Rendus de l'Académie Bulgarie des Sci. **57**(10), 9–18 (2004). ISSN: 1310-1331
21. Peneva, V., Popchev, I.: Aggregation of fuzzy preference relations with different importance. Comptes Rendus de l'Académie Bulgarie des Sci. **58**(5), 499–506 (2005). ISSN: 1310-1331
22. Peneva, V., Popchev, I.: Weighted transformation in aggregation operators uniting fuzzy relations. Adv. Stud. Contemp. Math. **10**(1), 25–44 (2005). ISSN: 1229-3067
23. Peneva, V., Popchev, I.: Models for weighted aggregation of fuzzy relations to multicriteria decision making problems. Cybern. Inf. Technol. **6**(3), 3–18 (2006). ISSN: 1311-9702
24. Peneva, V., Popchev, I.: Aggregation of fuzzy preference relations by composition. Comptes Rendus de l'Académie Bulgarie des Sci. **59**(4), 373–380 (2006). ISSN: 1310-1331
25. Peneva, V., Popchev, I.: Models of decision making support by multicriteria problems in fuzzy environment. Adv. Stud. Contemp. Math. **12**(2), 291–308 (2006). ISSN: 1229-3067
26. Peneva, V., Popchev, I.: Aggregation of fuzzy preference relations to multicriteria decision making. Fuzzy Optim. Decis. Mak. **6**(4), 351–365 (2007). ISSN: 1568-4539 (print version). ISSN: 1573-2908 (electronic version). https://doi.org/10.1007/s10700-007-9018-6
27. Peneva, V., Popchev, I.: Aggregation of fuzzy relations with fuzzy weighted coefficients. Adv. Stud. Contemp. Math. **15**(1), 121–132 (2007). ISSN: 1229-3067
28. Peneva, V., Popchev, I.: Aggregation of fuzzy relations using weighting function. Comptes Rendus de l'Académie Bulgare des Sci. **60**(10), 1047–1052 (2007). ISSN: 1310-1331
29. Peneva, V., Popchev, I.: Fuzzy criteria importance with weighting functions. Comptes Rendus de l'Académie Bulgarie des Sci. **61**(3), 293–300 (2008). ISSN: 1310-1331
30. Peneva, V., Popchev, I.: Fuzzy criteria importance depending on membership degrees of fuzzy relations. Comptes Rendus de l'Académie Bulgarie des Sci. **61**(5), 579–584 (2008). ISSN: 1310-1331

31. Popchev, I., Georgieva, P.: A fuzzy approach to solving multicriteria investment problems. In: Iskander M. Ph.D. (ed.) Innovative Techniques in Instruction Technology, E-Learning, E-Assessment, and Education, pp. 427–431. PE Polytechnic Institute of New York University, New York, Springer, Science+Business Media B.V., New York (2008). ISBN: 978-1-4020-8738-7, e-ISBN: 978-1-4020-8739-1, Library of Congress Control Number 2008931889

32. Peneva, V., Popchev, I.: Multicriteria decision making based on fuzzy relations. Cybern. Inf. Technol. **8**(4), 3–12 (2008). ISSN: 1311-9702

33. Angelova, V.: Investigation in the area of soft computing: targeted state of the art report. Cybern. Inf. Technol. **9**(1), 18–24 (2009). ISSN: 1311-9702

34. Peneva, V., Popchev, I.: Models for fuzzy multicriteria decision making based on fuzzy relations. Comptes Rendus de l'Académie Bulgarie des Sci. **62**(5), 551–558 (2009). ISSN: 1310-1331

35. Peneva, V., Popchev, I.: Models for decision making by fuzzy relations and fuzzy numbers for criteria evaluations. Comptes Rendus de l'Académie Bulgarie des Sci. **62**(10), 1217–1222 (2009). ISSN: 1310-1331

36. Peneva, V., Popchev, I.: Multicriteria decision making by fuzzy relations and weighting functions for the criteria. Cybern. Inf. Technol. **9**(4), 58–71 (2009). ISSN: 1311-9702

37. Peneva, V., Popchev, I.: Fuzzy multi-criteria decision making algorithms. Comptes Rendus de l'Académie Bulgarie des Sci. **63**(7), 979–992 (2010). ISSN: 1310-1331

38. Radeva, I.: Strategic Integration with MAP-CLUSTER software system. Cybern. Inf. Technol. **10**(2), 78–93 (2010). ISSN: 1311-9702

39. Ilieva, G.: Decision making methods in agent based modeling. In: Proceeding of the Workshop on Applications of Software Agents, pp. 8–17. Aachen (2011). ISSN: 1613-0073

40. Ilieva, G.: Bidding strategy selection via direct aggregation of fuzzy numbers. Manag. Sustain. Dev. **33**(2), 126–132 (2012). ISSN 1311-4506

41. Ilieva, G.: A fuzzy approach for bidding strategy selection. Cybern. Inf. Technol. **12**(1), 61–69 (2012). Print ISSN: 1311-9702 Online ISSN: 1314-4081

42. Georgieva, P., Popchev, I.: Application of Q-Measure in a real time fuzzy system for managing financial assets. Int. J. Soft Comput. (IJSC) **3**(4), 21–38 (2012). ISSN: 2229-6735 [Online]; 2229-7103 [Print]

43. Georgieva P., Popchev, I.: Fuzzy Q-measure model for managing financial investments. Comptes Rendus de l'Academie Bulgare des Sci. **66**(5), 651–658 (2013). ISSN 1310-1331. https://doi.org/10.7546/CR-2013-66-5-13101

44. Georgieva P., Popchev, I.: Cardinality Problem in Portfolio Selection. In: Lecture Notes in Computer Science 7824. Adaptive and Natural Computering Algorithms. Proceedings 11th International Conference, ICANNGA 2013, pp. 208–217. Lausanne, Switzerland, Springer-Verlag Berlin Heidelberg (2013). ISBN 978-3-642-37212-4. Last accessed 4–6 April 2013

45. Georgieva, P., Popchev, I., Stoyanov, S.: A multi-step procedure for asset allocation in case of limited resources. Cybern. Inf. Technol. **15**(3), 41–51 (2015). Print ISSN: 1311-9702, Online ISSN: 1314-4081. https://doi.org/10.1515/cait-2015-0040

46. Ilieva, G., Dimitrov, A.: Inter-criteria comparison of bulgarian construction companies using fuzzy relations. Int. J. Eng. Sci. Innov. Technol. **4**(2), 290–299 (2015). ISSN: 2319-5967

47. Georgieva, P.: Applying FSSAM for currency rates forecasting. Trans. Mach. Learn. Artif. Intell., Manchester, SSE UK, **4**(3), 30–40 (2016). ISSN 2054-7390

48. Georgieva, P.: Fuzzy rule-based systems for decision-making. Eng. Sci., BAS, **LIII**(1), 5–16 (2016). ISSN 1312-5702 (Print), ISSN: 2603-3542 (online)

49. Georgieva, P.: A genetic fuzzy system for asset allocation. In: Proceedings 19th International Symposium on Electrical Apparatus and Technologies (SIELA), pp. 1–6 (2016). https://doi.org/10.1109/SIELA.2016.7543004. IEEE: CFP1163N-CDR, Electronic ISBN: 978-1-4673-9522-9; Print on Demand (PoD) ISBN: 978-1-4673-9523-*6

50. Georgieva, P.: FSSAM: a fuzzy rule-based system for financial decision making in real time. In: Papakostas, G.A., Hatzimichailidis, A.G., Kaburlasos, V.G. (eds.) Handbook of Fuzzy Sets Comparison—Theory, Algorithms and Applications, Gate to Computer Science and Research, vol. 7, SCIENCE GATE PUBLISHING P.C., pp. 121–148 (2016). ISBN 2241-9055; ISSN 2241-9063

51. Ilieva, G.: TOPSIS Modification with Interval Type-2 Fuzzy Numbers. Cybern. Inf. Technol. **16**(2), 60–68. Print ISSN: 1311-9702 Online ISSN: 1314-4081. https://doi.org/10.1515/cait-2016-0020

52. Ilieva, G.: Group decision analysis with interval type-2 fuzzy numbers. Cybern. Inf. Technol. **17**(1), 31–44 (2017). Print ISSN: 1311-9702 Online ISSN: 1314-4081. https://doi.org/10.1515/cait-2017-0003

53. Radeva, I.: Multicriteria fuzzy sets application in economic clustering problems. Cybern. Inf. Technol. **17**(3), 29–46 (2017). Print ISSN: 1311-9702; Online ISSN: 1314-4081

54. Georgieva, P.: Genetic fuzzy system for financial management. Cybern. Inf. Technol. **18**(2), 20–35 (2018). Print ISSN: 1311-9702; Online ISSN: 1314-4081. https://doi.org/10.2478/cait-2018-0025

55. Georgieva, P.: Parameters of GFSSAM: coding the parameters of a hybrid genetic fuzzy system. In: Proceedings of the Seventh International Conference on Telecommunications and Remote Sensing, (ICTRS '18), pp. 85–92. Barcelona, Spain (2018); ACM New York, NY, USA ©2018, ISBN: 978-1-4503-6580-2. https://doi.org/10.1145/3278161.3278174

56. Ilieva, G.: Group decision analysis algorithms with EDAS for interval fuzzy sets. Cybern. Inf. Technol. **18**(2), 51–64 (2018). Print ISSN: 1311-9702 Online ISSN: 1314-4081. https://doi.org/10.2478/cait-2018-0027

57. Ilieva, G., Yankova, T., Klisarova-Belcheva, S.: Decision analysis with classic and fuzzy EDAS modifications. Comput. Appl. Math. **37**(5), 5650–5680 (2018). ISSN: 2238-3603 (Print) 1807-0302 (Online). https://doi.org/10.1007/s40314-018-0652-0

58. Ilieva, G.: Decision analysis for big data platform selection. Eng. Sci. **LVI**(2), 5–18 (2019). ISSN: 1312-5702 (Print) ISSN 2603-3542 (online). https://doi.org/10.7546/EngSci.LVI.19.02.01

59. Ilieva, G.: Fuzzy supervised multi-period time series forecasting. Cybern. Inf. Technol. **19**(2), 74–85 (2019). Print ISSN: 1311-9702 Online ISSN: 1314-4081. https://doi.org/10.2478/cait-2019-0016

60. Georgieva, P., Dolchinkov, R.: Fuzzy models for managing a micro grid PV system. In: Proceedings 42nd International Convention on Information and Communication Technology, Electronics and Microelectronics, (MIPRO 2019), pp. 1291–1255 (2019). ISSN: 1847-3946

The Contribution of the Academic Institute for Technical Cybernetics and Robotics (ITCR) to the Initiation, the Development and the Serial Production of Personal Computers in Bulgaria

Vassil Sgurev and Plamen Vachkov

Abstract At the end of 1978 in the Bulgarian Academy of Sciences (BAS) a state project was realized to create a powerful enough institute with an Experimental Base in the field of high technologies and in particular in robotics and also in computer cybernetics control systems. Not a big research group in this institute led by the experienced researcher, constructor and organizer eng. Ivan Marangozov early anticipated the appearance of the new perspective trend of the personal computers (PCs) and pretty fast and effectively developed the first Bulgarian 8-bit PC IMCO-2 totally compatible with the PC Apple II Plus of the Apple company. The produced in the ITCR Experimental Base small series of several hundreds of PCs IMCO-2 provoked a big interest in the potential users. The governing body of ITCR-BAS decides to help not a big plant in the town of Pravets (Bulgaria) so that the latter can realize the serial production of IMCO-2 under the trademark of PRAVETS-82. The state institutions develop 'a national program for PCs for the period 1982–1985' which strongly helps the activity of the development and the production of PCs. The team of eng. I. Marangozov develops a 16-bit PC totally compatible with IBM PC/XT the production of which also begins by the name of PRAVETS-16. An active work begins for the creation of a suitable periphery, software and also of systems for the design and control which are PC-based. At the end of the 80-s of the past century Bulgaria assumes a leading position among the Eastern-European countries in the field of PCs. The described events related to the Bulgarian PCs during the past century convincingly show what good results can be obtained when the academic science, the industry and the state institutions act together for a significant for the country goal.

Keywords Personal computers · Development · Serial production · Effective economy

V. Sgurev (✉) · P. Vachkov
Institute of Information, Communication Technologies in BAS, Sofia, Bulgaria
e-mail: vsgurev@gmail.com

© The Author(s), under exclusive license to Springer Nature Switzerland AG 2021   101
K. T. Atanassov (ed.), *Research in Computer Science in the Bulgarian Academy of Sciences*, Studies in Computational Intelligence 934,
https://doi.org/10.1007/978-3-030-72284-5_6

At the end of 1978 a state act subjects the Institute of Technical Cybernetics and Robotics in the Bulgarian Academy of Science (BAS) to a profound modernization the result of which is the creation of a new Institute of Technical Cybernetics and Robotics (ITCR) in BAS. The main goal of this reorganization is to direct the scientific and research activity to the solution of high-tech production problems in our country and most of all towards robotics and the computer cybernetic systems.

This takes place for years when small companies in USA (Apple, etc.) anticipate the market needs and they very fast develop and offer the first personal computers.

A small team is made in ITCR-BAS led by the experienced electronic constructor and organizer eng. Ivan Marangozov aimed at the assessment of strategically important for ITCR trends and also for pilot developments.

Eng. I. Marangozov reacts fast and effectively at the appearance of the first personal computers and together with the engineers of his team—Petar Petrov, George Zhelyazkov, Boris Vachkov, etc. develops the first 8-bit personal computer called IMCO (Individual Micro-COmputer). It has an Intel 8080 microprocessor, a DOS 3.2 hard disk drive, and a Pascal compiler. It is totally compatible with the Apple II Plus analogue of Apple, the company created in 1977 [1]. The choice of the product is very successful and allows for its massive use in our country. In ITCR's Experimental Base, 50 pieces of IMCO PC have been produced, which are welcomed by the consumers.

The Marangozov group is developing a newer version—IMCO-2, without compromising compatibility. This makes it possible to use the already developed Apple II Plus PC software. For the needs of the Ministry of Education the Experimental Base produces 200 pieces (pcs) IMCO-2, funded by the same ministry. I. Marangozov and his collaborators make successful presentations of IMCO-2 throughout the country—in schools, universities, institutions. Especially successful was their presentation to the military ministry. The popularity of IMCO-2 is growing incredibly among the public in our country [2].

The ITCR-BAS governing body since the very beginning of the initiation and imposition of PCs has been firmly behind the PC development by Marangozov and his group. The governing body knows very well that in the face of Marangozov it has found the right person in the right place at the right time. In this respect, the Director of ITCR-BAS, Prof. Angel Angelov, plays an extremely important role. He is at the base of the project 'ITCR' at BAS. His idea is to build an academic institute that has opportunities for a 'closed cycle'—from ideas and patents to prototyping, experimentation, the creation of tooling and complete working documentation for serial production. This is done with lots of efforts! Such an institute at BAS has not been there until now, and there is no such thing at present.

The ITCR at BAS is built with a maximal attraction of already established professionals from the industry and especially from the system of State Economic Holding (SEH) 'ISOT' and its Central Institute of Computing Equipment (CICE). This approach is very important and it allows for the fastest way to reach the industry and the serial production in it. It is also very helpful that most of the heads of the different institutional levels come from the industrial sector and departmental institutes; the director himself, Assoc. Prof. A. Angelov, is one of them. In addition, his position as a Vice Chairman of the State Committee for Science and Technical Progress (SCSTP) gives him the opportunity to actively support the work of ITCR and BAS.

On November 30, 1978, the Bureau of the Council of Ministers published its Order No 73, which established the Institute of Technical Cybernetics and Robotics (ITCR) on 1 December 1978 at BAS. It is based on the previously existing Institute of Technical Cybernetics (ITC) at BAS and three scientific-production laboratories of the Center for Accelerated Deployment (CAD) 'Progress' at the State Committee for Science and Technical Progress (SCSTP)—'Unique Electronics', 'Specialized Robots' and 'Mechanization and Automation of Welding Processes'. The order formulates the following main aim of the newly established institute: 'experimental and applied research in areas related to cybernetic systems for managing technical processes and objects, research and implementation in the field of robots and their accelerated application in the national economy, experimental production as well as training, preparation and post-graduate qualification of specialists in these fields' [2].

The ITCR-BAS activity in the field of robotics is significant and very successful, but it is not the subject of this article. The above-mentioned main objectives of ITCR explicitly do not mention personal computers. In 1978, their role and prospects have not yet been fully elucidated.

In 1980, ITCR received from the industry a site at Iskar Station together with the 'Sasho Kofardzhiev' Scales Plant. For a short time, they have been transformed into the Institute's Experimental Base. During the creation of the ITCR, target funds were earmarked—including 600 thousand lv. Some of them are used for new equipment, repairs and construction in the Experimental Base. It has an electrically-constructive and machine-constructive departments; technological documentation and archives; breeding base; electrical-assembly workshop for rapid implementation of prototype models; models and experimental installations; Automatic Design Unit for Testing Installations; a powerful enough line for PCBs. ITCR also has a specialized, up-to-date computer configuration for PCB design. Only the best departmental institutions have such opportunities. In other ways, it is not possible to directly—without intermediaries, implement the products of an academic institute in the industry!

In October 1982, the director of ITCR, Assoc. Prof. A. Angelov, was appointed ambassador of our country to Japan and went to Tokyo. In the period up to the beginning of 1987 his directorial functions were executed by Assoc. Prof., Doctor of Technical Sciences (DTSs) Vassil Sgurev. A. Angelov constantly maintains contact with the institute, he supports in many ways the institutional activities and during his holidays he actively participates in the work of the ITCR. During this time, no problems and misunderstandings were found with the management of the institute.

In 1983, with a view to staffing the Experimental Base, Assoc. Prof. Plamen Vachkov, who is also Deputy Director of ITCR, was appointed as Head of the Experimental Base. The numerical composition of the Institute and the Experimental Base reached 870 people. All of this gives the opportunity for small-scale production of the first IMCO personal computers.

When the success of the IMCO PCs has been convincingly confirmed by their great demand, the ITCR1 governing body has focused on choosing the right industrial base for their serial production. The Director, Assoc. Prof. A. Angelov tries to attract for this purpose the plant options of the State Economic Holding (SEH) ISOT, but it is not successful—the management of this business organization, albeit with a delay, realizes the importance of this class of computers and decides in parallel to make its own new development and production of PCs. The question of what prevents the efforts of SEH ISOT and its Central Institute of Computing Equipment (CICE) —on the one hand, and of BAS and its ITCR [12]—on the other, to unite in a noble case for Bulgaria, can hardly be answered now...

But it is clear to ITCR managers that a manufacturer of serial computers from the IMCO family needs to be found very quickly. The numerous and difficult discussions in a narrower format lead to the preference of the SEH 'Instrumentation and Automation' (I and A) over the other options mainly for two reasons: first, this organization is not the strong SEH and it has no leading high tech product, and secondly its plant in the town of Pravets could become a high-productivity plant for the serial production of the PCs, which is protected at the highest level—for understandable reasons. These considerations proved to be far-sighted, and the concept, though difficult, was realized to the end [3].

Assoc. Prof. A. Angelov asked Correspondent Member Ivan Popov and he helped in realizing this idea. The same was done by the President of BAS Acad. A. Balevski, who was well aware of the importance of serial production of the BAS products. Some ITCR collaborators who have accessed the highest state levels have also helped. Gradually and after numerous conversations, some of the key figures such as the Pravets leaders Naydenov and Zlatev, the regional secretary Dyulgerov, the General Director of the State Economic Holding (SEH) 'Instrumentation and Automation (IA)' Eng. Vasil Tsarevski and others were attracted as adherents. As a result of these efforts, it was decided to submit the technical documentation of IMCO-2 from the ITCR Experimental Base to the Instrumentation Plant in Pravets. It started production of this computer under the new legalized name of PRAVETS 82. At this plant in 1983 the first 500 pieces of PRAVETS 82 were produced. SEH I and A itself assigned to its Plovdiv plant 'Kocho Tsvetarov' the implementation of the floppy disk drives for PRAVETS 82, the Instrumentation Plant in Petrich—the acquisition of printers and printing devices and the 'Analytic' Plant in the town of Montana—of the monitors. On the agenda is the re-equipment of the Plant in Pravets with a view to its transformation into a basic plant for large-scale production [13].

In 1986, the plant in the town of Pravets was already a PRAVETS Microprocessor Equipment Complex (MEC). The Director of the Expert Base of ITCR, Assoc. Prof. Plamen Vachkov, was appointed as its Director, who maintained his position as Deputy Director of ITCR-BAS. His direct deputies in MEC were Eng. Hristo Hristov and Eng. Rumen Raychev. The three of them took on the difficult tasks of the first stage of the construction of the Pravets cimplex, as well as the implementation coordination of the new models of personal computers developed in ITCR and the documentation on them [4].

The governing bodies of ITCR and BAS are very interested in the success of the computer production at MEC PRAVETS. The management of the plant has been given direct effective help in several ways:

- Sending competent specialists from ITCR in the town of Pravets for temporary or permanent work there. Assoc. Prof. A. Angelov managed to negotiate with the above instances some very valuable specialists to get to work at MEC PRAVETS;
- Active participation of specialists from the ITCR and the Experimental Base in the design of new workshops and plants for PRAVETS, and which is very important, in the selection of suitable machines and equipment for them. Delegations for negotiations on the purchase of new foreign machinery and equipment are almost always attended by ITCR specialists;
- The Marangozov group developers are well aware of the technological capabilities of the PRAVETS plants and this is taken into account in the development process itself.

To this must be added the very good personal contacts between the management and working groups of the two organizations, which greatly helped their successful work together [3].

But not everything and not always goes painlessly and without great difficulties...

The incredibly rapid success of the PCs IMCO-PRAVETS led the governance of the State Committee for Science and Technical Progress (SCSTP) to develop a "program for the development of personal computers in the People's Republic of Bulgaria for the period 1982–1985" [5]. It envisages the production of 10 000 pieces of PCs by the end of 1989, as well as the necessary funds for this production—equipment and expansion of the capacity of the manufacturing plants. At that time SEH ISOT developed its competitive to IMCO-2 (PRAVETS 82) personal computer ISOT 1031. There is a problem—which of the two types of computers and which plants to produce. In order to solve it, the SCSTP appoints a panel of experts consisting of: P. Magerski, M. Krinkov and H. Karadjov. The report, signed by the experts, of 5 July 1983 noted that "according to the last instructions of comrade T. Chakarov, SEH P and A prepares a second version of the material for the production of 5 000 ISOT 1031 in the Plants for Instrumentation andAutomation (PIA) PRAVETS. The reason is the high percentage of equipment for IMCO-2 in the second direction, while ISOT 1031 is produced only with a socialist elements' base." The commission's inspection reveals other facts, that [5]:

1. The Soviet Elements' Base Set in ISOT 1031 is Inaccessible for Purchase in the Required Quantities and, Therefore, Both Computers Require the Same Currency for the Second Direction;
2. 'In the USSR and in our country, the only accessible computer IMCO-2, repeatedly demonstrated to specialists its advantages with graphics and color, and with a variety of software. It should be emphasized that the excited interest of the Soviet side and the inquiry made is based solely on the base of IMCO-2, and here it is not clear what will be the attitude towards the unknown with nothing, both in Bulgaria and in the international field ISOT 1031' [5];
3. 'ISOT 1031 has some significant disadvantages compared to IMCO-2: there is no graphic image and color management. This turns it into a computer with text messages on the screen and simple pseudo-graphs—squares, half-squares, dashes.' [5];

4.   'IMCO-2 with a ZET-80 PC Board (PCB) works with a CP/M operating system
      that makes it possible to use about 20 000 Apple programs and all CP/M
      programs while ISOT 1031 can not use the ones of Apple' [5].

In conclusion, the commission writes: 'We believe that with IMCO-2 (PRAVETS
82) the issues with our 8-bit PCs are solved and that the ISOT 1031 version at the
technical level does not exceed that of IMCO-2 and it is probably not to apply to
both ours and the markets of the socialist countries.' [5].

Attached to the report is a Protocol of 5 July 1983, signed by Eng. Petko Magerski
from SCSTP, and by the responsible designer of ISOT 1031—by CICE. The most
important finding in it is that Apple4s software can not be used by ISOT 1031, and
that of 500 imported software products work with IMCO-2 and only 4 programming
languages—with ISOT 1031.

With such categorical expert judgment, the decisions are also obvious in favor
of IMCO-2 (PRAVETS 82). The implementation of the program for the development of personal computers for the period 1982–1985 is assigned to SHE P and A,
respectively to its plant in the town of Pravets and to the institute of ITCR. For the
production of 10 000 pieces of PCs, SHE I and A envisaged 5, 6 million (mln) lv.,
including currency of 506 thousand lv. in the first direction and currency of 4,2 mln
lv. in the second direction. The limit of capital investment for this SHE is elevated
up to 16 mln lv.

The winning of the right to jointly implement the national PC program (1982–
1985) by ITCR-BAS and SHE I and A is a success that can not be reassessed. This is
a crucial moment because it leads to accelerated deployment of the development and
production of personal computers in Bulgaria. And even because for the first time
an institute of BAS manages to compete on the market for the serial production of
high-tech products and systems!

In 1982, IBM introduced IBM's 16-bit personal computer to world markets, and
in early 1983 IBM PC-XT. It works with Microsoft's MS DOS operating system
and a 10 megabyte hard disk drive. An ITCR delegation reviewed the many IBM
competing personal computers at the Hannover Fair. Some of them surpass the IBM
PC, which is a 'pseudo' 16-bit PC. But ITCR is definitely deciding to take IBM's
personal computer as a model because it is expected to become a 'de facto' standard
of 16-bit PCs. So it happens.

In the summer of 1984, the Marangozov group has already developed a 16-bit
personal computer IMCO-4 that is fully compatible with the IBM PC-XT. There is a
rapid development and transfer to the plant in Pravets of the full technical documentation of IMCO-4 and at the end of 1984 this computer is already produced under
the name of PRAVETS 16. This is only two years after the debut of IBM PC-XT.
The Marangozov group develops PRAVETS 16-M for professional applications and
PRAVETS 16-H for home use. The developer team is making a huge effort to create a
double-layer motherboard. At that time, in all the manufacturers of such computers,
these boards are multilayered, and this requires more sophisticated and expensive
technology. Another team of ITCR, led by K. Batmazyan, after many unsuccessful
attempts, manages to design a double-layer motherboard that does not affect signals

from tracks on signals from others. This is a significant technological success! Our country is already leading in the production of PCs among other socialist countries.

Meanwhile, with some delay after IMCO-4, the management of SHE ISOT is engaged in competing development of two 16-bit PCs, namely ISOT 1036 and ISOT 1037. Both computers are on more sophisticated and expensive multi-layer motherboards. It is stated that one of these computers will be produced at the Silistra plant and the other one at the Plants for Memory Devices (PMD in Veliko Tarnovo, but for one reason or another that is not happening—SEH ISOT refuses to develop the parallel line of the gathered speed and popularity family PCs IMCO-PRAVETS; ITCR-BAS remains the only developer of mass-produced PCs in our country. This was then marked in a categorical manner in the publication [6]. The production remains only in the system of SEH I and A, and its base plant in the town of Pravets. There is also the direct fruitful and effective relationship between BAS, its ITCR institute and шге Microprocessor Engineering Complex (MEC) PRAVETS—from the initiation to the development of this production. And, if we are right, the inconceivable sensation in the Bulgarian Academy of Sciences remains the significant contribution of the academy to the Bulgarian high-tech industry! This is not a small one...

At the initiative of Prof. V. Sgurev, acting as Director of ITCR-BAS in 1987, the first international meeting of representatives of the academies of sciences of the socialist countries in the professional field 'personal computers' took place in the same institute [10]. The Soviet delegation is led by Acad. Naumov—a prominent specialist and director of the Institute for Small Computer Machines. The meeting noted the importance of this direction and the need for its strengthening and development in the academic institutes.

Of particular interest in the development of the ICTR-BAS is Acad. E. Velihov, vice-president of the Academy of Sciences of the USSR. He closely follows the PC's development and often visits the ITCR. They are interested in the possibilities of their application in the instrumentation and the scientific experiment. Significant is his letter from 18.12.1989 to Corresponding member A. Angelov and Assoc. Prof. P. Vachkov as heads of 'Microprocessor Systems' [4].

At the initiative of the governing body of ITCR-BAS in 1985 and 1987 in Sofia two international conferences on the problems of the personal computers—PERSCOMP'85 and PERSCOMP'87 were organized and held. They play a significant role in coordinating the efforts of professionals in Bulgaria and abroad to develop personal computers. At that time, the authority of ITCR-BAS was too great and indisputable in this field of science and technology.

Serial PC production has caused a need for a variety of application software. In ITCR, the workstations CSY-8, CS-10 and CS-16 have been set up to collect laboratory data, process and visualize information, digital processing and image analysis. Through the "MicroScan" system, information from a significant number of thermocouples and thermoresistances is collected and processed. By means of PLOT-2 and 3 there is realized the automation of the design, and for the testing in the electronics is used the program-hardware system RISK. All this is based on the personal computers of the IMCO family—PRAVETS. On the same basis, the Ministry of Education creates types of training classes in informatics, robotics, biotechnology.

The ITCR-BAS actively supports such activities not only in the institutes but also in the other scientific and development units of BAS and outside the academy. The role of SEH 'Software Products and Systems', of the Interprogramme Institute, of HEEI 'Lenin' (Higher Engineering and Electrotechical Institute, now Technical University, TU-Sofia), of the United Center for Mathematics and Mechanics (UCMM) at BAS, of the academic physical institutes and not a small number of departmental institutes in different fields of material production. There is also specialized programming in the field of culture, public practice services.

There is also a significant activity in popularizing PCs by creating computer clubs, workshops, seminars, etc. An important role in this regard is played by the Institute of Highly Specialized Activities (IHSA)'Avangard' (i.e. Avant-Garde).

In 1986 there was a change in the structure of ITCR. At the insistenceof V. Sgurev, acting as director there was created a separate section 'Personal computers', led by I. Marangozov. I. Marangozov himself avoids taking administrative duties and prefers to work with a small circle of people. The new section performs the intense research and design work on the improvement of the PCs produced, as well as on the creation of new high-performance personal computers and modules for them. PRAVETS 16 must be assigned to this group with an Intel 8088 microprocessor and two built-in 500 KB diskette drives; PRAVETS 16I—portable computer with 3 pieces (pcs) inch disk drives; PRAVETS 16B—vertically oriented computer with 2 pcs inch disk drives and PRAVETS 16H—desktop PC with 2 pcs (and inch) disk drives. These computers are based on a standard euroformat and euro mechanic. Additional hardware devices have been developed: processor modules with a clock frequency of 10 MHz and using Intel-8088 and Intel-80286 compatible micropro-cessors; controllers for external storage devices; local area network modules; for connection to the IEEE standard; for memory up to 2 MB and more.

Overall, the efforts in the field of personal computers lead to substantial, some-what unexpected results. PCs begin to apply massively in various automated systems in which they are serially produced and used as computational cores. In many places they replace applied expensive minicomputers. Such processes are observed in training systems; design automation; the automation of scientific experimentation and research; automation of the establishment activity; process control in real-time mode, etc. In this activity, there was direct and effective coordination and interac-tion between ITCR and its divisions and SHE I and A and its plants in the town of Pravetz there was a direct and effective coordination and interaction. This stimulated and accelerated the overall activity. In the second half of the 1980s, a successful— let us call it—a 'second computer revolution' in our country. State leaders generally understand this and support this work. In 1986, to ITCR there were allocated targeted, including currency, funds to accelerate the development of computer equipment and technologies.

At the end of 1989, more than 1 300 associates were employed in ITCR, along with those from the affiliates in Veliko Tarnovo, Plovdiv, Pravets, Malko Tarnovo and in the village of Gramatikovo. There are 28 associate professors (then senior research associates), 5 professors (then senior research fellow 1st degree), one academician and one correspondent member of BAS. The number of candidates for science (currently doctors) is 40, and the doctors of science—5. ITCR built a separate block of flats for the needs of its collaborators and a creative base in the mountain above Vladaya. The construction of another such base near the Veleka river on the Black Sea coast starts. ITCR purchased two apartments in Tokyo to build a joint Bulgarian-Japanese lab for robotics and computers [7]. The average age of the institute is lower than the other institutes of BAS; the young people were striving to work at this institute.

Meanwhile, the production capabilities of SCSTP PRAVETS are developing very quickly—almost in the exponent. On 31.01.1987, with the decision № 1 of the Council of Ministers, based on SCSTP PRAVETS, the Economic Holding (EH) 'Microprocessor systems' (MS) was set up with the following activities: organization of scientific-applied, design-constructive, introductory, production—as well as commercial and engineering activities in the country and abroad, in the field of personal and professional microcomputers and the main electronic devices and their units, microprocessor modules and microprocessor systems, including microprocessor process control systems. The holding consists of: a microcomputer plant and systems; plant for mechanical assembly; plant for circuit boards (PCB plant) [11]. The following are also attached to EH MS: Machine-Experimental Plant in Gorna Malina; Analytic Plant in the town of Montana; Instrumentation Plant in the town of Petrich; Mechatronika Plant in the town of Gabrovo; Technological plant for microprocessor devices in Sofia city, as well as: Institute of Microprocessor Equipment in Sofia city, the bases for development and implementation for MEC PRAVETS, Analytic Plant in the town of Montana, Printed Devices in the town of Petrich, the software houses in Sofia and Pravets.

Assoc. Prof. Dr. Plamen Vachkov, was appointed as general director of EH MS, while at the same time he retained his place in the governing body of ITCR-BAS and his deputies were: Eng. Hristo Hristov and Eng. Rumen Raychev. The chairman of the Managing Board of EH MS has been appointed Corresponding Member A. Angelov [6, 8, 9].

The creation of this much-needed economic holding is preceded by considerable preparation at different levels, in which the management of ITCR plays a large part and too actively. The role of local leadership in Pravetz is also invaluable.

With the decision of the Council of Ministers to build the EH MS, a very important stage of the development of the personal computer industry in Bulgaria ends. A solid demolition line has been carried out: EH ISOT is primarily concerned with the big computing machines of the ES series, including peripheral devices for them, and the production and development of PCs and microprocessor equipment with their periphery is a task of EH MS. Later, EH ISOT assumes the very important tasks for the development of high-speed information and communication systems and technologies. In this way two centers for the production of computing equipment are formed in Bulgaria. It is true [3] that "the economic holding 'Microprocessor Systems' has arisen not as a continuation of ISOT, but as a result of the emergence of a new trend in computing technology—the distributed production of information that led to the current Internet network and to the ability of anyone with a medium degree of computer literacy to access limited information arrays. A computer on everyone's table is the idea that emerged in the world with the advent of Apple. Then the leadership of the computing technique within the Council for Mutual Economic Assistance (CMEA) missed this phenomenon."

These are strong and fair findings. The opinion of the three authors of [3] is key to understanding the logic of the development of computing equipment worldwide and in our country during the last 15 years before the changes in Bulgaria.

Until the establishment of EH MS, too many was made by his predecessor, SHE I and A, led by General Director Vassil Tsarevski, in the field of PCs. In 1983, the instrumentation plant in Pravets produced the first series of 500 pcs PRAVETS 82 and in the following 1984 5 350 pcs. 1985 marks a new success—15 000 pcs have been produced, including the upgraded PRAVETS 82XP.

At the same time, a major modernization and reconstruction was carried out at the instrumentation plant in Pravets, with a view to the production of 25 000 pcs. of PCs. From 8 mln lv. in 1983, the annual commodity production increased to 25 mln lv. in the end of 1984. It is projected to increase to 400 mln lv. in 1987, with an annual production of 100 000 pcs. of computers. This is about overcoming major difficulties—providing the necessary tooling, PCBs, keyboards, power supplies, and a product assembly and testing line. The goal is an annual production of 50 000 pcs. computers for one shift.

In 1985, ITCR handed over to the PRAVETS production plants the working documentation of IMCO-4, it was legalized as PRAVETS 16 and since April the same year starts a new stage in the development of the plant, namely, the production of 500 pcs. from this new 16-bit computer. With the Decision № 187 of 16.09.1985, the instrumentation plant in the town of Pravets was transformed into a Production Plant for Microprocessor Equipment (PPME under the Ministry of Transport, Information Technology and Communications or abbreviated as the Microprocessor Engineering Complex or MEC) PRAVETS consisting of three plants and a basis for development and deployment. The managerial staff remains the same.

In the same year 1985, the Complex Project for Investment Initiative (CPII) 'Electroproject' was assigned the design of the overall modernization of MEC PRAVETS, and from 1986 onwards the building and assembly work and staff training were carried out without disturbing the production rhythm. The commodity production of the plant increased from 54, 8 mln lv. in 1985 to 112, 8 mln lv. in 1986. A profit of 81, 8 mln lv. was achieved in the period from 1983 to 1986, while production funds amount to 24 mln lv.

At the beginning of 1988, the modernization project of the microprocessor plant was completed in essence. Firm measures are taken to provide software for production. For this purpose software houses have been created: 'Macrosystems' and 'Pravetz-Program', and the complex service throughout the country is taken over by 'SystemEngineering'. A branch of ITCR was established and, as the three directors of MEC PRAVETS, have already rightly noted in [3]: 'The cooperation between the newly-established plant and the Institute of Technical Cybernetics and Robotics of BAS is deepening. The institute is extensively expanding work on system applications. It is working in the areas of robotics, the control being carried out by the computers of the complex.' There is also a technical college with English language teaching—a division of HEEI 'Lenin'.

EH MS creates its foreign trade network, including the External Trade Holding (ETH) 'Comex' and two overseas companies—Octagon in Singapore and AEP in Germany. Together with the Academy of Sciences of USSR, the company 'Variant' in Tashkent was established for the installation of computers for educational purposes. With the same organization it participated in the preparation of the Soviet shuttle 'Buran' (snow storm in Russian). It is partnering with the Hungarian 'Videoton', the Ukrainian 'Electron' and the German 'Sherring'. For the Republic of Iraq, EH MS has built a PC plant in the town of town of Sallach al-Din.

For the successes of EH MS it is best written in [3], namely: "Only the profits of the 'Microprocessor Systems' complex from 1980 to 1991 amounts to 491, 4 mln lv.. Basic funds were introduced for 94, 9 mln lv. In the countries of the socialist camp, output of 407, 7 mln lv. and 16, 5 mln $ was exported. 288 mln lv. have been contributed to the state budget, and in the budget of the Pravets municipality—over 58 mln lv."

After all, there are people who deliberately strive to discard the achievements and the successes of the persistent devoted work of the hundreds and thousands of scientists, engineers, technicians and other specialists in this high-tech branch of Bulgarian material culture !?

At one of the Annual Reports of BAS in the 80 s of the last century, its President Acad. Angel Balevski, in the presence of senior officials, was proud to note that the economic effect in the Bulgarian economy, resulting from the implementation of the achievements of BAS, is much greater than the Academy's support. One of the positive examples is the fruitful link between ITCR and MEC PRAVETS. Everyone who listens to these words to the President of the Bulgarian Academy of Sciences understands whence his authority and self-confidence come to defend the best interests of the academy. Historically put the creation of ITCR is a significant success for BAS because it offers the possibility to initiate the development and the serial production in industry of a big number of personal computers for the inner market and also for export with all pluses and benefits for the Bulgarian economy. In this field in the 80-s of the past century Bulgaria was a confident leader among the rest of the Eastern-European countries.

References

1. Angelov, A., Sgurev, V., Petrov, P.: Some Aspects of the Automation and Informatics Development at the Bulgarian Academy of Sciences, Technicheska Myssal, XXXI, Anniversary Issue, pp. 14–19 (1994) (in Bulgarian)
2. Petrov, P.: The Contributions of BAS institutes for research and production of computer and communication systems. J. BAS **2**, 53–62 (2016). (in Bulgarian)

3. Vachkov, P., Boyanov, K., Yanev, K.: Prospects for the development of personal computers in the 'microprocessor systems' economic holding. In: Proceedings of PERSCOMP'87, vol. 1, S., Ed. BAS, pp. 51–63 (1987) (in Bulgarian)
4. Angelov, A., Kisyov, Y., Iliev, N.: Instrumental systems and computers. In: Proceedings of PERSCOMP '87, vol. 1, S., Ed. BAS, pp. 107–111 (1987) (in Bulgarian).
5. Marangozov, I., Petrov, P., Hristov, M.: Personal computer 'Pravets-8A'. In: Proceedings of PERSCOMP'87, vol. 1, S., Ed. BAS, pp. 167–171 (1987) (in Bulgarian).
6. Sgurev, V.: State, developed and perspective of personal computers of the 'Pravets' family. In: Proceedings of PERSCOMP'85, vol. 1, S., Ed. BAS (1985) (in Bulgarian).
7. Boyanov, K.: The role of BAS for the development of research and production of computer and communication systems. J. BAS **2**, 48–53 (2015). (in Bulgarian)
8. Angelov, A., Petrov, P., Zhelyazkov, G., Engineer Marangozov, I. (ed.): 'Technicheska Myssal' Magazine, vol. 1, pp. 9–31 (2004) (in Bulgarian)
9. Sgurev, V.: State of research on artificial intelligence in Bulgaria. J. BAS **3** (1986) (in Bulgarian)
10. Proceedings of the International Conference "RobCon'4", 20–23 Oct 1987, S., Ed. BAS (1987) (in Bulgarian)
11. Vachkov, P., Raichev, R., Hristov, H.: Microprocessor systems from pravets. In: The Golden Decades of Bulgarian Electronics, Trud (ed.), pp. 244–252 (2008) (in Bulgarian)
12. Sgurev, V.: The Achievements of the department of technical sciences of the bulgarian academy of sciences in automation, cybernetics, robotics and computers, Nauka (ed.) (2019) (under press, in Bulgarian)
13. Program for the development of PCs in the People's Republic of Bulgaria for the period 1982–1985, 6 documents, Archive ITCR-BAS (1983) (in Bulgarian)

Pattern Recognition Development in Bulgarian Academy of Sciences

Georgi Gluhchev, Dimo Dimov, and Atanas Uzunov

Abstract The development of Pattern Recognition (PR) theory in Bulgarian Academy of Sciences (BAS) is traced out from 1970. The major part of the activity has been carried out at the Institute of Information and Communication Technologies (IICT). Due to a few administrative transformations IICT is actually a follower of the ex-institutes ITC (Institute of Technical Cybernetics, 1970–1980), ITCR (Institute of Technical Cybernetics and Robotics, 1881–1989, Institute of Information Technologies, 1990–2010) and IICT (Institute of Information and Communication Technologies, 2010–). The research included image and speech processing and recognition. Attention has been paid to the following scientific problems: noise suppression, contrast enhancement contour delineation, object segmentation, feature selection, feature space minimization, methods, algorithms and software for object classification and identification. All this aimed the solution of different practical problems. The obtained theoretical results have been published as 370 papers and conference presentations, mainly international. 11 doctoral theses have been developed [1–11], 4 monographs [12–15], two of which in English, and a Handbook [16] have been written, and an international patent was obtained [17]. These results were connected to 51 projects, 4 of which in the frame of 4th, 5th, 6th and COST European programs. 14 software products have been implemented in the country and 2 abroad. This activity is further described in more detail for each of the predecessors.

Keyword Pattern recognition

1 Institute of Technical Cybernetics (ITC, 1970–1980)

The Pattern Recognition activity was mainly concentrated in the Department of Bionics headed by Minko Marinov. Theoretical approaches including threshold logics, fuzzy sets, deterministic methods, linear decision rules, statistical methods,

G. Gluhchev (✉) · D. Dimov · A. Uzunov
Institute of Information and Communication Technologies, 25A, Acad. G. Bonchev Str., 1113 Sofia, Bulgaria
e-mail: gluhchev1944@abv.bg

© The Author(s), under exclusive license to Springer Nature Switzerland AG 2021
K. T. Atanassov (ed.), *Research in Computer Science in the Bulgarian Academy of Sciences*, Studies in Computational Intelligence 934,
https://doi.org/10.1007/978-3-030-72284-5_7

structural-linguistic methods, heuristic rules, potential functions, system approach, random search with adaptation, experiment planning have been used. Obtained results in the area of image processing and pattern recognition were published in 62 papers and 6 doctoral theses were developed [1–6]. The major theoretical results were included in two monographs [7, 8].

The investigations in PR were carried out in the frame of the following projects.

1. "Methods and systems for image processing" (1971–1975, head M. Marinov, supported by ITC).

2. "Biotechnical systems, robots and ergonomics" (1977–1979), head Y. Marinov, supported by ITC.

3. "Automated system for early cancer diagnostic" (1973–1976), head M. Marinov, supported by the State Committee for Scientific and Technical Progress. The project was aimed to help oncologists for cancer screening on cell smears. Classification was based on the automatically measured morphometric cell parameters. Physicians and scientists from the Institute of Oncology, Medical Academy, Institute of Morphology at BAS and different hospitals in Sofia participated into the project development. Useful contacts were established with the Institute of Control Problems and Institute of Problems of Information Transfer in Moscow, Institute of Cybernetics (Kiev), and Institute of Electronics (Riga). This helped us to seriously update our equipment. The developed system was implemented in the Institute of Oncology.

4. "Ray". A system for the army need was developed together with specialists from the Military Medical Institute.
   A group heading by prof. Y. Marinov and including the scientists P. Venkov, M. Yancheva, L. Yankova, Z. Markov and prof. N. Naplatanov was doing investigations of the analysis and classification of physiological signals like ECG, VCG, phonocardiogram. Deterministic and statistical algorithms were developed for the analysis of heart activity. As aresult following projects have been developed.

5. "AORTA", project 1M-1 (1971–1974, supported by the Ministry of National Defense).
   A system for the objective detection of heart noise from phonocardiogram for the diagnostic of heart diseases in the military staff and recruits was developed. It was implemented in heart clinic at the High Military Medical Institute. A solution was suggested for the development of self-learning PR system for analysis of physiological signals, taking into account the patient time changes. A software system in FORTRAN was developed for the analysis of ECG/VCG signals and diagnostics of heart diseases. Automatic measurements of features, evaluation of their significance and classification of multidimensional feature vectors with Bayse decision rule were realized. The system was tested in the Department of heart signals at the Medish Physish Institut of Ultrecht, the Neterlands, using large data bases of ECG/VCG signals from the Veterans Administration Hospitals in USA.

6. "Electronic system for a diagnostics and prognoses of heart disorders", (1976–1980), Contract 2611/76 with the Ministry of Education.

7. "Development of electronized hospital" (1979–1980, contract 1501/79 with the Institute of Medical Technique and Ministry of Energy and Electronics).

   Statistical methods and algorithms for diagnostics of heart diseases based on a complex of symptoms were developed, modeling the physician's logics for the selection of informative combinations of symptoms. Linear prognostic rule for the evaluation of the issue of infarct of myocardium, and algorithms for the automatic evaluation of the prognostic thresholds were suggested.

Aside from the medical problems but actual and quite interesting and important were the investigations for the automatic classification of written and printed letters and ciphers which gave the possibility for the solution of different practical problems some time later.

During this period following results were obtained: two monographs [7, 8], 6 doctoral theses [1–6], 62 papers and conference presentations, 7 projects, one patent [9] and the organization of one International Conference on Pattern Recognition.

2 Institute of Technical Cybernetics and Robotics (ITCR, 1981–1989)

In 1980 ITC was transformed to ITCR. This together with the advance of microprocessing technique changed the direction of PR investigations. Research work was oriented to robotics. Additional difficulties were caused by the close down of the Department of Bionics and its staff leaved the new institute. However, very soon the administration understood the mistake and some of the researchers returned back. New specialist in electronics, TV and software development have been appointed. The new team, headed by D. Mutafov, consisted from the following researchers: H. Karaatanasov, K. Zagurski, S. Dimitrova, A. Uzunov, V. Shapiro and G. Gluhchev. Methods and algorithms for machine vision and robot control by speech were under development. During the next three years following systems have been developed.

1. "CSY-11"—a microprocessor system for processing and recognition of TV images (head V. Sgurev).

   This was the first such system in the East European countries. Nevertheless the difficulty with the import of electronic components the developed hardware and software were on a high level, comparable to the level of similar foreign products. In 1984 CSI-11 was presented at the Moscow Exhibition. On its basis a few products of practical and educational importance were further developed. The possibility to couple the system with a microscope enlarged its application in the area of biological and medical investigations. A number of systems were developed and implemented in different Institutes of BAS, Departments of Sofia University, and Institute for criminalistics and criminology.

   The interest to more objective evaluation of handwriting identification and signature verification for the purpose of the detection of fake documents and access control was realized with the development of the following systems.

2.   "Expert-1" and "Expert-2" (1986–1988, head D. Mutafov).
     The goal of the systems was to help the expert to select specific pieces in
     the scanned document and to automatically measure preselected features and
     their combination in a dialog mode. The decision making rules were giving the
     possibility for automatic comparison with similar documents and evaluation of
     the degree of similarity between them. This alleviated significantly the search of
     an unknown writer in a large data base of written documents. The systems were
     developed in collaboration with handwriting experts from the Scientific Institute
     for Criminology and Criminalistics (SICC) at the Ministry of the Interior and
     have been implemented there. The scientific achievements were described in
     the doctoral thesis of V. Shapiro.
3.   "Automatic recognition of car license plates" (head D. Mutafov).
     The automatic car license plates recognition is and continues to be a very compli-
     cated problem even for the modern PR systems because of the changing weather
     conditions, car speed, different plate design, noise and geometric transforma-
     tions. The system was tested at the station for technical control of the vehicles
     in Sofia and demonstrated excellent results. It is worth to note that this system
     was one of the first such systems in the world. The software update after a few
     years gave the possibility plates of different standard to be processed.
4.   'CSY-5' (head S. Ogorelkov).
     This TV system was oriented to engineering applications.
     In parallel with this investigations on the processing and analysis of speech
     signals oriented to robot control and speaker verification have been carried out.
     Following systems have been developed.
5.   "GraphSig" (1993, head A. Uzunov).
     The system was aimed at the analysis of speech signals. It was developed on the
     IBM PC AT platform under MS-DOS, has had a specific interactive graphical
     design and was implemented in SICC.
6.   "Software for digital processing of speech signals" (1989, head A. Uzunov).
     The software was a part of the expert system OBZOR at the Laboratory AI,
     Institute of Technical Cybernetics the Slovak Academy of Sciences.
7.   "ROBIK" (1991, head A. Uzunov).
     The purpose of this system was to control the educational robot ROBKO-1 using
     separate speech command. It was implemented in the Institute of Mechatronics
     at BAS.

   Due to the initiative and activity of V. Valev a Bulgarian Association for Pattern
Recognition (BDRO) was set up as a collective member of the International Association for Pattern Recognition (IAPR). Seminars and discussions of scientific problems
were organized.

   *The obtained results in ITCR could be summarized as follows: 1 dissertation, 37
papers, 8 projects (2 of them implemented abroad), 1 patent.*

3 Institute of Informatics (II, 1991–1993) and Institute of Information Technologies (IIT, 1994–2010)

ITCR was separated into 4 Institute in 1990. One of them was the Institute of Informatics which was renamed as IIT in 1994. Some of the researcher in PR moved to the new Institute and a few others attended it. They were: D. Dimov, S. Bonchev, E. Kalcheva, M. Mladenov, P. Veleva, D. Kamenov. The work was intensified, a regular seminar was organized. The possibility for new projects including European increased. The team was included in the following international projects.

1. "BIOSECURE—Network of Excellence", (2002–2006, a project from the 6th European Program, national leader G. Gluhchev).
   The project was aimed to person identification on the basis of biometric parameters. Our team was responsible for on-line handwriting identification/verification. Additional information related to the hand's characteristics, mutual disposition between the hand and the pen together with the speed of movement has been registered thus increasing the accuracy of the inference. The obtained results helped the development of the doctoral thesis of M. Mladenov.

2. "COST—2101 Biometric modalities for individuals identification" (2006–2007, national leader G. Gluhchev).
   Our team was responsible for the development of a system for signature identification and detection of fake signatures using a graphical tablet. Together with the signature features the dynamics of movement was registered. The obtained results were used by the development of the doctoral thesis of D. Bojadjieva.

3. "FIVES—Safer Internet Plus. Forensic Image and Video Examination Support" (2009–2010, 7th European Program, national leader G. Gluhchev).
   The goal was to analyze video clips for the detection of producers and propagators of pornographic and pedophiles materials. Our responsibility covered image enhancement, segmentation and detection of text regions. Also, search and comparison of similar images from large data base has to done.

4. "Processing of text information in cadastre maps" (4th European Program, national leader Z. Ilcheva).

5. "System for the identification of illegal CDs" (1997, head G. Gluhchev).

6. "System for facial individual identification" (1998, head G. Gluhchev).
   Experts from SICC offered a methodology for the evaluation of the similarity between photos. Some parameters related to face characteristic points are measured in an interactive mode. Thus similar faces could be extracted from a data base helping the expert. The system was implemented in SICC and a certificate was granted to the developers.

7. "System for person identification visualizing and evaluating the degree of pressure alongside the strokes of the signature" (2001, head G. Gluhchev).

8. "ABV—a system for graphometric identification of individuals using signatures" (2001, head G. Gluhchev).

The system implements the expert knowledge about sets of handwriting informative features. The expert selects pieces of handwriting and a set of features in a dialog mode. After that machine evaluates them and builds a "portrait" of the writer, compares it to other such "portraits", selecting them from a date base, and generates a conclusion.

9. "System for person identification using hand-pen position signing on a tablet" (2005, head G. Gluhchev).

Simultaneously with capturing the writing on a tablet a TV camera captures serious of hand-pen images. This allows for the evaluation of the changes in the disposition between the hand and the pen during the writing. The description of the movement together with signature parameters is used for more reliable comparison and objective conclusion. The obtained results were used for the development of the doctoral thesis of M. Mladenov.

10. "System for person identification using signature on a tablet" (2007, G. Gluhchev).

Using the coordinates from the tablet specific points are registered connected to the changes in the direction of movement, type of strokes and speed of movement between them. This information is used for writer's identification and detection of forgeries. The results were included in the doctoral thesis of D. Bojadjieva.

11. "Investigation of new methods for knowledge processing and machine learning" (1992–1995, contract НИ – И -203).

12. "Neural Networks. Application to information technologies" (1994–1996, contract НИ – И -411).

13. "Neural networks for classification: theoretical investigations and application aspects" (1996–2000, contract НИ – И -611).

14. "Biometric parameters for identification" (2003–2005, contract НИ – И - 1302).

15. "Methods and algorithms for the analysis of combined biometric information" (2006–2009, contract ВУ-ТН- 202).

16. "Analyse de la parole et reconnaissance du locuteur" (1996, head S. Hadjtodorov, Contract between CLMBI-BAS and CNRS, France).

17. "Microcomputer system for acoustical analysis of pathological voices-patient with laryngeal diseases" (1996, head B. Boyanov, contract between CLMBI-BAS and Phoniatric Clinic of Central University Hospital, Helsinki, Finland).

18. "Methods and means for image processing and Pattern Recognition" (1994–1995, head G. Gluhchev).

19. "Methods for image and speech recognition" (1997–1999, head G. Gluhchev).

20. "Identification with biometric parameters" (2006–2008, head G. Gluhchev, connected with European projects BIOSECURE and COST)

21. "Recognition with biometric parameters" (2009–2010, head G. Gluhchev, connected with European projects BIOSECURE and COST)

Following National and International projects have been developed under the leadership of D. Dimov.

1. "Recognition of neumen writing in historical documents".
   The neumen writing describes how religious texts have to be chanted. Neumens have been used in Early Christianity, some of them are preserved but their interpretation is lost. The goal was to help researchers of old neumen texts to restore their meaning by analogy to the modern neumen writing. The obtained results were used by L. Laskov in his doctoral thesis. Some results were used in the project "Astroinformatics" (2009–2012) for deciphering of handwritten numbers from archive photographic plates.

2. "Videostabilization and 3D recognition in real time" (2012–2016).
   This investigation was connected to some other projects as "AComiIn" (2012–2016, European project), "Stabilization of video from high speed camera" (ДО1-192р МОН), BG161PO 003-1.1.06-0038-C0001, 2012). Two problems had to simultaneously be solved—2D video stabilization and 3D person recognition. A generalization scheme using 3D reconstruction and evaluation of the movement parameters between regions of interest or characteristic points in image sequence for the evaluation of the camera and object coordinates connected them. The object stabilization was based on proper object separation from the background. The results were involved in the doctoral thesis of A. Nikolov.

3. "Astroinformatics: processing and analysis of digitized astronomical data and web-based application" (contract Do-02-275/2009 with Ministry of Education and Science).
   The project was aimed to the processing of scanned images from photographic plates and catalogs. Known and new methods, algorithms and techniques were used for the automatic processing of astronomical images of the following types: (i) plates of star chains obtained via successive movement and exposure, (ii) plates of type "Carte du Ciel" with additional measuring lines net, and (iii) plates of type "Lost in Space" with fully or partially lost information about the observation position. The WEB-based system for the preservation of and access to astronomical images was built and could be involved in European virtual observatory.

4. "ETN-FETCH—Future Education and Training in Computing: how to support learning at any time anywhere" (ETN. 539461-LLP-1-2013-1BG-ERASMUS-ENW, contract #2013-3862/001-001, coordinator Ruse University "A. Kanchev").
   The main goal of the project was to help and achieve intelligent growth and building of computer society based on knowledge and innovations via the quality increase in computer education, dissemination of knowledge, discussion of new teaching methodologies and support of good practices among all the countries.

5. "ETN TRICE: Teaching, Research and Innovation in Computer Education" (2009–2011, ETN 142399-LLP-1-2008-1-BG-ERASMUS-ENW, coordinator Ruse University "A. Kanchev").

Analysis of the problem was carried out: sharing of positive experience and dissemination of good practices among partners, and improvement of education quality using innovative educational technologies. Additionally, the integration between universities, research institutes and companies was searched for.

6. "Support of Ph.D. students, post-doctors and young scientist in the area in computer sciences", (2009–2011, Contract #BG 051 PO 001-3.3.04/13, financed by the European Social Fond, OP "Development of human resources", coordinator the university of V. Tarnovo).

    The goal of the project was bringing together the science and business trough the orientation of young researchers towards the scientific and innovative needs of the firms.

7. "ISSSE: Intelligent sensor systems for security enhancement" (2006–2009, Contract BY-MI-204/2006 with NFSI-Ministry of Education, coordinator Department of Mathematics and Informatics at Sofia University).

    New methods and means were tested and a prototype of a system was created. Possibilities as sensor calibration, 3D reconstruction, object recognition, and object data base support have been included.

8. "Study of biomedical data by the methods of multiresolution analysis. Polyspline wavelet analysis in immune computations and brain research", (Greek-Bulgarian project, Contract B-G-17/2005 with NFSI-Ministry of Education, coordinator Institute of Mathematics and Informatics at BAS).

9. ETN DEC—Doctoral education in computing" (2004–2007, Erasmus Thematic Network #114046-CP-1-2004-1-BG-ERASMUS-TN, Ministry of Education, coordinator Ruse University "A. Kanchev").

10. "Methods for fast access by content to multimedia data bases", (2003–2006, Contract И-1306 with NFSI-Ministry of Education).

    The project aimed the improvement of indexing techniques for effective, fast and noise-immune extraction by content of images from data bases.

11. "Development of method and software for effective search in large data bases of images by graphical content", (2004–2005, Contract ИД-6/2004 with the State Agency ITC).

    The goal of the project concerned the investigation of the world experience for content based image retrieval and effective search by graphical content from multimedia data bases in the aspect of speed and noise immunity.

12. "STEMB-Software system for verification of the State Emblem replicas" (2000–2001, assigned by the State Committee of Standards and Metrology).

    The goal was to compare objectively copies of the State Emblem to its original image—a problem that has generated disputes in the society. The system had to automatically compare colors and graphics in the two images, detect the discrepancies and create a document suggesting acception or rejection of the replica. It was successfully developed and implemented.

13. "Recognition of printed and handwritten text—methods and information technology for Cyrillic alphabet", (Contract И-524/1995 with NFSI-Ministry of Education).

The problems for the development of effective system for the recognition of printed and handwritten text images are delineated. A suggestion for the development of information technologies for the adaptation of methods, algorithms and means used for recognition of Latin letters was founded.

14. "Requirements and Framework for Environment and Transport Telematics", (1998–1999, CAPE Project TR 4101/IN 4101, from 5th European Program).

15. "NEW TO—Computer verification of replica images to sophisticated graphic standards (COVERIS)", (Ref. # OB-0086/2002, Technology offer, IRC-Sofia).

16. "NEW TO—Software technology for fast and noise tolerant image retrieval based on graphic content (FANTIR)", (Ref. # OB-0121/2004, Technology offer, IRC-Sofia).

The obtained results in IIT and partly in IIKT could be summarized as follows: 2 monographs (in English), 271 papers and conference presentations, 32 project, 5 Ph.D. dissertations, organization of 1 International conference.

References

1. Rangelova, E.: Fuzzy sets in pattern recognition. Ph.D. thesis
2. Venkov, P.: Machine methods and algorithms for the classification of physiological signals accompanying the heart activity. Ph.D. thesis
3. Yancheva, M.: Structural-linguistic approach for recognition and diagnostics. Ph.D. thesis
4. Mutafov, D.: System approach to pattern recognition. Ph.D. thesis
5. Gluhchev, G.: Tissue classification using cell smears. Ph.D. thesis
6. Tsonev, M.: Recognition of manufacture situations. Ph.D. thesis
7. Shapiro, V.: Computer investigation of handwritten text. Ph.D. thesis (1990)
8. Savov, M.: Person identification during signing. Ph.D. thesis (2007)
9. Laskov, L.: Recognition of neumen writing in historical documents. Ph.D. thesis (2010)
10. Bojadjieva, D.: Combined approach to on-line recognition of signatures. Ph.D. thesis (2014)
11. Nikolov, A.: Videostabilization and 3D real time recognition. Ph.D. thesis (2016)
12. Gluhchev, G., Venkov, P., Mutafov, D., Yancheva, M.: Elements of Pattern Recognition Theory. Publ. BAS (1982) (monograph)
13. Hristov, K., Gluhchev, G., Hinova, R.: Citophotometry. Publ. Medicine and physical education (1983) (monograph)
14. Atanassov, K., Gluhchev, G., Hadjitodorov, S., Shanon, A., Vassilev, V.: Generalized Nets in Pattern Recognition. KvB, Monograph #6, Visual Concepts, Pty. Ltd., Australia (2003)
15. Atanassov, K., Gluhchev, G., Hadjitodorov, S., Kasprzyk, J., Shanon, A., Schmidt, E., Vassilev, V.: Generalized Nets, Decision Making and Pattern Recognition. Warsaw School of Information Technology, Warsaw (2006)
16. Valev, V., Gluhchev, G.: Pattern Recognition Methods. Scientific and Technical Unions, Sofia (1986)
17. Sgurev, V., Mutafov, D., Gluhchev, G., et al.: A system for identification of registered vehicles. Patent NBR G06K9/00, 71029 (Switzerland)

On the History of Artificial Intelligence in Bulgaria

Vassil Sgurev

Abstract This work is dedicated to the history of emergence and the development of the research in the field of Artificial Intelligence (AI) in Bulgaria until the end of the past century. It is pointed out that first research activities are in the Bulgarian Academy of Sciences (BAS)—Institute of mathematics, and a little later in the Institute of technical cybernetics and robotics—BAS (ITCR-BAS). A stimulus for this is the creation of Bulgarian personal computers in ITCR. The beginning of a series of authoritative international conferences—AIMSA (Artificial Intelligence: Methodology, Systems, Applications) and IEEE-IS is laid in ITCR. In this connection, the Bulgarian association of artificial intelligence (BAAI) arises.

Keywords History of the AI in Bulgaria · IEEE-IS · ITCR-BAS · BAAI

The interest in Artificial Intelligence (AI) in Bulgaria originates almost simultaneously with other European countries and with a certain delay after the United States. Initially, AI is considered as a new advanced technology within the rapidly evolving computer information technology. To the AI direction there have been added also the expert systems, the logical methods for knowledge extraction, and the optimal solutions in various types of discrete computer games.

This research begins initially in the scientific units of the Bulgarian Academy of Sciences—Institute of Mathematics (IM), Institute of Technical Cybernetics (ITC), transformed in 1978 at the Institute of Robotics and Technical Cybernetics (ITCR), the Center for Scientific Information (CSI), the Center in Science and the Institute of Philosophy. This is because BAS is the most powerful Bulgarian scientific organization for basic research and it is assigned to deal with solving extremely complex and difficult problems being solved in a new scientific field such as the artificial intelligence. Later, already in the early 1980s of the previous century, such investigations gradually started to be carried out in HEEI 'Lenin' (Higher Engineering and Electrotechnical Institute, now Technical University, TU-Sofia), the Central Institute

V. Sgurev (✉)
Institute of Information and Communication Technologies (IICT), Bulgarian Academy of Sciences (BAS), Sofia, Bulgaria

© The Author(s), under exclusive license to Springer Nature Switzerland AG 2021
K. T. Atanassov (ed.), *Research in Computer Science in the Bulgarian Academy of Sciences*, Studies in Computational Intelligence 934,
https://doi.org/10.1007/978-3-030-72284-5_8

of Computing Engineering (CICE), the Central Institute of Complex Automation (CICA).

The intensification of this research greatly contributes to the successful long-term cooperation of the academies of the socialist countries. Within this framework, Working Group No. 18 on the Computing (Computer) Technique, led by Academician Germogen Pospelov, has been working for a long time, through which AI has been developed. Its activity has been coordinated by Professor Dmitry Pospelov. From the Soviet side, this international collaboration has been carried out through the Computing Center of the Academy of Sciences of the USSR (in Moscow). On the Bulgarian side, Prof. Dr. V. Sgurev from ITC, later ITCR at the Bulgarian Academy of Sciences and Prof. Math. Plander from the Czech Cybernetics Institute in Bratislava. During the work of WG No. 18, numerous international conferences, seminars, visits of scientists on multilateral cooperation, as well as specializations of young scientists in the field of AI have been realized.

In 1968, at the Institute of Mathematics at the Bulgarian Academy of Sciences, the "Mathematical Insurance" department started a computer-aided intellectual game development, heuristic programming and the acquisition of the Lisp and Prolog languages. This is where the Adles translator is used. In another group of the same department, under the guidance of Assoc. Prof. V. Tomov, there has been an activity of Analytical Transforms (SAP) [1, 2], which is specialized in different mathematical fields (indefinite integrals, rational functions, matrices, continued fractions). An expert system has been set up and language tools have been developed for the use of mathematical knowledge with the Reduce-2 system for analytical transforms. The same group developed a system for recognizing and classifying archaeological ceramics, which uses production rules. Attempts have been made to spread the results in the AI area to the humanitarian sciences. Steps have been realized to use L. Zadeh's newly created theory on fuzzy sets.

In 1965 a Machine Translation Group was set up at the Bulgarian Academy of Sciences, led by Dr. A. Ljudskanov [3]. In order to solve this complex problem, the group started to design algorithms for creative processes, optimal machine translation strategies, etc.

In 1978, a mathematical linguistics group led by Assoc. Prof. Radko Pavlov was formed at the Institute of Mathematics at the Bulgarian Academy of Sciences, whose efforts were focused on linguistically ensuring the human–computer dialogue in natural language [4]. The aim was to create a formal model for describing the Bulgarian language and to perform its computer processing. It relies on the use of advanced transition networks—the so-called ATN-grammars and on this basis algorithms for analysis and synthesis of phrases in Bulgarian language have been developed. An automated programming environment for algorithms based on ATH formalizations has been developed. As an instrumental tool, the M1-1 system is used in a language close to the Bulgarian language. Prof. R. Pavlov and Assoc. Prof. Dr. G. Angelova designed a linguistic processor with a basic software product [4].

The established systems of restricted natural language dialogue allow for the development of any matter, regardless of the specific subset of the natural language in a restricted subject area.

In the same laboratory, modeling and structural descriptions of the Bulgarian heroic epic were carried out, and a legal-information search system 'Sprint' was implemented. It can be used as a computer model of a system of legal relations.

In the 'Mathematical Logic' sector, activities are carried out on the formalization of logical and algorithmic processes related to the problems of AI. Special attention is paid to logical schemes that more realistically reflect the methods used in the intellectual activity to search for and reflect the truth. Different formal systems possess such quality. Classical schematics of predicate calculus with richer syntax, obtained by adding quantum substitution, were studied. Multiple isomorphisms are used in model theory and computational theory. Research has been done on machine checking and machine search issues, as well as on the possibilities of using fuzzy sets and some probabilistic structures. Results have been obtained in the creation of computer-based computer search-for-evidence systems [5].

In the 1980s, a number of studies on psychological issues of AI were carried out at the Institute for Foreign Students.

Under the guidance of Academician Azaria Polikarov at the Center for Scientific Information (SCI) at the Bulgarian Academy of Sciences for many years, a scientific group has been working on philosophical problems of AI and on the intensification of research through systems with AI. Academician A. Polikarov, in his research, draws attention to the exclusive role of AI in the philosophical conception of modern trends in the real, natural and social sciences. They have explored the logical structures in solving scientific problems and elements of heuristics. This is summed up in his divergent-convergent approach. In his fundamental work—'Methodology of Scientific Knowledge' of 1972, [6] he paid serious attention to the methodological problems of AI, as well as to the so called hypothetic-deductive models that advocate a multidisciplinary approach to heuristics.

The Center for Applied Mathematics, established in HEEI 'Lenin' (TU-Sofia), in the middle of the 1980s developed an intellectual programming system with the capabilities of adopting a restricted natural language, responding to 'how', availability of own knowledge base. This system provides for dialog with the user in the absence of sufficient information.

At the same center, a parallel computer with a hyper-structure is designed that can be used adequately in AI systems. Within the same center, under the leadership of Prof. Dr. Ludmil Dakovski formed an AI group developing different aspects of AI.

A number of other units of the Bulgarian Academy of Sciences gradually began to join the AI's problems. The Science Center focuses on the study of the heuristic and creative processes of man and the possibilities for their computer modeling. The Institute of Philosophy at the Bulgarian Academy of Sciences conducts research on some of the philosophical problems of mathematics and cybernetic modeling.

In the second half of the 70s of the last century at the Institute of Technical Cybernetics (ITC) of the Bulgarian Academy of Sciences, a department for automation of intellectual activity led by Assoc. Prof. V. Sgurev was established. In the department a well-known issue is endorsed covering the direction of artificial intelligence. In the reorganization of ITC in the ITCR at the Bulgarian Academy of Sciences, these problems go to the new direction 'Informatics' and the 'Cybernetics Systems' section, led by Assoc. Prof. V. Sgurev.

A significant part of the 'Cybernetic Systems' sectional research in the ITCR at BAS is related to various aspects of the diagnostic activity and to the creation of different intelligent systems. This is mainly due to the expanding development and implementation of complex technical systems requiring high-quality service. To solve such problems, the problem-independent expert system DIGS [7] has been developed. It incorporates instrumental tools that allow it to be quickly and efficiently reshaped for diagnostic activity from one technical system to another. This is done through a corresponding change in the system of some parameters, as well as through the exchange of the expert knowledge. A network model is developed that reflects the relationships and interactions between the different parameters of the diagnosed technical systems and the ways of realizing the whole diagnostic process. This allows build a proper structure of the developed expert system.

DIGS is platform-oriented for existing PCs and their operating systems and program tools. It contains two options. The first involves complete automation of the damage search strategy. In the second, an interactive mode is provided in which the user can construct hypotheses himself and introduce them into the expert system.

For the implementation of the expert system, appropriate instrumental tools have been selected and adapted, namely, some versions of the Prolog language. Some problems with the implementation of the Prolog operator have been solved and an extension of the same language has been carried out in order to improve its use in the PCs platforms [8]. Modules for modular organization of 'Prolog' were created through a corresponding tree structure. This allows processing with computers with limited RAM. The research carried out allows the extension of the possibilities of the standard 'Prolog' to the established expert system, as well as the creation of logical formalizations for the presentation of knowledge.

It has been shown that these results can be used in CAD/CAM/CAE systems that are widely used in various areas of material production and public practice [9–13].

In the same department, theoretical and applied research has been carried out and systems for identifying vehicle numbers were created for the needs of special services.

Other developments have been made, including:

- Intelligent document flow system with automatic text recognition and handwriting recognition;
- Industrial-robot diagnostic methods and methods for planning robot activity, as well as a medical expert system for diagnosing childhood diseases—DEDEX;
- Network-flow realization of causal processes for extracting new knowledge, as well as for making decisions in the knowledge base contradictions;
- General schemes and hypotheses for the information processes in cognitive activity;
- Distributed Intelligent Process Management Tools;
- Method for determining the brightness of an image histogram;
- Technical Hardware Diagnostic Method on a 16-bit PC.

The same department explores problems related to the possibilities and limitations of artificial intelligence. Primary attention is paid to the systems developed within the framework of the weak theses of AI, intelligent software systems and technologies for extraction and processing of knowledge. The prospect of building a natural human–machine interface, different cognitive aspects of the systems with AI, the instrumental tools of the AI, is discussed. Special attention is paid to modeling creative processes with computers. Different types of constraints are discussed which interfere with existing technologies with AI to perform full-scale modeling of cognitive processes, as well as the prospects for realization of AI's strong theses.

The technology of building expert systems for product reliability assessment is one of the core tasks of the System Research Department of the Institute of technical cybernetics and robotics, led by Assoc. Prof., Ph.D. Ivan Popchev. The aim is to provide methodological and expert assistance to the user in assessing the reliability of an item and making decisions based on the analysis. Two main blocks have been developed—computing and consulting. Appropriate research has been carried out on how to present and process knowledge. A significant part of the research focuses on decision-making methods and multi-criteria assessment for selection of elements from finite sets. Relevant software products and systems have been developed that can be used to model and support decision-making processes [14].

The problems of the 'Robot Management' section of the ITCR at the Bulgarian Academy of Sciences is mainly aimed at creating a common methodology and software for management systems containing AI elements [15]. The REX software system is designed to design real-time control systems. On its basis, a multi-level robot management system has been developed. REX is built on a modular principle and allows relatively independent design of the individual levels. It has the following modules, each of which is responsible for a separate management level: a robot environment model, an incremental compiler that implements a step-by-step user program where the program originally compiled is adjusted to the current situation. The planning system has two base blocks. One implements planning, and the other identifies the differences between the hierarchical principle model and the real state of the operative environment. The objects of the technological process have a 3D presentation. There is an orthogonal method for introducing new information. The technological object thus built is the basis for the training of the industrial robot, adapting to the specific conditions for realization.

The temporary executor of ITCR Director at BAS and the head of 'Informatics' Department Assoc. Prof. Vassil Sgurev and Assoc. Prof. B. Petkov from the CSI at BAS, initiated the idea to organize in Bulgaria an international conference on AI, at which to invite eminent foreign scholars. The aim of this conference is to stimulate the Bulgarian research in the field of AI. In September 1984, the first AIMSA'84 conference was held under the name AIMSA (artificial intellect, methodology, systems and applications) in Varna—Golden Sands [16–20]. It was actively assisted by Professor Wolfgang Bibel of the German Federal Republic, President of ECCAI (European Coordination Committee on Artificial Intelligence). The works of the conference were published by the Nord-Holland Publishing House, edited by V. Bibel and B. Petkov. The conference also sparked a broad response in the former socialist countries. It was decided that the AIMSA international conference would be held regularly in one year under the auspices of ECCAI.

The management of ITCR at BAS organizes a Bulgarian group on AI (BGAI), which under the leadership of Assoc. Prof. V. Sgurev made the necessary steps for membership in ECCAI. Due to the lack of legal status in Bulgaria for a non-governmental organization (NGO), the group was registered with the Ministry of Interior. At a consecutive ECCAI meeting in Brighton (UK), in the presence of a delegation from the ITCR at the Bulgarian Academy of Sciences, the Bulgarian AI group was officially accepted as a member of this organization. In the work of this group for years the most active participation was taken by Assoc. Prof. Danail Dochev, Assoc. Prof. Gennady Agre, Prof. Zdravko Markov and others. Their work in organizing the AIMSA conferences is too significant.

Particularly successful was the Second International Conference AIMSA '86, organized by ITCR at BAS, the Electronics Association, the Institute of Social Management and the Union of Mathematicians in Bulgaria. The conference took place from 16 to 19 September 1986 in Varna—Golden Sands, under the aegis of ECCAI. Chairman of AIMSA '86 was Scientific Secretary General of BAS acad. Blagovest Sendov and Chairman of the International Program Committee was Professor Philip Jorrand, Director of the Laboratory of Informatics and Artificial Intelligence in Grenoble, France. The conference was attended by 289 people from 17 countries, including 80 foreign participants. There were 91 reports grouped in six strands. The collection of works by AIMSA '86 was published by the Nord-Holland Publishing House, edited by F. Jorrand and V. Sgurev.

Of particular interest was the participation of the famous British scientist and lead researcher in Prolog Prof. Robert (Bob) Kowalski. He had been invited as a plenary lecturer and took an active part in the discussion on the AI future. The report was invited by Prof. R. Kowalski on 'Using Logic as a Formalism for Presenting Knowledge', by Professor V. Bibel for 'The Deductive Approach to Knowledge Generation' and by Prof. Alan Birman for 'Machine for Speech Processing for Joint Troubleshooting'.

An impressive group of scientists from the Computing Center of the Academy of Sciences (AS) of the USSR took part in the work of the conference: Academician Germogen Pospelov, Prof. Dmitry Pospelov, Assoc. Prof. Dr. Vladimir Horoshevski, Prof. Dr. Gennady Ossipov and others.

The AIMSA '86 conference took place under the strong dominance of the logical approach in AI, probably under the influence of the impressive presence of the founder of this scientific field—Prof. R. Kowalski.

Until and shortly after the changes in Bulgaria in 1989/1990, the AIMSA conferences were held regularly in one year, namely in 1988, 1990, 1992 and 1994 at the same place in Varna—the Golden Sands and at the same time—in September. These four conferences were organized by ITCR at BAS and after its reorganization by the Institute of Informatics (II) at the Bulgarian Academy of Sciences and under the aegis of ECCAI. The works of the three AIMSA conferences from 1988 to 1992 were published by the North Holland publishing house as follows: AIMSA '88, edited by

T. O'Shea and V. Sgurev, for AIMSA '90—edited by F. Jorrand and V. Sgurev, for ASIMSA '92, edited by B. du Boulay and V. Sgurev.

World Scientific, edited by F. Jorrand and V. Sgurev, published the papers at AIMSA '94.

The new legal possibilities in Bulgaria allowed for the official registration of the Bulgarian AI Group as the Bulgarian Association of Artificial Intelligence (BAAI) with President Prof. DSc. V. Sgurev. This association was the main organizer of the next 11 AIMSA conferences from 1996 to 2018, all under the aegis of ECCAI.

From 1994 to 2018, a total of 17 international AIMSA conferences were held without interruption. This series of conferences is currently the largest in the Bulgarian Academy of Sciences. It is a tremendous achievement for Bulgarian scientists because it creates an international AI site where young scientists meet with prominent Bulgarian and foreign researchers in this fast growing and promising scientific field.

During the same period of the previous century Bulgarian state funded a significant research program called 'Brain'. Its main task is a complex research of brain problems by specialists in neuroscience, surgeons, psychologists, experts in informatics, mathematicians, engineers and others. This program has played an important role in intensifying research on artificial intelligence in Bulgaria. A number of research teams from the Bulgarian Academy of Sciences, the Medical Academy, the universities and others took part in it.

In [21] there is a brief overview of the research in our country of Artificial Intelligence for the period from start to 1990–1994.

The period since 1994, and especially in the current 21st century, has been characterized by accelerating research into artificial intelligence within the framework of European programs and jointly with European research organizations—institutes and universities.

At the initiative of Acad. Vassil Sgurev and Acad. Mincho Hadjiyski, within the framework of the Union of Automation and Informatics (UAI) 'John Atanassov' and the IEEE Societies 'System, Man and Cybernetics' (SMC), 'Computational Intelligence' and 'Control Systems', a very successful series of IEEE-Conferences 'Intelligent Systems' (IS) was launched in 2002. The conferences are held every other year. So far, between 2002 and 2018, a total of 9 such international conferences have been held.

Valuable American scientist Prof. Lotfi Zadeh, who was elected a foreign member of the Bulgarian Academy of Sciences, was of value in organizing this series of artificial intelligence conferences. He personally attended the first IS'02 conference and was an honorary chair of other such conferences.

In addition to the IEEE database, the accepted and exported reports of the series of 'Intelligent Systems' conferences are published in separate paper volumes from the authoritative publishing house Springer, Germany.

The rows and inscriptions in Table 1 give data about the 10 IEEE conferences on intelligent systems, including the forthcoming in 2020, namely, the names of: First and Second Chairmen ('Chairs') of each conference, the Chairmen ('Chairs') of the International Programs Committees, the General Coordinators (Technical Program

Table 1 Coferences "Intelligent Systems – IEEE"

Conf. year	Chair 1	Chair 2	Int. progr. Committee Chair	General coord. Chair	Honorary Chair	Location
2002	Vassil Sgurev	Tariq Samad	Mincho Hadjiski	Vladimir Jotsov	Lotfi Zadeh	Varna
2004	Vassil Sgurev	Ronald Yager	Mincho Hadjiski	Vladimir Jotsov	Lotfi Zadeh	Varna
2006	Panagio tis Chountas	Ronald Yager, Janusz Kacprzyk	I. Smith	I. Petrounias, J. Kacprzyk	Lotfi Zadeh	London
2008	Vassil Sgurev	Ronald Yager	Mincho Hadjiski	Vladimir Jotsov	Lotfi Zadeh	Varna
2010	Panagio tis Chountas	Janusz Kacprzyk	Nadia Nedjah, Witold Pedrycz		Lotfi Zadeh	London
2012	Vassil Sgurev	Ronald Yager	Mincho Hadjiski, Janusz Kacprzyk	Vladimir Jotsov	Lotfi Zadeh	Sofia
2014	K. T. Atanassov, J. Kacprzyk	M. Hadjiski, V. Jotsov, L. Rutkowski, S. Zadrozny	P. Angelov, B. Bouchon-Meunier, D. Filev, W. Pedrycz	V. Duch, A. Hassanien, J. Korbicz, E. Szmidt	V. Sgurev, R. R. Yager, L. A. Zadeh, J. Zurada	Warsaw
2016	Vassil Sgurev, Vladimir Jotsov	Janusz Kacprzyk	Mincho Hadjiski, Plamen Angelov, Krassi, mir Atanass ov		Lotfi Zadeh, Ronald Yager	Sofia
2018	Ricardo Goncalves, Carlos Agostinho, Vladimir Jotsov	Maria Marques, Francisco Duarte, João Martins	João Pedro Mendonça, David Romero	Maria João Lopes, Vasco Delgado-Gomes, Raul Poler	Vassil Sgurev	Funshal-Madeira Island
2020	Rudolf Cruse, Vladimir Jotsov	Anatoly Sachenko, Vaclav Snasel	Plamen Angelov, Krassimir Atanassov		Vassil Sgurev, Ronald Yager	Varna

'Chairs'), the Honorary Chairmen (Honorary 'Chairs'). The last column indicates the venue for the individual conferences.

For a small country like Bulgaria, conducting two series of international AIMSA and Intelligent Systems (IS) conferences is a significant achievement of the professional community in these two closely intertwined scientific disciplines.

In conclusion, the following can be noted:

1. The 'Artificial Intelligence' direction originated in the Bulgarian Academy of Sciences very soon after its emergence in other countries and simultaneously with the penetration of computer and information technologies and robotics in the different spheres of science and economic and spiritual activity. A significant positive role played the initiation and development of the personal computers industry and industrial robots by the Institute of Technical Cybernetics and Robotics (ITCR) at the Bulgarian Academy of Sciences in the early 1980s.
2. Over the same period of time, the research in the field of artificial intelligence was closely related to the need to apply the theoretical and applied results obtained in industry and public practice.
3. The National 'Brain' Program, funded and assisted by the Bulgarian state in the 1980s, also played a positive role in stimulating the research in the field of AI.
4. Two series of international conferences AIMSA (since 1984) and later the IEEE series 'Intelligent Systems' (since 2002) initiated and continue to play a significant positive role for the exchange of ideas and programming tools for both Bulgarian AI specialists and their foreign counterparts. Their active support from ECCAI and IEEE has stimulated and continues to stimulate them in this direction. These series of conferences have their lasting place in the international calendar so far and hopefully in the future.
5. At the threshold of its 150th anniversary, the Bulgarian Academy of Sciences has every reason to take pride in its decisive role in the emergence and development of the 'Artificial Intelligence' scientific sphere and the 'Intelligent Systems' direction in Bulgaria.

In the literature, this monograph presents a list of publications on the AI mainly of the 'Informatics' Department of ITCR at the Bulgarian Academy of Sciences, but also of other Bulgarian researchers in this scientific field. Without claiming the completeness of this list.

References

1. Tomov, V., Spiridonova, M., Gergov, A.: Opportunities of modern systems for automatic transformations. ASU **1** (1983) (in Bulgarian)
2. Tomov, V., Sakhno, S.: Experimental recognition system and class of archaeological ceramics functions. In: Collection of Conference Reports on Systems for Automated Maintenance, Varna, 3–8 Oct. (1983) (in Russian)
3. Ljudskanov, A.: Is the generally accepted strategy of machine translation research optimization. Mech. Trans. N.Y. **11**(6), 33–42 (1968)

4. Pavlov, R., Angelova, G.: On an approach for designing linguistic processors. In: Proceedings of the World Conference 'Mathematics in the Service of Men', Spain, July 1982
5. Skordev, A.: Combinatorial Spaces and Recursiveness in them. S., BAN (1980) (in Russian)
6. Polikarov, A.: Methodology of Scientific Knowledge. S. (1972) (in Bulgarian)
7. Sgurev, V., Dochev, D., Dichev, H., Agre, G., Markov, Z.: DIGS: a domain-independent expert system for technical diagnostics. In: Bibel, W., Petkoff, P. (eds.) Artificial Intelligence, Methodology and Applications, North-Holland, pp. 137–144 (1985)
8. Dochev, D., Dichev, H., Markov, Z., Agre, G.: Problems in the implementation of prolog personal computers. In: AIMSA '84. Varna, Bulgaria, Sept. (1984)
9. Sgurev, V., Ivanov, S.: Electric welding process management system. Prob. Eng. Cybernet. Robot. BAS **14**, 56–61 (1982) (in Bulgarian)
10. Sgurev, V., Dochev, D., Dichev, H.: Possibilities of using memory in artificial automation in engineering automation. In: Aktualni Problemi na Naukata, No. 9, Center for Scientific Information, BAN, Sofia, p. 44 (1985) (in Russian)
11. Sgurev, V., Mutafov, D., Glukhchev, G.: Some issues of the development of discrimination systems. Prob. Eng. Cybernet. Robot. BAN **24**, 26–29 (1986). (in Russian)
12. Sgurev, V., Dochev, D., Dichev, H., Agre, G., Markov, Z.: An approach to building a technical diagnostic expert system. Comput. Artifi. Intell. **5**(2), 103–115 (1986) (Bratislava)
13. Sgurev, V.: Second international conference on artificial intelligence AIMSA '86. J. Bulgarian Acad. Sci. **2**, 82–85 (1987) (in Bulgarian)
14. Sgurev, V., Pavlov, R. (ed.): Expert Systems, Nauka, I., Izkustvo, C., 442 pp. (1990) (in Bulgarian)
15. Kuttov, O., Khristov, V.: On the use of artificial intelligence methods for creating robot control systems. In: Reports of the I-th International Conference on Artificial Intelligence, Smolenice (Czechoslovakia) (1980) (in Russian)
16. Jorrand, P., Sgurev, V. (eds.): Artificial Intelligence II—Methodology, Systems, Applications (AIMSA '86), North-Holland, p. 403 (1987)
17. O'Shea, T., Sgurev, V. (eds.): Artificial Intelligence III. Methodology, Systems, Applications (AIMSA '88), North-Holland, Amsterdam, p. 444 (1988)
18. Jorrand, P., Sgurev V. (eds.): Artificial Intelligence IV-Methodology, Systems, Application. (AIMSA '90), North-Holland, Amsterdam, p. 433 (1990)
19. du Boulay, B., Sgurev, V. (eds.): Artificial Intelligence V, Methodology, Systems (AIMSA '92), Applications, North-Holland, Amsterdam, p. 273 (1992)
20. Jorrand, P., Sgurev V. (eds.): Artificial Intelligence VI-Methodology, Systems, Application (AIMSA '94), World Scientific, p. 400 (1994)
21. Sgurev, V.: State of the research on artificial intelligence in the People's Republic of Bulgaria. J. Bulgarian Acad. Sci. **3**, 12–17 (1986). (in Bulgarian)

Computer Algebra Applications in the Institute of Mathematics and Informatics of the Bulgarian Academy of Sciences

Margarita Spiridonova

Abstract Research topics related to some Computer Algebra applications in the Institute of Mathematics and Informatics (IMI) of the Bulgarian Academy of Sciences (BAS) for the period of last 40 years are outlined. The paper consists in three parts. In the first part (Sect. 1) the main features of the Computer Algebra Systems are briefly described and some Computer Algebra developments and applications in IMI for the period of last 40 years are presented. In the second part (Sect. 2) a representative Computer Algebra application (for solving initial and boundary-value problems for some classes of differential equations with use of the Computer Algebra System Mathematica) is considered. The third part (Sect. 3) includes notes on some current and future Computer Algebra applications.

Keywords Computer algebra · Computer algebra system · Symbolic computation · Operational calculus · Convolution · Duhamel principle · Initial value problem · Boundary value problem

1 Computer Algebra Systems. Some Applications in IMI

1.1 Computer Algebra Systems

More than 40 years ago the terms symbolic algebraic computations and analytic transformations of mathematical expressions were used in IMI for describing the capabilities of the systems as REDUCE 2 for computations with symbolic expressions and obtaining solutions of mathematical, physical and technical problems in symbolic

M. Spiridonova (✉)
Bulgarian Academy of Sciences, Institute of Mathematics and Informatics,
1113 Sofia, Bulgaria
e-mail: mspirid@math.bas.bg

© The Author(s), under exclusive license to Springer Nature Switzerland AG 2021
K. T. Atanassov (ed.), *Research in Computer Science in the Bulgarian Academy of Sciences*, Studies in Computational Intelligence 934,
https://doi.org/10.1007/978-3-030-72284-5_9

form. Such program systems then were called symbolic and algebraic systems. Later the "international" term Computer Algebra Systems (CAS) was accepted in Bulgaria as well.

The first informal definition of Computer Algebra (CA) was given in 1982 by Loos [6]: "Computer algebra is that part of computer science which designs, analyses, implements and applies algebraic algorithms".

The Computer Algebra could be considered both as an intersection of computer science and mathematics and as a special form of scientific computation as well. The main features of Computer Algebra are considered, for example, in [6, 9, 116], etc.

In contrast to numerical computations, in CA there isn't necessity of assigning specific numerical values to the variables. The emphasis is on computing with symbols representing mathematical objects and concepts. Numbers can appear in the symbolic expressions and the numerical computations with them are made with high precision (exact integer and rational arithmetic and arbitrary precision real arithmetic are always provided).

The CA algorithms give exact results. They deal with polynomials, rational functions, trigonometric functions, algebraic numbers, logical formulas, equations, power series, matrices with symbolic entities, etc. and apply on them some transformations or solve mathematical problems.

The main sub-areas of CA are: algorithms, systems and applications. The considerations in this paper are related mainly to the systems and their applications.

The Computer Algebra Systems (CAS) are software tools providing implementation and application of algebraic algorithms. They manipulate with symbolic expressions and perform on them symbolic and numerical computations—simplifications, transformations, substitutions, differentiation and integration, finding symbolic or numerical solution of algebraic and differential equations and many others. Visualization of functions, generation of graphical images of different types and big number of other operations are also provided in many CAS.

Computer algebra is a relatively young but rapidly growing scientific field. In the notes on the beginning of CA, found in different papers, many authors cite the first successful programs for symbolic differentiation, developed in 1953 (see [45, 71]), or the development of the programming language for symbolic information processing Lisp in 1958 (first published in 1960 [56]), or the notes of Ada Lovelace in 1842 on the "analytic engine" of Charles Babbage [57], etc. There are authors (Kaltofen [46] , for example), finding the roots of CA in the Universal Arithmetic of Newton (1728), describing the principles of manipulation with mathematical expressions. It is important to note also, that approximately till 1971, the Symbolic Mathematical Computing (SMC) and the Artificial Intelligence(AI) had a common path of development (in papers [7, 8], for example, "the territory" common to AI and SMC and the possible future links between the SMC and AI are considered).

CAS began to appear in the 1960s. Dozens of CAS have been developed till now. Some of them include a large amount of mathematical knowledge and are called *general purpose* CAS, others are oriented to solving specific problems (in subfields of physics or mathematics, for example) and they are called *special purpose* CAS.

Each CAS has own user language allowing the use of the implemented capabilities of the system; often these languages are powerful programming languages.

The authors of each CAS have own conception about the language of "their" system. An example: the author of Mathematica Stephen Wolfram says : "The philosophy of the Wolfram Language is to build as much knowledge—about algorithms and about the world—into the language as possible" [117] .

Important feature of each CAS is its user interface, providing communication with files, packages, text processors, other systems, etc. Some of the powerful CAS provide sophisticated custom interfaces.

Not all features of the CAS will be commented here. But let's note the main application areas of the CAS: mathematics, physics, mechanics, chemistry, biology, computer science, technology, education, etc. (see [6, 85, 97, 106–109]).

Many innovative achievements and applications of CA have been presented at the international conferences (with short names ISSAC and ACA, for example) devoted to the main aspects of CA and published in journals (the Journal of Symbolic Computations, ACM SIGSAM Bulletin and others).

The combination of symbolic and numerical computations, based on powerful algorithms, with graphics capabilities, powerful programming language and advanced user interface is a reason the contemporary general purpose CAS to be considered as integrated computing environments for scientific and technical computations and for development of new application tools. In such environments the process of teaching and learning mathematics (in the universities and secondary schools) is also convenient.

The well known powerful general purpose CAS, such as Maple, Mathematica, Reduce, Matlab, Macsyma, Derive and some others are well known and used in Bulgaria.

In some CAS applications it is enough the built-in capabilities of the available systems to be used. Other applications need to be developed extensions of the capabilities of the used CAS. Such extensions are usually developed as program packages. Sometimes special purpose CAS are developed as well.

The research and developments in IMI considered bellow, include all mentioned forms of CAS applications.

1.2 Computer Algebra Applications in IMI—General Notes

The following aspects of research, developments and applications of CA in IMI will be outlined:

(i) Use of computer-aided symbolic computations for mathematical problem solving; use of CAS as an assistant of the researchers.
(ii) Development of special purpose CAS and program packages for solving specific problems, with use of the general purpose CAS Reduce, SAC-2, Maple and Mathematica.

(iii) Conceptual and Methodological Problems; a conception of Intelligent CAS—features and implementations; other conceptions.
(iv) Other aspects: applications in education; projects, including CA topics, supported by the Bulgarian National Science Fund for Research; national and international activities, etc.

1.3   Use of Computer-Aided Symbolic and Algebraic Computations in Research

The computer-aided symbolic and algebraic computations has been an important part of the research in IMI for many areas.

Some results, concerning research in the field of algebra, are presented below.

The first applications of computers to abstract algebra at IMI were in the 1980s in Group Theory, in the Ph.D. Thesis of Daniela Nikolova [61], see also [62]. In the beginning the programs were written in Fortran, but later in most part of the research in Group Theory the system GAP was used [63]. This is a system for Computational Discrete Algebra, elaborated with particular emphasis on Computational Group Theory, particularly useful for research and education.

Further applications of CAS to Group theory include the study of Fibonacci groups (see [64, 65]), the covering number of groups (see [33, 49]), etc.

Applications of computational methods were done also in Design Theory and Combinatorics [54].

Starting from the 1960s, theory of algebras with polynomial identities (PI-algebras) is one of the branches of algebra actively developed at IMI. In the 1980s new powerful methods were developed for the study of PI-algebras. These methods were applied to perform computations (by hand) which allowed to solve several important problems in the theory. An account on the work can be found in the survey articles [26–28] and also in the book [30].

In the beginning of 1990s these methods were computerized by a team including Bulgarian, Italian and American mathematicians with further applications, also in classical and in noncommutative invariant theory. An account is given in the surveys [3, 31].

In the 2000s classical results for solving linear Diophantine equations were computerized and applied for the needs of classical invariant theory and theory of symmetric functions. Later these methods were applied also to the study of PI-algebras (see [2]), adding new people from Bulgaria, Italy and Turkey to the research team. Other applications of this method are given in the recent survey [4].

Another result was an algorithm for fast computing of Schur functions obtained in [10], in cooperation with American and Canadian mathematicians. Computing of Schur functions is related with the important problem for the evaluation of hypergeometric functions of a matrix argument. The practical importance of computing the hypergeometric function of a matrix argument stems from far reaching applications in multitude of fields. For example, it is needed in applications ranging from genomics to wireless communications, finance, target classification, etc.

For some of the computations, presented in the cited papers, the CAS Maple is used. Fore some others C++ is applied using 64 processors on the Cray T3E at the Lawrence Berkeley Lab (about 8 h have been needed for the computations from start to finish).

Many useful Computer Algebra applications in Coding Theory, in Interval Computations and in other areas were made in IMI as well. Some of them were presented at the Conferences on Applications of Computer Algebra (see https://math.unm.edu/aca.html), in particular at those held in Bulgaria in 2006 (http://www.math.bas.bg/artint/mspirid/ACA2006/) and in 2012 (http://www.math.bas.bg/ACA2012/) .

1.4 Development of Special Purpose CAS and Program Packages for Solving Specific Problems

In 1980 a research group named "Artificial Intelligence in Software" in the framework of the Software Department of IMI (headed by Prof. Petar Barnev) has started studying and exploration of the CAS features. The established international IMI contacts with the Russian Joint Research Institute for Nuclear Research (JINR) in Dubna, the National Research Center of Italy, the Technical University of Dresden, the Institute of Physics of the Hungarian Academy of Sciences and some others were very helpful.

After foundation (in 1985) of the Artificial Intelligence Department of IMI (headed by Prof. Valentin Tomov), the CAS features and their applications became an important part of the research in this Department. Most of the briefly presented below program packages were developed in this Department). From 1995 Assoc. Prof. Alexander Gerov became a head of the AI Dept., till 2010, when some structural changes took place in IMI.

Some papers of the period 1975–1977 (see [80, 81], for example) are related to the languages for symbolic information processing.

The main subjects of the further research were directed to solving particular problems. In that time the applications of CAS in IMI were made with use of the CAS Reduce. At the beginning of the 80s it was the only available CAS in IMI. Later SAC 2 was delivered, then Maple and even later—Mathematica.

Some built-in capabilities of CAS were used for solving specific problems in mechanics [5], in geodesy [111], and in some other areas.

The following "groups" of developments could be considered:

– Development of special purpose CAS and packages. The special purpose CAS were based on the use of SAC-2 system, the packages have been developed as Reduce, Maple or Mathematica packages.
– Development of conceptions, related to the use of CAS.
– Some others.

The main developed CA program tools for solving particular problems [103] are mentioned below.

- Special Purpose CAS for linear algebra [38].
- Special Purpose CAS for Continued Fractions Manipulation [66].
- Reduce Package for Laplace Transformation [50]. It was developed after an application of Reduce system for solving a problem in Mechanics: modeling of building structures behaviour under seismic loading [5]. A Reduce program was developed and used in research. But a new "problem" was discovered: the well known Laplace transformation still was not implemented in Reduce. A program package for Laplace transformation for the CAS Reduce was developed (see [50, 82]) and later it was included in the Reduce library [43]. Some other applications in Mechanics were made later.
- Reduce Package for Building and Use of Formulae Bases (see [83, 84, 87]).
- Reduce Package for Power Series Manipulation [88].
- Reduce Package for Investigation of Rational Functions [89].
- Reduce Package for Symbolic and Numerical Computations in Mathematical Geodesy [11]. Main problem solved by the package: the geodesic coordinates of a point on the earthly ellipsoid to be determined if the coordinates of two or more points are given and the azimuths in those points to the new (wanted) point are observed. The problem is known as "Angle Intersection on the Earthly Ellipsoid". The package provides use of an especially developed algorithm for reversing of power series of two variables [112], good organization and convenient use of the symbolic and numerical computations and of the observation model, etc. (see [11, 111, 113]). Further use of the CAS Maple was discussed.
- Maple Package for Symbolic Derivation of Modified (Partial Differential) Equations, allowing convenient analysis of finite-difference schemes [93] (in collaboration with INRIA Sophia Antipolis, France).
- Mathematica package for experiments with the Poncelet theorem from the Projective Geometry (proposed by Prof. P. Kenderov; his formulae are used).
- Legendre Transformation packages (proposed also by Prof. P. Kenderov). These two program packages for a convex analysis transformation, known as Legendre or Legendre-Fenchel Transformation were developed in order this transformation to be used by the aid of a CAS. The systems Reduce and Maple were chosen, respectively for the Reduce package (described in [91]) and for the Maple package (described in [92]). The capabilities of the second package were a kind of extension of the capabilities of the first one; the graphics capabilities of Maple were used for visualization of the functions and their transforms.
- A collection of packages (with the use of the CAS Mathematica) for solving initial and boundary-value problems for some classes of differential equations, based on the Operational Calculus Approach [102]; a more detailed consideration is presented in Sect. 2 of this survey.
- An Environment for Learning Mathematics using a CAS [68, 95] was developed, as a part of the research project on application of CAS in mathematics education (see [60]).

Graduate and postgraduate students took place in the development of a big part of the mentioned above program packages.

The following general features of the above developments should be noted:

(1) The most part of the research topics were formulated in collaboration with researchers being experts in specific fields of mathematics, mechanics, geodesy, etc. (in Bulgaria or abroad).
(2) The work on all topics includes research, development of algorithms and software tools, experimental application, analysis, formulation of topics for further research and "real" applications.
(3) The development of a problem oriented software tool like the mentioned above can be considered as a definition and development of an applied problem oriented "microenvironment" in the environment of the used general purpose CAS. The main features of such "microenvironments" are considered below.
Some of the mentioned research and developments were included in projects, supported by the Bulgarian National Fund for Research (the abbreviation NSF is used later); some of them are:

- National Science Fund Project I-520/95 "Development and Application of Software Packages for Computer Algebra" (1995–1998);
- National Science Fund Project V Rp-I-5/99 "Representation, Processing and Exchange of Knowledge Using Computer Algebra Systems" (1999);
- National Science Fund Project I-1002/2000 "Computer Algebra Tools for Mathematics Education" (2000–2003).

1.5 Conceptual Aspects of the CA Applications

On the base of experience and some features of the used CAS, two conceptions were developed in 1983–84 (see [37, 85, 86, 105, 108]):

   (c1) Conception of **Intelligent Computer Algebra System (ICAS)**. It includes development of:

- language tools for more "natural" description of the problem to be solved;
- analysis of the problem description and choice of an appropriate algorithm from the "knowledge base" of the system;
- construction of a sequence of steps for full solution of the problem.

Some "additional" problems had to be discussed—about the mathematical knowledge representation and the knowledge bases support; how to choose an appropriate (or the best) algorithm, etc. Some experimental developments were made later (see [69]).

Similar ideas were published by other authors as well (see [8] and [47], for example).

(c2) Conception of **Symbolic-Numerical CAS**. It includes the following requirements: not only symbolic computations, but also "more advanced" numerical computations and effective "transition" between them to be developed in such a system in order both types of computation to be used in a convenient way for solving problems. This conception relates to 1984; it is important to note that the contemporary general purpose CAS have such properties.

Some features of the above two conceptions were implemented, for example, in a Special Purpose CAS for linear algebra [38] and a Special Purpose CAS for Continued Fractions Manipulation [66].

The research on *building, maintenance and use of bases of symbolic transformations* was related to the conception of ICAS. A way to represent and to add more mathematical knowledge to a CAS was considered. The CAS Reduce was used and bases consisting of sets of transformations rules (represented as substitution rules) and bases consisting of sets of defined procedures were suggested. Language tools for building, maintenance an use of such bases, as well as some experimental bases were developed (see [83, 84, 87]).

(c3) One more Conception (with methodological aspects) was developed—for **Applied Computer Algebra Microenvironments** in a general purpose CAS [100].

Shortly about this Conception:

An Applied Microenvironment (AM) can be considered as a framework within a CAS which provides specific tools for solving a problem or a class of problems. The adjective "Applied" points out that always such a microenvironment is developed for solving an applied problem. The particle "micro" indicates that it is a small framework in the large computational environment of the host general purpose CAS.

It is supposed that being in an AM, one should be able to solve the formulated problem or a class of problems without the necessity other tools than those defined in it to be used. The last feature may not be valid for an usual library package. An AM should provide also facilities for interaction with the user, for delivering information about the used method, for explanation of the results, etc.

More details and examples are presented in [100]. In general terms, this Conception is aimed to deliver comfortable use of each developed problem-oriented CA program tool.

1.6   Applications in Education

Education has become one of the fastest growing application areas for computers in general and for CAS in particular.

The contemporary CAS such as Mathematica, Maple, Reduce, Axiom, Macsyma, Derive and many others often are used as good assistants in teaching and learning mathematics in the schools and universities.

The use of CAS for educational purposes can help students to understand better the mathematical notions, to facilitate the most tedious computations, they to be concentrated on the content of the lessons, etc. As a whole, the mathematics can become more attractive for the students.

A possible way the use of CAS in education to be extended is the teachers to get more information about the CAS capabilities and their applications in education on all levels. Some appropriate presentations at Bulgarian conferences with topics related to the education, as well as seminars and workshops devoted to the CAS application in education were organized. Some of them were included in the scientific programs of the annual conferences of the Union of Bulgarian Mathematicians (see [53, 94, 98] and also [99]).

In the framework of the Project I-1002/2000, supported by the NSF, a special website was created (in Bulgarian) with useful information—papers, guides, web-addresses, etc.—about the use of CAS and algebraic calculators in mathematics education [60]. A critical analysis of the CAS from the point of view of the educational goals was also included (see [96]).

The practical use of CAS in mathematics education meets problems of different kind—the availability of an "appropriate" CAS, good tutorials, the necessity of reorganizing the work in the classroom, etc. These problems can be considered as a challenge to the educators and in most cases their solution needs elaboration and professional skils.

The above features of the "classical" CAS could be considered as a reason some versions of the well known CAS (Mathematica, for example) to be developed for application in education. The company WOLFRAM even suggests special programs for high-school students, college-students, graduate students, etc. [119].

Some CAS as Derive [52], and MuPAD [59], for example, were designed for the purposes of education.

The main reasons for problems like the mentioned above to arise is that the "classical" CAS have no mathematical knowledge represented in an explicit way. Their knowledge is embedded implicitly in the algorithms and the user can only see the solution of a given problem.

Therefore one of the possible solutions is the development of knowledge based CAS (called intelligent CAS). An "intelligent" CAS could be used successfully in teaching and learning mathematics and it would be a better assistant in such activities if it is integrated into an environment providing intelligent interface and many other facilities.

The features of an environment for learning mathematics are considered in [69, 95]. This environment is aimed to support learning mathematics from elementary schools up to university level. The first prototype covers algebra, geometry and first semester calculus.

Similar developments are described in [67, 68, 70].

Recently, the free dynamic mathematics software and, in particular, systems as GeoGebra [36] has been in use in many Bulgarian schools. Many new ideas about the use of CAS in education are under discussion.

Some Bulgarian contributions related to the use of CAS in education, were presented at the Special Sessions on Education of the International conferences on Applications of Computer Algebra. At the one held in Bulgaria in 2012 (http://www.math.bas.bg/ACA2012/), among the talks at the Special Session on Education, 7 interesting talks (of all 11) were presented by Bulgarian authors. In the talk of P. Kenderov and E. Sendova entitled "Spreading the Inquiry Based Mathematics Education in Bulgaria within the Fibonacci European Project", the idea of "Inquiry Based Mathematics" was considered. The importance of dynamic computer learning environment was also discussed [51].

Information about the mentioned Fibonachi Project as well as about some other projects related to the computer-aided mathematical education, can be found on the site of the so-called Virtual Mathematics Classroom [115].

1.7 National and International Activities, University Courses, Etc.

The National and International activities of IMI, related to CA, include organization of seminars, workshops, conferences, etc.

The regular participation in international conferences in the field has been used for presentation of the achieved rezults (as it is seen in some references), for useful discussions and for establishing new contacts.

The collaboration with Bulgarian and foreign research institutes has been important part of the research in IMI in the field of CA.

Some conferences organized in Bulgaria, were mentioned yet. Let's note once again the organization of two of them—the conferences on Applications of Computer Algebra (ACA)—respectively in 2006 in Varna and in 2012 in Sofia (see https://math.unm.edu/aca.html).

The activities of IMI related to workshops include, in particular, a workshop of the Working Group 24 named "Computer Analytics" of the Council for Science of Comecon (Mutual Economic Assistance Council) in Sofia in 1990 and a Summer School ALCCAL'2000 "Algebraic Combinatorics and Computer Algebra", in Varna in 2000 (see http://www.math.bas.bg/~ALCCAL/).

An important seminar: the National Seminar on CAS was organized and took place monthly in the period 1992–1996.

University courses on CAS took place in the Sofia University and in some other Bulgarian universities.

2 Application of the Operational Calculus Approach to Solving Initial and Boundary Value Problems for Some Classes of Differential Equations, in the Environment of a Computer Algebra System

2.1 Introduction

The main features of the operational calculus approach are presented. Its application for solving problems related to some classes of differential equations is considered.

The main characteristics of the program packages developed with use of the CAS Mathematica and supporting such application are briefly described. They provide the use of:

- the Heaviside algorithm for solving initial value problems for linear ordinary differential equations (LODE);
- an extension of the Heaviside algorithm to a class of boundary value problems for LODE with constant coefficients, connected with the problems of obtaining periodic solutions of LODE both in the non-resonance and the resonance cases; the obtaining of mean-periodic solutions of LODE with constant coefficients using such an approach is outlined;
- algorithms based on second order operational calculi for solving local and nonlocal boundary value problems, related to the heat, the wave and the beam equations.

The features of the program tools implemented with use of the CAS Mathematica, are presented.

About the Operational Calculus

The essence of the operational calculus consists in transformation of calculus problems to algebraic problems, treating the differentiation operator as an algebraic object.

Some ideas of "symbolic" operational calculus come from the works of Leibnitz, Euler, Cauchy and other mathematicians (see [79, 104], and also [114]). Nevertheless, Oliver Heaviside (1850–1925) is regarded to be the father of the operational calculus. He was the first who successfully applied this method in his research for solving initial value problems related to electromagnetic theory (see [44]). But Heaviside did not established a sound mathematical theory and his calculus was regarded by some scientists as inconsistent. The first justification of his approach was done by means of the Laplace transformation. Quite later—in the middle of the last century—the Polish mathematician Jan Mikusiński (1913–1987) made a return to the original operator viewpoint and developed a direct algebraic approach to the Heaviside operational calculus. He based his calculus on the notion of convolution quotient, without refering it to the Laplace transformation. His calculus is known as Mikusiński's operational calculus. From historical point of view, it is fair to call it as operational calculus of Heaviside–Mikusiński.

Scientists in many countries have published works related to the operational calculus of Mikusiński. Some of them are L. Berg, T. K. Boehme, I. H. Dimovski, V. A. Ditkin, A. P. Prudnikov, K. Yosida, etc. Other names are mentioned in some references, for example in [79]. Some recent results can be found in [74, 120] and others.

Mainly the results of I. H. Dimovski on development of operational calculi of Mikusiński's type are considered below.

The operational calculus has been widely used for solving problems in mathematics, physics, mechanics, electrical engineering, etc. The algorithms and the program tools described here are intended to facilitate the use of the operational calculus approach by means of computer.

2.2   *Heaviside Algorithm*

Since the main idea of the Operational Calculus (OC) of Oliver Heaviside is the conversion of differential equations to algebraic equations by treating of the differentiation operator as an algebraic object, an algorithm for doing that is needed.

The so called Heaviside algorithm based on the operational calculus approach is intended for solving initial value problems for linear ordinary differential equations with constant coefficients. We use it in the frames of Mikusiński's operational calculus. A description and implementation of the Heaviside algorithm using a computer algebra system are also considered. Special attention is paid to the features making this implementation efficient. Illustrative examples are included The Heaviside algorithm for solving initial value problems for linear ordinary differential equations with constant coefficients in the frames of the Mikusinski's operational calculus is described. The features of its program implementation is considered.

The most important role in the Mikusiński's operational calculus plays the classical Duhamel convolution (see [58]):

$$(f * g) = \int\limits_{0}^{t} f(t - \tau)g(\tau)d\tau, \tag{1}$$

in the space $\mathcal{C}[0, \infty)$ of the continuous functions on $[0, \infty)$. Mikusiński considers this space as a ring on $I\!R$ or $\mathcal{C}$. He uses the fact that due to a famous theorem of Titchmarsh the operation (1) has no divisors of zero and hence $(\mathcal{C}[0, \infty), *)$ is an integrity domain. In the same way, as the ring $Z\!\!\!Z$ of the integers is extended to the field $\mathcal{Q}$ of the rational numbers, he extends the ring $(\mathcal{C}[0, \infty)*)$ to the smallest field $\mathcal{M}$ containing the initial ring. This field is called Mikusiński's field and it is denoted by $\mathcal{M}$. The elements of $\mathcal{M}$ are convolution fractions $\dfrac{f}{g} = \dfrac{\{f(t)\}}{\{g(t)\}}$, called "operators".

In Mikusiński's calculus each function $f : [0, \infty) \to \mathbb{R}$ is considered as an algebraic object and the notation $f = s\{f(x)\}$ is used.

Basic operators in the Mikusiński approach are the integration operator $l: lf(t) = \int_0^t f(\tau)d\tau$, and the algebraic analogon $s = \dfrac{1}{l}$ of the differentiation operator $\dfrac{d}{dt}$.

The relation between the derivative $f'(t)$ and the product $s\{f(t)\}$ is presented by the basic formula of the Mikusiński operational calculus

$$\{f'(t)\} = s\{f(t)\} - f(0), \tag{2}$$

where $f \in C^1[0, \infty)$ and $f(0)$ is considered as a "numerical operator". If a function $f = \{f(t)\}$ has continuous derivatives to n-th order for $0 \le t < \infty$, a more general formula can be derived:

$$f^{(n)} = s^n f - \sum_{i=0}^{n-1} s^i f^{(n-1-i)}(0), \quad n = 1, 2, 3, \ldots \tag{3}$$

2.3 Solving Initial Value Problems for Linear Ordinary Differential Equation Using the Heaviside Algorithm

Let $P(\lambda) = a_0\lambda^n + a_1\lambda^{n-1} + \cdots + a_{n-1}\lambda + a_n$ be a non-zero polynomial of n-th degree with real or complex coefficients.

Consider the following initial value problem:

$$P\left(\frac{d}{dt}\right) y = f(t), \ y(0) = \gamma_0, \ y'(0) = \gamma_1, \ \ldots, \ y^{(n-1)}(0) = \gamma_{n-1}. \tag{4}$$

Using the main formula (2), (3) of the operational calculus of Mikusinski, an "algebraization" of the problem can be made. The problem (4) reduces to the following single algebraic equation of first degree:

$$P(s)\, y = f + Q(s), \tag{5}$$

with $\quad P(s) = \sum_{j=1}^{n} a_j s^j, \quad Q(s) = \sum_{j=1}^{n} \left(\sum_{k=j}^{n} a_{n-k}\, \gamma_{k-j} \right) s^{j-1}$, $\deg Q < \deg P$.

The formal solution has the form

$$y = \frac{1}{P(s)} f + \frac{Q(s)}{P(s)}. \tag{6}$$

It can be interpreted as a functional solution if we decompose $1/P(s)$ and $Q(s)/P(s)$ in elementary fractions and interpret these fractions as functions using the formula (see [58]):

$$\frac{1}{(s-\alpha)^n} = \left\{ \frac{t^{n-1}}{(n-1)!} e^{\alpha t} \right\}, \quad n = 1, 2, \ldots \tag{7}$$

Thus we represent $1/P(s)$ and $Q(s)/P(s)$ as functions:

$$G(t) = 1/P(s), \quad R(t) = Q(s)/P(s) \tag{8}$$

and the solution takes the form

$$y(t) = G(t) * f(t) + R(t) \tag{9}$$

At last the computation of the convolution product denoted by $*$ in (9) has to be performed.

The main steps of the Heaviside algorithm for solving initial value problems for linear ordinary differential equations with constant coefficients are shortly described.

The solution of an initial value problem for simultaneous ordinary linear differential equations with constant coefficients can be performed in a similar way: algebraization of the problem and reducing it to a system of linear algebraic equations; solving the obtained system using linear algebra methods and functional interpretation of the solution.

2.4 Program Implementation of the Heaviside Algorithm

A program implementation of the Heaviside algorithm would allow it to be used by means of computer. Having in mind the kind of the operations of this algorithm and the capabilities of the computer algebra system Mathematica, this system was chosen for development of a program package implementing the Heaviside algorithm.

Information about a full program implementation of the Heaviside algorithm was not found.

Main steps of the algorithm

Formulating once again the steps of the Heaviside algorithm, the features of their program implementation are considered below.

Step 1. *Algebraization of the problem.* The language tools of *Mathematica* allow the transformation of (4) into (5) to be made in a convenient way, using rules for presentation of the formulae (2), (3) and for the initial conditions of (4).

Step 2. *Solution of the algebraic equation* (5). A polynomial equation (or a system of such equations) has to be solved and *Mathematica* provides such capabilities. The result is obtained in the form (6).

Step 3. *Factorization of the polynomial* $P(s)$ *and partial fraction decomposition of* $\dfrac{1}{P(s)}$ *and* $\dfrac{Q(s)}{P(s)}$. The built-in function named *Factor* of *Mathematica* is used; a presentation of $P(s)$ as a product of factors, each of which is a polynomial of first or second degree, raised to an integer positive number, is obtained. More details are described in [21, 102]. This process may not finish with success if some of the coefficients of $P(s)$ are parameters and in the same time deg $P > 4$. In this case the solution of problem (4) is aborted. If the factorization of P is finished successfully, the *Mathematica* function *Apart* represents the rational expressions $\dfrac{1}{P(s)}$ and $\dfrac{Q(s)}{P(s)}$ as sums of terms with minimal denominators of minimal degrees.

Step 4. *Interpretation of the rational expressions* $\dfrac{1}{P(s)}$ *and* $\dfrac{Q(s)}{P(s)}$. Each fraction in these expressions has to be interpreted as a function by means of formulae, such as (7). The main part of the Mikusinski's table is used. The formulae are presented as *Mathematica* rules with appropriate pattern matching. An uniform interpretation of all fractions is obtained. As a result, the presentations (8), (9) are achieved.

Step 5. *Computation of the Duhamel convolution in the final form of the solution.* The *Mathematica* integrator is used.

Step 6. *Showing the result: solution or a message that the problem can not be solved.* It was mentioned above when the problem will not be solved in case of one equation. In case of solving initial value problem for a system of equations, similar situation may occur, but, in addition, the problem will not be solved if on Step 2 the algebraic system has not solution.

2.5 Program Package for the Heaviside Algorithm

An implementation of the Heaviside algorithm following the steps described above, is developed as a *Mathematica* program package. Its main function *DSolveOC* defines the performance of all steps of the Heaviside algorithm. The call of this function is similar to the call of the *Mathematica* function *DSolve*. The output also has similar form. The solution is presented as a rule or as a list of rules in case of several solutions. The use of options for visualization of the solution and for some additional capabilities is provided.

Illustrative examples
With the following two examples we illustrate the use of the main function *DSolveOC* of the package. The solutions of two initial value problems—for linear ordinary differential equation and for a system of two linear ordinary differential equations are shown.

Example 1 Initial value problem for one LODE with constant coefficients:

```
task1 = {x'''[t] + x'[t] == e^(2t), x[0] == 0, x'[0] == 0, x''[0] == 0}
{x'[t] + x^(3)[t] == e^(2t), x[0] == 0, x'[0] == 0, x"[0] == 0}
DSolveOC[task1, x[t], t]
DSolveOC[{x'''[t] + x'[t] == e^(2t), x[0] == 0, x'[0] == 0, x''[0] == 0}, x[t], t]
```

$$x[t] \rightarrow \frac{1}{10}\left(-5 + e^{2t} + 4\cos[t] - 2\sin[t]\right)$$

Example 2 Initial value problem for a system of LODE with constant coefficients; an option for visualization of the solution is used.

```
mysyst = {-x[t] + 2y[t] + x'[t] == -2 e^t, -2 x[t] - y[t] + y'[t] == 0, x[0] == 0, y[0] == 1};
DSolveOC[mysyst, {x[t], y[t]}, t, GraphInterval → {0, π}]
```

$$\{x[t] \rightarrow -4\, e^t \cos[t]\, \sin[t],$$
$$y[t] \rightarrow e^t\, (-1 + 2\cos[2t])\}$$

Concluding Remarks

- The Heaviside algorithm gives a closed form solution of an initial value problem for a linear ordinary differential equation or a system of such equations in a direct way, without trying to find partial and general solution.
- An uniform approach is used for homogenous and for non-homogenous equations.
- No special requirements to the right-hand side function are posed (as in the case of Laplace transformation).
- In the Heaviside algorithm the initial value conditions are supposed to be given in the point 0. It is easy to develop an extension of the algorithm allowing the initial value conditions to be given in point $t_0 \neq 0$.
- For solving an initial value problem for a system of ordinary linear differential equations with constant coefficients, all steps of the Heaviside algorithm can be performed in a similar way, as in case of solving initial value problem for one equation.

The presented implementation of the Heaviside algorithm is considered in more details in [21, 102].

2.6 Extension of the Heaviside Algorithm to a Class of Boundary Value Problems for LODE with Constant Coefficients. Periodic Solutions of Such Equations

An approach to obtaining periodic and mean-periodic solutions of LODE with constant coefficients is presented.

An Auxiliary Boundary Value Problem

An extension of the Heaviside—Mikusiński operational calculus is developed by Dimovski and Grozdev (see [13, 39]) and in the framework of this operational calculus an extension of the Heaviside algorithm is proposed. It is intended for solving nonlocal initial value problems for LODE with constant coefficients. This approach is used for obtaining periodic solutions of such equations.

Let's consider a non-zero polynomial with constant coefficients of degree n:
$$P(\lambda) = a_0\lambda^n + a_1\lambda^{n-1} + \cdots + a_{n-1}\lambda + a_n$$
and the following ordinary linear differential equation with constant coefficients:

$$P\left(\frac{d}{dt}\right)y = f(t), \quad -\infty < t < \infty \tag{10}$$

We are looking for a periodic solution $y(t)$ with period T of this equation, i.e. a solution satisfying the identity:

$$y(t + T) = y(t), \quad -\infty < t < \infty \tag{11}$$

An obvious necessary condition for the existance of a periodic solution of (10) with period T is the function $f(t)$ to be periodic with period T, i.e. for each $t \in \mathbb{R}$ the following condition to be satisfied:

$$f(t + T) = f(t). \tag{12}$$

The following **Theorem** could be proven: *A solution of* (10) *with periodic right-hand side* $f(t)$ *with period* T *is T-periodic if and only if the following "boundary" conditions are satisfied:*

$$y(T) - y(0) = 0, \; y'(T) - y'(0) = 0, \; \ldots \; y^{(n-1)}(T) - y^{(n-1)}(0) = 0 \tag{13}$$

This theorem allows the problem of obtaining periodic solutions of (10) to be reduced to the problem of finding a solution of this equation in the interval $(-\infty, \infty)$, satisfying the "boundary" conditions (13).

Further we reduce this problem to the following intermediate (auxiliary) boundary-value problem:

$$P\left(\frac{d}{dt}\right) y = f(t), \quad -\infty < t < \infty$$

$$\int_0^T y(\tau)\,d\tau = \alpha_0, \quad y^{(k)}(T) - y^{(k)}(0) = \alpha_{k+1}, \quad k = 0, 1, \ldots n - 2. \tag{14}$$

2.7 An Operational Method for Solving the Auxiliary Problem. Convolution of Dimovski

The Heaviside algorithm is developed for solving initial value problems for LODE with constant coefficients and it can not be used directly for finding periodic solutions of such equations.

Use of Fourier transform and Laplace transform for obtaining periodic solutions can be found in some works of Kaplan [48], Rosenvasser [75], Lurie [55] and some others. We use an alternative direct approach, similar to those of Mikusiński, but using another convolution, based on the operational calculus of Dimovski (see [13]) and related to the nonlocal boundary value problem in $C(\mathbb{R})$:

$$y' = f(x), \quad \int_0^T y(\tau)d\tau = 0,$$

where T is a constant.

The solution

$$L\,f(t) = \int_0^t f(\tau)d\tau - \frac{1}{T}\int_0^T\left(\int_0^\tau f(\sigma)d\sigma\right)d\tau$$

is an analogue of the integration operator $lf(t) = \int_0^t f(\tau)d\tau$ of Mikusiński's operational calculus.

The operational calculus of Dimovski for the operator L is an analogue of the operational calculus of Mikusiński, but the following convolution of Dimovski is used:

$$(f \overset{t}{*} g)(t) = \Phi_\tau\{\int_\tau^t f(t + \tau - \sigma)g(\sigma)d\sigma\},$$

with an arbitrary linear functional Φ in $C(\mathbb{R})$. In this case the functional

$\Phi\{f\} = \dfrac{1}{T}\displaystyle\int_0^T f(\tau)d\tau$ is used. The convolution

$$(f \overset{t}{*} g)(t) = \frac{1}{T} \int_0^T \left(\int_\tau^t f(t + \tau - \sigma) g(\sigma) d\sigma \right) d\tau$$

has the property $Lf(t) = \{1\} \overset{t}{*} f$.

Dimovski and Grozdev proposed a simpler convolution (without using of repeated integrals):

$$
\begin{aligned}
(f * g)(t) = {} & \frac{f(t)}{T} \int_0^T g(\tau) \, d\tau + \frac{g(t)}{T} \int_0^T f(\tau) \, d\tau \\
& - \frac{1}{T} \int_0^t f(t - \tau) \, g(\tau) \, d\tau - \frac{1}{T} \int_t^T f(t + T - \tau) \, g(\tau) \, d\tau,
\end{aligned}
\tag{15}
$$

for which $\{1\} \overset{t}{*} f = f$.

The constant function $\{1\}$ plays the role of a unity in the convolution algebra $(\mathcal{C}(\mathbb{R}), *)$. The operator L has the following representation:

$$L\{1\} = t - \frac{T}{2}, \text{ i.e. } Lf = \left\{ t - \frac{T}{2} \right\} \overset{t}{*} f.$$

Further, convolution fractions of the form f/g are considered (with $f, g \in C[0, T]$, g being a nondivisor of 0 of the operation (15)). The ring of the continuous functions on $(-\infty, \infty)$ is extended to the smallest ring $\mathcal{M}$, containing the convolution fractions $\dfrac{f}{g}$ with denominators which are nondivisors of 0. The most important convolution fraction

$$S = \frac{1}{L}$$

is considered as an algebraic analogue of d/dt.

The basic formula of the Operational Calculus of Dimovski is:

$$\{f'(t)\} = S\{f(t)\} - \frac{1}{T} \int_0^T f(\tau) d\tau. \tag{16}$$

Here $\dfrac{1}{T} \displaystyle\int_0^T f(\tau) d\tau$ is considered as a constant function.

For $f^{(n)}$ the following formula can be derived from (16):

$$f^{(n)} = S^n f - \frac{S^n}{T} \int_0^T f(\tau) d\tau - \sum_{k=1}^{n-1} \frac{S^k}{T} \left(f^{(n-1-k)}(T) - f^{(n-1-k)}(0) \right) \tag{17}$$

For the case $T = 1$, the integral operator L is called by Dimovski and Grozdev **Bernoullian integration operator** due to the following relation with the polynomials of Bernoulli:

$$L^n\{1\} = \frac{T^n}{n!} B_n\left(\frac{t}{T}\right), \quad n = 0, 1, 2, \ldots,$$

where $B_n(t)$ is the polynomial of Bernoulli of degree n.

Further the scheme of Mikusiński has to be followed, using the convolution (15) and taking into account the following differences:

(1) The operation (15) has a unit element.

(2) This operation has divisors of 0.

The eigenfunctions of L are divisors of 0 of (15). These functions have the form $\varphi_n(t) = C e^{\frac{2\pi i n t}{T}}, \quad n \in \mathbb{Z} \setminus \{0\}$.

For the application of the new operational calculus it is important we to have formulae for convolution fractions of the type $\dfrac{1}{(S - \lambda)^k}, \; k \in \mathbb{N}$. They exist iff $S - \lambda$ is a nondivisor of 0 and this is not truth iff $\lambda = \dfrac{2\pi i n}{T}$ and $n \in \mathbb{Z} \setminus \{0\}$.

Thus for each $\lambda \neq \dfrac{2\pi i n}{T}, \; n \in \mathbb{Z} \setminus \{0\}$ the following formulae hold:

$$\frac{1}{S - \lambda} = -\frac{1}{\lambda} + \frac{T\, e^{t\lambda}}{e^{\lambda T} - 1} \tag{18}$$

$$\frac{S}{S - \lambda} = \frac{T\lambda e^{t\lambda}}{e^{\lambda T} - 1} \tag{19}$$

Corollary *If* $\lambda \neq \dfrac{2\pi i n}{T}, \; n \in \mathbb{Z} \setminus \{0\}$, *more general formulae hold (for each integer* $k \geq 1$):

$$\frac{1}{(S - \lambda)^k} = \frac{1}{(k-1)!} \frac{\partial^{k-1}}{\partial \lambda^{k-1}} \left(-\frac{1}{\lambda} + \frac{T\, e^{t\lambda}}{e^{\lambda T} - 1} \right) \tag{20}$$

$$\frac{S}{(S - \lambda)^k} = \frac{1}{(k-1)!} \frac{\partial^{k-1}}{\partial \lambda^{k-1}} \left(\frac{T\lambda e^{t\lambda}}{e^{\lambda T} - 1} \right) \tag{21}$$

The formulae (18)–(21) are intended to be used for interpretation of rational expressions in the extended Heaviside algorithm. For the purposes of the program implementation of this algorithm additional formulae were derived–for the case when the denominator is an integer power of a second degree polynomial.

Non-resonance case

Let's apply the Operational Calculus of Dimovski for solving the auxiliary problem, formulated above:

$$P\left(\frac{d}{dt}\right)y = f(t), \quad -\infty < t < \infty$$

$$\int_0^T y(\tau)\,d\tau = \alpha_0, \quad y^{(k)}(T) - y^{(k)}(0) = \alpha_{k+1}, \quad k = 0, 1, \ldots n - 2. \tag{22}$$

Using the formulae (16), (17), we can make an "algebraization" of the problem, thus reducing it to one algebraic equation of 1st degree:

$$P(S)y = f + S\,Q(S), \tag{23}$$

where $P(S)$ and $Q(S)$ are polynomials of S and the degree of $Q(S)$ is less than the degree of $P(S)$.

The formal solution of the above equation has the form

$$y = \frac{1}{P(s)}f + S\frac{Q(s)}{P(s)}. \tag{24}$$

The above representation contains division by $P(S)$ and this is possible if $P(S)$ is not a divisor of 0 in $\mathcal{M}$, i.e. iff $P\left(\frac{2\pi i m}{T}\right) \neq 0$ for each $m \in \mathbb{Z} \setminus \{0\}$. This is the so–called **non-resonance case**.

Main steps of the extended Heaviside algorithm for solving the intermediate problem in the non-resonance case

(1) Finding the roots $\lambda_1, \lambda_2, \ldots, \lambda_n$ of the equation $P(\lambda) = 0$.
(2) Finding out that none of the roots have the form $\frac{2\pi i m}{T}$ with $m \in \mathbb{Z} \setminus \{0\}$.
(3) Finding the polynomial $Q(S)$.
(4) Expanding $\frac{1}{P(S)}$ and $\frac{Q(S)}{P(S)}$ into a sum of partial fractions.
(5) Interpretation of the fractions $w = \frac{1}{P(S)}$ and $v = S\frac{Q(S)}{P(S)}$ as functions.
(6) Representation of the solution in the form $u = w * f + v$.

A comparison with the classical Heaviside algorithm

- for algebraization of the problem the formulae (16), (17) are used now.
- we have here an additional step (step 2);
- new interpretation formulae (such as (18)–(21)) are used on step (5);
- the operation $*$ on step (6) is not the Duhamel convolution; it is the convolution (15).

Resonance case

If the above condition $\lambda \neq \frac{2\pi i n}{T}$, $n \in \mathbb{Z} \setminus \{0\}$ fails for one or more roots of P, we have the so-called **resonance case** and the corresponding roots are called resonance roots.

Let's denote with $n_1, n_2, \ldots, n_p$ all integer numbers, for which $P\left(\dfrac{2\pi i n_k}{T}\right) = 0$, $k = 1, 2, \ldots, p$, and let $C_{n_1, n_2, \ldots, n_p}$ be the subalgebra of $(C(\mathbb{R}), *)$, such that the convolution (15) plays the role of multiplication in it. It was mentioned above that the eigenfunctions of the operator L have the form $\varphi_n(t) = e^{\frac{2\pi i n t}{T}}$, $n \in \mathbb{Z} \setminus \{0\}$. It is shown in [40] that if $f \in C[0, T]$, then

$$f * \{e^{\frac{2\pi i n t}{T}}\} = \chi_n(f)\, e^{\frac{2\pi i n t}{T}}, \quad n = \pm 1, \pm 2, \ldots,$$

where

$$\chi_n(f) = \frac{1}{T}\int_0^1 (e^{\frac{2\pi i n t}{T}} - 1)f(t)dt, \quad n = \pm 1, \pm 2, \ldots, \tag{25}$$

is a complete system of multiplicative functionals. We call them Fourier coefficients of f with respect to $\left\{e^{\frac{2\pi i n t}{T}}\right\}$, $n \in \mathbb{Z} \setminus \{0\}$.

Due to a theorem proven in the above cited paper, at least one of the Fourier coefficients of the function f has to be equal to zero in order this function to be a divisor of 0 in the algebra $(C(\mathbb{R}), *)$. One can prove that this condition is necessary as well.

Let's denote by $\tilde{L}$ the restriction of the operator L to $C_{n_1, n_2, \ldots, n_p}$. Then instead of $Lf = r * f$, for $r(t) = t - \dfrac{T}{2}$ in $[0, T]$, the following presentation in $C_{n_1, n_2, \ldots, n_p}$ will hold: $\tilde{L}f = \tilde{r} * f$, where

$$\tilde{r}(t) = r(t) - \sum_{k=1}^{p} \chi_{n_k}(r)e^{\frac{2\pi i n_k t}{T}} = t - \frac{T}{2} - \sum_{k=1}^{p}\frac{T}{2\pi i n_k t}e^{\frac{2\pi i n_k t}{T}}.$$

We denote by $\mathcal{M}_{n_1, n_2, \ldots, n_p}$ the ring of the convolution fractions of $C_{n_1, n_2, \ldots, n_p}$, whose denominators are nondivisors of 0 of the convolution (15). Denote the algebraic inverse element of $\tilde{L}$ by $\tilde{S}$, i.e. $\tilde{S} = \dfrac{1}{\tilde{L}}$.

Two important theorems, proven by Dimovski and Grozdev, are denoted here by T1 and T2, respectively:

T1. The elements $\tilde{S} - \dfrac{2\pi i n_k}{T}$, $k = 1, 2, \ldots, p$ of the ring $\mathcal{M}_{n_1, n_2, \ldots, n_p}$ are reversible and

$$\frac{1}{(\tilde{S} - \frac{2\pi i n_k}{T})^m} = \left\{\frac{(-1)^{m-1}}{(\frac{2\pi i n_k}{T})^m} + \frac{e^{\frac{2\pi i n_k t}{T}}}{m!}B_m\left(\frac{t}{T}\right)\right\} * \tag{26}$$

for $m = 1, 2, \ldots$, where B_m is the polynomial of Bernoulli of degree m (the sign $*$ means a convolution operator).

T2. If $P(\dfrac{2\pi i n_k}{T}) = 0$ for $k = 1, 2, \ldots, p$ and $P(\dfrac{2\pi i n}{T}) \neq 0$ for all other integer numbers $n \neq 0$, an necessary and sufficient condition for solvability of (22) is:

$$\frac{1}{T} \int_0^1 f(t)(e^{\frac{2\pi i n_k t}{T}} - 1)dt = 0, \; k = 1, 2, \ldots, p, \tag{27}$$

i.e. the Fourier coefficients of $f(t)$ with numbers $n_1, n_2, \ldots, n_p$ to be equal to 0.

Let's formulate now **the algorithm for solving** (22) **in the resonance case**:

(1) As in the non-resonance case, we can make an algebraization of the problem, i.e. we can reduce it to a single equation but in $C_{n_1, n_2, \ldots, n_p}$:

$$P(\tilde{S})\,\tilde{y} = f + Q(\tilde{S}). \tag{28}$$

(2) We consider the homogenous BVP:

$$P\left(\frac{d}{dt}\right)y = 0, \; \int_0^T y(\tau)d\tau = 0, \; y^{(j)}(T) - y^{(j)}(0) = 0, \; j = 0, 1, \ldots n-2.$$

It is equivalent to the equation $P(\tilde{S})\,y = 0$ and its solutions have the form:

$$y = \left\{ C_1 e^{\frac{2\pi i k_1 t}{T}} + \ldots + C_m e^{\frac{2\pi i k_m t}{T}} \right\},$$

where $C_1, C_2 \ldots, C_m$ are constants.

(3) The solution of (22) has the form:

$$y = \tilde{y} + \left\{ C_1 e^{\frac{2\pi i k_1 t}{T}} + \cdots + C_m e^{\frac{2\pi i k_m t}{T}} \right\}, \tag{29}$$

where $\tilde{y}$ is the solution of (28).

The reducing of the problem for obtaining periodic solutions of LODE with constant coefficients to the auxiliary problem deserves special attention. This consideration is omitted here.

General algorithm for obtaining a periodic solution.

(1) Algebraization of the given problem and finding roots $\lambda_1, \lambda_2, \ldots, \lambda_n$ of the equation $P(\lambda) = 0$.

(2) (a) Finding out roots of the form $\dfrac{2\pi i m}{T}$ $(m \in \mathbb{Z} \setminus \{0\})$.

(b) Verifying whether the roots selected in 2 a) satisfy the conditions (27). If for some of the selected roots these conditions are not satisfied, periodic solutions do not exist.

(3) Forming the polynomial $Q(S)$.

(4) Partial fraction decomposition of $\dfrac{1}{P(S)}$ and $\dfrac{Q(S)}{P(S)}$ and separation of the resonance and non-resonance parts.

(5) Interpretation of the fractions $w = \dfrac{1}{P(S)}$ and $v = \dfrac{Q(S)}{P(S)}$ as functions. As was mentioned above, different groups of formulae are used for interpretation of the fractions from the resonance and the non-resonance parts.

(6) Presentation of the solution in the form:

$$u_{nr} = w_1 * f + v_1, \quad u_r = w_2 * f + v_2$$
$$u = u_{nr} + u_r, \tag{30}$$

where w_1 and w_2 are functions, obtained at (step 5) after interpretation of the non-resonance and resonance parts respectively of the partial fraction decomposition of w; v_1 and v_2 are functions obtained at (step 5) after interpretation of the non-resonance and resonance parts respectively of the partial fraction decomposition of v.

The general solution u is the sum of both parts of the solution—the non-resonance part u_{nr} and the resonance part u_r. It is possible, of course, for each of these parts to be equal to zero.

2.8 Program Implementation of the Algorithm

The program implementation of the general algorithm follows the successive steps formulated above. For obtaining both parts of the solution—the non-resonance and the resonance one, the above described extended algorithm of Heaviside is used. Its implementation is in fact a modified implementation of the classical algorithm of Heaviside. The main differences are as follows:

(i) For algebraization of the problem the formula (17) is used now.

(ii) Other interpretation formulae are used here. The main formulae mentioned above are (18)–(21) and (26). For practical applications more formulae based on them are derived (see [102]).

(iii) The operation denoted by $*$ in (30) is the convolution (15); convolution powers are computed as in case of use of Duhamel convolution.

(iv) The verification of condition (27) here is a part of the algorithm.

The implementation of the general algorithm considered above includes finding of periodic solutions of systems of linear ordinary differential equations with constant coefficients as well.

Main part of the interpretation formulae used by our program implementation

For the non-resonance case:

$$\frac{1}{(S-a)^m} = \frac{(-1)^m}{a^m} + \frac{T\,\vartheta_{\{a,-1+m\}}\frac{e^{ta}}{-1+e^{Ta}}}{(-1+m)\,!}$$

$$\frac{1}{S^2+a^2} = \frac{1}{a^2} - \frac{T\,\text{Cos}[a\,t-\frac{a\,T}{2}]\,\text{Csc}[\frac{a\,T}{2}]}{2\,a}$$

$$\frac{S}{S^2+a^2} = \frac{1}{2}\,T\,\text{Csc}[\frac{a\,T}{2}]\,\text{Sin}[a\,t-\frac{a\,T}{2}]$$

$$\frac{c\,S+d}{S^2+a^2} = c\left(\frac{1}{a^2} - \frac{T\,\text{Cos}[a\,t-\frac{a\,T}{2}]\,\text{Csc}[\frac{a\,T}{2}]}{2\,a}\right) + d\left(\frac{1}{2}\,T\,\text{Csc}[\frac{a\,T}{2}]\,\text{Sin}[a\,t-\frac{a\,T}{2}]\right)$$

$$\frac{1}{S^2+p\,S+q} = \frac{1}{q} + \frac{\left(\frac{e^{\frac{1}{2}\left(-p+\sqrt{p^2-4q}\right)t}}{-1+e^{\frac{1}{2}\left(-p+\sqrt{p^2-4q}\right)T}} - \frac{e^{-\frac{1}{2}\left(p+\sqrt{p^2-4q}\right)t}}{-1+e^{-\frac{1}{2}\left(p+\sqrt{p^2-4q}\right)T}}\right)T}{\sqrt{p^2-4q}}, \quad p^2-4\,q \neq 0$$

For $\sqrt{p^2-4\,q} = \delta$, $p+\delta = \alpha$, $-p+\delta = \beta$:

$$\frac{1}{S^2+p\,S+q} = \frac{1}{q} + \frac{T}{\delta}\,\frac{\left(-\frac{e^{-\frac{t\,\alpha}{2}}}{-1+e^{-\frac{T\,\alpha}{2}}} + \frac{e^{\frac{t\,\beta}{2}}}{-1+e^{\frac{T\,\beta}{2}}}\right)}{\sqrt{p^2-4\,q}}$$

$$\frac{S}{S^2+p\,S+q} = \frac{T}{\delta}\left(\frac{e^{-\frac{t\,\alpha}{2}}\,\alpha}{2\left(-1+e^{-\frac{T\,\alpha}{2}}\right)} + \frac{e^{\frac{t\,\beta}{2}}\,\beta}{2\left(-1+e^{\frac{T\,\beta}{2}}\right)}\right)$$

For the resonance case:

$$\frac{1}{S^m} = \frac{T^m}{m\,!}\,B[m,\,\frac{t}{T}]$$

$$\frac{1}{(S-a)^m} = \frac{(-1)^m}{a^m} + \frac{e^{at}}{m\,!}\,T^m\,B[m,\,\frac{t}{T}]$$

$$\frac{1}{S^2+a^2} = \frac{1}{a^2} + \frac{t\,\text{Sin}[a\,t]}{a} - \frac{T\,\text{Sin}[a\,t]}{2\,a}$$

$$\frac{S}{S^2+a^2} = t\,\text{Cos}[a\,t] - \frac{1}{2}\,T\,\text{Cos}[a\,t] + \frac{\text{Sin}[a\,t]}{a}$$

$$\frac{c\,S+d}{S^2+a^2} = c\left(\frac{1}{a^2} + \frac{t\,\text{Sin}[a\,t]}{a} - \frac{T\,\text{Sin}[a\,t]}{2\,a}\right) + d\left(t\,\text{Cos}[a\,t] - \frac{1}{2}\,T\,\text{Cos}[a\,t] + \frac{\text{Sin}[a\,t]}{a}\right)$$

$$\frac{1}{S^2+p\,S+q} = \frac{e^{-\frac{1}{2}t\,(p+\delta)}}{2\,q\,\delta}\left(e^{t\,\delta}\,q\,(2\,t-T) + q\,(-2\,t+T) + 2\,e^{\frac{1}{2}t\,(p+\delta)}\,\delta\right), \quad p^2-4\,q \neq 0$$

2.9 Program Package

The developed program package for Mathematica, provides application of all described operations of the general algorithm for obtaining periodic solutions of LODE with constant coefficients.

The main function of the package is called *DSolveOCP* and its use is similar to the use of the considered above function *DSolveOC*. An additional argument is the period T. Due to the above considerations, the boundary conditions have the form $\int_0^T y(\tau)\,d\tau = \alpha_0$, $y^{(k)}(T) - y^{(k)}(0) = \alpha_{k+1}$, $k = 0, 1, \ldots n - 2$. The use of an option for visualization of the solution, together with the right–hand side function is provided.

Some illustrative examples follow—for the resonance case and for the "mixed" case when the solution is a sum of two parts—resonance and non-resonance ones.

Example for the non-resonance case:

```
<< DSolveOCPpack`
```

```
Example1: {y(t) a² + y"(t) = sin(t), α(1) = 0}; T = 2π
```

```
DSolveOCP[{y''[t] +a^2 y[t] = Sin[t], α[1] = 0}, y[t], t, 2π]
```

$$y[t] \to \frac{1}{2\,a^3\,(-1+a^2)^2\,\pi}\,(2\,a^2\,\text{Cos}[t]\,\text{Sin}[a\,\pi]\,((-1+a^2)\,\pi\,\text{Cos}[a\,\pi] - 2\,a\,\text{Sin}[a\,\pi]) +$$
$$(a\,(1-3\,a^2+2\,a^4)\,\pi + a\,(-1+a^2)\,\pi\,\text{Cos}[2\,a\,\pi] + (1-3\,a^2)\,\text{Sin}[2\,a\,\pi])\,\text{Sin}[t])$$

Example for the resonance case (with option for visualization of the solution):

```
Example 2 : y''[t] + 4 y[t] == Cos[3 t]; the period T = 2 π
```

```
DSolveOCP [{y''[t]+4 y[t]==Cos[3t], α[1]==0}, y[t], t, 2π, Graph->True]
```

```
y[t] -> -1 / 5 Cos[3 t]
```

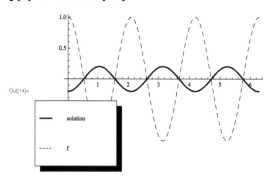

Example for a "mixed" case:

```
Example3: {4 y(t) + 4 y'(t) + y"(t) + y⁽³⁾(t) = cos(5t), α(1) = 0, α(2) = 0}; T = 2π;
```

```
de = y'''[t] + y''[t] + 4 y'[t] + 4 y[t] == Cos[5 t];
```

```
DSolveOCP[{de, α[1] = 0, α[2] = 0}, y[t], t, 2π]
```

$$y[t] \to \frac{1}{546}\,(-\text{Cos}[5\,t] - 5\,\text{Sin}[5\,t])$$

2.10 Mean-Periodic Solutions of LODE with Constant Coefficients

A more general approach to obtaining periodic solutions of LODE with constant coefficients is considered in the papers [20, 23].

Let $P(\lambda) = a_0\lambda^n + a_1\lambda^{n-1} + \cdots + a_{n-1}\lambda + a_n$ be a non-zero polynomial with constant coefficients of degree n and let us consider an ordinary linear differential equation of the form:

$$P\left(\frac{d}{dt}\right) y = f(t), \quad -\infty < t < \infty \tag{31}$$

Let Φ be a linear functional on $C(I\!R)$. We are looking for solutions of (31) which satisfy the relation

$$\Phi\{y(t+\tau)\} = 0 \tag{32}$$

for all $t \in I\!R$, i.e. for mean-periodic solutions of (32) with respect to the functional Φ.

Definition 1 The boundary value problem

$$P\left(\frac{d}{dt}\right) y = f(t), \; \Phi\{y^{(k)}\} = \alpha_k, \; k = 0, 1, \ldots, n-1, \; f \in C(\Delta) \tag{33}$$

is said to be a non-local Cauchy problem, associated with the functional Φ.

Definition 2 Let $\Phi \in [C(I\!R)]^*$ be a given linear functional on the space of the continuous functions on the real line. A function $f \in C(I\!R)$ is said to be mean-periodic [12] with respect to the functional Φ if

$$\Phi_\tau\{f(t+\tau)\} = 0 \text{ for } t \in I\!R.$$

The periodic functions with a period $T > 0$ are mean-periodic with respect to the functional

$$\Phi\{f\} = f(T) - f(0).$$

The antiperiodic functions with an antiperiod $T > 0$, i.e. the functions, satisfying the functional equation $f(T+t) = -f(t)$, are mean-periodic functions with respect to the linear functional

$$\Phi\{f\} = \frac{1}{2}\{f(0) + f(T)\}.$$

Further considerations related to the mean-periodic functions and the use of the Mikusiński type operational calculus of Dimovski (and the Heaviside algorithm with some modifications) for obtaining mean-periodic solutions of LODE with constant coefficients are presented in [20, 23].

For deriving some formulae and for practical application of some algorithms the CAS Mathematica is used.

2.11 Advantages of the Presented Approach for Obtaining Periodic Solutions of LODE with Constant Coefficients

The presented approach is more efficient than those in the above mentioned books of Kaplan [48], Rosenvasser [76] and Lurie [55].

The function $DSolve$ of Mathematica remains as undetermined the constants appearing in the solution in the resonance case.

In the classical methods for finding periodic solutions, originally the general solution is found and after that the periodicity conditions are used for determining the unknown constants in it. In the above guggested approach these conditions are taken into account at the level of algebraization of the problem.

In case of use the Laplace transformation for finding periodic solutions, the existence of Laplace transform of the right-hand side of the equation is needed.

The presented approach is more efficient (especially in the resonance cases) than the well known (and published) approaches.

A more general approach for the case of mean-periodic solutions is suggested.

All proposed algorithms are convenient for use in the program environment of a CAS.

2.12 Operational Calculus Approach for Solving Boundary Value Problems for Some Partial Differential Equations

If we are interested in application of the OC approach to PDEs, the Heaviside-Mikusiński's OC should be extended to multivariate functions. Such an extension for the Laplace transformation is proposed by Ditkin and Prudnikov in [25]. The principles of the applications of multivariate OC for solution of Cauchy problems for linear PDEs with constant coefficients are developed in Gutterman [42].

Another approach is considered here—a way of common use of a combination of two classical methods—the Fourier method [35] and the Duhamel principle [32], in the frames of a two-dimensional operational calculus, suggested by Dimovski [13]. It gives closed form solutions of boundary value problems for some equations of the mathematical physics.

As it is shown below, an extension of the Duhamel principle to the space variables enables a closed form solution of various boundary value problems for some partial differential equations to be obtained. To this end, as in the classical Duhamel principle, one special solution of the same problem, but for a simple and special choice of the initial value function should be obtained, using the Fourier method.

Then the solution for arbitrary initial value function can be obtained in the form of a non-classical convolution, using a two-dimensional operational calculus.

Duhamel formulated his principle in 1830 (see [32]). Due to this principle the solution of the boundary value problem

$$\frac{\partial u}{\partial t} = \frac{\partial^2 u}{\partial x^2}, \quad u(0, t) = 0, \quad u(1, t) = \varphi(t), \quad u(x, 0) = 0$$

can be obtained for arbitrary $\varphi(t)$, if a solution $U(x, t)$ of the same problem but for special choice of $\varphi(t)$, namely for $\varphi(t) \equiv 1$ is available. Then the general solution has the form:

$$u(x, t) = \frac{\partial}{\partial t} \int_0^t U(x, t - \tau)\varphi(\tau)d\tau \qquad (34)$$

for $0 \leq x \leq 1, \; 0 \leq t$.

The special solution can be obtained using the Fourier method – it has the form:

$$U(x, t) = x + \frac{2}{\pi} \sum_{n=1}^{\infty} \frac{(-1)^n}{n} e^{-n^2\pi^2 t} \sin n\pi x. \qquad (35)$$

An extension of the Duhamel principle for boundary value problems with non–homogenous initial conditions when the boundary value conditions are homogenous, will be considered.

2.13   A Two-Dimensional Operational Calculus for Boundary Value Problems

The following two types of convolutions, intended for one-variate OC, will be assembled to multivariate convolutions.

A. Convolutions for the differentiation operator
The basic BVP for the differentiation operator d/dt in the space $C\,[0, \infty)$ of the continuous functions $f(t), \; 0 \leq t < \infty$ is determined by an arbitrary linear functional χ on $C\,[0, \infty)$. It looks so: $y' = f(t), \; \chi(y) = 0$.

In order the solution y to exist it is necessary to assume $\chi\{1\} \neq 0$. For the simplicity sake, let $\chi\{1\} = 1$. Then the solution $y = lf(t)$ could be named a generalized integration operator. Evidently

$$l\, f(t) = \int_0^t f(\tau)\,d\tau - \chi_\tau \left\{ \int_0^t f(\tau)\,d\tau \right\}.$$

In [13] it is shown that the operation

$$(f * g)(t) = \chi_\tau \left\{ \int_\tau^t f(t - \sigma + \tau) \, g(\sigma) \, d\sigma \right\} \tag{36}$$

is a bilinear, commutative and associative operation such that

$$lf = \{1\} * f.$$

An one-variate operational calculus based on (36) is developed in [14].

B. Convolutions for the square of the differentiation operator

Let us consider the space $C[0, a]$ of the continuous functions on $[0, a]$.
The simplest non-local BVP for d^2/dx^2 in $C[0, a]$ is given by

$$y'' = f(x), \quad y(0) = 0, \quad \Phi\{y\} = 0$$

where Φ is a linear functional on $C^1[0, a]$. In order it to have a solution it is necessary to assume $\Phi\{x\} \neq 0$. For the simplicity sake it is assumed that $\Phi\{x\} = 1$. The solution $y = Lf(x)$ has the explicit form

$$Lf(x) = \int_0^x (x - \xi) \, f(\xi) \, d\xi - x \, \Phi_\xi \left\{ \int_0^\xi (\xi - \eta) \, f(\eta) \, d\eta \right\}.$$

In [13] it is proved that the operation

$$(f * g)(x) = -\frac{1}{2} \Phi_\xi \left\{ \int_0^\xi h(x, \eta) \, d\eta \right\} \tag{37}$$

where

$$h(x, \eta) = \int_x^\eta f(\eta + x - \zeta) \, g(\zeta) \, d\zeta - \int_{-x}^\eta f(|\eta - x - \zeta|) \, g(|\zeta|) \, \text{sgn}(\eta - x - \zeta) \, \zeta \, d\zeta$$

is a bilinear, commutative and associative operation and $Lf(x) = \{x\} * f$.
If the operations (36) and (37) are considered simultaneously, it can be written:

$$(f \overset{(t)}{*} g)(t) = \chi_\tau \left\{ \int_\tau^t f(t + \tau - \sigma) \, g(\sigma) \, d\sigma \right\},$$

$$(f \overset{(x)}{*} g)(x) = -\frac{1}{2} \Phi_\xi \left\{ \int_0^\xi h(x, \eta) \, d\eta \right\}.$$

2.14 Two-Dimensional Convolutions. Operational Calculi for l and L

The following idea of a multivariate operational calculus is used. Let $u = \{u(x, t)\}$ and $v = \{v(x, t)\}$ be arbitrary functions from the space $C = C([0, \infty) \times [0, a])$. We introduce a bilinear, commutative and associative operation $u * v$ in C such that the operators l and L are multipliers of the convolution algebra $(C, *)$ of the form $lu = \{1\} \overset{t}{*} u$ and $Lu = \{x\} \overset{x}{*} u$.

Theorem. *The operation*

$$\{u(x, t)\} * \{v(x, t)\} = -\frac{1}{2} \tilde{\Phi}_\xi \chi_\tau \{h(x, t; \xi, \tau)\} \tag{38}$$

with

$$h(x, t; \xi, \tau) = \int_\xi^x \int_\tau^t u(x + \xi - \eta, t + \tau - \sigma) v(\eta, \sigma) \, d\sigma d\eta -$$

$$- \int_{-\xi}^x \int_\tau^t u(|x - \xi - \eta|, t + \tau - \sigma) v(|\eta|, \sigma) \operatorname{sgn}[(x - \xi - \eta)\eta] \, d\sigma \, d\eta$$

and with the functional $\tilde{\Phi}_\xi = \Phi \circ \int_0^\xi$, *is a convolution of the operators* L *and* l *in* $C(\Delta)$ *(where* $\Delta = (0, a] \times [0, \infty)$*), for which* $L l u = \{x\} * u$. *The operators* $lu = \{1\} \overset{(t)}{*} u(x, t)$ *and* $Lu = \{x\} \overset{(x)}{*} u(x, t)$ *are multipliers of this operation.*

This theorem gives us an operation $(u * v)(x, t)$ in $C(\Delta)$, which is a convolution of each of both operators l and L.

The basic formulae of the operational calculus for l and L are

$$\frac{\partial u}{\partial t} = su - [\chi_\tau \{u(x, \tau)\}]_t \text{ and } \frac{\partial^2 u}{\partial x^2} = Su - [\Phi_\xi \{u(\xi, t)\}]_x \tag{39}$$

where the indices t and x mean that the corresponding functions of t and x, are considered as "partial" numerical operators.

2.15 Duhamel-Type Representations of Solutions of BVPs

Using multivariate operational calculi, algebraization of each BVP can be made. An explicit Duhamel-type representation of the solution is obtained, using one special solution satisfying simple boundary value conditions. The general solution is obtained as a multivariate convolution product of this solution with the right hand side function or with a given boundary value function (see [19, 102]). This representation can de used successfully for numerical computation of the solution. Representation formulae for the solutions of local and non-local BVPs for the heat equation, the wave equation and the equation of a supported beam are considered in [19, 102].

One of the problems solved by our approach, is the following non-local BVP for the string equation (called also "wave equation"):

$$u_{tt} = u_{xx}, \quad 0 < x < 1, \ 0 < t < \infty$$
$$u(0, t) = 0, \quad \int_0^1 u(\xi, t)d\xi = 0 \tag{40}$$
$$u(x, 0) = f(x), \quad u_t(x, 0) = g(x)$$

The second boundary condition is integral, i.e. nonlocal. Such (nonlocal) conditions arise mainly when the data on the boundary can not be determined directly.

The paper [77] is devoted to the numerical solution of a hyperbolic partial differential equation with a combination of classical and integral boundary conditions. The works of some other authors are cited in the paper.

S. Beilin (see [1]) considers problems of this type and formulates conditions for the existence and uniqueness of the solution.

Prof. Ivan Dimovski suggested an algebraic approach for solving linear nonlocal boundary value problems (see [13, 15]). In his common papers [18, 19] with the author of this survey, an explicit formula, convenient for practical applications was derived.

Case 1. $f(x) \equiv 0, \ g(x) \neq 0$.

The special solution $\Omega(x, t)$ for $g(x) \equiv x^3/6 - x/12$ and $f(x) \equiv 0$ was obtained using the Fourier method [17]:

$$\Omega(x, t) = \sum_{n=1}^{\infty} \left\{ \frac{x \cos(2n\pi x) \sin(2n\pi t)}{4n^3\pi^3} \right.$$

$$\left. + \left(\frac{t \cos(2n\pi t)}{4n^3\pi^3} - \frac{3\sin(2n\pi t)}{8n^4\pi^4} \right) \sin(2n\pi x) \right\}.$$

The solution $u(x, t)$ has the form (see [102])

$$u(x, t) = \frac{\partial^2}{\partial x^2} \left\{ \Omega(x, t) \overset{(x)}{*} g(x) \right\}$$

and after its simplification the following representation is derived:

$$u(x, t) = -2 \int_0^x \Omega_x(x - \xi, t)g'(\xi)d\xi \tag{41}$$

$$- \int_x^1 \Omega_x(1 + x - \xi, t)g'(\xi)d\xi + \int_{-x}^1 \Omega_x(1 - x - \xi, t)g'(|\xi|)d\xi.$$

Using this presentation, the problem was solved for

$$g(x) = 2\pi x \cos 2\pi x + \frac{3}{2} \sin 2\pi x.$$

The function $g(x)$ and the solution can be visualized (as it is shown by the example below).

In a similar way explicit formulae of the solutions of linear local and nonlocal boundary value problems for a free supported beam are derived.

All derived presentations of the solutions of BVPs for equations of mathematical physics are considered in more details in [24, 102].

2.16 Program Packages for the Considered BVPs

The representations of the solutions of BVPs considered above, are convenient for numerical computation of an arbitrary number of values of the solutions. A visualization of each solution can be made as well.

For practical application of these capabilities 3 program packages for the computer algebra system *Mathematica* were developed (for the 3 types of the considered equations). In each of these packages functions for solving local and nonlocal BVPs are implemented.

After loading the package for the srting equation, for example, the following call to the function for the considered above Case 1: $f(x) \equiv 0$, $g(x) \neq 0$ has to be performed:

DSolveOCStringNonl1[$g, u, x, t, \{x, 0, 1, 0.1\}, \{t, 0, 1, 0.1\}, 5$].

As a result, a table of numerical values of the solution in the given intervals is returned, followed by visualization of the boundary function and the solution (Figs. 1, 2, respectively); during the process of computation the visualization is shown right after the table of numerical solutions.

Table of numerical values of the solution:

```
{{0., 0., 0., 0., 0., 0., 0., 0., 0., 0., 0.},
 {0., 0.122599, 0.157754, 0.0669366, -0.115165, -0.293893, -0.360363,
  -0.248572, 0.0238813, 0.352929, 0.587785}, {0., 0.157754, 0.189536, 0.0425892,
  -0.226956, -0.475528, -0.542465, -0.336482, 0.104357, 0.611667, 0.951057},
 {0., 0.0669366, 0.0425892, -0.104357, -0.317774, -0.475528, -0.451647,
  -0.189536, 0.251303, 0.702484, 0.951057}, {0., -0.115165, -0.226956, -0.317774,
  -0.352929, -0.293893, -0.122599, 0.136138, 0.408592, 0.590693, 0.587785},
 {0., -0.293893, -0.475528, -0.475528, -0.293893, 3.33067×10⁻¹⁶, 0.293893, 0.475528,
  0.475528, 0.293893, 2.77556×10⁻¹⁵}, {0., -0.360363, -0.542465, -0.451647,
  -0.122599, 0.293893, 0.598127, 0.633283, 0.360829, -0.115165, -0.587785},
 {0., -0.248572, -0.336482, -0.189536, 0.136138, 0.475528, 0.633283, 0.483428,
  0.0425892, -0.520849, -0.951057}, {0., 0.0238813, 0.104357, 0.251303,
  0.408592, 0.475528, 0.360829, 0.0425892, -0.39825, -0.793302, -0.951057},
 {0., 0.352929, 0.611667, 0.702484, 0.590693, 0.293893, -0.115165, -0.520849,
  -0.793302, -0.828458, -0.587785}, {0., 0.587785, 0.951057, 0.951057, 0.587785,
  2.22045×10⁻¹⁶, -0.587785, -0.951057, -0.951057, -0.587785, -5.28466×10⁻¹⁴}}}
```

Fig. 1 Function $g(x)$

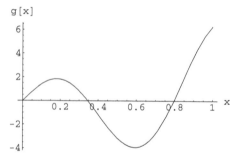

Fig. 2 Solution $u(x, t)$

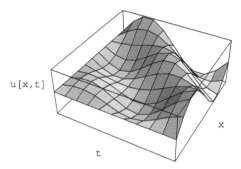

In the illustrative example a nonlocal BVP for the string equation, Case 1, is solved for $g(x) = 2\pi x \cos 2\pi x + \frac{3}{2} \sin 2\pi x$ and for degree of truncation equal to 5, of the series representing the partial solution $\Omega(x, t)$ (the last parameter of the call stands for that).

For the cases when the exact solution is known, a comparison of its values in a set of points with the values, computed by the respective function of the package in the same points could be easily made. A number of experiments were made and errors of computation of order 10^{-9}–10^{-13} were found.

As it is known, Mathematica system is not able to solve nonlocal BVPs.

2.17 Resonance Vibrations of String and Beam Under Integral Boundary Conditions

As an application of the presented approach to finding exact solutions of BVPs, a study of a real problem is considered.

The Tacoma Narrows bridge collapse on November 7, 1940 still has not obtained unanimous explanation. The physical phenomenon resonance is often pointed out as a possible explanation of the bridge failure. Many authors of studies of this disaster reject such an explanation.

If the bridge is considered as linear elastic system subjected to local boundary value conditions, it is not possible the resonance phenomenon to occur. Nevertheless, considering such a system as subjected to nonlocal boundary value condition of energetic type (integral boundary-value condition), there always occur resonances on all frequencies.

Let's consider again the linear nonlocal boundary value problems for the equations of a vibrating string and for the equations of a free supported beam, and their Duhamel—type representations of the solutions, derived by the presented approach (see [24]).

$$\frac{\partial^2 u}{\partial t^2} = \frac{\partial^2 u}{\partial x^2}$$

$$\frac{\partial^2 u}{\partial t^2} = -\frac{\partial^4 u}{\partial x^4}$$

$$0 < x < 1, 0 < t < \infty \qquad\qquad 0 < x < 1, 0 < t < \infty$$

$$u(0, t) = 0, \ \int_0^1 u(\xi, t)d\xi = 0 \qquad \int_0^1 u(\xi, t)d\xi = 0,$$

$$u(x, 0) = f(x), \qquad\qquad u_x(1, t) - u_x(0, t) = 0$$

$$u_t(x, 0) = g(x), \qquad\qquad u(0, t) = 0, u_{xx}(0, t) = 0$$

$$\qquad\qquad\qquad\qquad u(x, 0) = f(x), \ u_t(x, 0) = g(x).$$

For simplicity sake the cases when $f(x) \equiv 0$, $g(x) \neq 0$ for the string equation (the considered above Case 1) and when $f(x) \neq 0$, $g(x) \equiv 0$ for the beam equation are observed. The presentations of their solutions are as follows:

$$u(x, t) = -2 \int_0^x \Omega_x(x - \xi, t)g'(\xi)d\xi - \int_x^1 \Omega_x(1 + x - \xi, t)g'(\xi)d\xi +$$

$$+ \int_{-x}^1 \Omega_x(1 - x - \xi, t)g'(|\xi|)d\xi,$$

for the string equation (see the formula (41)), where

$$\Omega(x, t) = \sum_{n=1}^{\infty} \left\{ \frac{x \cos(2n\pi x) \sin(2n\pi t)}{4n^3 \pi^3} \right.$$

$$+ \left(\frac{t \cos(2n\pi t)}{4n^3 \pi^3} - \frac{3 \sin(2n\pi t)}{8n^4 \pi^4} \right) \sin(2n\pi x) \right\}$$

and

$$u(x, t) = -2 \int_0^x \tilde{\Omega}_{xx}(x - \xi, t) f^{iv}(\xi) d\xi + \int_x^1 \tilde{\Omega}_{xx}(1 + x - \xi, t) f^{iv}(\xi) d\xi$$

$$- \int_{-x}^1 \tilde{\Omega}_{xx}(1 - x - \xi, t) f^{iv}(|\xi|) \operatorname{sign}(\xi) d\xi + 2 f^{iv}(\xi)(\tilde{\Omega}_x(0, t) -$$

$$\tilde{\Omega}_x(1, t)) + f(x),$$

where

$$\tilde{\Omega}_{xx}(x, t) = \left\{ -\frac{x \cos(2 n \pi x) \sin(2 n^2 \pi^2 t)^2}{4 n^4 \pi^4} \right.$$

$$- \frac{\left(4 n^2 \pi^2 t \cos(2 n^2 \pi^2 t) - 3 \sin(2 n^2 \pi^2 t)\right) \sin(2 n^2 \pi^2 t) \sin(2 \pi x)}{4 n^7 \pi^5}$$

$$\left. - \frac{\sin(2 n^2 \pi^2 t)^2 \sin(2 n \pi x)}{4 n^5 \pi^5} \right\}$$

for the beam equation (see [24, 102]). For the purposes of simplification the presentation

$$\tilde{\Omega}(x, t) = \int_0^t \Omega_x(x, \tau) d\tau$$

is introduced.

Analizing all components of these presentations, a conclusion can be made that both formulae have the following general form:

$$u(x, t) = u_1(x, t) + t u_2(x, t),$$

where $u_1(x, t)$ and $u_2(x, t)$ are periodic with respect to t and hence bounded for fixed x. The resonance effect is due to the aperiodic term $t u_2(x, t)$. When its absolute value exceeds some fixed quantity, a demolition occurs.

Similar considerations are presented in more details in [22]. It is shown there that under a quite simple integral boundary condition, resonance inevitably occurs even in the case of absence of external forces. Conservation of the integral of displacement in time has a clear physical meaning: conservation of the bridge potential energy in the course of vibration.

The energy condition may be satisfied for a part of a bridge rather than for the entire bridge. It seems likely that it held for only one half of the Tacoma bridge (another half was not broken). It can be assumed that the role of wind in appearance of resonance vibration of bridges is to support the energy integral boundary condition at a strictly constant level.

By this phenomenon, an explanation of the resonance vibration of the Tacoma bridge can be given. This explanation is also applicable to the resonance vibration observed at the Volgograd bridge in May 2010, so that it was even closed for traffic for some time.

However, the prevention of such resonance vibration is not discussed in [22]. The design issues need special investigation.

2.18 Concluding Remarks

The considered operational calculus approach is convenient for obtaining closed form solutions and numerical solutions of initial and boundary value problems for some types of differential equationst.

The complete implementations of the Heaviside algorithm and the modified Heaviside algorithm are very useful tools for applications.

The suggested approach for finding periodic solutions od LODEs with constant coefficients has important features.

The Duhamel–type representations of the solutions of BVPs, mainly for the string and beam equations, deserve special attention.

The five developed Mathematica program packages for solving initial and boundary value problems by the operational calculus approach, were experimented and some positive conclusions about their features were made.

The choice of the computer algebra system Mathematica providing efficient symbolic and numerical computations, as well as convenient program language, was important precondition for achieving the expected results.

3 Notes on Some Current and Future Aspects of the Computer Algebra Applications in IMI

Further research and application of some of the presented results are under way. They are connected, in particular, with solving other boundary value problems (not solved in the previous developments), improvement of some algorithms, considering real phenomena, solving problems in other areas, etc. The conception of Applied CA Microenvironments [100] will be used as well.

Even though the theories of operational calculus and integral transforms are centuries old, their use in the fields of mathematics, physics, in electrical and radio engineering are constantly developing.

Some experimental application of the CAS Mathematica in research on a telecommunication problem was made. A problem related to the quality service prediction in overall telecommunication systems is presented in [72, 78]. As a part of the research a system of nonlinear algebraic equations has to be solved. Experiments with the use of CAS Mathematica are presented in [73]. Further developments of the accepted approach and some extended applications are to be made.

Some new applications and developments are under discussion.

The research and developments outlined in this paper, could be considered as a very small example of the variety of applications of CA. The CAS offers tools for extensive computations in science, in technical computing, etc.

CAS are more and more used in scientific computations, in mathematics education, in many industrial applications. By the aid of CAS the capacity of mathematical problem solving has been decisively improved.

The importance of the application aspect of the CAS is taken into account on the level of design and development of the systems. The real applications often pose research topics related to the design, development and improvement of the systems. New areas and new ways of use of CAS will be discovered, probably with broad impact.

Acknowledgements The author is very grateful to Prof. Vesselin Drensky for the support of this paper and for the useful information about the results in the field of algebra in IMI, presented in the paper.

The author expresses her gratitude to Prof.Ivan Dimovski for proposing the subject of our joint research and for the useful discussions.

This work is partially supported by the bilateral project DNTS-Russia 02/7, funded by Bulgarian National Science Fund.

This work is also partially supported by joint research grant of the project "Symbolic-Numerical Decision Methods for Algebraic Systems of Equations in Perspective Telecommunication Tasks" of the Institute of Mathematics and Informatics of the Bulgarian Academy of Sciences, Sofia, Bulgaria, and of the Joint Institute for Nuclear Research, Dubna, Russia.

References

1. Beilin, S.: Existence of solutions for one-dimensional wave equation with nonlocal conditions. El. J. Diff. Equ. **76**, 1–8 (2001)
2. Benanti, F., Boumova, S., Drensky, V., Genov, G.K., Koev, P.: Computing with rational symmetric functions and applications to invariant theory and PI-algebras. Serdica Math. J. **38**(1–3), 137–188 (2012)
3. Benanti, F., Demmel, J., Drensky, V., Koev, P.: Computational approach to polynomial identities of matrices—a survey. In: Giambruno, A., Regev, A., Zaicev, M. (eds.) Ring Theory: Polynomial Identities and Combinatorial Methods, Proceedings of the Conference in Pantelleria. Lecture Notes in Pure and Applied Mathematics, vol. 235, pp. 141–178, Dekker (2003)
4. Boumova, S., Drensky, V., Kostadinov, B.: A Diophantine transport problem from 2016 and its possible solution from 1903. In: Mathematics and Education in Mathematics Proceedings of the Forty-ninth Spring Conference of the Union of Bulgarian Mathematicians, pp. 89–113 (2020) http://www.math.bas.bg/smb
5. Brankov, G., Spiridonova, M., Ishtev, K., Philipov, Ph.: Modelling of building structures behavior under seismic loading with use of an algebraic manipulation system. In: Proceedings of the 7th European Conference on Earthquake Engineering, pp. 585–592, Athens(1982)
6. Buchberger, B., Collins, G., Loos, R.: Computer Algebra. Symbolic and Algebraic Computation. In: Computing Supplementum, vol. 4, Springer (1982)

7. Calmet, J., Campbel, J.: A Perspective On Symbolic Mathematical Computing and Artificial Intelligence (1997). https://www.researchgate.net/publication/36452068
8. Calmet, J., Tjandra, I.: On the Design of an artificial intelligence environment for computer algebra systems. In: Shirkov, D., Rostovtsev, V., Gerdt, V. (eds.) Computer Algebra in Physical Research, pp. 4–8. World Scientific, Singapore (1991)
9. Cohen, G.: Computer Algebra and Symbolic Computation: Mathematical Methods. CRC Press (2003)
10. Chan, C., Drensky, V., Edelman, A., Kan, R., Koev, P.: A linear-time algorithm for evaluating series of Schur functions. J. Algebraic Comb. **50**(2), 127–141 (2019)
11. Daskalova, M., Tomev, M., Spiridonova, M.: Symbolic and numerical computations for mathematical geodesy problem solving by MGCOMP. In: Proceedings of the First National Conference Informatics'94, pp. 75–82 Sofia (1994)
12. Delsarte, J.: Sur une Extension de la Formule de Taylor, J. de Math. 9, V. 17, 213–231 (1938)
13. Dimovski, I.H.: Convolutional Calculus. Kluwer Acad. Publishers, Dordrecht (1990)
14. Dimovski, I.H.: Non-local operational calculi. Proc. Steklov Inst. Math. **3**, 53–65 (1995)
15. Dimovski, I.: Nonlocal boundary value problems. In: Mathematics and Education in Mathematics, Proceedings of the 38th Spring Conference of UBM, pp. 31–40, Sofia (2009)
16. Dimovski, I., Spiridonova, M.: Numerical solution of boundary value problems for the heat and related equations. In: Gerdt, V.P. (ed.) Computer Algebra and its Application to Physics, Proceedings of International Workshop, Dubna, 28–30 June, 2001, pp. 32–42. JINR, Dubna (2002)
17. Dimovski I., Spiridonova, M.: The fourier method and the problem of its computer implementation. Mathematics and mathematics education. In: Proceedings of the 32nd Spring Conference of the Union of Bulgarian Mathematicians, pp. 117–127, Sofia (2003)
18. Dimovski, I., Spiridonova, M.: Computer implementation of solutions of BVP for finite vibrating systems. Mathe. Balkanica, Fasc. **18**(3–4), 277–285 (2004)
19. Dimovski, I., Spiridonova, M.: Computational approach to nonlocal boundary value problems by multivariate operational calculus. Math. Sci. Res. J. **9**(12), 315–329 (2005)
20. Dimovski, I., Spiridonova, M.: Operational calculus approach to nonlocal Cauchy problems. J. Math. Comput. Sci. **4**(2–3), 243–258 (2010)
21. Dimovski, M.: Spiridonova: an implementation of the heaviside algorithm. J. Phys. Part. Nucl. Lett. Pleiades Publishing Ltd. **8**(5), 491–493 (2011)
22. Dimovski, I.H., Spiridonova, M.N.: Construction of nonlocal linear vibration models using a computer algebra system. J. Program. Comput. Softw. **37**(2), 71–77, Pleiades Publication Ltd. (2011) Original Russian Text publications in Journal of Programmirovanie, pp. 20–28, vol. 37, no. 2 (2011)
23. Dimovski, I., Spiridonova, M.: Operational calculi for nonlocal Cauchy problems in resonance cases. LNCS **8372**, 83–95 (2014)
24. Dimovski, I., Spiridonova, M.: Exact solutions of boundary-value problems. Int. J. Appl. Math. **31**(3), 307–323 (2018)
25. Ditkin, V.A., Prudnikov, A.P.: Operational Calculus of Two Variables and Its Applications. GI Fiz.-mat. lit, Moscow (1958) (in Russian)
26. Drensky, V., Popov, A.P.: Prime varieties of associative algebras. In: Mathematics and Education in Mathematics Proceedings of the 14th Spring Conference of the Union of Bulgar. Mathematicians, pp. 35–52, Sofia (1987)
27. Drensky, V.: Computational techniques for PI-algebras. In: Banach Center Publications 26, Topics in Algebra, Part 1: Rings and Representations of Algebras, pp. 17–44. Polish Scientific Publishers, Warsaw (1990)
28. Drensky, V.: Polynomial identities for 2×2 matrices. Acta Appl. Math. **21**, 137–161 (1990)
29. Drensky, V.: Commutative and noncommutative invariant theory. In: Mathematics and Mathematics Education, Proceedings of the 24th Spring Conference of the Union of Bulgarian Mathematicians, Svishtov, 4–7 April 1995, pp. 14–50, Sofia (1995)
30. Drensky, V.: Free Algebras and PI-Algebras. Springer, Singapore (2000)

31. Drensky, V.: Computing with matrix invariants. Math. Balk., New Ser. **21**(1–2), 101–132 (2007)
32. Duhamel, J.M.-C.: Memoire sur le Métode generale relative au mouvement de la chaleur dans les corps solides plongés dans les mileaux dont la temperature varie avec le temps. J. de lÉc. Polyt. **14**, 20–77 (1830)
33. Epstein, M., Magliveras, S.S., Nikolova-Popova, D.: The covering numbers of A_9 and A_{11}. J. Comb. Math. Comb. Comput. **101**, 23–36 (2017)
34. Farlow S.: Partial Differential Equations for Scientists and Engineers. John Wiley and Sons, Inc. (1982)
35. Fourier, J.: The Analytical Theory of Heat. Dover, N.Y. (1955)
36. GeoGebra's Official website. https://www.geogebra.org/
37. Gerov, A., Kapitonova, Yu., Spiridonova, M., Tomov, V.: Intelligent computer algebra systems. In: Knowledge Representation in Man-machine and Robotic Systems, Vol. C: Applied Knowledge Based Man-machine Systems, Comp. Center of the Acad. of Sci. of USSR, VINITI, pp. 112–136. Moscow (in Russian) (1984)
38. Gerov, A.: Special purpose computer algebra system for matrix manipulation. Ph.D. thesis (in Bulgarian) (1987)
39. Grozdev, S.: Convolutional approach to abstract differential equations in the "special" case. PhD thesis (in Bulgarian), Sofia (1980)
40. Grozdev, S., Dimovski, I.: Bernoullian operational calculus, mathematics and mathematics education. In: Pcoceedings of the 9th Spring Conference of the UBS, Sunny Beach, 3–6 April 3–6, pp. 30–36, Sofia, BAS (in Bulgarian) (1980)
41. Grozdev, S.: A convolutional approach to initial value problems for equations with right invertible operators. C. R. Acad. Sci. Bulg. **33**(1), 35–38 (1980)
42. Gutterman, M.: An operational method in partial differential equations. SIAM J. Appl. Math. **17**(2), 468–493 (1969)
43. Hearn A.C.: REDUCE User's Manual, Version 3.8 (2004). http://www.reduce-algebra.com/reduce38-docs/reduce.pdf
44. Heaviside, O.: Electromagnetic Theory. London (1899)
45. Kahrimanian, H.G., Analytic differentiation by a digital computer. Master's Thesis, Temple University, Philadelphia, Pennsylvania (1953)
46. Kaltofen, E.: Computer algebra algorithms. Annu. Rev. Comput. Sci. **2**, 91–118 (1987)
47. Kapitonova, Y., Letichevsky, A., L'vov, M., Volkov, V.: Tools for Solving Problems in the Scope of Algebraic Programming. LNCS, 958, pp. 30–47. Springer (1995)
48. Kaplan, W.: Operational Methods for Linear Systems. Addison-Wesley Series in Mathematics, USA (1962)
49. Kappe, L.-C., Nikolova-Popova, D., Swartz, E.: On the covering number of small symmetric groups and some sporadic simple groups. Groups Complex. Cryptol. **8**(2), 135–154 (2016)
50. Kazasov, C.: Laplace Transformations in REDUCE 3. In: Proceedings of Eurocal '87, Lecture Notes in Computer Science, pp. 132–133. Springer (1987)
51. Kenderov, P., Chehlarova, T., Sendova, E.: A virtual mathematics laboratory in support of educating educators in inquiry-based style. In: Maa, K. (ed.) Conference Proceedings in Educating the Educators, 15–16 December 2014, pp. 167–176, Essen, Germany(2015)
52. Kutzler, B.: Introduction to DERIVE for Windows. Austria (1997)
53. Kuyumdjieva, B., Sendova, E., Spiridonova, M.: How to use graphics calculators in mathematics lessons. In: Proceedings of the 36th Spring Conference of the Union of Bulgaria Mathematicians, pp. 133–135 (2007)
54. Laue, R., Nikolova-Popova, D.: Design of designs. J. Comb. Des. **20**(1–2), 1–22 (2012)
55. Lurie, A.I.: Operational Calculus and Its Application to Problems of Mechanics. Publishing House of Technological Theory, Moscow, Leningrad (1950) (in Russian)
56. McCarthy, J.: Recursive functions of symbolic expressions and their computation by machine. Commun. ACM, April, Part I (1960)
57. Menabrea, L.F.: Sketch of the analytical engine invented by charles babbage, Bibliothque Universelle de Genve. October, **82**, 225–251 (1842). https://www.fourmilab.ch/babbage/sketch.html

58. Mikusinski, J.: Operational Calculus. Pergamon Press, Oxford (1965)
59. MuPAD System. https://www.mathworks.com/products/symbolic.html
60. National Science Fund Project I-1002/2000 "Computer Algebra Tools for Mathematics Education", 2000–2003. http://www.math.bas.bg/artint/mspirid/casined.htm
61. Nikolova, D.B.: Groups with a 2-variable commutator identity. Ph.D. thesis, Sofia (1983)
62. Nikolova, D.B.: Computation of a commutator identity in alternating groups on an electronic computer (in Russian). Serdica Bulg. Math. Publ. **10**, 28–40 (1984)
63. Nikolova, D.B.: Presentations of groups-computational methods. In: Zhong, L., Shum, K.P., Yang, C.C., Le, Y. (eds.) Proceedings of Asian Mathematical Conference, pp. 341–348, Hong Kong, 1990, World Scientific (1992)
64. Nikolova, D.B., Robertson, E.F.: One more infinite Fibonacci group. C.R. Acad. Bulg. Sci. **46**(3), 13–15 (1993)
65. Nikolova, D.B.: The Fibonacci groups—four years later. In: Shum, K.P., et al., (eds.) Semigroups Papers from the International Conference on Semigroups and its Related Topics, Kunming, China, 18–23 August 1995, pp. 251–255. Springer, Singapore (1998)
66. Nisheva, M.: Computer algebra system for continued fractions manipulation (in Bulgarian). PhD thesis (1987)
67. Nisheva-Pavlova, M.: A knowledge-based approach to building computer algebra systems. In: Proceedings of JCKBSE'96, Sozopol, pp. 222–225 (1996)
68. Nisheva-Pavlova, M.: KAM—a knowledge-based tool for developing computer algebra systems. In: Annuaire de l'Universit de Sofia "St. Kliment Ohridski", Facult de Mathmatiques et Informatique, vol. 90, no. 2, pp. 165–176 (1996)
69. Nisheva-Pavlova, M.: Knowledge representation and problem solving in the intelligent computer algebra system STRAMS. In: Annuaire de l'Universit de Sofia "St. Kliment Ohridski", Facult de Mathmatiques et Informatique, vol. 91, no. 2, pp. 193–202 (1997)
70. Nisheva-Pavlova, M.: An intelligent computer algebra system and its applicability in mathematics education. Int. J. Comput. Algebra Math. Educ. **6**(1), 3–16 (1999)
71. Nolan, J.F.: Analytic differentiation on a digital computer. Master's thesis, Mathematics Department, Massachusetts Institute of Technology, Cambridge, Massachusetts (1953)
72. Poryazov, S., Saranova, E.: Some general terminal and network teletraffic equations in virtual circuit switching systems. In: Ince, N., Topuz, E. (eds.) Modeling and Simulation Tools for Emerging Telecommunications Networks: Needs, pp. 471–505. LLC, Trends, Challenges, Solutions. Springer (2006)
73. Poryazov S., Saranova,E., Spiridonova, M.: Modeling of telecommunication processes in an overall complex system. Phys. Elem. Part. Atom. Nucl. Lett **12**(3)(194), 657–662 (2015)
74. Rosenkranz, M.: A new symbolic method for solving linear two-point BVPs on the operator level. J. Symbol. Comput. **39**, 171–199 (2005)
75. Rozenvasser, E.N.: Vibrations of nonlinear systems. Nauka, Moscow (1969) (in Russian)
76. Rozenvasser E.N., Volovodo, S.K.: Operator methods and vibration processes. Moscow "Nauka" (1985) (in Russian)
77. Saadatmandi, A., Dehghan, M.: Numerical solution of the one-dimensional Wawe quation with an integral condition. Numer. Methods Partial Diff. Eq. **23**, 282–292 (2007)
78. Saranova, E.T., Poryazov, S.A.: On the minimal number of easy-to-measure parameters, describing an overall telecommunication network state. In: International Conference on Distributed Computer and Communication Networks, pp. 79–88, Moscow, Russia, 26–28 October, pp. 79–88 (2011)
79. Skornik, K.: Professor Jan Mikusinski—life and work. Sci. Math. Jpn. **66**(1), 1–16 (2007)
80. Spiridonova, M.: Languages for symbolic information processing. J. Syst. Manage. (in Bulgarian) **3–4**, 29–34, Sofia (1975)
81. Spiridonova, M.: REFAL Language and Translator for Minsk 32 Computers. Bull. State Committee Sci. Res. **10**, 12–16, Sofia (in Bulgarian) (1977)
82. Spiridonova, M., Kazassov, Chr.: Computer-aided symbolic Laplace transformation. J. Autom. Control Syst. **3**, 90–94 (in Bulgarian) (1983)

83. Spiridonova, M., Djambazova, M.: Development and application of mathematical knowledge bases using REDUCE system in russian). In: Proceedings of the International Conference on Automated Research, Plovdiv, 15–20 October, pp. 463–466 (1984)
84. Spiridonova, M.: Design and implementation of symbolic transformations bases using REDUCE system (In Russian). In: Proceedings of the International Conference on Symbolic Computations and Their Application in Theory Physics, pp. 28–32, Dubna (1985)
85. Spiridonova, M., Nisheva, M.: Computer algebra system: state of the art. In: Proceedings of the International Workshop on Problems and Applications of Artificial Intelligence, pp. 184–188, Varna, 21–25 September (In Russian) (1987)
86. Spiridonova, M., Nisheva, M.: Intelligent computer algebra systems and their application in research and engineering (In Russian). In: Proceedings of the International Workshop on Theory and Applications of Artificial Intelligence, pp. 313–318, Sozopol, 29 May–2 June 2 (1989)
87. Spiridonova, M.: Some extensions and applications of REDUCE system. In: Proceedings of EUROCAL '87, Leipzig, 1987, LNCS 378, pp. 136–137. Springer (1989)
88. Spiridonova, M., Dokev, Ch. , Dikov, I.: Power series manipulation system. In: Proceedings of the 18th Spring Conference of the Union of Bulgarian Mathematicians, 6–9 April, pp. 494–497, Albena, (In Bulgarian) (1989)
89. Spiridonova, M., Dikov, I., Tzonkova, L.: Function limits computation using REDUCE system. Application to investigation of rational functions and power series manipulation. In: Proceedings of the 21st Spring Conference of the Union of Bulgarian Mathematicians, pp. 235–241, Sofia (in Bulgarian) (1992)
90. Spiridonova, M., Desideri, J.-A.:. Symbolic computations for analysis of finite-difference schemes by the modified equation approach. In: Extension Abstracts of the International Workshop on Computer Algebra Applications, pp. 46–49, Kiev, 9 July (1993)
91. Spiridonova, M., Baycheva, B.: Symbolic computation of the legendre transformation. J. Programm. **4**, 77–85 Russian Acad. of Sci., Nauka, Moscow (in Russian) (1994)
92. Spiridonova, M., Alamanov, L., Kenderov, P.: Implementation of Convex Analysis Operations Using Maple. In: W. Kuechlin (ed.), ISSAC'96 Posters, ETH, Zurich, 24–26 July, pp. 79–82. ACM Press (1996)
93. Spiridonova, M., Desideri, J.-A.: Symbolic derivation of linear and nonlinear modified (partial differential) equations. In: Mathematics and Education in Mathematics. Proceedings of the 27th Spring Conference of the Union of Bulgarian Mathematicians, pp. 234–238, Pleven (1998)
94. Spiridonova, M.: The computer algebra systems—a good assistant in mathematics education (Inv. paper, in Bulgarian). In: Proceedings of the 29th Spring Conference of the Union of Bulgarian Mathematics, pp. 67–78 (2000)
95. Spiridonova, M., Nisheva-Pavlova, M.: An environment for learning mathematics using computer algebra system. In: Summer School ALCCAL'2000 "Algebraic Combinatorics and Computer Algebra", Extended Abstracts, pp. 63–64 (2000)
96. Spiridonova, M., Nisheva, M.: Computer algebra program tools oriented to mathematics education. In: Projects of 1002 Report (2004). http://www.math.bas.bg/artint/mspirid/dog1002
97. Spiridonova, M.: Some application aspects of computer algebra systems. In: Proceedings of the Conference University of the Third Millenium Devoted to the 10th Anniversary of Burgas Free University, pp. 183–188, vol. 2, Burgas (2001)
98. Spiridonova, M., Nisheva-Pavlova, M., Nikolova, D., Kujumdzhieva, B.: Computer algebra systems in education of mathematics and computer science (In Bulgarian). In: Mathematics and Education in Mathematical, Proceedings of the 32nd Spring Conference of the Union of Bulgarian Mathematicians, pp. 416–420, Sunny Beach, 5–8 April 2003 (2003)
99. Spiridonova, M., Kuyumdjieva, B.: Computer algebra systems as element of the new technologies in education (In Bulgarian). In: Proceedings of the Second National Conference on the New Technologies in Education and Professional Learning, Plovdiv, 2004, Centre for European Education and Multicultural Communications, pp. 186–194, Sofia (2004)

100. Spiridonova, M.: Applied Computer Algebra Microenvironments, Mathematica Balkanica, New Series, vol. 19, Fasc. 3–4, pp. 445–452 (2005)
101. Spiridonova, M.: Computation of periodic solutions of differential equations by the operational calculus approach. In: Mathematics and Education in Mathematics. Proceedings of the 34th Spring Conference of the Union of Bulgarian Mathematicians, pp. 191–195, Sofia (2005)
102. Spiridonova, M.: Operational methods in the environment of a computer algebra system. Serdica J. Comput. Bulg. Acad. Sci., Inst. Math. Inform. 3(4), 381–424 (2009)
103. Spiridonova, M., Gerov, A., Nisheva-Pavlova, M.: Development and application of computer algebra systems in IMI of BAS (in Bulgarian). In: Proceedings of the National Conference on Informatics, 2015, Institute of Mathematical and Informatics, Bulgarian Academy of Science, pp. 30–38 Sofia (2016). http://www.math.bas.bg/infres/cib80/book/cib80-p03.pdf
104. Stephens, E.: The Elementary Theory of Operational Mathematics. McGraw Hill Book Co., Inc, New York and London (1937)
105. Tomov, V., Gerov A.: Intelligent applied program systems. In: Proceedings of the International Workshop on Problems and Applications of Artificial Intelligence, pp. 58–66, Varna, 21–25 September (In Russian) (1987)
106. Tomov, V., Spiridonova, M., Gerov, A.: Capabilities of the computer algebra systems. J. Autom. Control Syst. 1, 31–38 (in Bulgarian) (1983)
107. Tomov, V., Spiridonova, M., Gerov, A., Varbanov, S.: Systems for symbolic and numerical computations. In: Proceedings of 3rd International Summer School on Automation and Instrument-Making Industry, pp. 245–252, Varna, October 1984 (In Bulgarian) (1984)
108. Tomov, V., Spiridonova, M.. Gerov, G.: Development and application of computer algebra systems. In: Proceedings of the International Conference on Automated Research, Plovdiv, October 15–20, (Inv. paper, in Russian), pp. 58–64 (1984)
109. Tomov, V., Gerov, A., Spiridonova, M.: Computer algebra systems. In: Proceedings of the 14th Spring Conference of the Union of Bulgarian Mathematicians, Sunny Beach, pp. 92–103 (Inv. paper, in Bulgarian) (1985)
110. Tomov, V., Nisheva,M., Tonev, T.: Computer algebra system for continued fractions manipulation. In: LNCS, vol. 378, pp. 52–53. Springer (1989)
111. Trenkov, I., Spiridonova, M.: Symbolic Computation of the Solutions of High Geodesy Mathematical Problems. In: Proceedings of the International Workshop on Problems and Applications of Artificial Intelligence, pp. 206–210, Varna, September 21–25 (In Russian) (1987)
112. Trenkov, I., Spiridonova, M., Daskalova, M.: An application of REDUCE system for solving a mathematical geodesy problem. In: Proceedings of the International Symposium on Symbolic and Algebraic Computation ISSAC'91, July 15–17, pp. 448–449. ACM Press, Bonn, Germany (1991)
113. Trenkov, I., Spiridonova, M., Daskalova, M.: Bases of symbolic transformations for mathematical geodesy problems. In: Proceedings of the International Conference on Computer Algebra Systems in Physics Research, pp. 372–376, Dubna, New Scientist, Singapore (1991)
114. Vashchenko-Zaharchenko, M.E.: Symbolic computation and its application for integration of differential equations. Ms, Thesis (in Russian), Kiev (1862)
115. Virtual Mathematics Classroom. http://cabinet.bg/
116. Winkler, F.: Polynomial Algorithms in Computer Algebra. Springer, Wien (1996)
117. Wolfram Language. https://www.wolfram.com/language/
118. Wolfram Research. http://www.wolfram.com
119. Wolfram for Education. https://www.wolfram.com/education/
120. Yosida K., Okamoto, S.: A note on Mikusinski's operational calculus. Proc. Jpn. Acad. Ser. A Math. Sci. 56(1), 1–3 (1980)

Generalized Nets. An Overview of the Main Results and Applications

Dafina Zoteva and Nora Angelova

Abstract Generalized Nets (GNs) are a powerful tool for discrete event simulation and parallel processes flow representation. The apparatus of GNs is equally well suited for modelling simple systems, as well as large, complex systems. The major strength of a discrete event simulation is its ability to model random events and to predict the effects of complex interactions between these events. GN-models could be used as a quick method for analysing and solving complex problems. This article presents a brief overview of the evolution of the GNs theory and its various fields of application. The results discussed here are based on years of research by scientists of the Bulgarian Academy of Sciences.

Keywords Generalized nets · Petri nets extensions · Conceptual modelling

1 A Short Introduction to the Concept of Generalized Nets

The concept of Generalized Nets (GNs) as an extension of the Petri Nets [162] was defined in 1982 by Krassimir Atanassov. The introducing paper [39] however was published two years later in 1984. The first results on GNs were published in Bulgaria in 1983 in the papers [27, 38, 40, 41]. Their appearance is a natural continuation of the tendency of developing tools for modelling of discrete event systems. During the past 37 years, the GNs have become an object of extensive research, both in theoretical and applied aspect. The results are summarized in the books [25, 28, 70]. As of today, scientists from over 20 countries (Australia, Belgium, Germany, Great

D. Zoteva (✉)
Department of Bioinformatics and Mathematical Modelling, Institute of Biophysics and Biomedical Engineering, Bulgarian Academy of Sciences, Acad. G. Bonchev Str, Bl. 105, 1113 Sofia, Bulgaria
e-mail: dafy.zoteva@gmail.com

N. Angelova
Faculty of Mathematics and Informatics, Sofia University "St. Kliment Ohridski", 5 Blvd James Bourchier, 1164 Sofia, Bulgaria
e-mail: noraa@fmi.uni-sofia.bg

© The Author(s), under exclusive license to Springer Nature Switzerland AG 2021
K. T. Atanassov (ed.), *Research in Computer Science in the Bulgarian Academy of Sciences*, Studies in Computational Intelligence 934,
https://doi.org/10.1007/978-3-030-72284-5_10

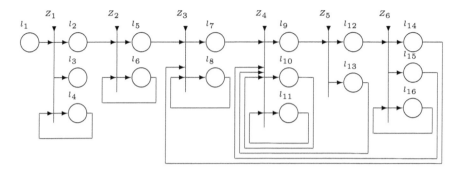

Fig. 1 Graphical representation of a GN

Britain, Morocco, Poland, Portugal, South Korea, Spain, etc.) have authored or co-authored with Bulgarian researchers more than 500 articles and over 25 monograhies [2, 166, 271].

Instead of summarizing the huge number of publications on the GNs theory and applications, the aim of this paper is to briefly outline the main notions of the GNs theory and to summarize the most important results in recent years in terms of GNs extensions, operator and programming aspect of the GNs theory and some of the GNs applications. The results are based on the work of research groups of the Bulgarian Academy of Sciences.

1.1 Definition of a Generalized Net

The GN is a relatively complex object. Before introducing its formal definition, we shall describe non-formally the elements used in the graphical representation of a GN. An example of a GN is shown in Fig. 1. GN's *places* are represented with $\bigcirc$. A GN's *transition* is a part of the net whose graphical representation looks like the object shown in Fig. 2. Every transition contains *transition's conditions* which are graphically represented by T

Like Petri nets, GNs contain tokens which are transferred from place to place through the *arcs* of the net. The arcs are denoted by arrows (see Fig. 1). They start at a place and end at the transition's condition or start at a transition's condition and end at a place. Each token enters the net with an initial characteristic. During their transfer in the net, the tokens receive new characteristics. This way, the tokens accumulate their "*history*". This is an essential difference from the other types of Petri nets.

At most one arc enters each place and at most one arc leaves it.

Some places have no entering arcs. They are called *input places* for the net. Such a place is, for example, the one denoted by l_1 in Fig. 1. Some places have no leaving

Fig. 2 Graphical
representation of a transition
of GN

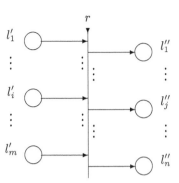

arc. They are called *output places* of the net. Such example is the place denoted by l_3 in Fig. 1. Usually, but not always, the input places stay to the left of the transition, while the output places are to its right. Some places can be both input and output places. Such places are l_4, l_6, l_8, l_{10}, l_{11} and l_{16} in Fig. 1.

When a token enters an input place of a transition, this transition becomes *potentially active*. The transition becomes actually *active* only when certain conditions for the transfer of the tokens from its input to its output places are met.

Besides the history stored in the tokens, another difference between the GNs and the Petri nets is the relation between the places and the transitions in a GN. The GNs transitions are more complex as the number of input and output places can be any natural number.

An important component of the formal description of the transitions is the Indexed Matrix (IM, see [34]). This is the third difference between the GNs and all other Petri nets. IMs are used to describe the conditions for tokens' transfer and the capacities of each arc in the net.

The time in GNs is discrete, i.e. it increases with discrete time steps. A GN starts functioning at a certain time moment, measured on some absolute scale, and it continues to function for a certain period of time. The overall state of a GN can be observed at any step of its functioning.

When different GN-models of connected parallel processes are used, not only different time scales (one for every net) can be used, but also a common time scale on which the time moments of activation and duration of the active states of each of the net can be measured. This is the fourth difference between the GNs and the other types of Petri nets.

A definition of a GN-transition should be given, before presenting the GN's formal definition.

The following notations are used in the formal definition of a GN-transition:

- $\mathcal{N} = \{0, 1, 2, \dots\} \cup \{\infty\}$;
- $pr_i X$ is the ith projection of the n-dimensional set, where $n \in \mathcal{N}, n \geq 1$, and $1 \leq i \leq n$.

– $card(X)$ is the cardinality of the set X.

Definition 1 Transition of a GN is the following ordered seven-tuple:

$$Z = \langle L', L'', t_1, t_2, r, M, \square \rangle$$

where

(a) L' and L'' are finite non-empty sets of the transition's input and output places, respectively. For the transition in Fig. 2 these sets are:

$$L' = \{l'_1, l'_2, \ldots, l'_m\},$$

$$L'' = \{l''_1, l''_2, \ldots, l''_n\}.$$

(b) t_1 is the current time moment at which the transition can be activated.
(c) t_2 is the duration of the active state of the transition.
(d) r is the IM of the transition's conditions which determine the output places to which the tokens in the input places can be transferred. It has the form:

$$r = \begin{array}{c|ccccc} & l''_1 & \ldots & l''_j & \ldots & l''_n \\ \hline l'_1 & & & & & \\ \vdots & & & & & \\ l'_i & & & r_{i,j} & & \\ & & & (r_{i,j} - \text{predicate}) & & \\ \vdots & & (1 \leq i \leq m, 1 \leq j \leq n) & & & \\ l'_m & & & & & \end{array}$$

where $r_{i,j}$ is the predicate which expresses the condition for transfer from the ith input place to the j-th output place.
When $r_{i,j}$ has truth-value "*true*", then a token from the i-th input place can be transferred to the j-th output place. Otherwise, this is impossible.

(e) M is an IM of the arcs' capacities. It has the form:

$$M = \begin{array}{c|ccccc} & l''_1 & \ldots & l''_j & \ldots & l''_n \\ \hline l'_1 & & & & & \\ \vdots & & & & & \\ l'_i & & & m_{i,j} & & \\ & & (m_{i,j} \geq 0 - \text{natural number or } \infty) & & & \\ \vdots & & (1 \leq i \leq m, 1 \leq j \leq n) & & & \\ l'_m & & & & & \end{array}.$$

(f) $\square$ is the transition type. It is an object having a form similar to a Boolean expression. It has as variables the same symbols that serve as labels for the transition's input places. It is constructed of these variables and the Boolean connectives $\wedge$ and $\vee$. For example:

$\wedge(l_{i_1}, l_{i_2}, \ldots, l_{i_u})$ — every place $l_{i_1}, l_{i_2}, \ldots, l_{i_u}$ must contain at least one token,

$\vee(l_{i_1}, l_{i_2}, \ldots, l_{i_u})$ — there must be at least one token in the set of places $l_{i_1}, l_{i_2}, \ldots, l_{i_u}$,

where $\{l_{i_1}, l_{i_2}, \ldots, l_{i_u}\} \subset L'$.
When the value of its type (calculated as a Boolean expression) is "$true$", the transition can become active. Otherwise, it cannot.

Definition 2 Generalized net is the ordered four-tuple

$$E = \langle\langle A, \pi_A, \pi_L, c, f, \theta_1, \theta_2\rangle, \langle K, \pi_K, \theta_K\rangle, \langle T, t^0, t^*\rangle, \langle X, \Phi, b\rangle\rangle,$$

where

(a) A is the set of GN-transitions.
(b) π_A is a function which gives the priorities of the transitions, i.e., $\pi_A : A \to \mathcal{N}$.
(c) π_L is a function which gives the priorities of the places, i.e., $\pi_L : L \to \mathcal{N}$, where

$$L = pr_1 A \cup pr_2 A$$

is the set of all GN-places.
(d) c is a function which gives the capacities of the places, i.e., $c : L \to \mathcal{N}$.
(e) f is a function which calculates the truth-values of the predicates of the transition's conditions. It obtains the values "$false$" or "$true$", or values from the set $\{0, 1\}$. If $\mathcal{P}$ is the set of the predicates used in a given GN-model, then f can be defined as $f : \mathcal{P} \to \{0, 1\}$.
(f) θ_1 is a function which gives the next time-moment when a given transition Z can be activated, i.e., $\theta_1(t) = t'$, where $pr_3 Z = t, t' \in [T, T + t^*]$ and $t \leq t'$. The value of this function is calculated at the moment when the transition terminates its functioning.
(g) θ_2 is a function which gives the duration of the active state of a given transition Z, i.e., $\theta_2(t) = t'$, where $pr_4 Z = t \in [T, T + t^*]$ and $t' \geq 0$. The value of this function is calculated at the moment when the transition starts functioning.
(h) K is the set of the GN's tokens. In some cases, it is convenient to consider this set in the form

$$K = \bigcup_{l \in Q^I} K_l,$$

where K_l is the set of tokens which enter the net from place l, and Q^I is the set of all input places of the net.
(i) π_K is a function which gives the priorities of the tokens, i.e., $\pi_K : K \to \mathcal{N}$.

(j) θ_K is a function which gives the time moment when a given token can enter the
     net, i.e., $\theta_K(\alpha) = t$, where $\alpha \in K$ and $t \in [T, T + t^*]$.
(k) T is the time-moment when the GN starts functioning. This moment is deter-
     mined with respect to a fixed (global) time-scale.
(l) t^0 is an elementary time-step, related to the fixed (global) time-scale.
(m) t^* is the duration of the GN's functioning.
(n) X is a function which assigns initial characteristics to each token when it enters
     an input place of the net.
(o) Φ is a characteristic function which assigns new characteristics to each token
     when it makes a transfer from an input to an output place of a given transition.
(p) b is a function which gives the maximum number of characteristics a given
     token can receive, i.e., $b : K \to N$.

   The input and output places of the transitions, the IMs of the capacities of the
arcs and the types of the transitions determine the static structure of a GN. The
dynamic nature of a GN is expressed in the tokens and the transitions' condition.
The temporal properties of a GN are presented through the time components T, t^0, t^*
and the elements of the set $pr_{3,4}A$, i.e., the functions θ_1 and θ_2 of the transitions. The
functions Φ, X and b serve as a GN's memory. The functions π_A, π_L, c are related
to the GN's static structure, f and π_K are related to the GN's dynamic elements and
θ_1, θ_2 and θ_K are related to the time components of a GN.

1.2 Algorithms for Transition Functioning and GN Functioning

The definition of a GN is significantly more complicated than the definition of a Petri
net. Therefore, it is naturally to expect that the algorithm for transfer of tokens is
also complicated.

   During the years the algorithms for transitions and GN's functioning have been
modified with the aim of optimizing them. Despite this, the later modifications
(see [73, 76, 90]) follow more or less the algorithms described in [28].

   The most general form of the algorithm of transition functioning, referred to
as *Algorithm A*, is presented here. In this form of the algorithm, the possibility
of transferring tokens to output places which have reached their capacities, if the
incoming token can merge with one of the tokens in the output place, is considered
(see [17]).

Algorithm A

(A01)   The input and output places of the transition are ordered according to their
        priorities.
(A02)   The tokens in the input places are ordered according to their priorities.
        The tokens in a given place are divided into two groups. The first group

consists of those tokens which can be transferred to output places at the current time step. The second group is initially empty.

(A03)   An empty IM R is generated which corresponds to the IM r of the predicates of the transition's conditions. Value "0" is assigned to all elements $R_{i,j}$ of R which are:

(a)   in a row corresponding to empty input place;
(b)   in a column corresponding to a full output place;
(c)   in a place with indexes (i, j) for which the predicate $r_{i,j}$ is given as *false* or $m_{i,j} = 0$.

Value "1" is assigned to all other elements of R which have indexes (i, j) for which the predicate $r_{i,j}$ is given as *true*.

(A04)   The ordered places are processed starting with the place having the highest priority in which there is at least one token and from which a token has not been transferred to output place at the current time step. The following steps are done one after the other:

(a)   The value of $R_{i,j}$ in R is checked and if $R_{i,j} = 1$, then go to **(A04c)**. If $R_{i,j}$ has no assigned value, then go to **(A04b)**. If $R_{i,j} = 0$, then check if in the corresponding output place there is a token which can merge with the current token. If there is such token, go to **(A04c)**, else go to **(A04d)**.

(b)   The truth-value of the predicate $r_{i,j}$ of r is checked. If this value is *true*, then go to **(A04c)**, else go to **(A04d)**.

(c)   The current token is transferred to the jth output place. If there is a suitable token there, it merges with it. If it cannot merge with another token in the j-th output place or if it has merged but there are no other tokens in the input places which can merge with it, the token obtains its next characteristic.
      If there is at least one token in the input places of the transition which can be merged with the current token, it will wait in the output place for this to happen and will obtain new characteristic either after the token is merged with the last of the suitable tokens from the input places or at the last time moment of the functioning of the transition.

(d)   If a transfer has been made and the current token cannot split, or if it can split and all predicates of the corresponding row of r have been checked, go to step **(A05)**, else go to **(A04a)**.

(A05)   If the token cannot be transferred during the current time step, it is moved to the second group of tokens of the input place.

(A06)   The current number of tokens in all places to which token has been transferred, without merging, is increased by 1.

(A07)   The current number of tokens in the input places from which a token has been transferred to the output places is decreased by one. If the number of tokens in the input place becomes 0, then 0 is assigned to the elements of the corresponding row of R.

(A08)   The capacities of all arcs through which a token has been transferred are decreased by 1. If the current capacity of an arc becomes 0, 0 is assigned to the corresponding element of the IM R.

(A09)   If there are input places with lower priority from which token has not been transferred to the output places, the algorithm continues with step (A04). Otherwise, go to step (A10).

(A10)   The current model time is increased by t^0.

(A11)   Check if the current model time is greater or equal to $t_1 + t_2$. If it is not, then the algorithm continues with step (A04). Otherwise, the algorithm continues with step (A12).

(A12)   End of the transition functioning.

The general algorithm for GN's functioning is denoted by *Algorithm B*. It uses the notion of an *abstract transition* which is a union of all activated transitions at a certain time moment.

Algorithm B

(B01)   All tokens α for which $\theta_K(\alpha) \leq T$ enter the input places of the GN.

(B02)   An abstract transition is constructed. It is initially empty.

(B03)   Check whether the current time moment is greater or equal to $T + t^*$.

(B04)   If the answer to the question in (B03) is "no", go to step (B12), else go to step (B05).

(B05)   Search for all transitions for which the value of t_1 is greater or equal to the current time.

(B06)   The types of all transitions found in step (B05) are checked in the following way:

   (a)   All identificators of the input places which are included in the type of the transition are substituted with 0, if the corresponding place is empty at the current time step, and with 1, otherwise.

   (b)   The truth-value of the boolean expression obtained in step (B06a) is evaluated.

(B07)   All transitions for which the truth-value of their types, evaluated according to step (B06), is 1 are added to the abstract transition.

(B08)   Algorithm A is applied to the abstract transition.

(B09)   All transitions which become non-active at the current time step are removed from the abstract transition.

(B10)   The current time moment is increased with t^0.

(B11)   Go to step (B03).

(B12)   The GN stops functioning.

For the functioning of each GN according to the algorithms above, the following important theorem is proved in [28]

Theorem 1 *No conflict situations appear in a GN which has no missing components.*

2   Extensions of the GNs and Reduced GNs

The GNs defined in Sect. 1 are also referred to as ordinary or standard GNs. In the theoretical studies of the GNs, since the very beginning, two important problems became apparent and were extensively researched.

The first problem is to study the properties of objects (nets) which have more components than the ordinary GNs and to specify the relation between them and the ordinary GNs with regards to the functioning and the results of their work. Such nets which have additional components are called *extensions of the GNs*.

The second problem is to study the properties of objects (nets) which have less components than the ordinary GNs and to specify the relation between them and the ordinary GNs in terms of the functioning and the results of their work. Such nets are called *reduced GNs*.

2.1   *Some Recent Extensions of the GNs*

During the years, more than 20 extensions of the GNs have been defined [2, 28, 36, 70].

First attempts for combining the GNs theory and fuzzy sets were published in 1985 [31]. First GN extensions are defined as a result: *Fuzzy GNs of first type* (FGN1) and *Intuitionistic Fuzzy GNs of first type* (IFGN1), respectively. Instead of {"*false*", "*true*"} (or {0, 1}), the results of the function f which evaluates the truth-values of the predicates in FGN1 are in the range [0; 1]. The values of the function f in the case of IFGN1 are in the set $\{\langle a, b \rangle \mid a, b, a + b \in [0, 1]\}$.

Also in [31], *IFGNs of second type* (IFGN2) are defined. The essential difference with IFGN1 is that the tokens in IFGN2 are regarded as "quantities" without initial or any other characteristics. Instead, the function Φ assigns new characteristics to the GN-places.

The GNs extensions defined after that are summarized first in [28].

Coloured GNs [26] define colours for the tokens and the arcs of a GN. The number of arcs leaving or entering a place of a coloured GN can be at most k. The movement of the tokens is restricted depending on their colour.

GNs where the time moment of transition firing is replaced by an interval are called *GNs with interval activation time*. *GNs with complex structure* are defined where more than one arc can leave and more than one arc can enter a place. GNs with a new component "global memory" are introduced. For the purpose of solving transportation problems, another extension is defined called *GNs with optimization components*.

A new class of GNs called *GNs with additional clocks* is introduced for processes with standard predefined procedure to deal with exceptional situations. Additional components are included in the definition of this new class GNs, so that in case the predicates in the transitions' conditions cannot be evaluated on time, the net uses a predefined set of truth-values.

GNs with stop conditions [52] define additional stop criteria for the transitions and the GN itself. In case a stop criterion is met, the functioning of the corresponding transition or the entire GN is terminated. *Opposite GNs* are a modification and extension of the standard GNs and the rest of the extensions where tokens move in the opposite direction, from the output to the input places.

The next group of extensions are included in the book [36].

GNs with tokens duration of life (GNTDL) add to the initial characteristics of each token a number corresponding to the number of time-steps for which the token can stay in the net. In the class *GNs with tokens possessing enhanced memory capabilities*, along with the initial or current characteristics, the tokens receive a real number that represents the degree of memorizing the corresponding characteristic. If the degrees of memorization of the characteristics are represented as "certainty" and "unreliability", the GNs extension is *GNs with intuitionistic fuzzy estimations of tokens memorizing capabilities*.

IFGN1 and IFGN2 are extended in [36] to *IFGN of type 3* (IFGN3) and *IFGN of type 4* (IFGN4), respectively. The tokens' characteristics in IFGN3 and the characteristics of the places in IFGN4 are estimated in intuitionistic fuzzy sense and only when their estimations satisfy the definition of intuitionistc fuzzy tautology, the corresponding tokens or places receive new characteristics.

Characteristics of the tokens in the above mentioned GNs extensions are always determined values, like symbols, numbers, strings, lists, etc. However, in [36] a new class GNs is defined with tokens which receive variables as characteristics. As a result of the functioning of such GNs linear systems can be obtained.

GNs are generalized into 3-dimensional objects in the extension *GNs with 3-dimensional structure* [36].

The fifth type of IFGN is the class of *GNs with places which have intuitionistic fuzzy capacities* defined in [48]. The algorithms for tokens' transfer in different types of IFGNs are discussed in [54].

In *GNs with volumetric tokens* (GNVT) [76] some of the GN-tokens exhibit a volumetric characteristic which has significant impact on the movement of the tokens inside the net. The volumetric characteristics of the tokens can be related also to other physical characteristics of the modelled objects, like weight, size, etc.

GNs with characteristics of the places (GNCP) [18] combine in one GN-tokens in the sense of the ordinary GNs with places of IFGN2. One justification for the introduction of this extension is that in many GN-models the places represent some real objects whose properties change with time. Apart from the simplification of the graphical representation of the models, GNCP can be used for evaluation of places, tokens, transitions and entire nets based on the characteristics of the tokens and the places [13, 19].

The connection between GNCP and IFGN1 and IFGN2 is studied in [11]. Possible applications of GNCP are discussed in [173].

IFGNs with characteristics of the places of type 1 (IFGNCP1) and *type 3* (IFGNCP3) [12] extend GNCP with the characteristics of IFGN1 and IFGN3.

The necessity of changing the tokens' priorities during simulation leads to the idea of *GNs with time dependent priorities* (GNTDP) [6]. The significant difference with the standard GNs is in the function which determines the priorities of the tokens at certain time moment during the GN's functioning. The idea of changing the priorities of the tokens during the GN's functioning is generalized in [5] with the definition of the class *GNs with dynamic priorities* (GNDP). The priorities of the tokens there are dependent on a certain expression.

GNs with characteristics of the arcs (GNCA) [15] define initial and current characteristics for the arcs of the GN-model. This class of GNs can be used to model transport networks where roads, railway lines or inland waterways can be represented in the model by arcs.

The following extensions can be successfully applied when the modelled process flow under uncertainty.

IFGN 1, ..., IFGN4 are extended once again in *Interval valued IFGN of type 1*(IVIFGN1) [74], ..., *type 4*(IVIFGN4) [24]. The evaluations of the predicates in the transitions' conditions are in the form of interval valued IF pairs $\langle M, N \rangle$, where $M, N \subseteq [0, 1]$, $M = [\inf M, \sup M]$, $N = [\inf N, \sup N]$ and $\sup M + \sup N \leq 1$ [35, 60].

In *GNs with additional IF conditions for tokens transfer* (GNAIFCTT) [273] an IF pair is assigned to each of the tokens. It is later exploited in the assessment of the transitions' conditions predicates. Since the additional IF conditions do not change during the GN functioning, it could be determined if a GN-token will pass through all the places of a certain route.

All GNs extensions are conservative extensions of the class of standard GNs, i.e. their functioning and the results of their work can be described by a standard GN. A different algorithm for tokens' transfer is applied for each of the GNs extensions taking into account the specifics of that extension.

2.2 Reduced GNs

Some of the components in the definition of the ordinary GNs may not be present. Such GNs form special classes called *reduced GNs*. The concept of reduced GNs is explained in details in [28].

Reduced GNs have great theoretical significance. They are important also for the modelling of real processes, since in the formal description of most of the GN-models developed so far, reduced GNs have been used.

The basic notations related to the reduced GNs are presented below.

Let

$$\Omega = \{A, \pi_A, \pi_L, c, f, \theta_1, \theta_2, K, \pi_K, \theta_K, T, t^0, t^*, X, \Phi, b\} \cup \{A_i | 1 \le i \le 7\},$$

where $A_i = pr_i A$ is the i-th projection of the set of GN-transitions A, i.e. $A_i \in \{L', L'', t_1, t_2, r, M, \square\}$.

Let $Y \in \Omega$. Σ^Y denotes the class of all GNs which do not have component Y. They are called Y- *reduced GNs* .

Theorems regarding the classes of reduced GNs are proved in [28].

The class $\Sigma^* = \Sigma^{A_3, A_4, A_6, A_7, \pi_A, \pi_L, c, \theta_1, \theta_2, \pi_K, \theta_K, T, t^0, t^*, b}$ is the class of the minimal reduced GNs (*-GNs). The minimal reduced GNs have the form

$$E' = \langle \langle A', *, *, *, *, *, *, * \rangle, \langle K, *, * \rangle, *, \langle X, \Phi, *, * \rangle \rangle,$$

where

$$A' = \{Z'|Z' = \langle L', L'', *, *, r, *, * \rangle \& Z = \langle L', L'', t_1, t_2, r, M, \square \rangle \in A\}.$$

Often, a minimal reduced GNs is denoted by $E' = \langle A', K, X, \Phi \rangle$.

The minimal elements of Σ^* are denoted by $E^* = \langle A^*, K^*, X^*, \Phi^* \rangle$.

Reduction of components can be applied to each of the existing GNs extensions (Sect. 2). Reduced GNCP are defined and studied in [14]. Reduced GNCA and a third class of minimal reduced GNs in which only the arcs obtain characteristics are defined in [16].

3 Operators over GNs

The idea of operators in the GNs' framework dates back to 1982. Different operators have been defined since then. When applied to a certain GN, each of them results in a new GN with specific properties. The diversity of the operators over GNs has motivated their different classifications.

The first classification is related to the time-moment when these operators can be applied to a given GN. The operators can be divided into those applied:

(a) before the start of the GN functioning;
(b) during the GN functioning;
(c) when the GN functioning is completed.

This way the operators are categorized into two groups depending on whether they can be or not applied during the functioning of a random GN. A GN which allows operators to be applied during its functioning is called a *Self-Modifying* GN.

The GN operators may be classified also according to the GN components to which they are applied. Thus they are divided into six different types as follows: *global*, *local*, *hierarchical*, *reducing*, *extending* and *dynamical*.

The basic properties and restrictions on the definitions of the operators are discussed in [25].

3.1   Global Operators

The *global operators* transform an entire GN or all of its components of a certain type according to a definite procedure. The first twenty global operators have been described in [28], while the last operator ($\mathcal{G}_{21}$) has been introduced in [71] and modified in [241] in the last years.

Depending on the objects of their action, the *global operators* are:

– *structural*, which change the entire structure of a given GN;
– *temporal*, which change the temporal components of a given GN;
– *dynamical* are connected to the changes in the dynamical components of the GN, like tokens, transition's conditions and the characteristics of the tokens;
– *auxiliary global operators*, which change the *auxiliary components* of a given GN E $\pi_A, \pi_L, c, \theta_1, \theta_2, \pi_K, \theta_K$.

The GNs' operators aspect has been developed at relatively slower pace after publishing the books [28, 36]. The abstract and overly formal definitions of some of the global operators are formalized in [88, 89].

3.2   Local Operators

The *local operators* transform single components of some of the transitions of a given GN. There are three types of local operators, according to the objects of their action:

– *temporal*, which change the temporal components of a given transition;
– *matrix*, which change some of the index matrices of a given transition;
– *other operators*, which alter the transition's type ($\square$), the capacities of some of the places in the net (c), the characteristic functions of the output places (Φ), the evaluation function associated with the transition condition predicates of a given transition (f).

3.3   Hierarchical Operators

The *hierarchical operators* fall into two groups according to their way of action: *expanding* or *shrinking* a given GN.

The *expanding* operators can be viewed as a mean of considering the modelled process' structure in more details, while the *shrinking* operators—as a mean of integration and ignoring the irrelevant details of the process.

By the object of their action, they can be divided into other three groups.

The first group of hierarchical operators either act upon a place (replace a certain place of the GN with an entire GN) or the result of their work is a place (replace a certain subset of a given GN with a single place).

The second group acts upon a transition (replace a certain transition with an entire GN) or the result of their work is a transition (replace a given subset of a GN with a single transition).

The last group of hierarchical operators replace a certain subnet with a new GN, or exchange a list of fixed tokens with corresponding GNs, or replace a list of predicates by a list of GNs at a particular time moment t. They are called hierarchical operators since they are related to the degree of details of the GN-models of given processes.

3.4 Reducing Operators

The next group of operators are called *reducing operators*, since they remove certain GN components from the model. Therefore, when applied on a certain GN, these operators produce a new GN, which is an element of a particular class of reduced GNs [28, 36].

3.5 Extending Operators

The operators of the fifth group are called *extending operators*. They extend a given GN E to one of the existing GN extensions. The extending operators need the consequent number of the respective GN extension according to the order in which they are defined in [36, 70] and the additional components which are necessary for the particular extension.

Additional extending operators should be constructed for the recently defined GN extensions like GNTDL [36], GNVT [76], GNCP [18], IFGNCP1 and IFGNCP3 [12], (GNTDP) [6], GNDP [5], GNCA [15], IVIFGN1, IVIFGN2, IVIFGN3 and IVIFGN4 [24, 74], and GNAIFCTT [273].

3.6 Dynamical Operators

The operators from the sixth and last group are related to the strategies of tokens transfer or the ways of the GNs functioning. Therefore, they are called *dynamical operators*.

Dynamical operators determine the different algorithms of token transfer depending on the capacities of the places and the arcs, and the number of time steps for which the GN-transitions are functioning. Some of them allow and prohibit the splitting and merging of the tokens, others allow and prohibit the transfer of the tokens from input to output places in a "package".

A group of *dynamical* operators is related to the ways of evaluating the transition condition predicates: standard predicate evaluation; expert estimations of the predicate values or predicates which depend on a solution of an optimisation (e.g., transportation) problem; based on statistical data or others.

The last group of operators are used to determine the direction of the tokens transfer from input to output places of the GN (in the standard case), from output to input places of the GN or transfer in both directions.

All these operators of different types, as well as the other that can be defined, have an important theoretical and practical value. They facilitate the modelling of different real processes. From theoretical point of view, they improve the studies of the GNs properties and behaviour.

4   Software Realization

During the years of GN's existence, the idea of its software realization has been discussed, developed and implemented.

There are a set of papers which describe different parts of the software implementation: current status, technologies, optimizations, new features, operators, simulations, etc.

Of course, the theory is being developed and its program realizations will also continue to be an object of interest.

The GNs' software realizations is associated with the base GN theory.

The last software realization is Generalized Nets Integrated Development Environment (GN IDE). GN IDE is a simulation tool for GN-models. At first, it has developed as a client of the GN's simulation server - GNTicker.

GNTicker is an interpreter for a certain kind of reduced GN. It works as a server for GN simulations and allows simultaneous execution of multiple GN-models [265]. It communicates with the client application via the GNTP protocol.

In the latest version of GN IDE, GNTicker has been removed. An EmbeddedSimulation class is implemented in GN IDE in order to replace the need of the GNTicker for the simulations.

Now, GN IDE allows the user to create, load and run GN-models. The GNs can be used as a part of more complex algorithms and products for optimization of the parallel processes, data security, data mining, etc.

GN IDE is a platform independent software tool written in Java. The software can run on any platform with Java 71 Runtime Environment (JRE) installed. The tool allows users to load and save GN XML files, to create models, to run, pause and resume simulation and to edit GN-models. Each model can be exported to tex file format [7, 85, 87, 91].

GN IDE is a tool that contains all model's elements, many features, operators, graphics, etc.

The GNs are described by objects that interact with one another.

Object-oriented programming (OOP) is a programming paradigm based on the concept of "objects" which can contain data in the form of fields (often known as attributes) and code in the form of procedures (often known as methods). GN IDE is designed in OOP style.

Each component of a GN-model is described like an object in the code with some properties to characterize it. The structure of the GN-model is defined in an XML file.

The approach has many advantages. XML format is platform independent. There are many tools for parsing, transforming and manipulating XML content. XML format is easily extensible. The visual and structural information of the models can be described easily. Furthermore, the structure of the XML definition allows GN IDE to extract information about the relations between those components [7].

There are two ways to create a GN-model. The first one is using the GN IDE menus and drawing functionality. The second is directly with the XML code.

The GN IDE can load a valid GN-model XML file. A GN-model includes description of its transitions, places, arcs, tokens, matrices (predicates), characteristic functions and data.

GN XML file has a strict hierarchy (see Fig. 3).

As each valid XML file, the GN XML file starts with a $<?xml ? >$ tag and its version. The model has one root node $< gn >$. It represents the GN itself. The root node has 4 children which describe the transitions, places, tokens and the functions of the model. The children of a transition are references to the input and output places, predicates and capacity matrices of the transition. The children of a place's reference are arcs.

Each token has a "Host" property which connects a token with a place. The functions are the final part of the GN's description. The latest version of GN IDE supports JavaScript characteristic and predicates functions. This is an important moment in the GN IDE development.

4.1   GN IDE Visual Representation

GN IDE has both Graphical and Tree view as shown in Fig. 4.

Graphical view

The Graphical view is the main view. It displays the graph of all transitions, places, tokens and arcs and all changes during the simulation. Each object on the scene is labelled with its name. Tokens are displayed in their "Host"—a place where they will enter when they become active. If a token has a colour characteristic, it is coloured according to its value. If there are more than four tokens in a place, only their count is displayed. The model can snap to grid or not if we choose the option from the *Edit* menu. The view has *Zoom In* and *Zoom Out* options. They are available from *View* menu and allow the user to view the model in detail.

```xml
<?xml version="1.0" ?>
<gn xmlns="http://www.clbme.bas.bg/GN" name="Logistics" time="256"
    timeStart="0" timeStep="1" language="JavaScript" root="true">
  <transitions>
    ...
    <transition id="Z4" name="Z_{4}" priority="0" startTime="0"
                lifeTime="-1" positionX="550" positionY="150" sizeY="140">
      <inputs>
        <input ref="19">
          <arc>
            <point positionX="450" positionY="230"/>
            <point positionX="550" positionY="230"/>
          </arc>
        </input>
        <input ref="113">
      </inputs>
      <outputs>
        <output ref="111">
        <output ref="112">
        <output ref="113">
      </outputs>
      <predicates>
        <predicate input="19" output="111">false</predicate>
        <predicate input="19" output="112">false</predicate>
        <predicate input="19" output="113">true</predicate>

        <predicate input="113" output="111">W13_11</predicate>
        <predicate input="113" output="112">W13_12</predicate>
        <predicate input="113" output="113">true</predicate>
      </predicates>
    </transition>
  </transitions>
```

Fig. 3 A part of a sample GN XML file

Tree View

The Tree view displays the hierarchical structure of a GN-model. The hierarchy is similar to that of the GN XML file but all token and places references are replaced with the real token and place objects.

The root node represents a GN. The children of the root are transitions and functions. The children of a transition are input and output places, predicates and capacity matrices of the transition. The children of a place are tokens. The children of the token are characteristic names, the children of the characteristic names are the values from the history. This is extremely useful for complex GN-models with many input and output places, many tokens in a place, etc. The nodes of the tree can be expanded or collapsed, allowing the user to display only specific information.

The bottom panel (Fig. 4) has 3 windows—*Properties, Problems, Run java script*.

Properties window describes the currently selected object. When clicking on an object on the Graphical view, the window displays its properties. When clicking on a property, the user can edit its value. The properties for the object are related with the structure of the component and the theory. Each GN component differs from the other by a unique id (identifier).

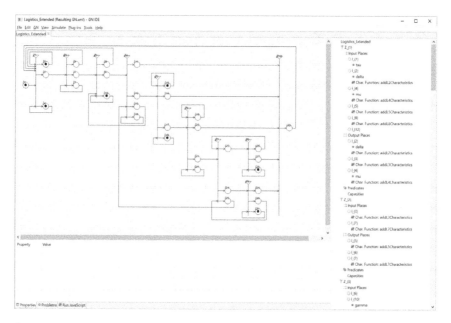

Fig. 4 GN IDE

Each transition is described by:

- Id—unique identifier;
- Friendly name—the name on the scene. For example: Z_2;
- Friendly name main part—the part of the Friendly name which contains the letter Z;
- Friendly name subscript—the part of Friendly name which contains the number identifier 2;
- Priority—the transition priority, part of the simulation algorithm;
- Start time—the time moment of transition activation;
- Life time—the time moment of the duration of the active state, for dynamic process or real time simulations it can be infinity;
- PositionX; PositionY—the position of the transition on the x-axis and y-axis;
- Height—the height of the transition.

Properties window can be used to set transition predicates and capacities matrices. If *Predicates* or *Capacities* option for some transition in the Tree view is selected, the *Properties* window opens the corresponding matrix.

Each place is described by:

- Id—unique identifier;
- Friendly name—the name on the scene. For example: l_1;
- Friendly name main part—the part of Friendly name which contains the letter l;

- Friendly name subscript—the part of Friendly name which contains the number identifier 1;
- Priority—the place priority, part of the simulation algorithm;
- Capacity—the place capacity (maximum number of tokens which can be in the place at the same time). It can be infinity;
- Characteristic Function—the characteristic function associated with the current place;
- Merge rule—the predicate which defines the merging of the tokens. If the predicate returns "true" the merging of the tokens is permitted;
- Output for transition—the id of the transition for which the place is output;
- Input for transition—the id of the transition for which the place is input;
- PositionX; PositionY—the position on x-axis and on y-axis.

In the place definitions and predicate matrix definitions, characteristic functions and predicates are only referred by their names. Their full definitions come in the second part of the GN code (*Functions view*).

Each token is described by:

- Id—unique identifier;
- Friendly Name—the name on the scene. For example: $alpha_2$;
- Friendly name main part—the part of Friendly name which contains the letters *alpha*;
- Friendly name subscript—the part of Friendly name which contains the number identifier 2;
- Priority—the token priority, part of the simulation;
- Host—the place where they will enter when they become active;
- Entering time—the time when the token enters the simulation;
- Default Characteristic—the start token's characteristic;
- Colour—the colour of the token.

The second window in the bottom panel is *Problems*. It displays the errors in the models. The last one is *Run java script* where the users can test code.

Another part of the simulation are the definitions of characteristic functions and predicates. In the definitions of the places and predicates matrices, characteristic functions and predicates are only referred by their names. Their full definitions are shown in Function view.

Function view opens in a new tab in the main window. It can be started from the Tree view or the *Properties* panel. When a place is selected, a characteristic function property can be added from the list of already defined functions or a new function can be created.

The characteristic functions can be written in JavaScript. This feature is now fully supported. GN IDE can be easily expanded to support characteristic functions in GNTCFL (GNTicker Characteristic Function Language). GNTCFL is a language with Lisp-like syntax developed specifically for GNTicker. Furthermore, each function should have access to the components of the GN-model. Thus, GN IDE exports a global variable *gn* which is a reference to the current GN-model. Each component of the model can be accessed thought it.

For example:

- gn.transitions—gives an array of the transitions of the model.
- gn.transitions['Z1']—gives the transition with id'Z1'.
- gn.transitions['Z1'].priority,
  gn.transitions['Z1'].inputs,
  gn.transitions['Z1'].outputs—give the transition with id'Z1' and its priority, input and output places, respectively.
- gn.places—gives an array with the places of the model.
- gn.places['place_t']—gives the place with id'place_t'.
- gn.places['place_t'].id,
  gn.places['place_t'].priority,
  gn.places['place_t'].capacity—give the place's id, its priority and capacity, respectively.
- gn.places['place_t'].tokens—gives an array of the tokens in the place with id'place_t' at the current time moment.
- gn.tokens—gives an array of the tokens of the model.
- gn.tokens['alpha1'].id,
  gn.tokens['alpha1'].priority,
  gn.tokens['alpha1'].host,
  gn.tokens['alpha1'].Default,
  gn.tokens['alpha1'].chars – give the token with id'alpha1' and return its id, priority, host, default characteristic and an array of all of its characteristics, respectively.
- gn.tokens['alpha1'].chars['charId'],
  gn.tokens['alpha1'].chars['charId'].value—give the characteristic with id'charId' for the token with id'alpha1' and its value.

Furthermore, the user can add and delete token's characteristic. GN IDE provides the following two methods: $addChar(id, type, history)$ and $delChar(id)$, where $history$ is the size of the array with the characteristic values.

Each token characteristic has a type. The types supported in the latest version of the software are:

- number—integers and float numbers.
  Example: 1, 1e-5, 0.00001.
- string—all string configurations.
  Example: "example string characteristic".
- vector—an array whose elements can be other arrays.
  Example: [], [30, 50.5, 23], [1, [2, 3], 4].
- object—a JSON object.
  Example: {name: "Example name", ISSN: 00000, year: 2016}.

For Example:

- gn.tokens["alpha1"].addChar("x", "number", 10);
  Adds new characteristic of type number.
- gn.tokens["alpha1"].chars["x"].value = 1000;
  Sets new value to the characteristic'x'.

4.2   GN API

GNs can be used in bigger algorithms and products for optimization of parallel processes, data security etc. Therefore, the GN-models should be created and updated only by code (JAVA). The results of each step of the algorithm execution should be available independently of GN IDE.

The Java API requires execution of Java predicates. For this purpose the *EmbededSImulation* class is extended.

The concept of the GN API is to implement the ability to create and modify GN-models, to create and control the simulation and to receive results back in convenient form using only Java code.

All of these features are available in two packages—GN API Builder and GN API.

GN API Builder contains one class *GeneralizedNetBuilder*.

GeneralizedNetBuilder is a class that implements all functions for creating and updating a GN-model. The basic steps of the GN's implementation include adding transitions, places, tokens and their properties.

GeneralizedNetBuilder implements the classes *TransitionBuilder*, *PlaceBuilder* and *TokenBuilder* which have an object property of the type (Transition, Place, Token) and methods for adding and setting their properties.

GeneralizedNetBuilder includes the following public features:

– Create a GN: *GeneralizedNetBuilder(String name)*.
  GeneralizedNetBuilder is a constructor with a parameter of *String* type. It creates an object of type *GeneralizedNetBuilder* by name.
– set duration of the GN functioning:
  GeneralizedNetBuilder setGnTime(int time).
  setGnTime is a setter method with a parameter of *int* type. It sets global GN time.
– set start simulation time:
  GeneralizedNetBuilder setGnTimeStart(int time)
  setGnTimeStart is a setter method with a parameter of *int* type. It sets the time moment (related to the gloabal time scale) when the GN starts functioning.
– set elementary time-step:
  GeneralizedNetBuilder setGnTimeStep(int timeStep).
  setGnTimeStep is a setter method with a parameter of *int* type. It sets an elementary time-step, related to the gloabal time scale.
– set token splitting property:
  GeneralizedNetBuilder setTokenSplittingEnabled(boolean enabled).
  setTokenSplittingEnabled is a setter method with a parameter of *boolean* type which enables or disables token splitting during the simulation.
  Each setter method returns an *GeneralizedNetBuilder* object, allowing the calls to be chained together in a single statement without requiring variables to store the intermediate results.
– add transition method:
  TransitionBuilder addTransition(String id).

addTransition is method with parameter of *String* type. It creates a transition by an id, adds it to the GN-model, creates and returns an object of type *TransitionBuilder*.

TransitionBuilder is a class part of *GeneralizedNetBuilder* which implements all methods of the transition setup.

- Create transitionBuilder: *TransitionBuilder(Transition transition)*.
  TransitionBuilder is a constructor with a parameter of *Transition* type. It creates an *TransitionBuilder* object by a transition object.
- set transition start time:
  TransitionBuilder setTransitionStartTime(int startTime).
  setTransitionStartTime is a setter method with a parameter of *int* type. It sets the current time-moment of the transition's firing.
- set transition life time:
  TransitionBuilder setLifeTime(IntegerInf lifeTime).
  setLifeTime is a setter method with a parameter of *IntegerInf* type. *IntegerInf* is a class which represents a natural number and allows the value "positive infinity". The method sets current value to the duration of transition's active state.
- set transition priority:
  TransitionBuilder setTransitionPriority(int priority).
  setTransitionPriority is a setter method with a parameter of *int* type. It sets the current transition priority during the simulation.
- set capacity:
  TransitionBuilder setCapacity(String fromId, String toId, IntegerInf value).
  setCapacity is a setter method with 3 parameters—two of *String* type (id of an input place and id of an output place) and one of *IntegerInf* type(capacity value). The method sets the capacity of a transition arc from an input place to an output place. The capacity value can be "positive infinity".
- set transition predicate:
  TransitionBuilder setPredicate(String fromId, String toId, JavaFunction predicate).
  setPredicate is a setter method with 3 parameters—two of *String* type(id of an input place and id of an output place) and one of *JavaFunction* type (predicate). The method sets the predicate which expresses the condition for transfer from the inplut place with id fromId to the output place with id toId.
- set transition type: *TransitionBuilder setType(String type)*.
  setType is a setter method with a parameter of String type. It sets the transition's type described above.
  Each TransitionBuilder setter method returns a *TransitionBuilder* object allowing the calls to be chained together in a single statement without requiring variables to store the intermediate results.
- add transition input place: *PlaceBuilder addInput(String id)*.
  addInput is a method with a parameter of *String* type. It creates a place by an id, adds it to the transition input places list, creates and returns an object of type *PlaceBuilder*.

– add transition output place: *PlaceBuilder addOutput(String id)*.

addOutput is a method with a parameter of *String* type. It creates a place, adds it to the transition output places list, creates and returns an object of type *PlaceBuilder*.

PlaceBuilder is a class, part of *GeneralziedNetBuilder* which implements all methods of a place setup.

– Create placeBuilder object:

PlaceBuilder(Place place, Transition lastTransition).

PlaceBuilder is a constructor with 2 parameters—one of *Place* type and one of *Transition* type. It creates a *PlaceBuilder* object for certain place and transition.

– set place capacity: *PlaceBuilder setPlaceCapacity(IntegerInf capacity)*.

setPlaceCapacity is a setter method with a parameter of *IntegerInf* type. It sets the place capacity. The capacity value can be—"positive infinity".

– set place priority: *PlaceBuilder setPlacePriority(int priority)*.

setPlacePriority is a setter method with a parameter of *int* type. It sets the current place priority during the simulation.

– set place merge rule: *PlaceBuilder setMergeRule(JavaFunction function)*.

setMergeRule is a setter method with a parameter of *JavaFunction* type. It sets a mergeRule which enables or disables tokens merge in the place during the simulation.

– set place merge boolean value: *PlaceBuilder setMergeTokens(boolean merge)*.

setMergeTokens is a setter method with a parameter of *boolean* type which enables or disables tokens merge in the place during the simulation.

– set place characteristic function:

PlaceBuilder setCharFunction(JavaFunction charFunction).

setCharFunction is a setter method with a parameter of *JavaFunction* type. It sets a characteristic function that assigns new characteristics to each token when it enters the place.

Each *PlaceBuilder* setter method returns a *PlaceBuilder* object, allowing the calls to be chained together in a single statement without requiring variables to store the intermediate results.

– add token: *TokenBuilder addToken(String id)*.

addToken is a method with a parameter of *String* type. It creates a token by an id, adds it to the model, creates and returns an object of type *TokenBuilder*.

– add periodic token generator:

TokenBuilder addPeriodicTokenGenerator(String id, int period).

addPeriodicTokenGenerator is a method with two parameters of *String* and *int* type. It creates a token generator by an id and time period which generates a token for this place for each period.

– add random token generator:

TokenBuilder addRandomTokenGenerator(String id).

addRandomTokenGenerator is a method with a parameter of *String* type. It creates a token generator which generated a token for this place randomly.

– add conditional token generator:

TokenBuilder addConditionalTokenGenerator(String id, JavaFunction condition).

addConditionalTokenGenerator is a method with two parameters of *String* and *JavaFunction* type. It creates a token generator which generates a token when the condition is *true*.

TokenBuilder is a class, part of *GeneralizedNetBuilder* which implements all methods of a token setup.

- Create tokenBuilder object:
  TokenBuilder(Token token, Place lastPlace, Transition lastTransition).
  TokenBuilder is a constructor with 3 parameters—one of *Token* type, one of *Place* type and one of *Transition* type. It creates a TokenBuilder object for a token, token's place and transition.
- set token priority: *TokenBuilder setTokenPriority(int priority)*.
  setTokenPriority is a setter method with a parameter of *int* type. It sets the current token priority during the simulation.
- set token entering time:
  TokenBuilder setTokenEnteringTime(int enteringTime).
  setTokenEnteringTime is a setter method with a parameter of *int* type. It sets a time-moment when a given token can enter into the GN-model.
- set token leaving time:
  TokenBuilder setTokenLeavingTime(IntegerInf leavingTime).
  setTokenLeavingTime is a setter method with a parameter of *IntegerInf* type. It sets a time-moment when a given token can leave the GN-model.
- add token characteristic:
  TokenBuilder addCharacteristic(String name, String type, int history).
  addCharacteristic is a method with 3 parameters—two of *String* type and one of *int* type. The method adds a token's characteristic with name, type and history.
- add token characteristic with value:
  TokenBuilder addCharacteristic(String name, String type, int history, String value).
  addCharacteristic is a method with 4 parameters—three of *String* type and one of *int* type. The method adds a token characteristic with name, type and history and sets its value with the given string.
  Each *TokenBuilder* method returns a *TokenBuilder* object allowing the calls to be chained together in a single statement without requiring variables to store the intermediate results.

Once the GN is completed, it should be built and prepared to start. *GeneralizedNetBuilder* includes *JavaGeneralizedNet build()* method witch finally creates a *JavaGeneralizedNet* object from a *GeneralizedNetBuilder*.

GN API implements the ability to create and control the simulation and to receive results back in convenient form. GN API includes the following classes:

- GeneralizedNetFacade—*GeneralizedNetFacade* is a class which implements start simulation functionality. It works only with GN objects of *JavaGeneralizedNet* type created by the *GeneralizedNetBuilder*. *GeneralizedNetFacade* includes 2 methods to start a simulation.

– Start simulation:
JavaSimulation startSimulation(JavaGeneralized-Net gn).
startSimulation method accepts one parameter of *JavaGeneralizedNet* type and
returns an object of *JavaSimulation* type. The method creates a simulation events
listener and calls startSimulation method with two parameters.
– Start simulation with events listener:
JavaSimulation startSimulation(JavaGeneralizedNet gn,
SimulationEventsListener listener).
The method has 2 parameters—one of type *JavaGeneralizedNet* type and one
of type *SimulationEventsListeners*. It creates a simulation object, adds observer,
starts simulation and returns a simulation object.

– JavaSimulation—*JavaSimulation* is a class which extends EmbededSimulation
class [87]. *JavaSimulation* class implements the ability to create and control the
simulation. It has one constructor: *JavaSimulation(GeneralizedNet gn)* with a
parameter of *GeneralizedNet* type. The simulation is carried out in steps. Steps
are initiated by *void step(int count)* method. It has a parameter of type *int* which
says how many steps of the simulation to be released.
– SimulationEventsListener—*SimulationEventsListener* is a class which implements a simulation events observer.

  – Create SimulationEventsListener: *SimulationEventsListener().*
  SimulationEventsListener is a default constructor that creates an observer
  with an update method. The update method handles 3 type of events—
  JavaEnterEvent, JavaMoveEvent and *JavaLeaveEvent*.
  – Get events observer: *BaseObserver getObserver().*
  getObserver is a method with no parameters that returns the events observer.

– JavaGnEvent—*JavaGnEvent* is a class which implements base event functionality.
It has a protected *GnEvent* object and two methods.
– JavaEnterEvent—*JavaEnterEvent* is a class which implements a place entering
event. The event is fired when a token enters the GN in a place. *JavaEnterEvent*
class extends *JavaGNEvent* class described above and retains all its methods, adds
a constructor and *getPlace* method.
– JavaLeaveEvent—*JavaLeaveEvent* is a class which implements a place leaving
event. The event is fired when a token leaves the GN from a place. *JavaLeaveEvent*
class extends *JavaGNEvent* class described above and retains all its methods, adds
a constructor and *getPlace* method.
– JavaMoveEvent—*JavaMoveEvent* is a class which implements a place moving
event. The event is fired when a token moves from one place to another. *Java-
MoveEvent* class extends *JavaGNEvent* class described above and retains all its
methods, adds a constructor, *getStartPlace* and *getEndPlace* methods.

GN API implements classes which allow to create Java objects from engine objects
and to get their properties. The API differs the following objects—GN, transition,
place, token, characteristic or function.

All classes are described below.

- JavaGeneralizedNet—*JavaGeneralizedNet* is a class with one private property of *GeneralizedNet* type. It implements constructor and getter methods associated with the Java(API) GN-model.
- JavaTransition—*JavaTransition* is a class with one private property of *Transition* type. It implements constructor and getter methods associated with the Java(API) Transition.
- JavaPlace—*JavaPlace* is a class with one private property of *Place* type. It implements constructor and getter methods associated with the Java(API) Place.
- JavaToken—*JavaToken* is a class with one private property of *Token* type. It implements constructor and getter methods associated with the Java(API) Token.
- JavaCharacteristic—*JavaCharacteristic* is a class with one private property of *Characteristic* type. It implements constructor, getter and setter methods associated with the Java(API) Characteristic.
- JavaFunction—*JavaFunction* is a class that implements constructor and runs a method for Java function objects. Java Functions are important for the predicates definitions.
- JavaFunctionReference—*JavaFunctionReference* is a class that implements a reference to a function. It has one private property—*JavaFunction*, a constructor and a get method.
- JavaFunctionRunner—*JavaFunctionRunner* is a class that implements function runner functionality. It has one private and static instance property, get and run methods.

4.3 GNs Operators Implementation in GN IDE

The main functionality implemented in GN IDE involves creating GN-models and running simulations of such. The merging and splitting of tokens implemented in the algorithm of GN's functioning can be regarded as an integration of some of the dynamical operators in GN IDE. The further improvements of GN IDE have been also related to the GNs' operators theory.

GN-models are repeatedly changed and improved during their design process. The need to optimize (or simplify) a model often occurs after the simulation of the designed process has been completed. The reducing operators can be used at this point when certain GN components are found redundant for the GN-model, and thus can be omitted.

Two features related to the reducing operators have been integrated in GN IDE. The first one allows reduction of components from an already created GN-model without any loss of information, so that the user can revert the action of the applied reducing operators. The second feature allows identifying the class of reduced GNs to which a particular GN-model may be referred.

The integration of the reducing operators in GN IDE has been described in details in [23]. Their implementation in GN IDE makes it possible for the user to explore different test scenarios by creating a single GN-model. In addition, the reduction of certain GN-components may lead to optimisation of the entire GN's performance.

Decomposition is a common approach when modelling a complex system. Different subprocesses or states of the system are first described by simpler models and integrated later into one composite and detailed model. This approach is often easier than directly creating the detailed version of the model.

The states or the subprocesses of a system are often depicted by GN-places or GN-transitions in one rough simplified model. When they are presented in more details, entire GNs are commonly used. To include the detailed models into the overall model of a system, two expanding hierarchical operators can be exploited, replacing respectively a place or a transition of a given GN-model with an entire GN.

In order to present or analyse certain GN-models in more details, these two hierarchical operators have been integrated in GN IDE too. The requirement and restrictions observed during the implementation of the two expanding hierarchical operators are discussed in [269]. The possibility to easily extend a given GN-model by replacing a single GN-transition with an entire GN has been demonstrated there with an example.

Reverting operators to restore the original GN-model to its state before applying the hierarchical operators have not been implemented in GN IDE. However, the model of the original GN is not changed unless the modifications of the operators $\mathcal{H}_1$ or $\mathcal{H}_3$ are stored in the same XML file with its description. If the resulting GN is stored in a new XML file, the two GNs (original and resulting) remain independent.

4.4   GN Web Simulator

The current version of GN IDE implements a big part of the theory and many features as - merging and splitting of the tokens, JavaScript functions, drawing functionality, load and export xml and tex files, Matlab integration, simulation, step by step simulation, graphics, operators, relations, etc. These features are very helpful but they make the tool hard to use. Furthermore, the GN simulations and their components should change on real time. For this purpose, a new version of the GNs Web Simulator is developed. The GN Web Simulator use the synchronization system for real time changes. The web version should be implemented on the JavaScript language and has simplified functionality. The designers and analysts sections pay a special attention to the user interface. The aim is to allow easy creation of simulations and tests in different areas.

5 Applications of Generalized Nets

GNs find application in a wide number of areas. For the last 30 years, GNs have found applications in the areas of Artificial Intelligence (AI) and Data Mining (DM), medicine, biology and ecology, chemistry and economics, university administration processes and many others. Part of the research discussed in this paper is within the framework of scientific projects at the Bulgarian Academy of Sciences.

GNs have been proven to be an easy and comprehensible way to describe the logic of each modelled process [32, 42, 84, 101, 126, 132, 208, 240].

The first GN-models are created for the needs of the lorry-load transportation. They solve the problem of tracing and controlling the cars. GNs as a tool for modelling complex systems of urban, railway and waterborne transport are used in [99, 136, 243, 266]. The proposed models describe the current state of the networks. They can be used to explore the possibilities for optimizing the networks and to identify a feasible solution in case of a problem.

Other GN-models are constructed for communication companies and telephone exchanges [163, 249]. A GN-model of the switching stage of an overall telecommunication system is proposed in [164] and it is compared to a model based on Service Systems Theory. The extensive use of GNs in the modelling of telecommunication systems has led to the development of GN constructions corresponding to the base virtual conceptual devices from Service Systems Theory in [21]. These results allow easier construction of GN-models of telecommunication systems based on already existing conceptual models. The next step in this direction is the paper in [22], where GN representations of control structures in Service Systems Theory are described, such as information feedback and feedforward and requests feedbacks. An important aspect of the GNs applications has become the modelling of queuing system in service networks. GN-models of queuing systems with various queuing disciplines are described in [165, 261, 262].

GN-models with important practical application are developed for the Petrochemical Combine, in Burgas, Bulgaria. These models proved to be useful for the control of the production processes. The work of oil refinery is modelled in [247]. The GN-model of the process of evaluation of the impact of refinery activity on the environment is shown in [138] and enhanced in terms of IFS in [151].

A number of GN-models are related to ecology and in particular to the problem of wildfires. A GN-model of the process of wildfire extinguishing by a fire service is presented in [20]. Clearing as a prevention measure is modelled in [81]. The models can be used to control the available resources and can help in the decision making process in different simulated situations. A GN-model of the work of a typical wastewater treatment plant and its simulation is shown in [106]. Other models concerning the wastewater treatment process, like mechanical water pre-treatment [107], physics-chemical purification [108], purification in "Biological Reservoir—Sedimentor" [103, 149] are designed. A comparison of wastewater treatment modelling with partial differential equations and GNs is shown in [150]. Part of the mentioned GN-models are collected in [202].

The administrative and academic functions of electronic information processing and exchanges within a university have been studied in series of research. The initial studies in GN modelling of the basic processes in a university (flow of information [214], finance, organization of timetable schedules, administrative services and electronic archives), models of e-learning systems (electronic learning, evaluation, intelligent training systems), university Intranet [215] and Internet flow models are included in a book [216].

An extended model of the information flow in an abstract university is presented in [217]. The models of processes of producing a timetable [218], updating [219] and evaluating existing ones [220] are extensions of previous models. A GN-model of the process of administrative servicing in a digital university with IF estimations is proposed in [222]. GN-models of a training system are explored in [223, 224], the precise order of the university subjects which is directly related to the training process is modelled in [225]. The models for electronic archives are enhanced with IF estimations of the searched information in [226]. A GN-model of an electronic system for the student-teacher interaction is shown in [227].

GN-models are created to describe the process of selection and usage of an intelligent e-learning system [139, 228], as well as the process of personalization of the e-learning environment [140].

An attempt to introduce relatively objective methods for evaluating each of the objects is made in the following studies. The evaluation of the students is made on the basis of assessments of students' problem solving and lecturers' evaluation of student work [133, 229–231]. The standardization and reliability of the assessments of the students problem solving is modelled in [232]. University subject ratings are modelled and explored in [234]. The evaluated objects can be lecturers, students, Ph.D. candidates, problems solved by students, courses, etc. The ICrA method is used to calculate the estimates, some possible correlations between pairs of criteria are detected.

Another group of articles is connected to processes of progressing through the hierarchy in universities and scientific institutions. A GN-model with IF estimations is designed for the process of obtaining scientific titles and degrees [235]. The administrative information processes are modelled in [122]. A GN-model of the process of PhD preparation is shown in [236]. A new perspective of the last two processes is shown in [237]. The presented GN-models can be used to optimize the functioning of a particular university, for monitoring and management of the quality of university education.

GNs are successfully applied to modelling processes concerning security issues. A GN-model model simulating different situations during social protests is proposed in [117]. The model can be used for preventing conflicts as well as for training police officers. GN-models applied to verification and identification of a person based on biometrics [49, 58, 59, 116, 238] can be also related to this field. Another example is the GN-model constructed to describe processes for protection patients' personal e-data, transmitted between healthcare information systems over the Internet [267]. Computer scene analysis in which both objects and the scene background are taken into consideration are described in [110, 111]. The use of IF contrast intensification operators in enhancing images is shown in [141].

The initial studies regarding GNs as a means of describing pattern recognition processes, such as face [114], signature, handwriting [115, 239] and speaker [118–120] recognition, are summarized in [62]. Some of these studies are enhanced with IFL in order to take into account the fact that the information may be imprecise and there might be inherent uncertainty [109, 121]. The book that summarizes the studies in this field up to that point is [63].

The preprocessing phase of the digital image recognition process is modelled in terms of GNs in [142]. The focus is mainly on describing the major step in image preprocessing concerning the image smoothing and contrast enhancement.

As a complement to the GN-models for writer verification and identification [61], a GN-model describing the process of on-line/off-line signature verification is proposed in [112]. The model explores the parallel processing of two signature models: digital image and tablet signal. Another model of signature verification is investigated in [160]. A GN-model for a combined method for on-line signature verification is later proposed in [82].

The process of handwritten Arabic word recognition is modelled in [113].

In general, all the above models can be included as sub-GNs in a larger scheme for biometric authentication. Verification and identification of a person based on biometrics, including iris recognition, are modelled with GNs in [49, 58, 59, 116, 238].

GN-models of algorithms for aggregation of estimates derived by the application of two or more classification procedures solving a particular pattern recognition problem are presented in [104, 257]. A GN-model of an algorithm for pattern classification based on the k-nearest neighbours rule modified for the case of intuitionistic fuzziness is proposed in [268].

A GN-model for simultaneous calculation of estimates for pattern recognition problems in medicine is shown in [251].

A GN-model used to distinguish different social groups, to compare their behaviour and to investigate the information propagation patterns, analysing data from Twitter social network is proposed in [125].

The common feature shared by these models is that they describe general phases of the processes, namely the stages of processing, feature extraction and classification. The models can be used not only for graphical representation of the processes, but also for their subsequent analysis.

Metaheuristics are mathematical optimization algorithms that have been successfully applied in various real-life problems. A large part of the research on the applicability of the GNs as a modelling tool focuses on genetic algorithms (GAs), as one of the most popular metaheuristics inspired by the natural selection.

The main studies regarding modelling GA functions, operators and results of standard GAs with the means of GNs are summarized in [208] or discussed in [152, 154, 155, 157, 204–207, 209]. The GN-models are analysed and modified in order to present some modifications of the standard GA regarding the order of the basic GA operators. An estimation of the calculation time of the modifications is also performed and analysed. The basic operators of GA, namely selection, crossover and mutation, as well as different functions for selection (roulette wheel selection method [157] and stochastic universal sampling [158]), different crossover techniques [156] (one-point crossover, two-point crossover and "cut and splice" technique) and a mutation operator (mutation operator of the Breeder genetic algorithm [203]) are described with GN-models. These models can be considered as separate modules, but they can also be accumulated into a single GN-model to describe an entire GA. There are GN-models elaborated to present modelling of GA functions and operators [197], as well as GN-models that present the possible usage of different modifications of the standard GA regarding the order of the basic GA operators [67, 185]. Other results on modelling GA with GNs are found in [153, 159, 186, 205, 206].

Ant colony optimization (ACO) is another metaheuristic method inspired by the foraging behaviour of the ants. A GN-model of the functioning of a standard ACO algorithm is first described in [95]. It is used to identify the weak points in the algorithm as well as some possibilities for improvements. IF estimations are added to the model in [96]. The GN model in [95] is extended in [100]. The aim is to use the experience of the ants from previous iterations to choose a better starting node. The estimations are based on IF logic. The model shows that the use of start strategies can lead to better results than the random start.

There have been many attempts to enhance the performance of metaheuristics with local search procedures or some exact methods. As a result new algorithms called hybrids are created. They are more efficient and flexible when dealing with large scale problems. An example of such a hybrid ACO algorithm combined with local search procedures is modelled with a GN in [97].

In addition to the above mentioned studies, others like [77, 79, 98, 102], concerning the research of the GN-models and ACO are summarized in [101].

Later, a GN-model with ACO algorithm optimization component is described in [79]. Hybridizations of ACO and GA are considered in [80]

Other metaheuristic algorithms explored with the means of GNs are Cuckoo Search algorithm, Firefly algorithm, Artificial Bee Colony optimization and Bat algorithm [190, 201, 210].

Enhanced models of the Cuckoo Search and the Firefly algorithms are presented in [212]. A Universal GN-model of population-based metaheuristic algorithm is constructed. Numerical experiments are performed on the Universal GN-model, adapted for the Cuckoo Search and the Firefly algorithm, using a predefined set of benchmark functions. The results are compared with the results of the two standard algorithms and their GN-models simulated on the same set of benchmark functions.

One of the main challenges regarding the metaheuristic algorithms is to choose appropriate parameter values for the algorithm run. GNs are used as a tool to describe the process of parameters' control while trying to improve the algorithm performance. In order to tune dynamically the parameters of a metaheuristic algorithm a GN-model of IFL controller is shown in [211].

The GN-model of the work-flow of the Artificial Bee Colony presented in [270] is enhanced with an IF logic controller to determine the magnitude of perturbation, depending on the current iteration of the algorithm in [272].

The algorithm of neuro-dynamic programming is studied in [123].

Examples of modelling real-life problems in which some optimization algorithms are applied are explored in [86, 92, 93]. The proposed GN-models are used for comparing different mathematical models of an *E. coli* fed-batch cultivation process obtained by a GA, a Tabu search algorithm and a Firefly algorithm. The best process model is determined based on a predefined criteria. Simulations of the GN-models are performed.

A series of studies is concerned with describing the dynamics of some bioprocesses [105, 137, 147, 193, 195, 213], as well as their monitoring and control, based on GNs [187–189, 191, 192, 194, 196, 198–200]. GNs are also applied to modelling of the advisory system for yeast cultivation on-line control [143], further expanded and upgraded with IFL [144, 145].

Expert Systems (ES) are designed to model the knowledge of experts and to apply this knowledge into assessing and defining solutions to well-structured problems.

A GN representation of the basic properties of a knowledge base is shown in [94].

Ten models of ES are described with the means of GNs. Models of ordinary ES [44–47], ES with priorities of their Database facts and Knowledge Base rules [29, 57], ES with metafacts [55], IF ES [56] and ES with temporal facts and answering to temporal questions [37], as well as a new type of ES [43], are collected in [32]. Later, ES are extended into ES with temporal components, the process of their functioning and the results of their works are described in [50]. This tenth model of ES is included also in [126].

GNs are used as an instrument for describing the process of ES construction in [161]. A methodology for modelling open, hybrid and closed systems by using GNs is presented in [242]. The validity testing process of an ES is modelled in [65].

A GN model of an ES, in which new facts and rules with IF degree of validity and non-validity can be introduced during its functioning, is proposed in [51]. A GN model of ES with Frame-Type Data Bases is introduced in [72]. The model is enhanced with IF estimations of the frames in [68]. An extension of the last model with truth-values of the hypotheses is the model described in [66]. Three different estimation types are discusses there as possibilities for evaluation of the truth-values of the hypotheses: optimistic, average and pessimistic. A GN model based on IFS theory, that gives the possibility to evaluate the time of the occurrence or completion of the events, is shown in [83]. The model can answer questions such as: "Is the fact valid?", "How many facts does it contradict?", "How many facts does it confirm?".

The above mentioned GN models of ESs can be regarded as a proof that ES can be expanded and each of their extensions can be described in terms of GNs.

GNs can be used for representing any abstract system (statical, dynamical, stochastic, etc.). Examples of such models are given in [69, 167]. An IF system capable of self-modifying is represented with a GN-model in [1]. Some properties of the connections and the events of the abstract systems, interpreted in terms of GNs, are discussed in [30, 33, 64]. Other applications of GNs can be found in describing the standard commands in robotic systems.

A GN representation of production system interpreters is shown in [248]. The presented GN-model includes all phases, all possible work cycles in the process.

The model illustrates the capabilities of GNs for modelling and simulation of large-scale technological processes.

GNs as a tool for modelling multi-agent systems are presented in [263, 264]. One of the reasons GNs are chosen as a modelling tool for multi-agent systems is the possibilities they offer for modelling parallel processes.

GN-models of an industrial robot [260] and a modular robotic system [135] are designed. The different units of the robotic systems are regarded as elements (agents) that are part of an abstract multi-agent system.

GN-models of conflict resolution approaches in version control systems are proposed in [75, 78].

Many GN-models have been developed in the area of medicine, initially through a partnership between a medical doctor Joseph Sorsich, Krassimir Atanassov and Anthony Shannon. As a result, more than 800 GN-models of processes of diagnostics are constructed.

A large part of the initial studies of the applications of GNs in medicine in general are collected in [233]. This research is later extended with models of diagnoses of neurological deceases [53]. The GN applications in general and internal medicine are collected in [221].

A GN-model considering carnitine role in human diseases is presented in [124]. Some investigations related to GNs application for description of genetic maps are shown in [127–131, 148].

The latest studies are related to modelling the processes of diagnosing different deceases and pathological conditions, like shoulder pain [174], osteoarthritis [175], muscle pain [176], adhesive capsulitis [177] and proximal humeral fractures [178]. The process of diagnosing of asymptomatic osteoporosis is modelled with an ordinary GN-model in [179]. A novel approach for early detection of adolescent idiopathic scoliosis (AIS) and its categorization using a GN-model is proposed in [180]. The combined means of GNs and IFS are used for classification and evaluation of the curve progression probability in patients with confirmed AIS in [181]. GN-models for diagnosis of multiple sclerosis [252], assessment [253] and monitoring the degree of disability in patients with multiple sclerosis [254] are constructed.

The mechanical ventilation process is modelled in terms of GNs in [255]. A GN-model of the treatment of mechanically ventilated patients with nosocomial pneumonia is presented in [250], while the transition to spontaneous breathing after long term mechanical ventilation is described in [256]. An enhanced GN-model for evaluating the objective condition of ventilated patients in order to determine if they are ready for weaning off the mechanical ventilation support is given in [258]. A GN-model for describing the diagnostic method before implantation of cerebrospinal fluid draining shunt in infants is proposed in [259]. The models in medicine can be used for simulation of real processes with educational purpose. They can guide specialists in studying the logic of the processes related to diagnoses and help in acquiring knowledge and diagnostic skills.

Another direction of the description of medical processes is the modelling of the human body and particularly some of its organs and systems. In [168–173], for example, are presented GN-models of the logical relations between the structures of the upper limb.

Some of the recent investigations are related to GNs application in medical mechatronic engineering [182–184], particularly in the development of orthotic devices.

The information system "GP Help" is modelled in terms of GNs in [146]. In the past few years, the processes between health care units have been also described with GN-models [3, 4]. The automated planning of the resources (necessary equipment and specialists), medical staff working schedules, patients reception, vastly facilitate the health care units administration. The store and the processing of the patients personal data and examination results in data bases and expert systems aid the decision making. The appearance of the electronic health care improves the people informativeness in a low price.

During the past decade, GNs have been used in the modelling of telecare/ telemedicine processes. As a starting point, the communication process between patients with life sensors attached to their bodies and remote specialized medical center via GSM network is modelled by GNs [8, 9]. At the next stage, a GN-model of the signal classification and the reaction of the medical staff at the hospital unit is constructed in [134]. These models are later used in [10] as a base for construction of a GN-model of telemedicine based on a specific sensor, i.e., body temperature sensor.

GN-model of processes, related to tracking the changes in health status of adult patients with diabetes is described in [245]. The contemporary state of the art of the telecommunications and navigation technologies allows this model to be extended to the case of active and mobile patients. This requires the inclusion of patients current location as a new and significant variable of the model. In [244], a detailed GN-model of telemedicine for people with diabetes is proposed. The sensors included in the model are blood pressure monitor, weight scale, pulse oximeter and blood glucose monitor. Smart filtering of false positive alarm messages is included which reduces the number of events for which the health care person has to take a decision. The model can be used in the development of a decision support tool for telemedicine for people with diabetes and can be easily extended to include estimations of the costs of the telemedicine center.

A review of the GN-models in medicine with focus on telehealth is shown in [246].

In general, the GN-models developed in medicine can be useful in early detection of pathological abnormalities, may facilitate diagnosis or point to the most likely diagnosis, direct patients to appropriate specialists, as well as to identify the bottlenecks of the processes for providing health care services.

6 Conclusion

GNs have been proven to be a very convenient and proper tool for the modelling complex algorithms and processes. The benefits of using GNs can be found in:

- managing the modelled object or process;
- finding relations with other objects and processes;
- generating hypotheses for the improvement of the process.

Another GNs advantage is that the created models can be easily extended or generalized when necessary. The existence of a software tool to construct and simulate GN-models with real experimental data increases GNs applicability.

Acknowledgements The work research was supported by the National Scientific Fund of Bulgaria under the Grants DN02/10 "New Instruments for Knowledge Discovery from Data, and their Modelling".

References

1. Aladjov H., Atanassov, K., Kim, S.K., Chang, O.B., Kim, Y.S.: Generalized net-based machine learning model of an intuitionistic fuzzy abstract system. In: Proceedings of the 1999 IEEE International Fuzzy Systems Conference, Seoul, Aug. 22–25, vol. II, pp. 645–649 (1999)
2. Alexieva, J., Choy, E., Koycheva, E.: Review and bibloigraphy on generalized nets theory and applications. In: Choy, E., Krawczak, M., Shannon, A., Szmidt, E. (eds.) A Survey of Generalized Nets, vol. 10, pp. 207–301. Raffles KvB Monograph, Sydney (2007)
3. Andreev, N., Vassilev, P., Atanassova, V., Roeva, O., Atanassov, K.: Generalized net model of the cooperation between the departments of transfusion, haematology and the national centre of transfusion haematology. In: Mladenov, V., Slavova, A., Sgurev, V., Hadjiski, M., Boshnakov, K. (eds.) IEEE Proceedings Advances in Neural Networks and Applications, pp. 82–85 (2018)
4. Andreev, N., Atanassov, K., Sotirova, E., Atanassova, V., Roeva, O., Zoteva, D., Vasilev, P.: Generalized net models of the processes in and between centers for transfusion haematology. In: Advances in Intelligent Systems and Computing. Springer, in press (2020)
5. Angelova, N.: Generalized nets with dynamic priorities. Issues in Intuitionistic Fuzzy Sets and Generalized Nets **11**, 15–22 (2014)
6. Angelova, N.: Generalized nets with time dependent priorities. In: Proceedings of 15th International Workshop on Generalized Nets, Burgas, pp. 17–22 (2014)
7. Angelova, N., Todorova, M., Atanassov, K.: GN IDE: implementation, improvements and algorithms. Comptes rendus de l'Academie Bulgare des Sciences, Tome **69**(4), 411–420 (2016)
8. Andonov, V., Stojanov, T., Atanassov, K., Kovachev, P.: Generalized net model for telecommunication processes in telecare services. In: Proceedings of the First International Conference on Telecommunications and Remote Sensing, Sofia, 2012, pp. 158–162 (2012)
9. Andonov, V., Stefanova-Pavlova, M., Stojanov, T., Angelova, M., Cook, G., Klein, B., Atanassov, K., Vassilev, P.: Generalized net model for telehealth services. In: Proceedings of the 6th IEEE International Conference "Intelligent Systems". Sofia, pp. 221–224 (2012)
10. Andonov, V., Stephanova, D., Esenturk, M., Angelova, M., Atanassov, K.: Generalized net model of telemedicine based on body temperature sensors. In: Proceedings of the 14th International Workshop on Generalized Nets, Burgas, pp. 78–89 (2013)
11. Andonov, V.: Connection between generalized nets with characteristics of the places and intuitionistic fuzzy generalized nets of type 1 and type 2. Notes on Intuitionistic Fuzzy Sets **19**(2), 77–88 (2013)
12. Andonov, V.: Intuitionistic fuzzy generalized nets with characteristics of the places of type 1 and type 3. Notes on Intuitionistic Fuzzy Sets **19**(3), 99–110 (2013)

13. Andonov, V.: Intuitionistic fuzzy evaluation of places in generalized nets and generalized nets with characteristics of the places. In: Proceedings of the 15th International Workshop on Generalized Nets, Burgas, pp. 8–16 (2014)
14. Andonov, V.: Reduced generalized nets with characteristics of the places. International Journal "Information Models and Analyses" **3**(2), 113–125 (2014)
15. Andonov, V.: Generalized nets with characteristics of the arcs. Comptes Rendus de l'Academie Bulgare des Sciences **70**(10), 1341–1347 (2017)
16. Andonov, V.: Reduced generalized nets with characteristics of the arcs. Issues in Intuitionistic Fuzzy Sets and Generalized Nets **14**, 25–35 (2018/19)
17. Andonov, V., Angelova, N.: Modifications of the algorithms for transition functioning in GNs, GNCP, IFGNCP1 and IFGNCP3 when merging of tokens is permitted. Springer Series Studies in Fuzziness and Soft Computing, pp. 275–288 (2015)
18. Andonov V., Atanassov, K.: Generalized nets with characteristics of the places. Comptes Rendus de l'Academie Bulgare des Sciences **66**(12), 1673–1680 (2013)
19. Andonov, V., Shannon, A.: Intuitionistic fuzzy evaluation of the behavior of tokens in generalized nets. Adv. Intell. Syst. Comput. **322**, 633–644 (2015)
20. Andonov, V., Atanassov, K., Shannon, A., Sotirova, E., Velizarova, E.: Generalized net model of the process of wildfire extinguishing by a fire service. In: Proceedings of the 15th International Workshop on Generalized Nets, pp. 23–28 (2014)
21. Andonov, V., Poryazov, S., Saranova, E.: Generalized net representations of elements of service systems theory. Adv. Stud. Contemp. Math. **29**(2), 179–189 (2019)
22. Andonov, V., Poryazov, S., Saranova, E.: Generalized net representations of control structures in service systems theory. Adv. Stud. Contemp. Math. (Kyungshang), **30**(1), 49–60 (2020)
23. Angelova, N., Zoteva, D.: Implementation of the reducing operators over generalized nets in GN IDE. Issues in Intuitionistic Fuzzy Sets and Generalized Nets, **12**, 80–92 (2015/2016)
24. Angelova, N., Zoteva, D., Atanassov, K.: Interval-valued intuitionistic fuzzy generalized nets with fluids instead of tokens. In: Proceedings of the 12th International Conference Information Systems and Grid Technologies (ISGT 2018), CEUR Workshop Proceedings vol. 2464, pp. 15–28 (2019). ISSN 1314–4855
25. Atanassov, K.: Applications of Generalized Nets. World Scientific, Singapore (1993)
26. Atanassov, K.: Colour Generalized nets. AMSE Review, **3**(1), 7–11 (1986)
27. Atanassov, K.: Conditions in Generalized nets. In: Proceedings of the XIII Spring Conference of the Union of Bulgaria Math., Sunny Beach, pp. 219–226 (in Bulgarian) (1984)
28. Atanassov, K.: Generalized Nets. World Scientific, Singapore (1991)
29. Atanassov, K.: Generalized nets and expert systems VII. Adv. Modelling Anal. A AMSE Press **21**(2), 15–22 (1994)
30. Atanassov, K.: Generalized Nets and Systems Theory. "Prof. M. Drinov", Academic Publishing House, Sofia (1997)
31. Atanassov, K.: Generalized nets and their fuzzings. AMSE Review **2**(3), 39–49 (1985)
32. Atanassov, K.: Generalized Nets in Artificial Intelligence. Vol. 1: Generalized nets and Expert Systems. "Prof. M. Drinov". Academic Publishing House, Sofia (1998)
33. Atanassov, K.: Generalized nets models in systems theory. In: Atanassov, Kacprzyk, J., Krawczak, M., Szmidt, E. (Eds.)Issues in Intuitionistic Fuzzy Sets and Generalized Nets, vol. 4. Wydawnictwo WSISiZ, Warszawa, pp. 35–42 (2007)
34. Atanassov, K.: Index Matrices: Towards an Augmented Matrix Calculus. Springer, Cham (2014)
35. Atanassov, K.: Interval-Valued Intuitionistic Fuzzy Sets. Studies in Fuzziness and Soft Computing book series, vol. 388 (2020). ISBN 978-3-030-32089-8, ISBN 978-3-030-32090-4 (eBook)
36. Atanassov, K.: On Generalized Nets Theory. Sofia, "Prof. Marin Drinov", Publishing House (2007)
37. Atanassov, K.: Temporal Intuitionistic Fuzzy Sets. Comptes Rendus de l'Academie Bulgare des Sciences, Tome **44**(7), 5–7 (1991)

38. Atanassov, K.: Theory of Generalized nets (An algebraic aspect. II). In: Proceedings of II International Symposium "Automation and Scientific Instrumentation", Varna, pp. 391–396 (in Bulgarian) (1983)

39. Atanassov, K.: Theory of generalized nets (an algebraic aspect). Adv Modelling Simul 1(2), 27–33 (1984)

40. Atanassov, K.: Theory of generalized nets (a functional aspect). In: Proceedings of the IX National Youth School "Mathematical methods in Informatics", Varna, pp. 112–113 (in Bulgarian) (1983)

41. Atanassov, K.: Theory of generalized nets (a logical aspect). In: Proceedings of Summer School on Math. Logic and its Applications, Primorsko, pp. 26–29 (in Bulgarian) (1983)

42. Atanassov, K., Aladjov, H.: Generalized Nets in Artificial Intelligence, vol. 2: Generalized Nets and Machine Learning. "Prof. M. Drinov". Academic Publishing House, Sofia (2000)

43. Atanassov, K., Aladjov, H.: Generalized nets model of a new type of expert systems. Adv. Stud. Contemp. Math. 3(1), 43–58 (2001)

44. Atanassov, K., Atanassova, L., Dimitrov, E., Gargov, G., Kazalarski, I., Marinov, M., Petkov, S.: Generalized nets and expert systems. In: Methods of Operations Research, vol. 59, Proceeding of the 12th Symposium on Operations Research of the Gesellschaft fur Mathematik, Okonomie und Operations Research, 1987, pp. 301–310. Passau, Frankfurt a.M: Athenaeum (1989)

45. Atanassov, K., Atanassova, L., Dimitrov, E., Gargov, G., Kazalarski, I., Marinov, M., Petkov, S.: Generalized nets and expert systems II. In: Proceedings of International Conference "Networks Information Processing Systems", Sofia, vol. 2, pp. 54–67 (1988)

46. Atanassov, K., Atanassova, L., Dimitrov, E., Gargov, G., Kazalarski, I., Marinov, M., Petkov, S.: Generalized nets and expert systems IV. In: Proceedings of the 19th Spring Conference of the Union of Bulgaria Math., Sunny Beach, pp. 155–161 (1990)

47. Atanassov, K., Atanassova, L., Dimitrov, E., Gargov, G., Kazalarski, I., Marinov, M., Petkov, S., Stefanova-Pavlova, M.: Generalized nets and expert systems III. In: Methods of operations research, vol. 63. Proceedings of the 14th Symposium on Operations Research, Ulm, pp. 417–423 (1989)

48. Atanassov, K., Atanassova, V., Chountas, P., Shannon, A.: Generalized nets with places having intuitionistic fuzzy capacities. Notes Intuitionistic Fuzzy Sets 17(4), 21–28 (2011). ISSN: 1310-4926

49. Atanassov, K., Boumbarov, O., Gluhchev, G., Hadjitodorov, S., Shannon, A., Vassilev, A.: A generalized net model of biometric access–control system. In: Proceedings of the 9th WSEAS International Conference on Automatic Control, Modeling & Simulation, Istanbul, pp. 77–80 (2007)

50. Atanassov, K., Chountas, P., Kolev, B., Sotirova, E.: Generalized net model of an expert system with temporal components. Adv. Stud. Contemp. Math. 12(2), 255–289 (2006)

51. Atanassov, K., Chountas, P., Kolev, B., Sotirova, E.: Generalized net model of a self-developing expert system. In: Atanassov, K., Kacprzyk, J. (Eds.) Proceedings of the Tenth International Conference on Intuitionistic Fuzzy Sets. Notes on Intuitionistic Fuzzy Sets, Sofia, 28–29 Oct 2006, vol. 12, No. 3, pp. 35–40 (2006)

52. Atanassov, K., Christov, R.: New conservative extensions of the generalized nets. Adv. Modelling Anal. AMSE Press 14(2), 27–34 (1993)

53. Atanassov, K., Daskalov, M., Georgiev, P., Kim, S., Kim, Y., Nikolov, N., Shannon, A., Sorsich, J.: Generalized Nets in Neurology. "Prof. M. Drinov". Academic Publishing House, Sofia (1997)

54. Atanassov, K., Dimitrov, D., Atanassova, V.: Algorithms for tokens transfer in different types of intuitionistic fuzzy generalized nets. J. Cybern. Inform. Technol. 10(4), 22–35 (2010). ISSN: 1311-9702 (print), 1314-4081 (online)

55. Atanassov, K., Georgiev, P.: Generalized nets and expert systems VI. Adv. Modelling Anal A 21(2), 1–14 (1994)

56. Atanassov, K., Georgiev, P.: Generalized nets and expert systems VIII. Adv. Modellin Anal. A 25(3), 53–64 (1995)

57. Atanassov, K., Georgiev, P., Tetev, M.: Generalized nets and expert systems V. In: Atanassov, K. (Ed.) Applications of Generalized Nets, pp. 96–105. World Scientific Publ. Co., Singapore (1993)

58. Atanassov, K., Gluhchev, G., Hadjitodorov, S., Shannon, A., Vasilev, V.: Application of generalized nets in biometrics. Comptes Rendus de l'Academie Bulgare des Sciences, Tome 56, No. 5, pp. 13–18 (2003)

59. Atanassov, K., Gluhchev, G., Hadjitodorov, S., Savov, M., Szmidt, E.: Signature based person verification: a generalized net model. In: de Baets, B., De Caluwe, R., De Tre, G., Fodor, J., Kacprzyk, J., Zadrozny, S. (Eds.) Current issues in data and knowledge engineering. Academicka Oficyna Wydawnicza EXIT, Warszawa, pp. 33–41 (2004)

60. Atanassov, K., Gargov, G.: Interval valued intuitionistic fuzzy sets. Fuzzy Sets Syst. **31**(3), 343–349 (1989)

61. Atanassov, K., Gluhchev, G., Hadjitodorov, S., Shannon, A., Vasilev, V.: Generalized nets in image processing and pattern recognition. In: Proceedings of the Sixth International Workshop on Generalized Nets, Sofia, pp. 47–60 (2005)

62. Atanassov, K., Gluhchev, G., Hadjitodorov, S., Shannon, A., Vassilev, V.: Generalized nets and pattern recognition. KvB Visual Concepts Pty Ltd, Monograph No. 6, Sydney (2003)

63. Atanassov, K., Gluhchev, G., Hadjitodorov, S., Kacprzyk, J., Shannon, A., Szmidt, E., Vassilev, V.: Generalized nets decision making and pattern recognition. Warsaw School of Information Technology, Warszawa (2006)

64. Atanassov, K., Hadjyisky, M., Kazprzyk, J., Radeva, V., Szmidt, E.: Generalized nets and systems theory. In: IX. Proceedings of the Second International Workshop on Generalized Nets, Sofia, 26–27 June, pp. 62–72 (2001)

65. Atanassov, K., Orozova, D., Sotirova, E., Chountas, P., Tasseva, V.: Generalized net model of expert system validity testing process. In: Proceedings of the international conference of BFU, pp. 165–173 (2007)

66. Atanassov, K., Orozova, D., Sotirova, E.: Generalized net model of an intuitionistic fuzzy expert system with frame-type data bases and different forms of hypotheses estimations. Annual of Burgas Free Univ. **XXIII**, 257–262 (2010)

67. Atanassov, K., Pencheva, T.: Generalized net model of simple genetic algorithm modifications. Issues Intuitionistic Fuzzy Sets Generalized Nets **10**, 97–106 (2013)

68. Atanassov, K., Peneva, D., Tasseva, V., Sotirova, E., Orozova, D.: Generalized net model of an expert system with frame-type data bases with intuitionistic fuzzy estimations. In: First International Workshop on Intuitionistic Fuzzy Sets, Generalized Nets and Knowledge Engineering, London, 6–7 Sept, pp. 111–116 (2006)

69. Atanassov, K., Radeva, V.: On the abstract systems with properties. Adv. Stud. Contemp. Math. **4**(1), 55–71 (2001)

70. Atanassov, K., Sotirova, E.: Generalized nets. Sofia,"Prof. Marin Drinov", Publishing House (2017)

71. Atanassov, K., Sotirova, E.: On global operator G_{21} defined over generalized nets. Cybern. Inform. Technol. **4**(1), 30–40 (2004)

72. Atanassov, K., Sotirova, E., Orozova, D.: Generalized net model of an expert system with frame-type data bases. Proc. Jangjeon Math. Soc. **9**(1), 91–101 (2006)

73. Atanassov, K., Tasseva, V., Trifonov, T.: Modification of the algorithm for token transfer in generalized nets. Cybern. Inform. Technol. **7**(1), 62–66 (2007)

74. Atanassov, K., Zoteva, D., Angelova, N.: Interval-valued intuitionistic fuzzy generalized nets. Notes Intuitionistic Fuzzy Sets **24**(3), 111–123 (2018)

75. Atanassova, V.: Generalized net models of conflict resolution approaches in version control systems. Part 1: Lock-Modify-Unlock. Developments in Fuzzy Sets, Intuitionistic Fuzzy Sets, Generalized Nets and Related Topics, vol. II: Applications, Warsaw, pp. 13–24 (2010). ISBN: 978-8389475305

76. Atanassova, V.: Generalized nets with volumetric tokens. Comptes rendus de l'Academie Bulgare des Sciences **65**(11), 1489–1498 (2012). ISSN: 1310-1331

77. Atanassova, V., Atanassov, K.: Ant colony optimization approach to tokens' movement within generalized nets. In: Numerical Methods and Applications. Lecture Notes in Computer Science. No 6046, pp. 240–247. Springer, Germany (2011). ISSN 0302-9743

78. Atanassova, V., Georgiev, P.: Generalized net models of conflict resolution approaches in version control systems. Part 2: Copy-Modify-Merge. In: Proceedings of 10th International Workshop on Generalized Nets, 5 Dec, pp. 14–21 (2009), ISSN: 1313-6860

79. Atanassova, V., Fidanova, S., Chountas, P., Atanassov, K.: A generalized net with an ACO-algorithm optimization component. In: Lecture Notes in Computer Science, vol. 7116, pp. 190–197 (2012). ISSN: 0302-9743

80. Atanassova, V., Fidanova, S., Popchev, I., Chountas, P.: Chapter 5. Generalized nets, ACO algorithms, and genetic algorithms. In: Sabelfeld, K., Dimov, I. (Eds.) Monte Carlo Methods and Applications, Proceedings of the 8th IMACS Seminar on Monte Carlo Methods, August 29–September 2, 2011, Borovets, Bulgaria, De Gruyter, pp. 39–46 (2012). eBook ISBN: 9783110293586

81. Atanassova, V., Marinov, E., Velizarova, E., Sotirova, E., Atanassov, K.: Clearcutting as a forest fire prevention measure. A generalized net model. In: 14th International Workshop on Generalized Nets, Burgas, 29 Nov, pp. 11–16 (2013)

82. Boyadzieva, D., Gluhchev, G.: A GN model for on-line signature verification. In: 14th International Workshop on Generalized Nets, Burgas, 29–30 Nov, pp. 65–70 (2013)

83. Chountas, P., Atanassov, K., Sotirova, E., Bureva, V.: Generalized net model of an expert system dealing with temporal hypothesis. In: Flexible Query Answering Systems 2015, Advances in Intelligent Systems and Computing, vol. 400, pp. 473–481 (2016)

84. Chountas, P., Kolev, B., Rogova, E., Tasseva, V., Atanassov, K.: Generalized Nets in Artificial Intelligence. vol. 4, Generalized Nets, Uncertain Data and Knowledge Engineering. "Prof. M. Drinov" Academic Publishing House, Sofia (2007)

85. Dimitrov, D.G.: A graphical environment for modeling and simulation with generalized nets. Ann. "Informatics" Sect. Union Sci. Bulgaria **3**, 51–66 (2010) (in Bulgarian)

86. Dimitrov, D., Roeva, O.: Development of a generalized net for comparison of different models obtained using metaheuristic algorithms. Issues IFS GNs **13**, 109–118 (2017)

87. Dimitrov, D.G.: GN IDE—a software tool for simulation with generalized nets. In: Proceedings of Tenth International Workshop on Generalized Nets, Sofia, 5 Dec, pp. 70–75 (2009)

88. Dimitrov, D.G.: On the global operator G_2 over generalized nets. In: Proceedings of 12th International Workshop on Generalized Nets, Burgas, Bulgaria, 17 June, pp. 1–5 (2012)

89. Dimitrov, D.G.: On the global operator G_6 over generalized nets. Ann. "Informatics" Sect. Union Sci. Bulgaria **5**, 43–52 (2012)

90. Dimitrov, D.: Optimized algorithm for token transfer in generalized nets. In: Recent Advances in Fuzzy Sets, Intuitionistic Fuzzy Sets, Generalized Nets and Related Topics, vol. 1, Warsaw, SRI PAS, pp. 63–68 (2010)

91. Dimitrov, D.G.: Software products implementing generalized nets. Ann. Sect. "Informatics" Union Sci. Bulgaria **3**, 37–50 (2010) (in Bulgarian)

92. Dimitrov, D., Roeva, O.: Comparison of different mathematical models of an *E. coli* fed-batch cultivation process using generalized net model. In: 13th International Workshop on Generalized Nets, 29 Oct, London, UK, pp. 15–23 (2012)

93. Dimitrov, D., Roeva, O.: Development of generalized net for testing of different mathematical models of *E. coli* cultivation process. In: Angelov, P., Atanassov, K.T., Doukovska, L., Hadjiski, M., Jotsov, V., Kacprzyk, J., Kasabov, N., Sotirov, S., Szmidt, E., Zadrożny, S. (Eds) Advances in Intelligent Systems and Computing (Intelligent Systems'2014), vol. 322, pp. 657–668 (2015)

94. Dimitrov, E., Kazalarski, I., Atanassov, K.: Generalized net representing the basic properties of knowledge base. In: Atanassov, K., (Ed.) First Scientific Session of the "Mathematical Foundation of Artificial Intelligence" Seminar, Sofia, Oct 10: Preprint IM-MFAIS-7-89, Sofia, 1989, 12–14; in Applications of generalized nets. World Scientific Publ. Co. Singapore 1993, pp. 106–111 (1989)

95. Fidanova, S., Atanassov, K.: Generalized net models of the process of ant colony optimization. Issues Intuitionistic Fuzzy Sets Generalized Nets **7**, 108–114 (2008)
96. Fidanova, S., Atanasov, K.: Generalized net models of the process of ant colony optimization with intuitionistic fuzzy estimations. In: Proceedings of the 9th International Workshop on Generalized Nets, Sofia, Bulgaria, pp. 41–48 (2008)
97. Fidanova, S., Atanassov, K.: Generalized net models for the process of hybrid ant colony optimization. Comptes Rendus de l'Academie Bulgare des Sciences, Tome 61, No. 12, pp. 1535–1540 (2008)
98. Fidanova, S., Atanassov, K.: Generalized net models and intuitionistic fuzzy estimation of the process of ant colony optimization. Issues Intitionistic Fuzzy Sets Generalized Nets **8**, 109–124 (2010). ISBN 978-83-61551-00-3
99. Fidanova, S., Atanassov, K., Dimov, I.: Generalized nets as a tool for modelling of railway networks. In: Advanced Computing in Industrial Mathematics: Revised Selected Papers of the 10th Annual Meeting of the Bulgarian Section of SIAM, Dec 21–22, Sofia, Bulgaria, pp. 23–35 (2015)
100. Fidanova, S., Atanassov, K., Marinov, P.: Start Strategies of ACO Applied on Subset Problems. Lecture Notes in Computer Science, vol. 6046. Springer, Berlin (2011)
101. Fidanova, S., Atanassov, K., Marinov, P.: Generalized Nets in Artificial Intelligence, vol. 5, Generalized Nets and Ant Colony Optimization. "Prof. M. Drinov" Academic Publishing House, Sofia (2011)
102. Fiodanova, S., Marinov, P., Atanassov, K.: Generalized net models of the process of ant colony optimization with different strategies and intuitionistic fuzzy estimations. In: Proceedings Jangjeon Math. vol. 13, No 1, pp. 1–12 (2010). ISSN 1598-7264
103. Georgiev, P., Roeva, O., Pencheva, T., Szmidt, E.: Generalized net model of wastewater treatment process in system "Biological Reservoir - Sedimentor". Issues Intuitionistic Fuzzy Sets Generalized Nets **3**, 11–16 (2006)
104. Georgiev, P., Todorova, L., Vassilev, P.: Generalized net model of algorithm for intuitionistic fuzzy estimations. Notes Intuitionistic Fuzzy Sets **13**(2), 61–70 (2007). ISSN 1310-4926
105. Georgieva, O., Pencheva, T., Krawczak, M.: An application of generalized nets with intuitionistic fuzzy sets for modelling of biotechnological processes with distributed parameters. Issues Intuitionistic Fuzzy Sets Generalized Nets **3**, 5–10 (2006)
106. Georgieva, V., Angelova, N., Roeva, O., Pencheva, T.: Simulation of parallel processes in wastewater treatment plant using generalized net integrated development environment. Comptes rendus de l'Academie Bulgare des Sciences, Tome **69**(11), 1493–1502 (2016)
107. Georgieva, V.: Generalized net model of mechanical wastewater pre-treatment. Int. J. Bioautom. **21**(1), 133–144 (2017)
108. Georgieva, V., Roeva, O., Pencheva, T.: Generalized net model of physics-chemical wastewater treatment. J. Int. Sci. Publ. Ecol. Saf. **9**, 468–475 (2015)
109. Gluhchev, G., Atanassov, K., Hadjitodorov, S., Szmidt, E., Shannon, A.: Intuitionistic fuzzy generalized net model of the process of handwriting analysis. In: Proceedings of the 3rd Conference of the European Society for Fuzzy Logic and Technology EUSFLAT' 2003, Zittau, 10–12 Sept, pp. 218–222 (2003)
110. Gluhchev, G., Atanassov, K., Vassilev, V.: A generalized net model of scene analysis process. In: Advances in Fuzzy Sets, Intuitionistic Fuzzy Sets, Generalized Nets and Related Topics, vol. II: Applications, pp. 63–68. Academic Publishing House EXIT, Warszawa (2008)
111. Gluhchev, G., Atanassov, K., Hadjitodorov, S., Shannon, A.: A generalized net model of the process of scene analysis. Cybern. Inform. Technol. **9**(1), 13–17 (2009)
112. Gluhchev, G., Atanassov, K., Hadjitodorov, S., Szmidt, E.: Generalized net model for signature verification. In: Atanassov, K., Shannon, A. (Eds.) Proceedings of the 10th International Workshop on Generalized Nets, Sofia, 5 Dec, pp. 27–30 (2009)
113. Gluhchev, G., Baccour, L., Alimi, M., Atanassov, K.: Generalized nets for handwritten arabic word recognition. In: International Conference Automatics and Informatics, Sofia, 1–4 Oct, II-41–II-43 (2008)

114. Gluhchev, G., Atanassov, K., Hadjitodorov, S., Vasilev, V., Shannon, A.: Face recognition via generalized nets. In: Atanassov, K., Kacprzyk, J., Krawczak, M. (Eds.) Issues in Intuitionistic Fuzzy Sets and Generalized Nets. Wydawnictwo WSISiZ, Warszawa, pp. 57–60 (2004)

115. Gluhchev, G., Atanassov, K., Hadjitodorov, S.: Handwriting analysis via generalized nets. Proceedings of the International Scientific Conference on Energy and Information Systems and Technologies, vol. III, Bitola, June 7–8, pp. 758–763 (2001)

116. Gocheva, E., Sotirov, S.: Modelling of the verification by Iris scanning by generalized nets. In: 9th International Workshop on Generalized Nets, Sofia, 4 July, pp. 9–13 (2008)

117. Ismaili, S., Fidanova, S.: Representation of civilians and police officers by generalized nets for describing software agents in the case of protest. Studies of Computational Intelligence, vol. 728. Springer (2018). https://doi.org/10.1007/978-3-319-65530-7_7. ISSN:1860-949X

118. Hadjitodorov, S., Atanassov, K., Mitev, P., Shannon, A., Gluhchev, G., Vasilev, V.: Generalized net representing the process of speaker classification—identification and verification. Adv. Stud. Contemp. Math. **6**(2), 129–140 (2003)

119. Hadjitodorov, S., Mitev, P., Atanassov, K., Kolev, B., Gluhchev, G., Vasilev, V.: A generalized net description for laryngeal pathology detection without refusal option. In: Proceedings of the 3rd International Workshop on Generalized Nets, Sofia, 1 Oct, pp. 14–17 (2002)

120. Hadjitodorov, S., Mitev, P., Atanassov, K., Gluhchev, G., Vassilev, V., Shannon, A.: A generalized net description for laryngeal pathology detection excluding the refusal from classification option. Cybern. Inform. Technol. **2**, 27–32 (2002)

121. Hadjitodorov, S., Szmidt, E., Atanassov, K., Shannon, A., Gluhchev, G., Vasilev, V.: Generalized net models in speaker verification and identification with intuitionistic fuzzy estimations. In: Atanassov, K., Kacprzyk, J., Krawczak, M., Szmidt, E. (Eds.) Issues in the Representation and Processing of Uncertain Imprecise Information: Fuzzy Sets, Intuitionistic Fuzzy Sets, Generalized Nets, and Related Topics, pp. 127–140. Akademicka Oficyna Wydawnictwo EXIT, Warszawa (2005)

122. Hadjitodorov S., Orozova, D., Sotirova, E., Atanassov, K.: Generalized net model of administrative information processes in the Bulgarian academy of sciences. In: 16th International Symposium on Electrical Apparatus and Technologies, pp. 122–130. SIELA, Proceedings, 1 (2009)

123. Ilkova, T., Roeva, O., Petrov, M., Vanags, J.: Generalized net model of neuro-dynamic programming algorithm. In: 13th International Workshop on Generalized Nets, 17 June, Burgas, pp. 17–26 (2012)

124. Ilkova, T., Petrov, M., Roeva, O.: Carnitine role in human diseases. Pharmaceutical ways, optimization and generalized net description. J. Int. Sci. Publ. Mater. Methods Technol. **9**, 585–597 (2015)

125. Kapanova, K., Fidanova, S.: Generalized nets: a new approach to model a htags linguistic network on Twitter. Stud. Comput. Intell. **793**, 211–221 (2019), ISBN:978-3-319-97277-0

126. Kolev, B., El-Darzi, E., Sotirova, E., Petronias, I., Atanassov, K., Chountas, P., Kodogianis, V.: Generalized Nets in Artificial Intelligence. Vol. 3: Generalized nets, Relational Data Bases and Expert Systems. "Prof. M. Drinov" Academic Publishing House, Sofia (2006)

127. Kosev, K., Roeva, O.: Generalized net model of E. coli glicolysis control. Ann. "Informatics" Sect. Union Sci. Bulgaria **4**, 53–61 (2011)

128. Kosev, K., Roeva, O.: Generalized Net Model of the lac Operon in Bacterium E. coli. In: IEEE 6th International Conference IS 2012, Sofia, Bulgaria, vol. 2, pp. 237–241 (2012)

129. Kosev, K., Roeva, O., Atanassov, K.: Generalized Net Model Cytokinin/Auxin Interactions for Plant Root Formation, New Developments in Fuzzy Sets, Intuitionistic Fuzzy Sets, Generalized Nets and Related Topics, Volume II: Applications, IBS PAN—SRI PAS, (Systems Research Institute, Polish Academy of Sciences), pp. 91–99. Warsaw (2012)

130. Kosev, K., Roeva, O., Atanassov, K.: Generalized net model of cytokinin-auxin signalling interactions. In: Recent Advances in Fuzzy Sets, Intuitionistic Fuzzy Sets, Generalized Nets and Related Topics, Volume II: Applications, IBS PAN—SRI PAS (Systems Research Institute, Polish Academy of Sciences), pp. 93–100. Warsaw (2011)

131. Kosev, K., Ivanov, V., Ananiev, A., Denev, P., Roeva, O.: Generalized net model of interval mapping QTL analysis. Issues IFSs GNs **10**, 136–142 (2013)

132. Krawczak, M., Sotirov, S., Atanassov, K.: Multilayer Neural Networks and Generalized Nets. Warsaw School of Information Technology, Warsaw (2010)

133. Melo-Pinto, P., Kim, T., Atanassov, K., Sotirova, E., Shannon, A., Krawczak, M.: Generalized net model of E-learning evaluation with intuitionistic fuzzy estimations. In: Atanassov, K., Kacprzyk, J., Krawczak, M., Szmidt, E. (Eds.) Issues in the Representation and Processing of Uncertain Imprecise Information: Fuzzy Sets, Intuitionistic Fuzzy Sets, Generalized Nets, and Related Topics, pp. 241–249. Akademicka Oficyna Wydawnictwo EXIT, Warszawa (2005)

134. Matveev, M., Andonov, V., Atanassov, K., Milanova, M.: Generalized Net Model for Telecommunication Processes in Telecare Services. In: Proceeding of the 2013 International Conference on Electronics and Communication Systems, pp. 142–145. Rhodes Island (2013)

135. Minchev, Z., Atanassov, K.: On the possibility for generalized nets modelling of modular robotic system. Adv. Stud. Contemp. Math. **10**(2), 169–174 (2005)

136. Nedev, S., Atanassova, L.: GNs for simulation of booking systems for passenger transport. In: Atanassov, K. (Ed.) Applications of Generalized Nets, pp. 139–147. World Scientific, Singapore, New Jersey, London (1993)

137. Nikolova, M., Pencheva, T., Roeva, O.: Generalized nets model of methanization process. In: Atanassov, K., Kacprzyk, J., Krawczak, M. (Eds.) Issues in Intuitionistic Fuzzy Sets and Generalized Nets, vol. 4, pp. 95–103. Wydawnictwo WSISiZ, Warszawa (2007)

138. Novachev, N., Marinov, I., Stratiev, D., Pencheva, T., Atanassov, K.: Generalized net model of the process of evaluation of the environmental impact of refinery activity. In: Proceedings of the 13th International Workshop on Generalized Nets, London, UK, pp. 56–61, October 29 (2012). ISSN 1313-6860

139. Orozova, D., Atanassov, K.: Generalized net model of the process of selection and usage of an intelligent E-learning system. Comptes Rendus de l'Academie Bulgare des Sciences, Tome **65**(5), 591–598 (2012)

140. Orozova, D., Atanassov, K., Todorova, M.: Generaized net model of the process of personalization and usage of an E-learning environment. In: Proceedings of the Jangjeon Mathematical Society, vol. 19, No. 4, pp. 615–624 (2016)

141. Parvathi, R., Gluhchev, G., Atanassov, K.: Generalized net model of face recognition. In: Atanassov, K., Shannon, A. (Eds.) Proceedings of the 9th International Workshop on Generalized Nets, Sofia, vol. 2, pp. 102–105, 4 July (2008)

142. Parvathi, R., Sotirov, S., Gluhchev, G., Atanassov, K.: A Generalized Net Model of Intuitionistic Fuzzy Image Preprocessing. Comptes Rendus de l'Academie Bulgare des Sciences, Tome 64, No. 3, pp. 333–338 (2011)

143. Pencheva, T.: Generalized net model of an advisory system for on-line control of yeast fedbatch cultivation. In: Atanassov, K., Homenda, W., Hryniewicz, O., Kacprzyk, J., Krawczak, M., Nahorski, Z., Szmidt, E., Zadrożny, S. (Eds.) Developments in Fuzzy Sets, Intuitionistic Fuzzy Sets, Generalized Nets and Related Topics. Volume II: Applications, pp. 217–231 (2010)

144. Pencheva, T.: Intuitionistic fuzzy logic in generalized net model of an advisory system for yeast cultivation on-line control. Notes Intuitionistic Fuzzy Sets **15**(4), 45–51 (2009)

145. Pencheva, T.: Modelling of expanded advisory system for yeast cultivation on-line control using generalized nets and intuitionistic fuzzy logic. Issues Intuitionistic Fuzzy Sets Generalized Nets **9**, 101–115 (2011)

146. Pencheva, T., Ljakova, K., Roeva, O.: Modelling of an information system "GP Help" using generalized nets. In: Gorunescu, F., El-Darzi, E., Gorunescu, M. (Eds.) First East European Conference on Health Care Modelling and Computation, August 31–September 2, Romania, pp. 224–230 (2005)

147. Pencheva, T., Roeva, O., Bentes, I., Barroso, J.: Generalized Nets Model for Fixed-bed Bioreactors. In: Proceedings of the 10th ISPE International Conference on Concurrent Engineering Advanced Design, Production and Management Systems, Madeira, July 26–30, pp. 1025–1028 (2003)

148. Pencheva, T., Roeva, O., Atanassova, V., Angelova, M.: Generalized net model of lac operon. Issues IFSs GNs **10**, 183–192 (2013)

149. Pencheva, T., Georgieva, O.: Modelling of waste water purification in system "Biological Reservoir–Sedimentor" on the basis of generalized nets. In: International Symposium "Bioprocess Systems'2002 - BioPS'02", Sofia, Bulgaria, October 28–29, pp. III.23–III.26 (2002)

150. Pencheva, T., Georgiev, P., Roeva, O.: A comparison of wastewater treatment modelling with partial differential equations and generalized nets. In: Proceedings of the 1st International Workshop on Intuitionistic Fuzzy Sets, Generalized Nets and Knowledge Engineering, University of Westminister, London, UK, 6–7 Sept, pp. 105–110 (2006)

151. Pencheva, T., Novachev, N., Stratiev, D., Atanassov, K.: Generalized net model of the process of evaluation of the environmental impact of refinery activity using intuitionistic fuzzy estimations. Notes Intuitionistic Fuzzy Sets **18**(4), 32–39, 1310–4926 (2012)

152. Pencheva, T.: Generalized nets model of crossover technique choice in genetic algorithms. Issues Intuitionistic Fuzzy Sets Generalized Nets **9**, 92–100 (2011)

153. Pencheva, T., Atanassov, K., Shannon, A.: Generalized nets model of offspring reinsertion in genetic algorithms. Ann. "Informatics" Sect. Union Sci. Bulgaria **4**, 29–35 (2011)

154. Pencheva, T., Atanassov, K., Shannon, A.: Generalized net model of selection function choice in genetic algorithms. In: Recent Advances in Fuzzy Sets, Intuitionistic Fuzzy Sets, Generalized Nets and Related Topics. Vol. II: Applications, pp. 193–201. Warsaw, Systems Research Institute, Polish Academy of Sciences (2011)

155. Pencheva, T., Atanassov, K., Shannon, A.: Generalized nets model of rank-based fitness assignment in genetic algorithms. In: New Trends in Fuzzy Sets, Intuitionistic Fuzzy Sets, Generalized Nets and Related Topics. Vol. II: Applications, pp. 127–136. Warsaw, Systems Research Institute, Polish Academy of Sciences (2013)

156. Pencheva, T., Roeva, O., Shannon, A.: Generalized net models of crossover operators in genetic algorithms. In: Proceedings of 7th International Workshop on Generalized Nets, Sofia, 4 July, vol. 2, pp. 64–70 (2008)

157. Pencheva, T., Atanassov, K., Shannon, A.: Modelling of a roulette wheel selection operator in genetic algorithms using generalized nets. Int. J. Bioautom. **13**(4), 257–264 (2009)

158. Pencheva, T., Atanassov, K., Shannon, A.: Modelling of a stochastic universal sampling selection operator in genetic algorithms using generalized nets. In: Proceedings of the 10th International Workshop on Generalized Nets, Sofia, 5 Dec, pp. 1–7 (2009)

159. Pencheva, T., Roeva, O., Shannon, A.: Generalized net models of basic genetic algorithm operators. Stud. Fuzziness Soft Comput. **332**, 305–325 (2016)

160. Peneva, D., Atanassov, K., Chountas, P.: Generalized net model for signature verification. In: Atanassov, K., Shannon, A. (Eds.) Proceedings of the 10th International Workshop on Generalized Nets, Sofia, 5 Dec, pp. 59–62 (2009)

161. Peneva, D., Tasseva, V., Kodogiannis, V., Sotirova, E., Atanassov, K. (Eds.) Generalized nets as an instrument for description of the process of expert system construction. In: Proceedings of 3rd International IEEE Conference "Intelligent Systems" IS06, London, 4–6 Sept, pp. 755–759 (2006)

162. Petri, C.-A.: Kommunication mit Automaten. Ph.D. Thesis, Univ. of Bonn, 1962.; Schriften des Inst. fur Instrument. Math., No. 2, Bonn (1962)

163. Poryazov, S., Atanassov, K.: Generalized net subscribers' traffic model of communication switching systems. Adv. Modelling Anal. B **37**(1–2), 27–35 (1997)

164. Poryazov, S., Andonov, V., Saranova, E.: Comparison of conceptual models of overall telecommunication systems with QoS guarantees. In: Christiansen, H., Jaudoin, H., Chountas, P., Andreasen, T., Legind Larsen, H. (eds.) Flexible Query Answering Systems, 10333, pp. 260–268. Springer, LNCS (2017)

165. Poryazov, S., Andonov, V., Saranova, E.: Comparison of four conceptual models of a queuing system in service networks. In: Proceedings of the 26th National Conference with International Participation-TELECOM, Sofia, 25–26 Oct, 71–77 (2018)

166. Radeva, V., Krawczak, M., Choy, E.: Review and bibliography on generalized nets theory and applications. Adv. Stud. Contemp. Math. **4**(2), 173–199 (2002)

167. Radeva, V., Atanassov, K., Kim, S.K., Chang, O.B., Kim, Y.S.: Generalized net-models of an intuitionistic fuzzy abstract system. In: Proceedings of the 1999 IEEE International Fuzzy Systems Conference, Seoul, August 22–25, vol. II, pp. 1039–1044 (1999)
168. Ribagin, S.: Generalized net model of age-associated changes in the upper limb musculoskeletal structures. Comptes rendus de l'Academie bulgare des Sciences **67**(11), 1503–1512 (2014)
169. Ribagin, S., Chakarov, V., Atanassov, K.: Generalized net model of the Scapulohumeral rhythm. Recent Contributions in Intelligent Systems, pp. 229–247. Springer (2017). [book chapter]
170. Ribagin, S., Chakarov, V., Atanassov, K.: Generalized net model of the upper limb in relaxed position. In: New Developments in Fuzzy Sets, Intuitionistic Fuzzy Sets, Generalized Nets and Related Topics, vol. 2, Applications, pp. 201–210 (2014)
171. Ribagin, S., Chakarov, V., Atanassov, K.: Generalized Net Model of the Upper limb Vascular System, vol. 2 of Proceedings of IEEE 6th Conference on Intelligent Systems, 6–8 Sept Sofa, pp. 221–224 (2012)
172. Ribagin, S., Chakarov, V., Atanassov, K.: Generalized net model of the upper limb withdrawal reflex. New Developments in Fuzzy Sets, Intuitionistic Fuzzy Sets, Generalized Nets and Related Topics, vol. 2: Applications, pp. 71–82 (2014)
173. Ribagin, S., Andonov, V., Chakarov, V.: Possible applications of generalized nets with characteristics of the places. A medical example. In: Proceedings of 14th International Workshop on Generalized Nets, 29–30 Nov, Burgas, pp. 56–64 (2013)
174. Ribagin, S., Atanassov, K., Shannon, A.: Generalized net model of shoulder pain diagnosis. Issues in Intuitionistic Fuzzy Sets and Generalized Nets, vol. 11, Warsaw School of Information Technology, Warsaw, pp. 55–62 (2014)
175. Ribagin, S.: Generalized net model of osteoarthritis diagnosing. Adv. Stud. Contemp. Math. Jangjeon Math. Soc. **27**(4), (2017) (in press)
176. Ribagin, S., Chountas, P., Pencheva, T.: Generalized Net Model of Muscle Pain Diagnosing. Lecture Notes on Artificial Intelligence, 10333, pp. 269–275 (2017). ISSN 0302-9743, ISSN 1611-3349 (electronic), ISBN 978-3-319-59692-1
177. Ribagin, S., Sotirova, E., Pencheva, T.: Generalized Net Model of Adhesive Capsulitis Diagnosing. Lecture Notes in Computer Science (2017) (in press). ISBN 978-3-662-43879-4
178. Ribagin, S., Zaharieva, B., Pencheva, T.: Generalized net model of proximal humeral fractures diagnosing. Int. J. Bioautom. **22**(1), 11–20 (2018). ISSN 1314-2321 (on-line), 1314-1902 (print)
179. Ribagin, S., Roeva, O., Pencheva, T.: Generalized net model of asymptomatic osteoporosis diagnosing. IEEE 8th International Conference on Intelligent Systems, Sofia, 4–6 Sept, pp. 604–608 (2016). ISBN 978-1-5090-1353-1, IEEE Catalog Number CFP16802-USB
180. Ribagin, S., Atanassov, K., Roeva, O., Pencheva, T.: Generalized net model of adolescent idiopathic scoliosis diagnosing. In: Atanassov, K., Kacprzyk, J., Kałuszko, A., Krawczak, M., Owsiński, J., Sotirov, S., Sotirova, E., Szmidt, E., Zadrożny, S. (Eds.) Uncertainty and Imprecision in Decision Making and Decision Support: Cross-fertilization, New Models and Applications, vol. 559 of the Advances in Intelligent Systems and Computing, pp. 333–348 (2018). ISBN 978-3-319-65544-4, ISBN 978-3-319-65545-1 (eBook)
181. Ribagin, S., Vassilev, P., Pencheva, T., Zadrożny, S.: Intuitionistic fuzzy generalized net model of adolescent idiopathic scoliosis classification and the curve progression probability. Notes Intuitionistic Fuzzy Sets **23**(3), 88–95 (2017). Print ISSN 1310–4926, Online ISSN 2367-8283
182. Ribagin, S., Atanassov, K.: Generalized net model for user-oriented control of active elbow orthosis device. In: Proceedings of ADP'2018, Sozopol, Bulgaria, pp. 228–232 (2018)
183. Ribagin, S., Pencheva, T., Shannon, A.: Generalized net model of surface EMG data processing for control of active elbow orthosis device. In: Proceedings of ANNA '18, St. Konstantin and Elena Resort, Bulgaria, pp. 86–89 (2018)
184. Ribagin, S., Vassilev, P., Zoteva, D.: Generalized net model of an active elbow orthosis prototype. In: Proceedings of IWIFSGN'2018, Warsaw, Poland, September 27–28, (2018) (in press)

185. Roeva, O., Shannon, A., Pencheva, T.: Description of simple genetic algorithm modifications using generalized nets. In: IEEE 6th International Conference IS 2012, Sofia, Bulgaria, vol. 2, pp. 178–183 (2012)

186. Roeva, O., Michalíková, A.: Generalized net model of intuitionistic fuzzy logic control of genetic algorithm parameters. In: 9th International Workshop on IFSs, Banská Bystrica, 8 October 2013. Notes Intuitionistic Fuzzy Sets **19**(2), 71–76 (2013)

187. Roeva, O.: Generalized net for optimal feed rate control of fed-batch fermentation processes. In: Proceedings of the Fifth International Workshop on Generalized Nets, Sofia, Nov 10, pp. 6–12 (2004)

188. Roeva, O.: Generalized net model for foam monitoring control systems. In: Proceedings of the Eighth International Workshop on Generalized Nets, Sofia, June 26, pp. 6–10 (2007)

189. Roeva, O.: Generalized nets model for functional state recognition. In Book Series "Challenging Problems of Sciences" Computer Sciences, pp. 233–240 (2008)

190. Roeva, O.: Pedro Melo-Pinto. Generalized net model of Firefly algorithm. In: Proceedings of 14th International Workshop on Generalized Nets, Burgas, 29 Nov, pp. 22–27 (2013)

191. Roeva, O., Tzonkov, S.: Fermentation processes monitoring and control using generalized nets. Ann. "Informatics" Sect. Union Sci. Bulgaria **2**, 38–45 (2009)

192. Roeva, O., Pencheva, T.: Generalized net for control of temperature in fermentation processes. Issues in Intuitionistic Fuzzy Sets and Generalized Nets, Wydawnictwo WSISiZ, Warszawa, vol. 4, pp. 49–58 (2007)

193. Roeva, O., Pencheva, T.: Generalized net model of Brevibacterium flavul 22LD fermentation process. Int. J. Bioautom. **2**, 17–23 (2005)

194. Roeva, O., Pencheva, T.: Generalized net model of pH control system in biotechnological processes. In: Proceedings of the Seventh International Workshop on Generalized Nets, Sofia, July 14–15, pp. 20–24 (2006)

195. Roeva, O., Pencheva, T., Bentes, I., Barroso, J.: Modelling of *Escherichia coli* cultivation process on the basis of generalized nets. In: Proceedings of the 10th ISPE International Conference on Concurrent Engineering—Advanced Design, Production and Management Systems, Madeira, July 26–30, pp. 1039–1042 (2003)

196. Roeva, O., Pencheva, T., Bentes, I., Nascimento, M.M.: Modelling of temperature control system in fermentation processes using generalized nets and intuitionistic fuzzy logics. In: Notes on Intuitionistic Fuzzy Sets, Proceedings of the Ninth International Conference on Intuitionistic Fuzzy Sets, Sofia, May 7–8, vol. 11, 4, pp. 151–157 (2005)

197. Roeva, O., Pencheva, T., Atanassov, K.: Generalized net of a genetic algorithm with intuitionistic fuzzy selection operator. In: New Developments in Fuzzy Sets, Intuitionistic Fuzzy Sets, Generalized Nets and Related Topics, Volume I: Foundations, IBS PAN—SRI PAS, (Systems Research Institute, Polish Academy of Sciences), pp. 167–178. Warsaw (2012)

198. Roeva, O., Pencheva, T., Tzonkov, S.: Generalized net for carbon dioxide monitoring of fermentation. In: Atanassov, K., Boeva, V., Boyanov, K., Denchev, S., Sotirov, S., Todorova, M. (Eds.) Annual of "Informatics" Section of Union of Scientists in Bulgaria, vol. 1, pp. 93–97 (2008). ISSN 1313-6852

199. Roeva, O., Pencheva, T., Tzonkov, S.: Generalized net for proportional-integral-derivative controller. In: In Book Series "Challenging Problems of Sciences"—Computer Sciences, pp. 241–247 (2008)

200. Roeva, O., Slavov, T.: PID controller tuning of glucose control using generalized nets. In: New Developments in Fuzzy Sets, Intuitionistic Fuzzy Sets, Generalized Nets and Related Topics, Volume II: Applications, IBS PAN—SRI PAS, (Systems Research Institute, Polish Academy of Sciences), Warsaw, pp. 211–218 (2012)

201. Roeva, O.: Bat algorithm in terms of generalized net. In: Proceedings of 15th International Workshop on Generalized Nets, Burgas, pp. 1–6 (2014)

202. Roeva, O., Pencheva, T., Melo-Pinto, P.: A survey of generalized nets implementation for modelling in ecology. Chapter in: Choy, E., Krawczak, M., Shannon, A., Szmidt, E. (Eds.) A Survey of Generalized Nets, Raffles KvB Monograph 10, pp. 166–197. Raffles KvB Institute Pty Ltd, North Sydney, Australia (2007). ISBN 0-9578457-8-2

203. Roeva, O., Shannon, A.: A generalized net model of mutation operator of the breeder genetic algorithm. In: Proceedings of the 9th International Workshop on Generalized Nets, Sofia, Bulgaria, 4 July, vol. 2, pp. 59–63 (2008)
204. Roeva, O., Atanassov, K.: Generalized net model of a modified genetic algorithm. Issues Intuitionistic Fuzzy Sets Generalized Nets 7, 93–99 (2008)
205. Roeva, O., Atanassov, K., Shannon, A.: Generalized net for evaluation of the genetic algorithm fitness function. In: Proceedings of the 8th International Workshop on Generalized Nets, Sofia, Bulgaria, 26 June, pp. 48–55 (2007)
206. Roeva, O., Atanassov, K., Shannon, A.: Generalized net for selection of genetic algorithm operators. Ann. "Informatics" Sect. Union Sci. Bulgaria 1, 117–126 (2008)
207. Roeva, O., Pencheva, T.: Generalized net model of a multi-population genetic algorithm. Issues Intuitionistic Fuzzy Sets Generalized Nets 8, 91–101 (2010)
208. Roeva, O., Pencheva, T., Shannon, A., Atanassov, K.: Generalized Nets in Artificial Intelligence, vol. 7: Generalized Nets and Genetic Algorithms. "Prof. M. Drinov" Academic Publishing House, Sofia (2013)
209. Roeva O., Pencheva, T., Atanassov, K., Shannon, A.: Generalized net model of selection operator of genetic algorithms. In: Proceedings of the IEEE International Conference on Intelligent Systems, 7–9 July 2010, pp. 286–289. University of Westminster, London, UK (2010)
210. Roeva, O., Atanassova, V.: Generalized net model of cuckoo search algorithm. In: Proceedings of the 2016 IEEE 8th International Conference on Intelligent Systems, Sofia, 4–6 Sept 2016, pp. 589–592 (2016). ISBN 978-1-5090-1353-1, IEEE Catalog Number CFP16802-USB
211. Roeva, O., Michalíková, A.: Intuitionistic fuzzy logic control of metaheuristic algorithms' parameters via a generalized net. Notes Intuitionistic Fuzzy Sets 20(4), 53–58 (2014)
212. Roeva, O., Zoteva, D., Atanassova, V., Atanassov, K., Castillo, O.: Cuckoo search and firefly algorithms in terms of generalized net theory. Soft Comput. 24(2020), 4877–4898 (2020). https://doi.org/10.1007/s00500-019-04241-7
213. Shannon, A., Roeva, O., Pencheva, T., Atanassov, K.: Generalized Nets Modelling of Biotechnological Processes. "Marin Drinov" Publishing House of Bulgarian Academy of Sciences, Sofia (2004)
214. Shannon, A., Sotirova, E., Atanassov, K., Krawczak, M., Ralev, N.: A generalized net model of the informational process in an abstract university. In: Atanassov, K., Kacprzyk, J., Krawczak, M. (Eds.) Issues in Intuitionistic Fuzzy Sets and Generalized Nets. Wydawnictwo WSISiZ, Warszawa, pp. 41–47 (2004)
215. Shannon, A., Langova-Orozova, D., Sotirova, E., Petrounias, I., Atanassov, K., Melo-Pinto, P., Kim, T.: Generalized net model of information flows in intranet in an abstract iniversity. Adv. Stud. Contemp. Math. 8(2), 183–192 (2004)
216. Shannon, A., Atanassov, K., Langova-Orozova, D., Krawczak, M., Sotirova, E., Melo-Pinto, P., Petrounias, I., Kim, T.: Generalized Net Modelling of University Processes. KvB Institute of Technology (2005). ISBN 0-9578457-5-8
217. Shannon, A., Langova-Orozova, D., Sotirova, E., Petrounias, I., Atanassov, K., Krawczak, M., Melo-Pinto, P., Kim, T., Tasseva, V.: A generalized net model of the separate information flow connections within a university. Proceedings of 3rd International IEEE Conference "Intelligent Systems" IS06, London, 4–6 Sept, pp. 760–763 (2006)
218. Shannon, A., Orozova, D., Sotirova, E., Atanassov, K., Krawczak, M., Melo-Pinto, P., Nikolov, R., Sotirov, S., Kim, T.: Towards a model of the digital university: a generalized net model for producing course timetables. In: Proceedings of the 4th International IEEE Conference Intelligent Systems, Varna, 6–8 Sept, vol. 2, pp. 16-25–16-28 (2008)
219. Shannon, A., Orozova, D., Sotirova, E., Atanassov, K., Krawczak, M., Chountas, P., Georgiev, P., Nikolov, R., Sotirov, S., Kim, T.: Towards a model of digital university: a generalized net model of update existing timetable. In: Atanassov, K., Shannon, A. (Eds.) Proceedings of the 9th International Workshop on Generalized Nets, Sofia, 4 July, vol. 2, pp. 71–79 (2008)

220. Shannon, A., Orozova, D., Sotirova, E., Hristova, M., Atanassov, K., Krawczak, M., Melo-Pinto, P., Nikolov, R., Sotirov, S., Kim, T.: Towards a model of the digital university: a generalized net model for producing course timetables and for evaluation the quality measurements. In: Sgurev, V., Hadjiski, M., Kacprzyk, J. (Eds.) Intelligent System: From Theory to Practice, Studies in Computational Intelligence, vol. 299, pp. 373–382. Springer, Berlin (2010)

221. Shannon, A., Sorsich, J., Atanassov, K., Nikolov, N., Georgiev, P.: Generalized Nets in General and Internal Medicine. Academic Publishing House "Marin Drinov" (2000). ISBN 954-430-703-6

222. Shannon, A., Sotirova, E., Atanassov, K., Krawczak, M., Melo-Pinto, P., Sotirov, S., Hadjitodorov, S., Kim, T.: Generalized net model of the process of administrative servicing in a digital university with intuitionistic fuzzy estimations. In: 14th International Conference on Intuitionistic Fuzzy Sets, Sofia, 15–16 May 2010, pp. 50–56, NIFS 16 (2010)

223. Shannon, A., Riecan, B., Orozova, D., Sotirova, E., Petrounias, I., Atanassov, K., Krawczak, M., Melo-Pinto, P., Kim, T.: A generalized net model of a training system. In: Choy, E.Y.H., Krawczak, M., Shannon, A., Szmidt, E. (Eds.) A Survey of Generalized Nets, pp. 198–206. Raffles KvB Monograph No. 10, Sydney (2007)

224. Shannon, A., Orozova-Langova, D., Sotirova, E., Trelewicz, J., Atanassov, K., Melo-Pinto, P., Kim, T.: Generalized net model of a training system. Adv. Stud. Contemp. Math. 10(2), 175–179 (2005)

225. Shannon, A., Sotirova, E., Atanassov, K., Krawczak, M., Melo-Pinto, P., Sotirov, S., Kim, T.: Generalized net model of the process of ordering of university subjects. In: Proceedings of the 7th International Workshop on Generalized Nets, Sofia, 14–15 July, pp. 25–29 (2006)

226. Shannon, A., Sotirova, E., Petrounias, I., Atanassov, K., Krawczak, M., Melo-Pinto, P., Kim, T.: A generalized net model of the university electronic archive with intuitionistic fuzzy estimations of the searched information. In: Atanassov, K., Kacprzyk, J., Krawczak, M., Szmidt, E. (Eds.) Issues in Intuitionistic Fuzzy Sets and Generalized Nets, vol. 4, pp. 75–88. Wydawnictwo WSISiZ, Warszawa (2007)

227. Shannon, A., Orozova, D., Sotirova, E., Atanassov, K., Krawczak, M., Melo-Pinto, P., Kim, T.: System for electronic student-teacher interaction with intuitionistic fuzzy estimations. Notes Intuitionistic Fuzzy Sets 13(2), 81–87 (2007)

228. Shannon, A., Rieca, B., Orozova, D., Sotirova, E., Atanassov, K., Krawczak, M., Melo-Pinto, P., Parvathi, R., Kim, T.: Generalized net model of the process of selection and usage of an intelligent E-learning system. In: IS'2012 6th IEEE International Conference Intelligent Systems, Proceedings. Art. No. 6335223, pp. 233–236 (2012)

229. Shannon, A., Orozova, D., Sotirova, E., Atanassov, K., Krawczak, M., Melo-Pinto, P., Kim, T.: Generalized net model of E-learning evaluation with intuitionistic fuzzy estimations. Issues in Intuitionistic Fuzzy Sets and Generalized Nets, vol. 5, pp. 46–53. Warsaw School of Information Technology, Warsaw (2007)

230. Shannon, A., Sotirova, E., Atanassov, K., Krawczak, M., Melo-Pinto, P., Kim, T., Jang, L.C., Kang, D.-J., Rim, S.H.: A note on generalized net model of E-learning evaluation associated with intuitionistic fuzzy estimations. Int. J. Fuzzy Logic Intell. Syst. 6(1), 6–9 (2006)

231. Shannon, A., Langova-Orozova, D., Sotirova, E., Atanassov, K., Melo-Pinto, P., Kim, T.: Generalized net model for adaptive electronic assessment, using intuitionistic fuzzy estimations. In: Reusch, B. (Ed.) Computational Intelligence, Theory and Applications, pp. 291–297. Springer, Berlin (2005)

232. Shannon, A., Sotirova, E., Atanassov, K., Krawczak, M., Melo-Pinto, P., Kim, T.: Generalized net model for the reliability and standardization of assessments of student problem solving with intuitionistic fuzzy estimations. In: Developments in Fuzzy Sets, Generalized Nets and Related Topics. Applications. vol. 2, pp. 249–256. System Research Institute, Polish Academy of Science (2008)

233. Shannon, A., Sorsich, J., Atanassov, K.: Generalized Nets in Medicine. Sofia, Academic Publishing House "Marin Drinov", (1996). ISBN 954-430-501-7

234. Shannon, A., Sotirova, E., Inovska, G., Atanassov, K., Krawczak, M., Melo-Pinto, P., Kim, T.: A generalized net model of university subjects rating with intuitionistic fuzzy estimations. Notes on Intuitionistic Fuzzy Sets 18(3), 61–67 (2012)

235. Shannon, A., Langova-Orozova, D., Sotirova, E., Atanassov, K., Melo-Pinto, P., Kim, T.: Generalized net model with intuitionistic fuzzy estimations of the process of obtaining of scientific titles and degrees. In: Proceedings of the International Conference on Intuitionistic Fuzzy Sets, vol. 11, No. 3, Sofia, pp. 95–114 (2005)

236. Shannon, A., Riečnan, B., Sotirova, E., Krawczak, M., Atanassov, K., Melo-Pinto, P., Kim, T.: Modelling the process of PhD preparation using generalized nets. In: 14th International Workshop on Generalized Nets, Burgas, 29 Nov, pp. 34–38 (2013)

237. Shannon, A., Riečnan, B., Sotirova, E., Atanassov, K., Krawczak, M., Melo-Pinto, P., Parvathi, R., Kim, T.: Generalized net models of academic promotion and doctoral candidature. In: Sgurev, V., Yager, R., Kacprzyk, J., Atanassov, K. (Eds.) Recent Contributions in Intelligent Systems, pp. 263–277. Springer International Publishing Switzerland (2017)

238. Shannon, A., Gluhchev, G., Aanassov, K., Hadjitodorov, S., Vassilev, V.: A generalized net model of biometric-based access permission. Comptes Rendus de l'Academie Bulgare des Sciences Tome 60(11), 1157–1162 (2007)

239. Shannon, A., Gluhchev, G., Atanassov, K., Hadjitodorov, S.: Generalized net representing process of handwriting identification. Cybern. Inform. Technol. 1, 71–80 (2001)

240. Sotirov, S., Atanassov, K.: Generalized Nets in Artificial Intelligence. Vol. 6: Generalized Nets and Supervised Neural Networks. "Prof. M. Drinov" Academic Publishing House, Sofia (2012)

241. Sotirova, E.: On a modification of the global operator G_{21} over generalized nets. Proc. Jangjeon Math. Soc. 9(1), 65–78 (2006)

242. Sotirova, E., Atanassov, K., Chountas, P.: Generalized nets as tools for modelling of open, hybrid and closed systems: an example with an expert system. Adv. Stud. Contemp. Math. 13(2), 221–234 (2006)

243. Stefanova-Pavlova, M., Atanassov, K.: Generalized net models for flexible manufacturing systems. In: Atanassov, K. (Ed.) Applications of Generalized Nets. World Scientific Publ. Co., Singapore, pp. 172–207 (1993)

244. Stefanova-Pavlova, M., Andonov, V., Tasseva, V., Gateva, A., Stefanova, E.: Generalized Nets in Medicine: An Example of Telemedicine for People with Diabetes, pp. 327–357. Springer series Studies in Fuzziness and Soft Computing, Springer (2015)

245. Stefanova-Pavlova, M., Andonov, V., Stoyanov, T., Angelova, M., Cook, G., Klein, B., Vassilev, P., Stefanova, E.: Modeling telehealth services with generalized nets. In: Sgurev, V., Yager, R., Kacprzyk, J., Atanassov, K. (eds.) Recent Contributions in Intelligent Systems. Studies in Computational Intelligence, vol. 657, pp. 279–290. Springer, Cham (2017)

246. Stefanova-Pavlova, M., Andonov, V., Tasseva, V., Gateva, A., Stefanova, E.: Generalized nets in medicine: an example of telemedicine for people with diabetes. In: Angelov, P., Sotirov, S., (Eds.) Imprecision and Uncertainty in Information Representation and Processing. Studies in Fuzziness and Soft Computing, vol. 332. Springer, Cham

247. Stratiev, D., Marinov, I., Pencheva, T., Atanassov, K.: Generalized net model of an oil refinery. In: Proceedings of the 12th International Workshop on Generalized Nets, Burgas, Bulgaria, June 17, pp. 10–16 (2012). ISSN 1313-6860

248. Stoeva, S., Atanassov, K.: Generalized nets representation of production systems interpreters. In: Proceedings of the 15th Spring Conference of the Union of Bulg. Math., Sunny Beach, pp. 456–464 (1986)

249. Tashev, T.: Generalized net model of algorithm for non-conflict switch in packet communication node. Proceedings of the Fifth International Workshop on Generalized Nets, Sofia, 10 Nov, pp. 19–24 (2004)

250. Todorova, L., Bentes, I., Barroso, J., Temelkov, A.: A generalized net model of the treatment of mechanically ventilated patients with nosocomial pneumonia. Proceedings of the 10th ISPE International Conference on Concurrent Engineering, 26–30 July, Madeira, pp. 1057–1063 (2003)

251. Todorova, L., Georgiev, P., Vassilev, P., Szmidt, E.: A GN model for simultaneous calculation of estimates for pattern recognition problems in medicine. In: Choy, E.Y.H., Krawczak, M., Shannon, A., Szmidt, E. (Eds.) A Survey of Generalized Nets, pp. 141–165. Raffles KvB Monograph No.10, Sydney (2007). ISBN 0-9578457-8-2

252. Todorova, L., Ignatova, V., Haralanov, L.: Generalized net model for diagnosis of multiple sclerosis. In: Proceedings of 12th International Workshop on Generalized Nets, 17 June, Burgas, pp. 32–38 (2012)

253. Todorova, L., Vassilev, P., Ignatova, V.: A generalized net model for assessment of the degree of disability in patients with multiple sclerosis based on the abnormalities of visual evoked potentials. Issues Intuitionistic Fuzzy Sets Generalized Nets **10**, 173–182 (2013). ISBN 978-83-61551-08-9

254. Todorova, L., Ignatova, V., Hadjitodorov, S., Vassilev, P., Model, G.N., for Monitoring the Degree of Disability in Patients with Multiple Sclerosis Based on Neurophysiologic Criteria. Studies in Fuzziness and Soft Computing, pp. 289–303. ISSN 1434–9922 ISSN 1860–0808 (electronic) Studies in Fuzziness and Soft Computing ISBN 978-3-319-26301-4 ISBN 978-3-319-26302-1 (eBook). DOI (2016). https://doi.org/10.1007/978-3-319-26302-1

255. Todorova, L., Sorsich, J.: Generalized net model of the mechanical ventilation process. In: Proceedings of the Second International Workshop on Generalized Nets, Sofia, 26–27 June, pp. 34–39 (2001)

256. Todorova, L., Temelkov, A., Antonov, A.: GN model of transition to spontaneous breathing after long term mechanical ventilation. First East European Conference on Health Care Modelling and Computation, Romanis, 31 August–2 September, pp. 300–308 (2005). ISBN 973-7757-67-X

257. Todorova, L., Vassilev, P., Georgiev, P.: Generalized net model of aggregation algorithm for intuitionistic fuzzy estimates of classification. Issues Intuitionistic Fuzzy Sets Generalized Nets **5**, 54–63 (2007). ISBN 987-83-88311-90-1

258. Todorova, L., Vassilev, P., Matveev, M., Krasteva, V., Jekova, I., Hadjitodorov, S., Georgiev, G., Milanov, S.: Generalized net model of a protocol for weaning from mechanical ventilation. Comptes rendus de l'Academie bulgare des Sciences **66**(10), 1385–1393 (2013)

259. Todorova, L., Vassilev, P., Surchev, J.: Diagnostic method before implantation of cerebrospinal fluid draining shunt in infants. A generalized net model. In: 13th International Workshop on Generalized Nets, London, 29 Oct, pp. 77–82 (2012). ISSN 1313-6860

260. Toleva, S., Atanassov, K., Fukuda, T.: A generalized net model of an industrial robot. In: Atanassov, K., Kacprzyk, J., Krawczak, M. (Eds.) Issues in Intuitionistic Fuzzy Sets and Generalized Nets, vol. 2, pp. 29–36. Wydawnictwo WSISiZ, Warszawa (2004)

261. Tomov, Z., Krawczak, M., Andonov, V., Dimitrov, E., Atanassov, K.: Generalized net models of queuing disciplines in finite buffer queuing systems. In: Proceedings of the 16th International Workshop on Generalized Nets, Sofia, 9–10 February, pp. 1–9 (2018)

262. Tomov, Z., Krawczak, M., Andonov, V., Atanassov, K., Simeonov, S.: Generalized net models of queueing disciplines in finite buffer queueing systems with intuitionistic fuzzy evaluations of the tasks. Notes Intuitionistic Fuzzy Sets **25**(2), 115–122 (2019)

263. Trifonov, T., Atanassov, K.: On some generalized net models of multiagent systems (part 1). In: Atanassov, K., Kacprzyk, J., Krawczak, M. (Eds.) Issues in Intuitionistic Fuzzy Sets and Generalized Nets, vol. 2, pp. 59–66. Wydawnictwo WSISiZ, Warszawa (2004)

264. Trifonov, T., Atanassov, K.: On some generalized net models of multiagent systems. Ministerului Educatiei si Cercetarii **VII**(37), 49–52 (2002)

265. Trifonov, T., Georgiev, K.: GNTicker: a software tool for efficient interpretation of generalized net models. Issues Intuitionistic Fuzzy Sets Generalized Nets **3**, 71–78 (2005)

266. Valkov, I., Atanassov, K., Doukovska, L.: Generalized nets as a tool for modeling of the urban bus transport. In: Christiansen, H., Jaudoin, H., Chountas, P., Andreasen, T., Larsen, H.L. (Eds.) Flexible Query Answering Systems, Lecture Notes in Artificial Intelligence, vol. 10333, pp. 276–285. Springer, Cham (2017)

267. Vardeva, I., Atanassov, K.: Generalized net model for protection of patients' personal E-data, transmitted between healthcare information systems over the internet. Comptes Rendus de l'Academie Bulgare des Sciences, Tome **65**(3), 291–298 (2012)

268. Vassilev, P., Todorova, L.: Generalized net model of a K-nearest neighbor rule pattern recognition algorithm for the case of intuitionistic fuzzy sets. In: Atanassov, K., Shannon, A. (Eds.) Proceedings of the Tenth International Workshop on Generalized Nets, pp. 8–13. Sofia, 5 Dec. (2009). ISSN 1313-6852

269. Zoteva, D.: Implementation of hierarchical operators in GN IDE. In: Atanassov, K.T., Kacprzyk, J., et al. (Eds.) Uncertainty and Imprecision in Decision Making and Decision Support: Cross fertilization, New Models and Applications. Springer, Cham (2019) (in press)

270. Zoteva, D., Atanassova, V., Roeva, O., Szmidt, E.: Generalized net model of Artificial Bee Colony optimization algorithm. In: ANNA'18; Advances in Neural Networks and Applications 2018, pp. 53–58. VDE VERLAG GMBH Berlin Offenbach (2018). ISBN: 978-3-8007-4756-6

271. Zoteva, D., Krawczak, M.: Generalized nets as a tool for the modelling of data mining processes. A survey. Issues Intuitionistic Fuzzy Sets Generalized Nets **13**, 1–60 (2017)

272. Zoteva, D., Roeva, O., Atanassova, V.: Generalized net model of artificial bee colony optimization algorithm with intuitionistic fuzzy parameter adaptation. Notes Intuitionistic Fuzzy Sets **24**(3), 79–91 (2018). https://doi.org/10.7546/nifs.2018.24.3.79-91

273. Zoteva, D., Szmidt, E., Kacprzyk, J.: Generalized nets with additional intuitionistic fuzzy conditions for tokens transfer. Notes Intuitionistic Fuzzy Sets **25**(2), 104–114 (2019)

On Intuitionistic Fuzziness

Peter Vassilev, Lyudmila Todorova, and Evgeniy Marinov

Abstract In the present work we try to provide an overview of the main results related to intuitionistic fuzzy sets and related concepts contributed by scientists from the Bulgarian Academy of Sciences from the very birth of the concept to present day. Of course, it is impossible not to mention briefly the contributions of other authors but we will try to keep the focus of our exposition on the results of the groups that have worked within the Academy.

Keywords Intuitionistic fuzzy sets · Interval-valued intuitionistic fuzzy sets · Intuitionistic fuzzy logic · Operators · Extensions

1 Introduction

The idea of intuitionistic fuzziness appeared for the first time in 1983 in [14]. This concept relates two rather different but in some sense connected ideas—the intuitionism of L. Brouwer firstly introduced in a speech in 1912 [33] and the fuzzy sets of L. Zadeh introduced in 1965 [82]. Both ideas deal with the train of thought which tries to do away with the law of excluded middle (LEM) which dates back at least to the great Aristotle, who gave its first surviving recorded formulation. The fact that LEM has been crucial in the development of modern science is completely undeniable. However, since our aim in this work is not to provide an extensive review of the philosophical and logical consequences of surrendering the certainty provided by LEM (and the related subtleties involved), we shall not delve deeper into this topic. The interested reader is referred to e.g. [38, 59]. We will restrict ourselves to a brief overview of the historical facts and will keep our attention on the relevant concepts, insomuch and insofar as they are useful, necessary and relevant to our exposition.

P. Vassilev (✉) · L. Todorova · E. Marinov
Bulgarian Academy of Sciences, Institute of Biophysics and Biomedical Engineering,
Acad. G. Bonchev Str. bl. 105, 1113 Sofia, Bulgaria
e-mail: lpt@biomed.bas.bg

© The Author(s), under exclusive license to Springer Nature Switzerland AG 2021
K. T. Atanassov (ed.), *Research in Computer Science in the Bulgarian Academy of Sciences*, Studies in Computational Intelligence 934,
https://doi.org/10.1007/978-3-030-72284-5_11

227

Before proceeding any further we have to mention several notions required before the introduction of the idea of intuitionistic fuzziness.

Historically speaking, the first notion related to intuitionistic fuzziness introduced was the intuitionistic fuzzy set. What essentially distinguishes it from the concept of fuzzy set is the existence of the logical modal operators "necessity" and "possibility", which have no analogue in the fuzzy case.

2 Basic Definitions and Notions

The original definition of the concept intuitionistic fuzzy set (IFS) (see [14]) is the following:

Definition 1 (*cf.* [17]) An intuitionistic fuzzy set A in E is defined as an object of the following form:

$$A = \{\langle x, \mu_A(x), \nu_A(x) \rangle | x \in E\} \tag{1}$$

where the functions:

$$\mu_A : E \to [0, 1] \tag{2}$$

and

$$\nu_A : E \to [0, 1] \tag{3}$$

define the degree of membership and the degree of non-membership of the element $x \in E$, respectively, and for every $x \in E$:

$$0 \le \mu_A(x) + \nu_A(x) \le 1. \tag{4}$$

The quantity

$$\pi_A(x) := 1 - \mu_A(x) - \nu_A(x) \tag{5}$$

is called hesitancy or indeterminacy degree of the element $x \in E$.

Remark 1 In a sense a fuzzy set may be thought of, and in the scientific literature is usually given, as an IFS A such that

$$A = \{\langle x, \mu_A(x), 1 - \mu_A(x) \rangle | x \in E\} \tag{6}$$

with μ_A given by (2).

Here we propose a slightly more general definition of the concept of intuitionistic fuzziness which includes not only the above but also several other related concepts.

Definition 2 A set of intuitionistic fuzzy nature (SIFN) A in E is defined as an object of the following form:

$$A = \{\langle x, \widehat{M}_A(x), \widehat{N}_A(X)\rangle | x \in E\} \qquad (7)$$

where the functions:

$$\widehat{M}_A(x) : E \rightarrow M^*([0, 1]) \qquad (8)$$

($M^*([0, 1])$ being a set representable as union of subintervals of the unit interval) and

$$\widehat{N}_A(x) : E \rightarrow N^*([0, 1]) \qquad (9)$$

($N^*([0, 1])$ being a set representable as a union of subintervals of the unit interval) define the degree of membership and the degree of non-membership of the element $x \in E$, respectively, and for every $x \in E$:

$$0 \leq f(\widehat{M}_A(x), \widehat{N}_A(x)) \leq 1, \qquad (10)$$

where f is a suitably chosen function.

Remark 2 The above definition includes in itself IFS, interval-valued intuitionistic fuzzy sets (IVIFS) [25], intuitionistic fuzzy sets of p-th type (IFSp) [29], and d-intuitionistic fuzzy sets (d-IFS) [68], for instance, as particular cases for suitable choices of M^*, N^* and f. Note, that for all these sets we invariably have

$$f(\widehat{M}_A(x), \widehat{N}_A(x)) = f(\widehat{N}_A(x), \widehat{M}_A(x)). \qquad (11)$$

One may be tempted to consider functions that do not have this property, but this is beyond the scope of this work, and is left to the inquisitive reader for possible investigations.

We are also obliged to mention that the apparent generality of Definition 2 is a bit misleading—in order to investigate the particular properties of each of the above sets, one has to investigate the particular behaviour of the function f and of the sets M^* and N^* in detail. However, here we have chosen it in order to highlight just how powerful the concept of intuitionistic fuzziness is and how general and encompassing it can easily be.

3   General Overview of the Main Results on Intuitionistic Fuzziness

3.1   Theoretical Results

3.1.1   Intuitionistic Fuzzy Sets—Basic Facts and Geometrical Interpretations

Most of the results on intuitionistic fuzzy sets may be found collected in several monographs—see e.g. [17, 21, 28]. We will provide a brief and concise summary outlying the main results.

As our starting point we will consider the different geometric interpretations which have been employed for the visualization of the intuitionistic fuzzy sets. The first and somewhat simplest one is the following (Fig. 1).

The second widely used geometric interpretation may be derived from the first by simple using the values of $1 - \nu$ instead of the values of ν (Fig. 2).

The third geometrical interpretation predominantly in use mostly by the Bulgarian researchers is the so-called interpretation triangle (Fig. 3).

We are obliged to mention that several other interpretations exist and are used. The most notable among them are a three dimensional geometric interpretation discussed

Fig. 1 The first geometrical interpretation using μ-membership and ν-non-membership degrees

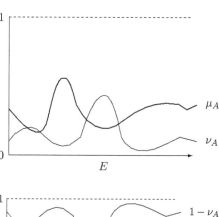

Fig. 2 The interpretation of IFS with μ-the membership and $1 - \nu$-complementary non-membership degrees

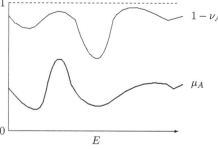

Fig. 3 The interpretation
triangle for IFSs

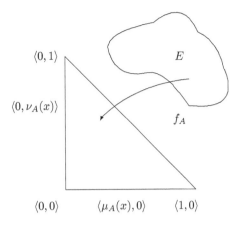

Fig. 4 A radar chart
interpretation for an IFS

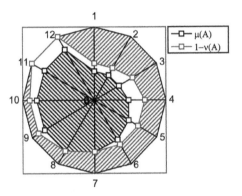

in [66]. Another interesting idea is the spherical interpretation proposed in [80]. And
last, but not least, the idea of V. Atanassova—radar chart interpretation as given in
[30]—see Fig. 4.

Clearly, the choice of the type of the most appropriate interpretation depends
on the problem under consideration. Currently, in most practical applications the
interpretation from Fig. 3 is preferred for the data visualization.

Considering other ways to interpret intuitionistic fuzzy sets, it is possible to think
of them as sets for which each element of the universe set is labeled by an *intuitionistic
fuzzy pair* (IFP). The formal definition goes as follows:

Definition 3 (*cf.* [27]) An *intuitionistic fuzzy pair* is an object having the form $\langle a, b \rangle$,
where $a, b \in [0, 1]$ and $a + b \leq 1$.

The values a and b may be interpreted as degrees of membership and non-
membership, or degrees of validity and non-validity, or degree of correctness and
non-correctness, etc.

In some sense, an IFP may be thought of an IFS with universe set consisting of
exactly one element x, which, for simplicity, is omitted in the denotation. In this

view, we can represent all relations, operations, etc. with the ordering and algebraic operations defined over IFPs. In such treatment all properties of IFSs are derived from the structure of the lattice defined by the different underlying IFPs, and this allows results valid for these lattices to be "mechanically" transferred to the theory of IFSs.

Similar reasoning has been adopted, for instance, in the implementations of the InterCriteria Analysis (see [26]) and the study of different types of assignment problems in intuitionistic fuzzy environment [78, 79]. This also allows to easily compare IFSs with other extensions of fuzzy sets, by finding a suitable bijective correspondence between the IFPs and the membership-function values (possibly piece-wise for disjoint cases) and establishing whether there is an equipollence between the two concepts.

3.1.2 Basic Operations and Relations. Modal and Extended Modal Operators

Returning, to the classical point of view on IFSs, we remind and highlight some of the most important aspects that differentiate them from other extensions of fuzzy sets.

Before we continue to the modal and extended modal operators, that have no analogues in the case of fuzzy sets, we briefly present the basic relations and operations between two IFS two IFSs A and B the following operations and relations can be defined (everywhere below as customary "iff" is a shorthand for "if and only if"):

$$A \subset B \text{ iff } (\forall x \in E)((\mu_A(x) < \mu_B(x) \text{ \& } \nu_A(x) \geq \nu_B(x))$$
$$\lor (\mu_A(x) \leq \mu_B(x) \text{ \& } \nu_A(x) > \nu_B(x))$$
$$\lor (\mu_A(x) < \mu_B(x) \text{ \& } \nu_A(x) > \nu_B(x)));$$

$$A \supset B \text{ iff } B \subset A;$$

$$A \subseteq B \text{ iff } (\forall x \in E)(\mu_A(x) \leq \mu_B(x) \text{ \& } \nu_A(x) \geq \nu_B(x));$$

$$A \supseteq B \text{ iff } B \subseteq A;$$

$$A = B \text{ iff } (\forall x \in E)(\mu_A(x) = \mu_B(x) \text{ \& } \nu_A(x) = \nu_B(x));$$

$$\overline{A} = \{\langle x, \nu_A(x), \mu_A(x)\rangle | x \in E\};$$

$$A^c = \{\langle x, \nu_A(x), \mu_A(x)\rangle | x \in E\};$$

$$A \cap B = \{\langle x, \min(\mu_A(x), \mu_B(x)), \max(\nu_A(x), \nu_B(x))\rangle | x \in E\};$$

$$A \cup B = \{\langle x, \max(\mu_A(x), \mu_B(x)), \min(\nu_A(x), \nu_B(x))\rangle | x \in E\};$$

$$A + B = \{\langle x, \mu_A(x) + \mu_B(x) - \mu_A(x).\mu_B(x), \nu_A(x).\nu_B(x)\rangle | x \in E\};$$

$$A.B = \{\langle x, \mu_A(x).\mu_B(x), \nu_A(x) + \nu_B(x) - \nu_A(x).\nu_B(x)\rangle | x \in E\};$$

$$A @ B = \left\{ \left\langle x, \frac{\mu_A(x) + \mu_B(x)}{2}, \frac{\nu_A(x) + \nu_B(x)}{2} \right\rangle | x \in E \right\}.$$

We remark that in the above notations the conjunction (&) and disjunction ($\vee$) are to be understood in the sense of classical logic. Also, the complement (denoted by A^c) and the (classical) negation (denoted by $\overline{A}$) coincide. Other types of negations, have also been introduced but we will discuss them later.

> We emphasize that the operations + and . above have nothing to do with addition and multiplication between intuitionistic fuzzy sets, *as has been wrongly assumed by some authors*, but since these are the original denotations used when they were introduced we have kept them as such here to avoid further confusion.

Six versions of Cartesian products of two IFSs have been defined (see [1, 21]). Let E_1 and E_2 be two universes and let

$$A = \{\langle x, \mu_A(x), \nu_A(x)\rangle | x \in E_1\},$$
$$B = \{\langle y, \mu_B(y), \nu_B(y)\rangle | y \in E_2\},$$

be two IFSs over E_1 and over E_2, respectively.

$$A \times_1 B = \{\langle\langle x, y\rangle, \mu_A(x).\mu_B(y), \nu_A(x).\nu_B(y)\rangle | x \in E_1 \& y \in E_2\},$$
$$A \times_2 B = \{\langle\langle x, y\rangle, \mu_A(x) + \mu_B(y) - \mu_A(x).\mu_B(y), \nu_A(x).\nu_B(y)\rangle$$
$$|x \in E_1 \& y \in E_2\},$$
$$A \times_3 B = \{\langle\langle x, y\rangle, \mu_A(x).\mu_B(y), \nu_A(x) + \nu_B(y) - \nu_A(x).\nu_B(y)\rangle$$
$$|x \in E_1 \& y \in E_2\},$$
$$A \times_4 B = \{\langle\langle x, y\rangle, \min(\mu_A(x), \mu_B(y)), \max(\nu_A(x), \nu_B(y))\rangle | x \in E_1 \& y \in E_2\},$$
$$A \times_5 B = \{\langle\langle x, y\rangle, \max(\mu_A(x), \mu_B(y)), \min(\nu_A(x), \nu_B(y))\rangle | x \in E_1 \& y \in E_2\},$$
$$A \times_6 B = \{\langle\langle x, y\rangle, \frac{\mu_A(x) + \mu_B(y)}{2}, \frac{\nu_A(x) + \nu_B(y)}{2}\rangle | x \in E_1 \& y \in E_2\}.$$

We are now ready to discuss the modal operators and their extensions. As we stated previously, the modal operators have no counterparts in the ordinary fuzzy set theory. What they do is transform a given IFS to a fuzzy set in certain way. Following [14, 15, 17], we give the definition of the modal operators, whose name was chosen since they are similar to the operators *"necessity"* and *"possibility"* defined in some modal logics (cf. [39]).

Definition 4 (*cf.* [17]) Let for every IFS A, defined over the universe set E.

$$\Box A \overset{\text{def}}{=} \{\langle x, \mu_A(x), 1 - \mu_A(x)\rangle | x \in E\}, \tag{12}$$

$$\Diamond A \overset{\text{def}}{=} \{\langle x, 1 - \nu_A(x), \nu_A(x)\rangle | x \in E\}. \tag{13}$$

Obviously, when A is an ordinary fuzzy set, then $\Box A = A = \Diamond A$, and hence these operators are meaningless.

Fig. 5 The geometrical
interpretations of the
operators necessity □ and
possibility ◊

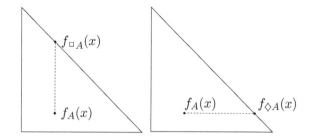

The geometrical interpretation of the two operators are the following (Fig. 5).

These operators can in turn be extended further by the operator D_α, with $0 \leq \alpha \leq 1$ and its generalization $F_{\alpha,\beta}$, where $0 \leq \alpha, \beta \leq 1$ and $0 \leq \alpha + \beta \leq 1$.

Definition 5 (*cf.* [17]) For any IFS A and $0 \leq \alpha \leq 1$, the operator D_α is defined as follows:

$$D_\alpha(A) = \{\langle x, \mu_A(x) + \alpha.\pi_A(x), \nu_A(x) + (1 - \alpha).\pi_A(x)\rangle | x \in E\}.$$

From this definition it is easily seen that that $D_0(A) = \square A$ and $D_1(A) = \lozenge A$, also in the general case the result of the application of D_α is a fuzzy set since:

$$\mu_A(x) + \alpha.\pi_A(x) + \nu_A(x) + (1 - \alpha).\pi_A(x) = \mu_A(x) + \nu_A(x) + \pi_A(x) = 1.$$

Definition 6 (*cf.* [17]) For any IFS A, and IFP $\langle \alpha, \beta \rangle$ the operator $F_{\alpha,\beta}$ is defined as follows:

$$F_{\alpha,\beta}(A) = \{\langle x, \mu_A(x) + \alpha.\pi_A(x), \nu_A(x) + \beta.\pi_A(x)\rangle | x \in E\}.$$

We see that D_α coincides with $F_{\alpha,1-\alpha}$

These may be generalized further to point-wise operators, which is achieved in most concise form as follows:

Definition 7 (*cf.* [21]) For any IFS A, and a fixed IFS B the operator F_B is defined as follows:

$$F_B(A) = \{\langle x, \mu_A(x) + \mu_B(x).\pi_A(x), \nu_A(x) + \nu_B(x).\pi_A(x)\rangle | x \in E\}.$$

Besides these modal and extended operators, there have been other operators defined in a way similar to them.

We present some of them below

$$G_{\alpha,\beta}(A) = \{\langle x, \alpha.\mu_A(x), \beta.\nu_A(x)\rangle | x \in E\},$$

with $\alpha, \beta \in [0, 1]$.

$$H_{\alpha,\beta}(A) = \{\langle x, \alpha.\mu_A(x), \nu_A(x) + \beta.\pi_A(x)\rangle | x \in E\},$$
$$H^*_{\alpha,\beta}(A) = \{\langle x, \alpha.\mu_A(x), \nu_A(x) + \beta.(1 - \alpha.\mu_A(x) - \nu_A(x))\rangle | x \in E\},$$
$$J_{\alpha,\beta}(A) = \{\langle x, \mu_A(x) + \alpha.\pi_A(x), \beta.\nu_A(x)\rangle | x \in E\},$$
$$J^*_{\alpha,\beta}(A) = \{\langle x, \mu_A(x) + \alpha.(1 - \mu_A(x) - \beta.\nu_A(x)), \beta.\nu_A(x)\rangle | x \in E\},$$

with $\langle \alpha, \beta \rangle$ an IFP.

The operator, that generalizes all the above operators is the following one:

$$X_{a,b,c,d,e,f}(A) = \{\langle x, a.\mu_A(x) + b.(1 - \mu_A(x) - c.\nu_A(x)),$$
$$d.\nu_A(x) + e.(1 - f.\mu_A(x) - \nu_A(x))\rangle | x \in E\}$$

where $a, b, c, d, e, f \in [0, 1]$ and $\langle a, e(1 - f)\rangle$, $\langle d, b(1 - c)\rangle$, and $\langle b, e\rangle$ are IFPs.

We conclude our outline on modal and extended modal operators by mentioning some similar to them operators.

In [32] two new operators bearing resemblance to the modal operators are proposed and their properties are studied. The first one is given by:

$$CSC_F(A) = \left\{\langle x, 1 - u_A(x), u_A(x)\rangle \,\middle|\, x \in E\right\},$$

where

$$u_A(x) \overset{def}{=} \frac{\sqrt{2((1 - \mu_A(x))^2 + \nu_A^2(x))}}{2}$$

and the second by:

$$CSC_T(A) = \left\{\langle x, w_A(x), 1 - w_A(x)\rangle \,\middle|\, x \in E\right\},$$

where

$$w_A(x) \overset{def}{=} \frac{\sqrt{2(\mu_A^2(x) + (1 - \nu_A(x))^2)}}{2}$$

The following inclusions are also stated in [32]:

$$\Box A \subseteq CSC_T(A) \subseteq CSC_F(A) \subseteq \Diamond A.$$

In [70] two new modal-like operators are introduced based on the power mean M_p given for any two non-negative numbers x and y by:

$$M_p(x, y) = \left(\frac{x^p + y^p}{2}\right)^{\frac{1}{p}}, \tag{14}$$

where $p \in [-\infty, +\infty]$ is a parameter. The following special cases deserve mentioning:

$$M_{-\infty}(x, y) \overset{\text{def}}{=} \min(x, y)$$

$$M_{+\infty}(x, y) \overset{\text{def}}{=} \max(x, y)$$

and

$$M_0(x, y) \overset{\text{def}}{=} \sqrt{xy}$$

Since we have

$$\square A = \left\{ \langle x, \mu_A(x), M_{+\infty}(\nu_A(x), 1 - \mu_A(x)) \rangle \,\middle|\, x \in E \right\},$$

and

$$\Diamond A = \left\{ \langle x, M_{+\infty}(\mu_A(x), 1 - \nu_A(x)), \nu_A(x) \rangle \,\middle|\, x \in E \right\},$$

the authors have introduced by analogy the operators

$$\square_{M_p} A = \left\{ \langle x, \mu_A(x), M_p(\nu_A(x), 1 - \mu_A(x)) \rangle \,\middle|\, x \in E \right\},$$

and

$$\Diamond_{M_p} A = \left\{ \langle x, M_p(\mu_A(x), 1 - \nu_A(x)), \nu_A(x) \rangle \,\middle|\, x \in E \right\},$$

in the general case. The following properties are preserved for the so-defined operators:

$$\overline{\square_{M_p} \overline{A}} = \Diamond_{M_p} A$$
$$\overline{\Diamond_{M_p} \overline{A}} = \square_{M_p} A$$
$$\square_{M_p} A \subseteq A \subseteq \Diamond_{M_p} A$$

When $p \neq +\infty$ however, in general, we have:

$$\square_{M_p}\square_{M_p} A \neq \square_{M_p} A$$
$$\Diamond_{M_p}\Diamond_{M_p} A \neq \Diamond_{M_p} A$$

Another modal-like operator also defined in [70] is the following:

$$\boxed{\Diamond}_p A = \left\{ \langle x, M_p(\mu_A(x), 1 - \nu_A(x)), M_p(\nu_A(x), 1 - \mu_A(x)) \rangle \,\middle|\, x \in E \right\},$$

which is well-defined for $p \leq 1$.

3.1.3 Other Types of Operators. Topological and Level Operators

We will briefly mention some of the other types of operators defined over IFSs. The topological operators *closure* (C) and *interior* (I) and their variations thereof.

In [21], nine different topological operators are presented and their basic properties are given. They are analogous to the classical topological operators of closure and interior (see, e.g. [44, 81]). The first two of them have been introduced in 1983 some months after the appearance of the IFSs (see [15]).

Definition 8 (*cf.* [21]) For every IFS A,

$$C(A) = \{\langle x, K, L \rangle | x \in E\}, \tag{15}$$

where

$$K \stackrel{\text{def}}{=} \sup_{y \in E} \mu_A(y), \tag{16}$$

$$L \stackrel{\text{def}}{=} \inf_{y \in E} \nu_A(y) \tag{17}$$

and

$$I(A) = \{\langle x, k, l \rangle | x \in E\}, \tag{18}$$

where

$$k \stackrel{\text{def}}{=} \inf_{y \in E} \mu_A(y), \tag{19}$$

$$l \stackrel{\text{def}}{=} \sup_{y \in E} \nu_A(y). \tag{20}$$

The following operators (see [21]) are extensions of the two topological operators C and I:

$$C_\mu(A) = \{\langle x, K, \min(1 - K, \nu_A(x)) \rangle | x \in E\};$$
$$C_\nu(A) = \{\langle x, \mu_A(x), L \rangle | x \in E\};$$
$$I_\mu(A) = \{\langle x, k, \nu_A(x) \rangle | x \in E\};$$
$$I_\nu(A) = \{\langle x, \min(1 - l, \mu_A(x)), l \rangle | x \in E\},$$

where K, L, k, l are defined by (16), (17) and (19) and (20), respectively.

In [21], the two new topological operators C_μ^* and I_ν^* are introduced as

$$C_\mu^*(A) = \{\langle x, \min(K, 1 - \nu_A(x)), \min(1 - K, \nu_A(x))\rangle | x \in E\},$$

$$I_\nu^*(A) = \{\langle x, \min(1 - l, \mu_A(x)), \min(l, 1 - \mu_A(x))\rangle | x \in E\},$$

where K, L, k, l are defined by (16), (17) and (19) and (20), respectively.

In [21, 24], *"weight-center operator"* over a given IFS A is introduced by:

$$W(A) = \left\{ \left\langle x, \frac{\sum\limits_{y \in E} \mu_A(y)}{\text{card}(E)}, \frac{\sum\limits_{y \in E} \nu_A(y)}{\text{card}(E)} \right\rangle | x \in E \right\},$$

with $\text{card}(E)$ denoting cardinality of the set E. For the continuous case, the summation is replaced by integration (the membership and non-membership functions are assumed measurable) over E and $\text{card}(E)$ in the above formula is to be understood as the measure of E.

Another two operators of topological nature are T and U, defined as follows:

Definition 9 *(cf.* [19] *)*

$$T(A) = \left\{ \left\langle x, \frac{\mu_A(x)}{\sup\limits_{y \in E}(\mu_A(y) + \nu_A(y))}, \frac{\nu_A(x)}{\sup\limits_{y \in E}(\mu_A(y) + \nu_A(y))} \right\rangle | x \in E \right\}. \tag{21}$$

Definition 10 *(cf.* [20])

$$U(A) = \left\{ \left\langle x, \frac{\mu_A(x) - \inf\limits_{y \in E} \mu_A(y)}{u_{\text{inf}}(A)}, \frac{\nu_A(x) - \inf\limits_{y \in E} \nu_A(y)}{u_{\text{inf}}(A)} \right\rangle | x \in E \right\}, \tag{22}$$

where $u_{\text{inf}}(A) = 1 - \inf\limits_{y \in E} \mu_A(y) - \inf\limits_{y \in E} \nu_A(y) - \inf\limits_{y \in E} \pi_A(y)$.

We next turn our attention to the so-called level and level type operators and then to some other operators defined over IFSs.

First we remind the definition of the level operator $N_{\alpha,\beta}$.

Definition 11 *(cf.* [21]) For any IFS A and an IFP $\langle \alpha, \beta \rangle$ we can define the level operator $N_{\alpha,\beta}$ as follows:

$$N_{\alpha,\beta}(A) = \{\langle x, \mu_A(x), \nu_A(x)\rangle | x \in E \ \& \ \mu_A(x) \geq \alpha \ \& \ \nu_A(x) \leq \beta\}. \tag{23}$$

Thus, the result of the application of $N_{\alpha,\beta}$ is a set defined over a specific subset of the universe E.

This provides problems with practical application as the result of the application of the level operator over two different IFS may produce sets defined over completely different elements. In order to mitigate this, level type operators were introduced.

Before we proceed with them, however, we mention that in [31] a new level operator is defined. Here we provide a brief summary of the main results.

Definition 12 (*cf.* [31]) An IFS A is called v-positive if for each $x \in E$, $v_A(x) > 0$. For such sets the following operator may be defined:

$$N_\gamma(A) = \left\{ \langle x, \mu_A(x), v_A(x) \rangle \mid x \in E \And \frac{\mu_A(x)}{v_A(x)} \geq \gamma \right\}. \tag{24}$$

The following property is satisfied by the N_γ operator:

$$N_\gamma(\Box A) \subseteq A \subseteq N_\gamma(\Diamond A)$$

In contrast to the level operators which reduce the underlying elements from the universe set, level type of operators are special type of operators which preserve the universe set but diminish or increase the membership/non-membership of the elements to a preliminary defined threshold level. The formal definition is as follows:

Definition 13 (*cf.* [21]) For any IFS A and an IFP $\langle \alpha, \beta \rangle$ we can define the level type operators $P_{\alpha,\beta}$ and $Q_{\alpha,\beta}$ as follows:

$$P_{\alpha,\beta}(A) = \{\langle x, \max(\alpha, \mu_A(x)), \min(\beta, v_A(x)) \rangle | x \in E\}, \tag{25}$$

$$Q_{\alpha,\beta}(A) = \{\langle x, \min(\alpha, \mu_A(x)), \max(\beta, v_A(x)) \rangle | x \in E\}, \tag{26}$$

It is clear that the above two operators may be extended by replacing the IFP $\langle \alpha, \beta \rangle$ with another IFS B defined over the same universe E, thus replacing α with $\mu_B(x)$ and β with $v_B(x)$ in the above definition.

Oftentimes in practical applications one needs a method for ranking alternatives. When the said alternatives are evaluated in the form of IFPs, there exists a partial ordering between them. The classical partial ordering between two IFPs $x = \langle a, b \rangle$ and $y = \langle c, d \rangle$ is the following:

$$x \preceq y \Leftrightarrow a \leq c \And b \geq d.$$

In [72] a new ordering is proposed, which is later generalized with the help of power mean (see (14)) in [73]. In its general form it is said that $\preceq_{\mu;M_p} y$ iff:

$$\begin{cases} u_1 \leq v_1 \\ M_p(u_2, 1 - u_1) \geq M_p(v_2, 1 - v_1) \end{cases} \tag{27}$$

Since, $x \preceq_{\mu;M_{-\infty}} y$ coincides with the classical ordering it can be established that

Fig. 6 Triangular
representation of the the
intuitionistic fuzzy sets A
and $B \in$ IFS(E) in a
particular point $x \in E$

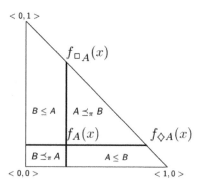

$$x \preceq y \Rightarrow x \preceq_{\mu;M_0} y \Rightarrow x \preceq_{\mu;M_1} y \Rightarrow x \preceq_{\mu;M_2} y$$

i.e. the $\preceq_{\mu;M_2}$ ordering allows more elements to be compared. When $\preceq_{\mu;M_{+\infty}}$ is considered the ordering becomes since the two conditions in (27) degenerate into one

In [46] two main notions—the π-ordering $\preceq_{\pi}$ and the index of indeterminacy were introduced. As a starting point were employed the modal operators on IFSs and their properties in respect to the π-ordering were investigated. It has been shown that (IFS(E), $\preceq_{\pi}$) is a left semi-lattice but not a right semi-lattice, i.e. it is not a complete lattice. This ordering has a very good geometrical representation in terms of the triangular form of the IFSs as shown on Fig. 6. Finally, the new ordering is used for the definition of the index of indeterminacy over IFSs, which is supposed to satisfy three corresponding axioms. A few versions of the index of indeterminacy were proposed according the the structure of the underlying universe set E, over which the IFSs are considered. This index measures how far (close) an IFS is from (to) the family of the usual FSs on the same universe set E.

In [48] a new measure of object similarity has been introduced—the modified weighted Hausdorff distance and it was applied for intuitionitic fuzzy sets. A generalization of this method is also proposed where one takes distance between two collections of denumerable subsets satisfying certain conditions. The methods employ elements of the distance measures between the intuitionistic fuzzy sets. In [45] further research extensions of [48] were proposed. Many new interesting properties valid for any metric space have been investigated focusing on the applications on various operations stemming from the theory of intuitionistic fuzzy sets. Some examples include the notions "degree of friendship" and various methods for algorithmic determination of weights. Ideas for algorithms for the generation of the weights (priorities) assigned to every object have also been discussed. The degree of isolation (distance) from the other objects of the group is also considered to augment intuitionistic fuzzy decision making procedures.

The concept of an intuitionistic fuzzy neighbourhood was introduced in [49]. This notion turns out to be more sensitive to variations in the different points of the universe compared to the standard neighbourhood definition of metric space.

The intuitionistic fuzzy neighbourhood have been applied to the extended modal operators, defined over intuitionistic fuzzy sets. This lead to the introduction of the concepts of IF-neighbourhood and IF-ball. Many relations between them and the standard notions of neighbourhood and ball in metric spaces have also been investigated. Further research on directed intuitionistic fuzzy neighbourhood may be found in [47].

First we remind the classical notion of open and closed balls in metric spaces.

Definition 14 *(cf. [49])* Let (X, d) be a metric space. Then the d-open ball (briefly, open ball) of radius ε, centered at $x \in X$, is defined by:

$$\mathcal{B}(x, d, \varepsilon) = \{y \mid y \in X \ \& \ d(x, y) < \varepsilon\}$$

and the corresponding closed ball (centered at $x \in X$) is defined by:

$$\overline{\mathcal{B}}(x, d, \varepsilon) = \{y \mid y \in X \ \& \ d(x, y) \le \varepsilon\}.$$

Then, we can by analogy define open IF-ball and closed IF-ball:

Definition 15 *(cf. [49])* Given a universe set E, and $A, B \in \text{IFS}(E)$ and a real number $\varepsilon \in [0, 1]$. We shall say that:

- $\mathcal{B}_{\text{IF}}(A, d_j, \varepsilon) = \{B \mid B \in \text{IFS}(E) \ \& \ (\forall x \in E)(d_j(f_A(x), f_B(x)) < \varepsilon)\}$ is **open IF-ball** with center A and radius ε.
- $\overline{\mathcal{B}}_{\text{IF}}(A, d_j, \varepsilon) = \{B \mid B \in \text{IFS}(E) \ \& \ (\forall x \in E)(d_j(f_A(x), f_B(x)) \le \varepsilon)\}$ is **closed IF-ball** with center A and radius ε.

Analogically, can be defined the open and closed IF-neighbourhoods.

Definition 16 *(cf. [49])* Let us be given a universe set E, $A, B \in \text{IFS}(E)$ and $\xi, \eta \in [0, 1]$.

- $\mathcal{B}_{\text{IF}}(A, \xi, \eta) = \{B \mid B \in \text{IFS}(E) \ \& \ (\forall x \in E)(f_A(x) \in \mathcal{B}(f_B(x), \xi, \eta))\}$ is **open (ξ, η)-IF-neighbourhood** of A
- $\overline{\mathcal{B}}_{\text{IF}}(A, \xi, \eta) = \{B \mid B \in \text{IFS}(E) \ \& \ (\forall x \in E)(f_A(x) \in \overline{\mathcal{B}}(f_B(x), \xi, \eta))\}$ is **closed IF-(ξ, η)-neighbourhood** of A

It turns out that this IF-neighbourhood is more sensitive to large variations in points from the universe (when considering two IFSs) in comparison to the neighbourhoods defined by the distances in IFSs. That is, in the case of the IF-neighbourhood or IF-ball, if in only a few points (at least one) $x_0 \in E$ for the intuitionistic fuzzy sets $A, B \in \text{IFS}(E)$ we have that $d(f_A(x_0), f_B(x_0))$ (where d is some metric in $\mathbb{R}$) is large, then the minimal IF-neighbourhood centered in A and containing B would be large, too. Whereas, suppose that only $d(f_A(x_0), f_B(x_0))$ is large for some point x_0 from the universe, but for the rest of the points from E, $d(f_A(x), f_B(x))$ is very small, in some sense, it may compensate the value of the large scale of $d(f_A(x_0), f_B(x_0))$ and $L(A, B)$ could be not so large (L is some metric for IFSs).

Following [49] in [50], relying on the notions of IF-neighbourhoods and IF-balls defined, the concepts of IF-separability in different points of the universe E were introduced.

Definition 17 ([50]) For any $\varepsilon, \xi, \eta \in (0, 1]$ such that $\xi + \eta < 1$, the IFSs A and $B \in \text{IFS}(E)$ are called:

1. (ε, d_j)**-IF-separable** ($j \in \{0, 1, \infty\}$) in $E_1 \subseteq E$ iff $B \in C_{\text{IF}}^{E_1}(A, d_j, \varepsilon)$, where

$$C_{\text{IF}}^{E_1}(A, d_j, \varepsilon) = \{D \mid D \in \text{IFS}(E) \ \& \ (\forall x \in E_1)(d_j(f_A(x), f_D(x)) \geq \varepsilon)\}$$

   is a subfamily of $\text{IFS}(E)$.
2. $(\xi - \eta)$**-IF-separable** in $E_1 \subseteq E$ iff $B \in C_{\text{IF}}^{E_1}(A, \xi, \eta)$, where

$$C_{\text{IF}}^{E_1}(A, \xi, \eta) = \{D \mid D \in \text{IFS}(E) \ \& \ (\forall x \in E_1)(|\mu_A(x) - \mu_D(x)| \geq \xi$$
$$\& \ |\nu_A(x) - \nu_D(x)| \geq \eta)\}$$

   is a subfamily of $\text{IFS}(E)$.

The weakest form of separability between A and $B \in \text{IFS}(E)$ in the standard sense happens when we have separability in at least one point of the universe E. Whereas a stronger form, i.e. separability in all the points of E, corresponds to the concept of complete IF-separability.

Definition 18 ([50]) A and $B \in \text{IFS}(E)$ are **completely** (ε, d_j)**-IF-separable** iff

$$B \in C_{\text{IF}}^{E}(A, d_j, \varepsilon).$$

Analogically, A and B are **completely** $(\xi - \eta)$**-IF-separable** iff

$$B \in C_{\text{IF}}^{E}(A, \xi, \eta).$$

Lastly the notion of total IF-separability is proposed again in [50].

Definition 19 ([50]) Let us consider again A and $B \in \text{IFS}(E)$. The two IFSs are called:

1. **totally** (ξ, η)**-IF-separable** iff

$$\begin{cases} K_A < k_B \text{ and } k_B - K_A \geq \xi \\ l_B < L_A \text{ and } L_A - l_B \geq \eta. \end{cases}$$

   or

$$\begin{cases} K_B < k_A \text{ and } k_A - K_B \geq \xi \\ l_A < L_B \text{ and } L_B - l_A \geq \eta. \end{cases}$$

where the respective K, L, k, l are given by (16), (17), (19) and (20), respectively, for the considered set.

2. **totally** $\pi - (\xi, \eta)$**-IF-separable** or briefly $\overline{(\xi, \eta)} - \pi$**-separable** iff

$$\begin{cases} K_A < k_B \text{ and } k_B - K_A \geq \xi \\ l_A < L_B \text{ and } L_B - l_A \geq \eta. \end{cases}$$

or

$$\begin{cases} K_B < k_A \text{ and } k_A - K_B \geq \xi \\ l_B < L_A \text{ and } L_A - l_B \geq \eta. \end{cases}$$

where the respective K, L, k, l are given by (16), (17), (19), (20), respectively, for the considered set.

All the forms of separability introduced in [50] provide a tool to measure how far or "distinct" two IFSs are from each other in a more sensitive way compared to using only the standard properties of metric spaces (that is employing only the three axioms for distances).

In somewhat a similar vein of research in [74] new equivalence relation called (μ, ν)-coherence is defined over the class of all IFSs over a fixed universe E, and some of its properties are studied. It may be regarded as another tool for measuring similarity between two IFSs, which is different from one obtained using some type of distance between them.

Other results related to the practical aspect when considering decision making problems is how to deal with inconsistent or "unconscientious" expert evaluations in the form of ordered pairs which fail to be IFPs. Some possible ways for altering their estimates are given in [69], where the existing correspondenx ce between d_{φ_p}-IFS (for $p > 1$) and IFS is used. This approach can also be applied to define other operators over IFSs, see e.g. [67]. Other ways for altering expert evaluations may be found in [21], and another two are proposed in works by P. Dworniczak [36, 37].

3.1.4   Intuitionistic Fuzzy Logic

We will provide only a very concise outline of some new results in this area and we refer the interested reader to [23] for a more detailed and substantial overview of the results.

Currently there are around 190 intuitionistic fuzzy implications defined, the first several being analogues to fuzzy implications. Some of the others have no counterpart in the fuzzy case. Each of these implications generates certain negation. Since many of the negations coincide, there are less negations somewhere around 54.

Below we provide a table showing what negations are generated by which implication(s) (Table 1).

Based on the defined implications and negations N. Angelova and M. Stoenchev in a series of papers (with V. Todorov as coauthor in one of them) [11–13] have

Table 1 Negations generated by specific implications

$\neg 1$	$\to 1$, $\to 4$, $\to 5$, $\to 6$, $\to 7$, $\to 10$, $\to 13$, $\to 61$, $\to 63$, $\to 64$, $\to 66$, $\to 67$, $\to 68$, $\to 69$, $\to 70$, $\to 71$, $\to 72$, $\to 73$, $\to 78$, $\to 80$, $\to 124$, $\to 125$, $\to 127$, $\to 166$
$\neg 2$	$\to 2$, $\to 3$, $\to 8$, $\to 11$, $\to 16$, $\to 20$, $\to 31$, $\to 32$, $\to 37$, $\to 40$, $\to 41$, $\to 42$, $\to 172$, $\to 173$, $\to 181$, $\to 182$, $\to 183$
$\neg 3$	$\to 9$, $\to 17$, $\to 21$
$\neg 4$	$\to 12$, $\to 18$, $\to 22$, $\to 46$, $\to 49$, $\to 50$, $\to 51$, $\to 53$, $\to 54$, $\to 91$, $\to 93$, $\to 94$, $\to 95$, $\to 96$, $\to 98$, $\to 134$, $\to 135$, $\to 137$, $\to 169$, $\to 170$, $\to 179$, $\to 180$
$\neg 5$	$\to 14$, $\to 15$, $\to 19$, $\to 23$, $\to 47$, $\to 48$, $\to 52$, $\to 55$, $\to 56$, $\to 57$, $\to 171$, $\to 174$, $\to 175$, $\to 184$, $\to 185$
$\neg 6$	$\to 24$, $\to 26$, $\to 27$, $\to 65$
$\neg 7$	$\to 25$, $\to 28$, $\to 29$, $\to 62$
$\neg 8$	$\to 30$, $\to 33$, $\to 34$, $\to 35$, $\to 36$, $\to 38$, $\to 39$, $\to 76$, $\to 82$, $\to 84$, $\to 85$, $\to 86$, $\to 87$, $\to 89$, $\to 129$, $\to 130$, $\to 132$, $\to 167$, $\to 168$, $\to 177$, $\to 178$
$\neg 9$	$\to 43$, $\to 44$, $\to 45$, $\to 83$
$\neg 10$	$\to 58$, $\to 59$, $\to 60$, $\to 92$
$\neg 11$	$\to 74$, $\to 97$
$\neg 12$	$\to 75$
$\neg 13$	$\to 77$, $\to 88$
$\neg 14$	$\to 79$
$\neg 15$	$\to 81$
$\neg 16$	$\to 90$
$\neg 17$	$\to 99$
$\neg 18$	$\to 100$
$\neg 19$	$\to 101$
$\neg 20$	$\to 102$, $\to 108$
$\neg 21$	$\to 103$
$\neg 22$	$\to 104$
$\neg 23$	$\to 105$
$\neg 24$	$\to 106$
$\neg 25$	$\to 107$
$\neg 26$	$\to 109$, $\to 110$, $\to 111$, $\to 112$, $\to 113$
$\neg 27$	$\to 114$, $\to 115$, $\to 116$, $\to 117$, $\to 118$
$\neg 28$	$\to 119$, $\to 120$, $\to 121$, $\to 122$, $\to 123$
$\neg 29$	$\to 126$
$\neg 30$	$\to 128$

(continued)

Table 1 (continued)

$\neg 31$	$\rightarrow 131$
$\neg 32$	$\rightarrow 133$
$\neg 33$	$\rightarrow 136$
$\neg 34$	$\rightarrow 138$
$\neg 35$	$\rightarrow 139$
$\neg 36$	$\rightarrow 140$
$\neg 37$	$\rightarrow 141$
$\neg 38$	$\rightarrow 142, \rightarrow 143$
$\neg 39$	$\rightarrow 144, \rightarrow 145$
$\neg 40$	$\rightarrow 146, \rightarrow 147$
$\neg 41$	$\rightarrow 148, \rightarrow 149$
$\neg 42$	$\rightarrow 150$
$\neg 43$	$\rightarrow 151$
$\neg 44$	$\rightarrow 152$
$\neg 45$	$\rightarrow 153$
$\neg 46, \lambda$	$\rightarrow 154, \lambda, \rightarrow 155, \lambda$
$\neg 47, \lambda$	$\rightarrow 156, \lambda, \rightarrow 157, \lambda$
$\neg 48, \gamma$	$\rightarrow 158, \gamma, \rightarrow 159, \gamma$
$\neg 49, \gamma$	$\rightarrow 160, \gamma, \rightarrow 161, \gamma$
$\neg 50, \alpha, \beta$	$\rightarrow 162, \alpha, \beta, \rightarrow 163, \alpha, \beta$
$\neg 51, \alpha, \beta$	$\rightarrow 164, \alpha, \beta, \rightarrow 165, \alpha, \beta$
$\neg 52$	$\rightarrow 167$
$\neg 53$	$\rightarrow 176$
$\neg 54$	$\rightarrow 190$

taken up the challenge to calculate in explicit form the conjunctions and disjunctions generated from several types of formulas involving implications and negations. Their important results may be used with regards to different algebraic structures such as monoids, groups and lattices and the study of their properties.

In [71] an attempt is made to classify the intuitionistic fuzzy implications not according to the Axioms they satisfy, as is usually done (see e.g. [18], but by by verifying whether they can be represented in certain equivalent form.

4 Applications of Intuitionistic Fuzzy Sets

The main applications of IFSs are related to areas dealing with imprecise information, where they can either play an auxiliary role in augmenting a certain deterministic approach or provide a framework for selection or decision making of its own.

Currently there is a software being developed by E. Marinov (using Python with mat-plotlib library) to help users better visualize and use the intuitionistic fuzzy sets. It requires as an input a universe set, and the membership and non-membership degrees. Currently it can calculate how the membership and non-membership degrees of the original set are transformed under the application of different operators and provides the user with additional information. The resulting sets can be exported for further use. It also supports visualization of different version of histograms (see Figs. 7 and 8) and some of the orderings defined over IFSs, allowing one to experiment, construct and check validity of properties concerning the investigated set. There is a strong focus on the topological operators and the software highlights the different types of inclusions and ranges where they are valid. This can help one design in a very user-friendly manner a particular IFS as a starting point in a decision making task which has desirable properties.

The most common areas where the apparatus of IFSs has found application are considered below.

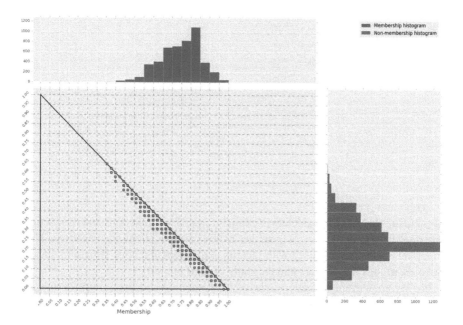

Fig. 7 A 2D histogram visualization of an IFS

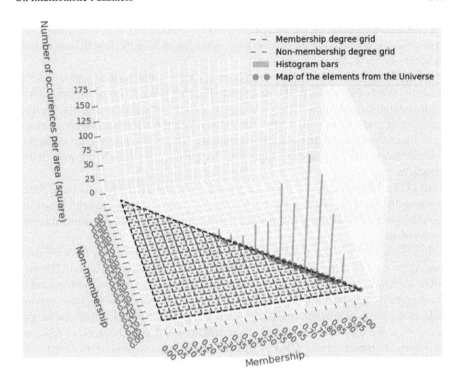

Fig. 8 A 3D histogram visualization of an IFS

4.1 Practical Applications of the Intuitionistic Fuzzy Sets to Medicine and Classification

In this regard the considered methods are either used to augment an existing method or to provide a simpler or more objective tool for a decision making process. Below, we briefly summarize some of the results in this areas.

In [41] An F-operator intuitionistic fuzzy version of one of the basic statistical nonparametrical methods, the nearest prototype (respectively nearest neighbor—NN) classification method, is proposed. It is based on a procedure of adjusting the degrees of membership, nonmembership and indeterminacy to the nearest class by means of the F-operator. The parameter values of that operator iteratively take into account the distances to the prototype (mean) vector of the class, to the 1-NN which belongs to that nearest class, to the 3-NN, to the 5-NN, etc. The procedure stops when the degree of indeterminacy is considerably diminished and the degree of membership reaches a sufficiently high value and in this way increases the confidence in the classification decision.

In [42] a version for increasing of the classification accuracy of one of the basic statistical nonparametrical methods, the K-NN method, is proposed. Here the idea of including fuzzy information in K-NN method is applied in a new way. The distances are modified by means of the pattern degrees of membership and nonmembership to the classes to which the pattern belongs. Thus for each of the labeled samples its typicality is taken into consideration.

In [43] Intuitionistic fuzzy versions of one of the basic statistical non-parametrical methods, the K-NN method, are proposed. The inclusion of fuzzy information is made through modification of the distances by means of the pattern degrees of membership and non-membership to the classes to which the reference pattern belongs. Thus for each of the labeled samples its typicality and non-typicality are taken into consideration. The versions are applied to Respiratory Distress Syndrome (RDS) detection.

In [76] a classification of a set of patterns with the aid of intuitionistic fuzzy Voronoi diagrams, in view of the possibility for more adequate description of the considered objects, allowing for recognition of patterns with non-strict membership is considered.

In [75], a new way for determining intuitionistic fuzzy estimates for a decision making process in medicine based on objective observations is proposed. The proposed algorithm provides an objective and measurable way for the determination of the degrees of membership, non-membership and hesitancy in the decision making tasks in the medical practice. An illustrative example is provided, which shows that decisive for the classification of the pattern (patient) is the combination of features, for which the patient is classified as sick and those that assign her/him to the class "healthy." The more a feature is expressed, the greater impact it would have on the total value for the respective class. In the considered example it turns out that this are the time interval to the first revision, the age at the shunt implantation, and the type of complication which led to the first revision. According to the obtained results the type of the shunt system and the kind of implanted valve are not indicative for the number of following surgical interventions. This coincides with similar findings in the scientific findings in the literature.

In [77] a scheme for more objective comparison of patients, represented by Kaplan-Meier graphs, by classification separating them in groups (m in number) with "graded survivability" is proposed. The classification uses the apparatus of the intuitionistic fuzzy sets. The final result is a clear and easy to interpret table. To avoid loss of information, as well as for the obtainment of graded, sound and real evaluation of the information, contained in the Kaplan-Meier graphs the authors recommend the consideration of all possible cases, for which it is fulfilled $n < m$, where n is the number of the patient groups, represented by Kaplan-Meier survivability curves. As an illustrative example a particular case from the neurosurgical practice is considered and analyzed.

In the paper [64], two topological operators T and U over intuitionistic fuzzy sets are considered and applied. As a case study a parameter identification problem of *E. coli* fed-batch cultivation process model using genetic algorithms is investigated. A new theoretical result linking the operators T and U is established. The results obtained by the application of the topological operators over data processed by InterCriteria Analysis are discussed.

4.2   Practical Application of Intuitionistic Fuzzy Logic

In this regard the application is usually in evaluating the quality of certain procedure or to provide an extended logical inference schemes.

A step-wise procedure based on an intuitionist fuzzy logic (IFL) has been developed for cross-evaluation of the performance of different kinds of genetic algorithms (GA), both simple (single-population) GA (SGA) and multi-population GA (MpGA) [55]. The performance of different kinds of GA is assessed based on the average values of the fitness function, the algorithm convergence time and the values of the estimated parameters resulted from model parameter identification procedures. The algorithms quality assessment is realized by an assignment of estimates formed by the application of IFL.

The developed procedure for the assessment of the quality of GA performance has been successfully applied to parametric identification of yeast cultivation process model. The designed procedure has assisted in the assessment of the performance of: different kinds of SGA by pairwise comparison [53], different kinds of MpGA by pairwise comparison [55], different kinds of SGA by triple comparison [55], different kinds of MpGA by triple comparison [10, 54], as well as for the comparison of SGA and MpGA [5, 55]. This procedure has been also successfully applied over the results of another procedure, namely for the purposeful model parameters genesis in GA [6]. Both procedures have been consequently applied to SGA [2, 4] and to MpGA [3] for the purposes of model parameter identification of fed-batch yeast cultivation process.

The efficacy of the developed procedure for the quality assessment of GA performance has been additionally demonstrated through its application at evaluation of the influence of main operators of GA, namely selection, crossover and mutation, on the performance of different kings of GA. The developed procedure has assisted in the assessment of the performance of: SGA at two different values of GA parameters, namely generation gap, crossover and mutation rates [9]; MpGA at two different values of GA parameters, namely generation gap, crossover and mutation rates [8], SGA at three different values of generation gap [55], as well as SGA and MpGA at three different values of generation gap [7].

Both aforementioned procedures, for purposeful model parameters genesis in GA and for the quality assessment of GA performance can be applied for either parametric identification of different objects or to other optimization algorithms.

Intuitionist fuzzy logic has been also implemented to the models developed by the application of the theory of generalized nets (GN), developed by Atanassov [16, 22]. Applying the IFL to the GN models by the introduction of degrees of validity/agreement and non-validity/disagreement instead of the transition predicates values "true" and "false" leads to so called Intuitionistic Fuzzy Generalized Nets (IFGN). Such approach has been successfully applied to GN models of: adolescent idiopathic scoliosis classification [60], the process of evaluation of the environmental impact of refinery activity [56], advisory system [52] and expanded advisory system for yeast cultivation on-line control [51], biotechnological processes with distributed parameters [40], modelling of temperature control system in fermentation processes [63], etc. Some of the researches are related with the implementation of IFL in various metaheuristic techiques, such as GN model of GA with intuitionistic fuzzy selection operator [65], GN-models of artificial bee colony optimization algorithm [83] and water cycle algorithm augmentation [34] with intuitionistic fuzzy parameter adaptation, intuitionistic fuzzy logic control of metaheuristic algorithms' parameters via a GN [61, 62]. Based on the investigation in [58] a parameter adaptation of the bat algorithm, using type-1, interval type-2 fuzzy logic and intuitionistic fuzzy logic is proposed in [57].

In the paper [35] the implementation of a scalable flexible relational algebra in MapReduce based on IFS set theory is reported. The IFS operations discussed are the IFR selection, projection, union, difference and equal join. The cost of each algorithm is discussed in terms of the cost of the map and reduces functions as well as the communication cost. The developed MapReduce algorithms enhance the standard relational operations with IFS predicates.

5 Conclusion

Our brief overview highlights mostly the achievements of the colleagues from the Bulgarian Academy of Sciences in areas related to intuitionistic fuzziness. We have tried to mention some other results which are related to them but have been obtained by scientists outside the Academy. We have certainly neither exhausted the topic, nor have we gone in sufficient detail regarding very important theories, contributions and consideration proposed by colleagues from Poland, South Korea, China, India, Spain, Mexico, etc.; which deal both with practical and theoretical aspects of the intuitionistic fuzziness. However, the vast amount of publications and monographs on the topic make it virtually impossible to write a concise and balanced survey of the main results in the field in something shorter than a book. With this final comment we end our journey in the domain of the intuitionistic fuzziness and we hope to have given the reader an enlightening and useful, even if limited in scope, information.

References

1. Andonov, V.: On some properties of one Cartesian product over intuitionistic fuzzy sets. Notes Intuit. Fuzzy Sets **14**(1), 12–19 (2008)
2. Angelova, M., Atanassov, K., Pencheva, T.: Intuitionistic fuzzy logic based quality assessment of simple genetic algorithm. In: Proceedings of the 16th International Conference on System Theory, Control and Computing (ICSTCC), Electronic edn., vol. 2, Sinaia, Romania, 12–14 October (2012)
3. Angelova, M., Atanassov, K. & Pencheva, T.: Multipopulation genetic algorithm quality assessment implementing intuitionistic fuzzy logic. In: Proceedings of the Federated Conference on Computer Sciences and Information Systems—FEDCSIS 2012, Workshop on Computational Optimization—WCO'2012, Wrocław, pp. 365–370, Poland, 9–12 September (2012)

4. Angelova, M., Atanassov, K., Pencheva, T.: Intuitionistic fuzzy estimations of purposeful model parameters genesis. In: Proceedings of the IEEE 6th International Conference on Intelligent Systems, pp. 206–211, Sofia, Bulgaria, 6–8 September (2012)
5. Angelova, M., Pencheva, T.: Quality assesment procedure for genetic algorithms performance using intuitionistic fuzzy logics. In: 10th National Young Scientific-Practical Session, pp. 244–249, Sofia, Bulgaria, 23–25 April (2012) (in Bulgarian)
6. Angelova, M., Atanassov, K., Pencheva, T.: Purposeful model parameters genesis in simple genetic algorithms. Comput. Math. Appl. **64**, 221–228 (2012)
7. Angelova, M., Atanassov, K., Pencheva, T.: Intuitionistic fuzzy logic as a tool for quality assessment of genetic algorithms performances. Stud. Comput. Intell. **470**, 1–13 (2013)
8. Angelova, M., Pencheva, T.: Genetic operators' significance assessment in multipopulation genetic algorithms. Int. J. Metaheur. **3**(2), 162–173 (2014)
9. Angelova, M., Pencheva, T.: Genetic operators significance assessment in simple genetic algorithm. In: Lecture Notes Computer Science, vol. 8353, pp. 223–231 (2014)
10. Angelova, M., Pencheva, T.: How to assess multi-population genetic algorithms performance using intuitionistic fuzzy logic. Adv. Comput. Ind. Math. Stud. Comput. Intell. **793**, 23–25 (2018)
11. Angelova, N., & Stoenchev, M.: Intuitionistic fuzzy conjunctions and disjunctions from first type. Annu. Inf. Sect. Union Sci. Bulg. **8**, 1–17 (2015–2016)
12. Angelova, N., Stoenchev, M., Todorov, V.: Intuitionistic fuzzy conjunctions and disjunctions from second type. Issues IFSs GNs **13**, 143–170 (2017)
13. Angelova, N., Stoenchev, M.: Intuitionistic fuzzy conjunctions and disjunctions from third type. Notes Intuit. Fuzzy Sets **23**(5), 29–41 (2017)
14. Atanassov, K.: Intuitionistic fuzzy sets, VII ITKR's Session, Sofia, June 1983 (Deposed in Central Sci.—Techn. Library of Bulg. Acad. of Sci., 1697/84) (in Bulg.). Reprinted: Int. J. Bioautom. **20**(S1), S1–S6 (in English) (1983 & 2016)
15. Atanassov, K.: Intuitionistic fuzzy sets. Fuzzy Sets Syst. **20**(1), 87–96 (1986)
16. Atanassov, K.: Generalized Nets. World Scientific, Singapore (1991)
17. Atanassov, K.: Intuitionistic Fuzzy Sets. Springer, Heidelberg (1999)
18. Atanassov, K.: On the intuitionistic fuzzy implications and negations. In: Part 1. 35 Years of Fuzzy Set Theory pp. 19–38. Springer, Berlin, Heidelberg (2010)
19. Atanassov, K.: New Topol. Oper. Over Intuit. Fuzzy Sets. Notes Intuit. Fuzzy Sets **21**(3), 90–92 (2015)
20. Atanassov, K.: Errata or a new form of the uniformly expanding intuitionistic fuzzy operator. Notes Intuit. Fuzzy Sets **23**(1), 100–103 (2017)
21. Atanassov, K.: On Intuitionistic Fuzzy Sets Theory. Springer, Berlin (2012)
22. Atanassov, K.: On Generalized Nets Theory. Prof. M. Drinov Academic Publishing House, Sofia (2007)
23. Atanassov, K.T.: Intuitionistic Fuzzy Logics. Springer International Publishing, Cham (2017)
24. Atanassov, K., Ban, A.: On an operator over intuitionistic fuzzy sets. Comptes Rendus de l'Academie bulgare des Sciences, Tome **53**(5), 39–42 (2000)
25. Atanassov, K., Gargov, G.: Interval valued intuitionistic fuzzy sets. Fuzzy Sets Syst. **31**(3), 343–349 (1989)
26. Atanassov, K., Mavrov, D., Atanassova, V.: Intercriteria decision making: a new approach for multicriteria decision making, based on index matrices and intuitionistic fuzzy sets. Issues Intuit. Fuzzy Sets Gener. Nets **11**, 1–8 (2014)
27. Atanassov, K., Szmidt, E., Kacprzyk, J.: On intuitionistic fuzzy pairs. Notes Intuit. Fuzzy Sets **19**(3), 1–13 (2013)
28. Atanassov, K., Vassilev, P., Tsvetkov, R. (2013). Intuitionistic Fuzzy Sets, Measures and Integrals, "Prof. Marin Drinov" Academic Publishing House, Sofia, (2013)
29. Atanassov, K.T., Vassilev, P.: On the intuitionistic fuzzy sets of n-th type. In: Gaweda, A., Kacprzyk, J., Rutkowski, L., Yen, G. (eds.) Advances in Data Analysis with Computational Intelligence Methods. Studies in Computational Intelligence, vol. 738, pp. 265–274 (2018)

30. Atanassova, V.: Representation of fuzzy and intuitionistic fuzzy data by Radar charts. Notes Intuit. Fuzzy Sets **16**(1), 21–26 (2010)
31. Atanassova, V.: New modified level operator N_γ over intuitionistic fuzzy sets. Lect. Notes Comput. Sci. **10333**, 209–214 (2017)
32. Atanassova, V., Doukovska, L.: Compass-and-straightedge constructions in the intuitionistic fuzzy interpretational triangle: two new intuitionistic fuzzy modal operators. Notes Intuit. Fuzzy Sets **23**(2), 1–7 (2017)
33. Brouwer, L.E.J.: Intuitionism and formalism. Bull. Am. Math. Soc. **20**(2), 81–96 (1913)
34. Castillo, O., Ramirez, E., Roeva, O.: Water cycle algorithm augmentation with fuzzy and intuitionistic fuzzy dynamic adaptation of parameters. Notes Intuit. Fuzzy Sets **23**(1), 79–94 (2017)
35. Chountas, P., Atanassov, K., Atanassova, V., Sotirova, E., Sotirov, S., Roeva, O.: Big data, intuitionistic fuzzy sets and MapReduce operators. Notes Intuit. Fuzzy Sets **24**(2), 129–135 (2018)
36. Dworniczak, P.: A note on the unconscientious experts' evaluations in the intuitionistic fuzzy environment. Notes Intuit. Fuzzy Sets **18**(3), 23–29 (2012)
37. Dworniczak, P.: Further remarks about the unconscientious experts' evaluations in the intuitionistic fuzzy environment. Notes Intuit. Fuzzy Sets **19**(1), 27–31 (2012)
38. Feferman, S.: In the Light of Logic. Oxford University Press, Oxford (1998)
39. Feys, R.: Modal Logics. Gauthier, Paris (1965)
40. Georgieva, O., Pencheva, T., Krawczak, M.: An application of generalized nets with intuitionistic fuzzy sets for modelling of biotechnological processes with distributed parameters. Issues Intuit. Fuzzy Sets Gener. Nets **3**, 5–10 (2006)
41. Hadjitodorov, S.T.: A F-operator intuitionistic fuzzy version of the nearest neighbor classifier. Notes Intuit. Fuzzy Sets **6**, 1–6 (2000)
42. Hadjitodorov, S.: An intuitionistic fuzzy sets application to the k-NN method. Notes Intuit. Fuzzy Sets **1**(1), 66–69 (1995)
43. Hadjitodorov, S.T.: Intuitionistic fuzzy versions of k-nn method and their application to respiratory distress syndrome detection. Notes Intuit. Fuzzy Sets **4**(4), 62–67 (1998)
44. Kuratowski, K., Topology, Vol. 1, New York, Acad. Press (1966)
45. Marinov, E.: On the algorithmic aspect of the modified weighted hausdorff distance. Inf. Models Anal. **126–135** (2012)
46. Marinov, E.: π-ordering and index of indeterminacy for intuitionistic fuzzy sets. In: Proceedings of 12th International Workshop on IFS and GN, IWIFSGN'13, Warsaw, Oct. 2013, Modern Approaches in Fuzzy Sets, Intuitionistic Fuzzy Sets, Generalized Nets and Related Topics. Volume I: Foundations, IBS PAN-SRI PAS, Warsaw, pp. 129–138 (2014)
47. Marinov, E., Atanassov, K., Vassilev, P., Su, J.: Directed intuitionistic fuzzy neighbourhoods. In: Proceedings of the IEEE 8th International Conference on Intelligent Systems (IS), pp. 544–549, Sofia, Bulgaria (2016)
48. Marinov, E., Szmidt, E., Kacprzyk, J., Tcvetkov, R.: A modified weighted Hausdorff distance between intuitionistic fuzzy sets. In: Proceedings of the 6th IEEE International Conference on Intelligent Systems, pp. 138–141 (2012)
49. Marinov, E., Vassilev, P., Atanassov, K.: On intuitionistic fuzzy metric neighbourhoods. In: Proceedings of the Conference of the International Fuzzy Systems Association and the European Society for Fuzzy Logic and Technology (IFSA-EUSFLAT-15), Gijon, Spain
50. Marinov, E., Vassilev, P., Atanassov, K.: On separability of intuitionistic fuzzy sets. In: Atanassov, K., et al. (eds) Novel Developments in Uncertainty Representation and Processing. Advances in Intelligent Systems and Computing, Vol. 401, pp. 111–123 (2016)
51. Pencheva, T.: Modelling of expanded advisory system for yeast cultivation on-line control using generalized nets and intuitionistic fuzzy logic. Issues Intuit. Fuzzy Sets Gener. Nets **9**, 101–115 (2011)
52. Pencheva, T.: Intuitionistic fuzzy logic in generalized net model of an advisory system for yeast cultivation on-line control. Notes Intuit. Fuzzy Sets **15**(4), 45–51 (2009)

53. Pencheva, T., Angelova, M.: Intuitionistic fuzzy logic implementation to assess purposeful model parameters genesis. Stud. Comput. Intell. **657**, 179–203 (2017)
54. Pencheva, T., Angelova, M., Atanassov, K.: Quality assessment of multi-population genetic algorithms performance. Int. J. Sci. Eng. Res. **4**(12), 1870–1875 (2013)
55. Pencheva, T., Angelova, M., Atanassov, K.: Genetic algorithms quality assessment implementing intuitionistic fuzzy logic, Chapter 11. In: Vasant, P. (ed.) Handbook of Research on Novel Soft Computing Intelligent Algorithms: Theory and Practical Applications, pp. 327–354. Hershey, Pennsylvania (USA), IGI Global (2013)
56. Pencheva, T., Novachev, N., Stratiev, D., Atanassov, K.: Generalized net model of the process of evaluation of the environmental impact of refinery activity using intuitionistic fuzzy estimations. Notes Intuit. Fuzzy Sets **18**(4), 32–39 (2012)
57. Perez, J., Valdez, F., Roeva, O., Castillo, O.: Parameter adaptation of the Bat Algorithm, using type-1, interval type-2 fuzzy logic and intuitionistic fuzzy logic. Notes Intuit. Fuzzy Sets **22**(2), 87–98 (2016)
58. Perez, J., Valdez, F., Castillo, O., Roeva, O.: Bat Algorithm with parameter adaptation using interval type-2 fuzzy logic for benchmark mathematical functions. In: 2016 IEEE 8th International Conference on Intelligent Systems, Sofia, 04–06 September, pp. 120–127 (2016)
59. Priest, G.: An Introduction to Non-classical Logic: From if to is. Cambridge University Press (2008)
60. Ribagin, S., Vassilev, P., Pencheva, T., Zadrożny, S.: Intuitionistic fuzzy generalized net model of adolescent idiopathic scoliosis classification and the curve progression probability. Notes Intuit. Fuzzy Sets **23**(3), 88–95 (2017)
61. Roeva, O., Michalíková, A.: Intuitionistic fuzzy logic control of metaheuristic algorithms' parameters via a generalized net. Notes Intuit. Fuzzy Sets **20**(4), 53–58 (2014)
62. Roeva, O., Michalíková, A.: Generalized net model of intuitionistic fuzzy logic control of genetic algorithm parameters. Notes Intuit. Fuzzy Sets **19**(2), 71–76 (2013)
63. Roeva, O., Pencheva, T., Bentes, I., Manuel Nascimento, M.: Modelling of temperature control system in fermentation processes using generalized nets and intuitionistic fuzzy logics. Notes Intuiti. Fuzzy Sets **11**(4), 151–157 (2005)
64. Roeva, O., Vassilev, P., Chountas, P.: Application of topological operators over data from intercriteria analysis. In: Christiansen, H., et al. (eds.): FQAS 2017, Lecture Notes in Artificial Intelligence, vol. 10333, pp. 215–225 (2017)
65. Roeva, O., Pencheva, T., Atanassov, K.: Generalized net of a genetic algorithm with intuitionistic fuzzy selection operator, new developments in fuzzy sets, intuitionistic fuzzy sets, generalized nets and related topics. In: Atanassov, K.T., Baczyński, M., Drewniak, J., Kacprzyk, J., Krawczak, M., Szmidt, E., Wygralak, M., Zadrożny, S. (eds.) Foundations, vol. 1, IBS PAN-SRI PAS, pp. 167–178, Warsaw (2012)
66. Szmidt, E., Kacprzyk, J.: Distances between intuitionistic fuzzy sets. Fuzzy Sets Syst. **114**(3), 505–518 (2000)
67. Vassilev, P.: Operators similar to operators defined over intuitionistic fuzzy sets **18**(4), 40–47 (2012)
68. Vassilev, P.: Intuitionistic fuzzy sets generated by archimedean metrics and ultrametrics. In: Sgurev, V., Yager, R., Kacprzyk, J., Atanassov, K. (eds.) Recent Contributions in Intelligent Systems. Studies in Computational Intelligence, vol. 657, pp. 339–378 (2017)
69. Vassilev, P.: On reassessment of expert evaluations in the case of intuitionistic fuzziness. Adv. Stud. Contemp. Math. **20**(4), 569–574 (2010)
70. Vassilev, P., Ribagin, S.: A Note on intuitionistic fuzzy modal-like operators generated by power mean. Adv. Intell. Syst. Comput. **643**, 470–475 (2018)
71. Vassilev, P., Ribagin, S., Kacprzyk, J.: A remark on intuitionistic fuzzy implications. Notes Intuit. Fuzzy Sets **24**(2), 1–7 (2018)
72. Vassilev, P., Stoyanov, T.: On a new ordering between intuitionistic fuzzy pairs. In: Proceedings of the 8th European Symposium on Computational Intelligence and Mathematics, pp. 77–80, Sofia (Bulgaria), 5–8 October (2016)

73. Vassilev, P., Stoyanov, T.: On power mean generated orderings between intuitionistic fuzzy pairs. In: Kacprzyk, J., Szmidt, E., Zadrożny, S., Atanassov, K., Krawczak, M. (eds.) Advances in Fuzzy Logic and Technology 2017. EUSFLAT 2017, IWIFSGN 2017. Advances in Intelligent Systems and Computing, vol. 643, pp. 476–481 (2018)

74. Vassilev, P., Todorova, L., Kosev, K.: Note on the (μ, ν)-coherence relation, defined over intuitionistic fuzzy sets. Notes Intuit. Fuzzy Sets **20**(4), 7–9 (2014)

75. Vassilev, P., Todorova, L., Surchev, J.: Determining intuitionistic fuzzy estimates for decision making in medical tasks. Notes Intuit. Fuzzy Sets **20**(5), 62–68 (2014)

76. Todorova, L.: Determining the specificity, sensitivity, positive and negative predictive values in intuitionistic fuzzy logic. In: Twelfth International Conference on IFSs, pp. 73–79, Sofia, 17–18 May 2008, Notes on Intuitionistic Fuzzy Sets, vol. 14, no. 2 (2008)

77. Todorova, L., Vassilev, P., Hadjistoykov, P., Surchev, J.: Application of intuitionistic fuzzy sets for more objective comparison of Kaplan-Meier curves. In: Intelligent Systems (IS), Proceedings of the 6th International IEEE Conference Intelligent Systems, Sept. 2012, pp. 212–215 (2012)

78. Traneva, V., Atanassova, V., Tranev, S.: Index matrices as a decision-making tool for job appointment. In: Nikolov, G., Kolkovska, N., Georgiev, K. (eds.) Numerical Methods and Applications. NMA 2018. Lecture Notes in Computer Science, Vol. 11189, pp. 158–166 (2019)

79. Traneva, V., Tranev, S., Atanassova, V. (2019) An intuitionistic fuzzy approach to the Hungarian Algorithm. In: Nikolov, G., Kolkovska, N., Georgiev, K. (eds) Numerical Methods and Applications. NMA 2018. Lecture Notes in Computer Science, vol. 11189, pp. 167–175 (2019)

80. Yang, Y., Chiclana, F.: Intuitionistic fuzzy sets: spherical representation and distances. Int. J. Intell. Syst. **24**(4), 399–420 (2009)

81. Yosida, K.: Functional Analysis. Springer, Berlin (1965)

82. Zadeh, L.: Fuzzy sets. Inf. Control **8**, 338–353 (1965)

83. Zoteva, D., Roeva, A., Atanassova, V.: Generalized net model of artificial bee colony optimization algorithm with intuitionistic fuzzy parameter adaptation. Notes Intuit. Fuzzy Sets **24**(3), 79–91 (2018)

Acoustic Analysis of Voices

Stefan Hadjitodorov

Abstract The research and results in the field of acoustic analysis of voices for different purposes—laryngeal pathology detection, speaker recognition and evaluation of the emotional state at the Center on Biomedical Engineering, Bulgarian Academy of Sciences (now Institute of Biophysics and Biomedical Engineering) has been described. A lot of novelties have been proposed and introduced regarding: acoustical parameters and approaches; methods for evaluation of well-known acoustical parameters; methods for classification and recognition; computer (software) systems for acoustic analysis of voices and speaker recognition.

Keywords Acoustic analysis of voices · Acoustical parameters · Classification methods · Computer system · Speaker recognition

1 Introduction

The research in the field of acoustic analysis of voices for different purposes—laryngeal pathology detection, speaker recognition and evaluation of the emotional state has started in 1987 at the Center on Biomedical Engineering, Bulgarian Academy of Sciences (now Institute of Biophysics and Biomedical Engineering). Since then a number of projects supported by various national and international funds and programs have been accomplished. Within these projects a lot of novelties have been proposed and introduced:

- acoustical parameters and approaches;
- methods for evaluation of well-known acoustical parameters;
- methods for classification and recognition;
- computer (software) systems for acoustic analysis of voices and speaker recognition.

S. Hadjitodorov (✉)
Institute of Biophysics and Biomedical Engineering, Bulgarian Academy of Sciences, Acad. G. Bonchev Str. Bl. 105, 1113 Sofia, Bulgaria
e-mail: sthadj@bas.bg

K. T. Atanassov (ed.), *Research in Computer Science in the Bulgarian Academy of Sciences*, Studies in Computational Intelligence 934,
https://doi.org/10.1007/978-3-030-72284-5_12

All novelties have been published in a lot of papers in national and international journals and conference proceedings [1–25]. Bellow specific details about some new proposals are described.

2 New Parameters, Methods and Systems

1. *New acoustical parameters and approaches*

   – A new acoustic parameter is introduced and it is shown that it may serve as an indicator of laryngeal function. It is named Turbulent Noise Index (*TNI*) and is defined as $100(1 - \bar{R}_{\max})$, where $\bar{R}_{\max}$ is the mean value of the maximum correlation coefficient between each pair of consecutive glottal cycles in the voiced signal. A method for its calculation is given. Experiments with synthetic and natural voice signals show that *TNI* is almost independent from frequency modulation noise and amplitude modulation noise. *TNI* is compared with *HNR* (Harmonic-to-Noise Ratio) and *NNE* (Normalized Noise Energy) which require high stationarity of the voice signal and are substantially affected by slow changes of frequency and amplitude. When the parameters *HNR* and *NNE* are used for discrimination between normal and pathological voices, the overlap area contains 21.5% and 23.5% of the total number of pathological voices, respectively. Using *TNI*, the normal and pathological voices overlap in 14.8% of the total number of pathological voices, i.e. compared to the other noise parameters *TNI* has a significant advantage as a diagnostic parameter [1];

   – A robust hybrid pitch period (To) detector characterized by parallel analyses of the speech signal in temporal, spectral and cepstral domains and preprocessing for periodic/aperiodic (unvoiced) separation (PAS) is proposed. The preprocessing is realized by analyses in these three domains and PAS by multilayer perceptron neural network. *To* evaluation is carried out by analyses in: *1. Time domain*: 1.1 Preprocessing by adaptive central clipping (ACC). 1.2. Calculation of the short-term autocorrelation function ($Rst(\tau)$). 1.3. Calculation of *To* by analysis of the distances between the maximas in $Rst(\tau)$. *2. Spectral domain*: 2.1. Spectral ACC. 2.2. Calculation of $Rst(\tau)$ over the power and logarithmic spectra. 2.3. *To* detection by algorithm like 1.3. *3. Cepstral domain*: 3.1. Liftering the cepstra to eliminate the formant structure. 3.2 Calculation of $Rst(\tau)$. 3.3. *To* detection by algorithm like 1.3. Final *To* evaluation by a logical analysis of the results from the three domains [2–6];

   – The following new measures (correlated with the laryngeal pathology) are proposed [5, 7, 8]:
     (a) Degree of stability of pitch generation (STAB);
     (b) Degree of dissimilarity of the shape of the pitch pulses (DISS);

(c)   Degree of unvoiceness (DUV) during the phonation of sustained vowels;

(d)   Ratio {energy concentrated in the cepstral pitch impulse}-to-{total cepstral energy} named PECM.

2.   *New methods for evaluation of well-known acoustical parameters*
For the pathology detection, new methods for evaluation of the following voice parameters are proposed:

- pitch period (To), by means of the methods [2, 4, 5, 9–14]. Using the *To* the stable zones are determined;
- deviations of the pitch periods (PPQ-jitter) and of the amplitudes of the pitch pulses (APQ-shimmer) [10–16]. PPQ and APQ are determined for the entire phonation and they retain the same values for all the input vectors corresponding to the different segments analyzed in the speaker's phonation;
- stability of *To* generation (STAB) during vowel phonation [4, 5, 7, 10, 15]. STAB is determined for the entire phonation and it retains the same value for all the input vectors of the speaker's phonation;
- the degree of dissimilarity of the shape (DISS) of the pitch pulses [4, 5, 10, 15]. DISS is determined for each stable zone and it retains the same value for all the segments in the stable zone;
- low-to-high energy ratio (LHER) [4, 7, 17]. LHER is determined for every segment analyzed in the given stable zone;
- harmonics-to-noise ratio (HNR) in time domain by means of a modification (described in [15]) of the method of Yumoto. HNR is determined for each stable zone and it retains the same value for all the input vectors corresponding to the different segments analyzed in the given stable zone;
- ratio (energy concentrated in the pitch impulse–in cepstra)-to-total cepstral energy [9]. This parameter is determined for each segment analyzed in the given stable zone.

3.   *New methods for classification and recognition*

- One of the basic problems at the stage of classification is the approximation and estimation of the probability density functions of the given classes. In order to increase the accuracy of laryngeal pathology detection and to eliminate the most dangerous error—classification of a patient with laryngeal disease as a normal speaker, an approach based on modeling of the probability density functions (*pdf*) of the input vectors of the normal and pathological speakers by means of two prototype distribution maps (PDM) respectively is proposed. The *pdf* of the input vectors of an unknown normal or pathological speaker is also modeled by such a prototype distribution neural map—PDM(X) and the pathology detection is done by means of a ratio of specific similarities rather than by a direct comparison of some type of distance/similarity with a threshold [10];
- A two-level scheme for speaker identification is proposed. The first classifier level is based on the self-organizing map (SOM) of Kohonen. LP-derived

cepstrum coefficients (LPCC) are used as input vectors for this classifier. LPCC coefficients are passed again through the already trained SOMs and as result the prototype distribution maps (PDMs) are obtained. The PDMs are the input for the second classifier level. The second level consists of multilayer perceptron (MLP) networks for each speaker. The first level of the classifier is a preprocessing procedure for the second level, where the final classification is made. The goal of the proposed approach is to combine the advantages of the two type of networks into one classification scheme in order to achieve higher identification accuracy. The experiments show an increased accuracy of the proposed two-level classifier, especially in the case of noise corrupted signals [18, 19].

4. *Computer (software) systems for acoustic analysis of voices and speaker recognition*
A number of computer systems for acoustic analyses of pathological voices and speaker recognition have been developed and used as a consulting system for clinical practice and speaker identification and verification [4, 7, 8, 13, 14, 19, 20].

References

1. Mitev, P., Hadjitodorov, S.: A method for turbulent noise estimation in voiced signals. Med. Biol. Eng. Comput. **38**, 625–631 (2000)
2. Boyanov, B., Ivanov, T., Hadjitodorov, S., Chollet, G.: Robust hybrid pitch detection. Electron. Lett. (IEE Publ., UK) **29**(22), 1924–1926 (1993)
3. Boyanov, B., Hadjitodorov, S., Chollet, G.: Robust periodicity/aperiodicity detector. Comptes Rendus de L'Academie Bulgare des Sciences (Ann. Bulg. Acad. Sci.) **50**(1), 43–46 (1997)
4. Boyanov, B., Hadjitodorov, S.: Acoustic analysis of pathological voices. IEEE Eng. Med. Biol. Mag. **16**(4), 74–82 (1997)
5. Boyanov, B., Hadjitodorov, S., Teston, B., Doskov, D.: Software system for pathological voice analysis. In: Proceedings of the Larynx'97, Marseille, 16–18 June, 1997, pp. 139–142 (1997)
6. Boyanov, B., Doskov, D., Mitev, P., Hadjitodorov, S., Teston, B.: New cepstral parameters for description of pathologic voice. Comptes Rendus de L'Academie Bulgare des Sciences (Ann. Bulg. Acad. Sci.) **53**(3), 41–44 (2000)
7. Boyanov, B., Doskov, D., Ivanov, T., Hadjitodorov, S.: PC based system for analysis the voice of patients with Laryngeal diseases. ENT J. **70**(11), 767–772 (1991)
8. Hadjitodorov, S., Boyanov, B.: PC-based system for robust speaker recognition. J. Comput. Inf. Technol. **6**(4), 415–423 (1998)
9. Boyanov, B., Chollet, G.: Pathological voice analysis using cepstra, bispectra and group delay functions. In: Proceedings of the ICSLP'92, Banff, Alberta, Canada, Oct. 1992, pp. 1039–1042 (1992)
10. Hadjitodorov, S., Boyanov, B., Teston, B.: Laryngeal pathology detection by means of class-specific neural maps. IEEE Trans. Inf. Technol. Biomed. **4**(1), 68–73 (2000)
11. Mitev, P., Hadjitodorov, S.: Fundamental frequency estimation by means of the spectrum maximum. In: Proceedings of the International Conference BIOSIGNAL 2000, Brno, Czech Republic, June 2000, pp. 159–161 (2000)

12. Mitev, P., Hadjitodorov, S.: Cepstrum method for fundamental frequency evaluation in pathological voices. In: Proceedings of the VIII National Conference on Biomedical Physics and Engineering (with international participation), Sofia, 12–14 Oct. 2000, pp. 54–57 (2000)

13. Hadjitodorov, S., Mitev, P.: A computer system for acoustic analysis of pathological voices and laryngeal diseases screening. Med. Eng. Phys. **24**(6), 419–429 (2002)

14. Mitev, P., Hadjitodorov, S.: Fundamental frequency estimation of voice of patients with laryngeal disorders. Inf. Sci. **156**(1–2), 3–19 (2003)

15. Boyanov, B.: Analysis of Pathological Voice. Research report under contract ERB-CIPA- CT-92-0170 with the European Community (1992)

16. Hadjitodorov, S., Ivanov, T., Boyanov, B.: Analysis of disphony using objective voice parameters. In: Proceedings of the 2nd Balkan Conference on Operational Research, Thessaloniki, Greece, 18–21 Oct. 1993, pp. 911–917 (1993)

17. Boyanov, B., Hadjitodorov, S., Baudoin, G., Doskov, D.: Method for evaluation the energy in the singer formant. Comptes Rendus de L'Academie Bulgare des Sciences(Ann. Bulg. Acad. Sci.) **48**(8), 25–28 (1995)

18. Hadjitodorov, S., Boyanov, B., Ivanov, T., Dalakchieva, N.: Text—independent speaker identification using neural nets and AR-vector models. Electron. Lett. (IEE Publ., UK) **30**(11), 838–840 (1994)

19. Hadjitodorov, S., Boyanov, B., Dalakchieva, N.: A two-level classifier for text-independent speaker identification. Speech Commun. **21**, 209–217 (1997)

20. Boyanov, B., Hadjitodorov, S., Baudoin, G., Vilkman, E.: Acoustical analysis of pathological voice. In: Proceedings of the of 3rd Slovenian-German and 2nd SDRV Workshop on Speech and Image Understanding, 24–26 Apr. 1996, Ljubljana, Slovenia, pp. 157–165 (1996)

21. Boyanov, B., Hadjitodorov, S., Krumov, O.: Analysis of voiced speech by means of bispectrum. In: Proceedings of IEEE 1990 South African Symposium on Communications and Signal Processing, COMSIG'90, Johannesburg, 29 June 1990

22. Hadjitodorov, S., Ivanov, T.: Analysis of objective voice parameters in time domain with respect to the Degree of Disphony. In: Proceedings of MEDICOMP'92, Sceged, Hungary, 29 Nov.–2 Dec. 1992, pp. 135–139 (1992)

23. Boyanov, B., Hadjitodorov, S., Ivanov, T.: Analysis of voiced speech by means of bispectrum. Electron. Lett. (IEE Publ., UK) **27**(24), 2267–2268 (1991)

24. Boyanov, B., Hadjitodorov, S., Baudoin, G.: Stress detection by means of speech analysis. In: Proceedings of the of 3rd Slovenian-German and 2nd SDRV Workshop on Speech and Image Understanding, 24–26 Apr. 1996, Ljubljana, Slovenia, pp. 149–156 (1996)

25. Boyanov, B., Hadjitodorov, S.: Evaluation of the emotional state by means of voice analysis and neural networks. In: Proceedings of NATO ASI "Computational Models of Speech Pattern Processing", Jersey, 14–18 July 1997, pp. 1–4 (1997)

The Mathematical Aspects of Some Problems from Coding Theory

Peter Boyvalenkov and Ivan Landjev

Abstract We briefly review the history and main research directions of the research in department "Mathematical Foundations of Informatics" (MFI) at the Institute of Mathematics and Informatics (IMI) of the Bulgarian Academy of Sciences, founded and headed by Professor Stefan Dodunekov (1945–2012). We describe two major themes which stay in the focus of MFI for many years. We present results and pose some open problems, as well as directions for future research.

Keywords Finite geometry · Universal bounds for codes

1 Introduction

The department MFI was founded in 1988 as a result of deep changes in Bulgarian science which saw the strong Department of Algebra of the Joint Center of Mathematics and Mechanics divided into three parts—the section "Algebra" of IMI, the department "Algebra" at the Faculty of Mathematics and Informatics of Sofia University and MFI at IMI. All three parts survived the changes and are still significant research centers in their institutions, thus at national and international levels.

Researchers from MFI cover classical themes from Coding theory—optimal linear codes, one of the favourite topics of Dodunekov (see his DSc Thesis [34] and references therein), bounds for codes in a very broad sense, finite geometries, designs and other combinatorial configurations, covering radius, practical application of codes, etc. Recent and modern topics include applications in Cryptography, network cod-

Peter Boyvalenkov—The research of this author was supported, in part, by a Bulgarian NSF contract DN02/2-2016. Ivan Landjev—This research was supported by the Research Fund of Sofia University under contract No 80-10-81/15.04.2019.

P. Boyvalenkov · I. Landjev (✉)
Institute of Mathematics and Informatics, Bulgarian Academy of Sciences, 8 Acad. G. Bonchev str., 1113 Sofia, Bulgaria
e-mail: i.landjev@nbu.bg

© The Author(s), under exclusive license to Springer Nature Switzerland AG 2021
K. T. Atanassov (ed.), *Research in Computer Science in the Bulgarian Academy of Sciences*, Studies in Computational Intelligence 934,
https://doi.org/10.1007/978-3-030-72284-5_13

ing, relations to Analysis and Potential theory, deep investigations of combinatorial structures, etc.

We present below the state of the art in two major areas of research of groups in MFI—universal bounds for codes and finite geometries. Both themes were highly influenced by Stefan Dodunekov by his invaluable guidance and research contributions.

2 Universal Linear Programming Bounds for Spherical Codes and Designs

One of the central and traditional themes for a group in MFI is about obtaining bounds for codes and designs in polynomial metric spaces (PM-spaces). The research was started by the first author as a university student under the supervision of Stefan Dodunekov. Realizing the importance and perspectives of this area, Stefan recommended creation of the group which included, over the years, more than 20 researchers.

We describe below investigations on bounds for spherical codes and designs, where results obtained by our group in 1990s received recently recognition and quite interesting developments.

2.1 Linear Programming on $\mathbb{S}^{n-1}$

We consider the unit Euclidean sphere $\mathbb{S}^{n-1}$ in $\mathbb{R}^n$ with the Euclidean distance between $\mathbf{x} = (x_1, x_2, \ldots, x_n)$ and $\mathbf{y} = (y_1, y_2, \ldots, y_n)$ in $\mathbb{R}^n$ defined by

$$d(\mathbf{x}, \mathbf{y}) := \sqrt{(x_1 - y_1)^2 + (x_2 - y_2)^2 + \cdots + (x_n - y_n)^2}.$$

We work with the inner product

$$\langle \mathbf{x}, \mathbf{y} \rangle := x_1 y_1 + x_2 y_2 + \cdots + x_n y_n.$$

On $\mathbb{S}^{n-1}$ one has the identity

$$\langle \mathbf{x}, \mathbf{y} \rangle = 1 - \frac{d^2(\mathbf{x}, \mathbf{y})}{2}.$$

A spherical code $\mathcal{C}$ is a non-empty subset of $\mathbb{S}^{n-1}$. The main parameters of a spherical code are the dimension n, the cardinality $|\mathcal{C}| = M$ and the maximal inner product

$$s = s(\mathcal{C}) := \max\{\langle \mathbf{x}, \mathbf{y} \rangle : \mathbf{x}, \mathbf{y} \in \mathcal{C}, \mathbf{x} \neq \mathbf{y}\}$$

(equivalently, the minimum distance $d = d(C) = \min\{d(\mathbf{x}, \mathbf{y}) : \mathbf{x}, \mathbf{y} \in C, \mathbf{x} \neq \mathbf{y}\}$). We will denote C as (n, M, s)-code.

Investigations of different parameters of spherical codes is interesting in mathematical analysis, discrete geometry, coding theory, optimization, numerical analysis, and information theory. In particular, coding theory applications are described in the books [15, 29, 40, 78] and in many of the papers cited in these books.

A classical problem in coding theory and discrete geometry asks for estimation of the quantity

$$A(n, s) := \max\{|C| : C \text{ is a spherical } (n, M, s)\text{-code}\}$$

for fixed n and s. Indeed, in some important models the (constant-energy) signals can be considered as points on $\mathbb{S}^{n-1}$ and this explains why codes with many points (signals) and good separation (small maximal inner product) are interesting. Moreover, evaluation of the decoding capabilities of codes requires good bounds for $A(n, s)$ and other code parameters.

Another popular interpretation is the Tammes problem[1] about packing a given number of circles (spherical caps) on the surface of $\mathbb{S}^{n-1}$ such that the minimum distance between circles is maximized. In other words, we have to distribute M dictators on $\mathbb{S}^{n-1}$ in such a way that the minimum possible distance between them is maximal (otherwise they may start a war). Anyway, we estimate the quantity

$$s(n, M) := \min\{s(C) : C \subset \mathbb{S}^{n-1}, |C| = M\},$$

the minimum maximal inner product of a code on $\mathbb{S}^{n-1}$ with fixed cardinality M.

The linear programming (LP) as explained below is a powerful tool for estimating $A(n, s)$ and $s(n, M)$. LP was introduced in coding theory and combinatorics by Delsarte [30] in the beginning of 1970s and developed later by many mathematicians (see, for example [32, 57] and the references therein).

For $a, b \in \{0, 1\}$ denote by $\{P_i^{a,b}(t)\}_{i=0}^{\infty}$ the Jacobi polynomials $\{P_i^{\alpha,\beta}(t)\}_{i=0}^{\infty}$ (see, for example [3, 72]) with parameters

$$(\alpha, \beta) = \left(a + \frac{n-3}{2}, b + \frac{n-3}{2}\right),$$

normalized by $P_i^{\alpha,\beta}(1) = 1$ (if $(a, b) = (0, 0)$, then we get Gegenbauer polynomials, which will be denoted by $P_i^{(n)}(t)$). The Gegenbauer polynomials are orthogonal on the interval $[-1, 1]$ with respect to the measure

$$d\mu(t) := \gamma_n(1 - t^2)^{\frac{n-3}{2}}\, dt, \quad t \in [-1, 1],$$

---

[1] Named after a Dutch botanist who posed the problem in 1930 while studying the distribution of pores on pollen grains.

where $\gamma_n := \Gamma(\frac{n}{2})/\sqrt{\pi}\Gamma(\frac{n-1}{2})$ is a normalizing constant, and satisfy the three-term recurrence relation

$$(i + n - 2)P_{i+1}^{(n)}(t) = (2i + n - 2)t P_i^{(n)}(t) - i P_{i-1}^{(n)}(t), \quad i = 1, 2, \ldots, \quad P_0^{(n)}(t) = 1, \quad P_1^{(n)}(t) = t.$$

Every real polynomial is expanded in terms of Gegenbauer polynomials uniquely, say

$$f(t) = \sum_{i=0}^{k} f_i P_i^{(n)}(t).$$

A real polynomial $f(t)$ is called positive-definite if all coefficients in its Gegenbauer expansion are nonnegative; i.e. $f_i \geq 0$ for $i = 0.5, \ldots, k$.

Theorem 2.1 (LP for spherical codes [31, 49]) *Let $n \geq 2$, $s \in [-1, 1)$, and the polynomial $f(t) \in \mathbb{R}[t]$ be positive definite and such that $f(t) \leq 0$ for $-1 \leq t \leq s$. Then $A(n, s) \leq f(1)/f_0$.*

2.2 Universal LP Bounds for the Size of Spherical Codes and Designs

Using suitable polynomials, Levenshtein [54] (see also [55–57]) obtained the following universal upper bound for $A(n, s)$. Set $r_i := \frac{2i+n-2}{i+n-2}\binom{i+n-2}{i}$ and denote by $t_i^{a,b}$ the largest root of the polynomial $P_i^{a,b}(t)$ (as $t_0^{1,1} = -1$ by definition).

Theorem 2.2 (Levenshtein bound) *Let $n \geq 2$ and $s \in I_\tau$, where I_τ is the interval*

$$I_\tau := \left[t_{k-1+\varepsilon}^{1,1-\varepsilon}, t_k^{1,\varepsilon} \right], \quad \tau = 2k - 1 + \varepsilon, \quad \varepsilon \in \{0, 1\}.$$

Then

$$A(n, s) \leq L_\tau(n, s) := \left(1 - \frac{P_{k-1+\varepsilon}^{1,0}(s)}{P_k^{0,\varepsilon}(s)} \right) \sum_{i=0}^{k-1+\varepsilon} r_i. \quad (1)$$

Example 2.3 The first three Levenshtein bounds (obtained for $\tau = 1, 2, 3$) are $A(n, s) \leq (s - 1)/s$ for $s \in I_1 = [-1, -1/n]$,

$$A(n, s) \leq \frac{2n(1 - s)}{1 - ns}$$

for $s \in I_2 = [-1/n, 0]$, and

$$A(n, s) \leq \frac{n(1 - s)(2 + (n + 1)s)}{1 - ns^2}$$

for $s \in I_3 = \left[0, \frac{\sqrt{n+3}-1}{n+2} \right]$.

Investigations of the conditions for attaining the bound (1) lead naturally to a special class of spherical codes—combinatorial objects, called spherical designs. The spherical designs were introduced by Delsarte et al. [31] as counterparts of the classical combinatorial t-designs.

Definition 2.4 *(Delsarte et al. [31])* A spherical code $C \subset \mathbb{S}^{n-1}$ is called *spherical τ-design* if the quadrature formula

$$\frac{1}{\mu(\mathbb{S}^{n-1})} \int_{\mathbb{S}^{n-1}} f(x) d\mu(x) = \frac{1}{|C|} \sum_{x \in C} f(x)$$

$(\mu(x)$ is the Lebesque measure) holds true for every polynomial $f(x) = f(x_1, x_2, \ldots, x_n)$ of total degree at most τ. The maximum number τ such that C is a spherical τ-design is called *strength* of C.

A naturally arising problem is to minimize the cardinality provided the dimension n and the strength τ are fixed, that is, to determine

$$B(n, \tau) := \min\{|C| : C \subset \mathbb{S}^{n-1} \text{ is a spherical } \tau\text{-design}\}.$$

The next theorem gives the LP bound on $B(n, \tau)$.

Theorem 2.5 (LP for spherical designs [31]) *Let $n \geq 2$ and $\tau \geq 1$ be fixed and the polynomial $f(t) \in \mathbb{R}[t]$ be such that $f(t) \geq 0$ for $-1 \leq t \leq 1$ and the coefficients in the Gegenbauer expansion $f(t) = \sum_{i=0}^{r} f_i P_i^{(n)}(t)$ satisfy $f_0 > 0$, $f_i \leq 0$ for $i = \tau + 1, \ldots, r$. Then $B(n, \tau) \geq f(1)/f_0$.*

Using suitable polynomials, Delsarte et al. [31] obtained a universal upper bound for $B(n, \tau)$.

Theorem 2.6 (Delsarte-Goethals-Seidel bound) *Let $n \geq 2$ and $\tau \geq 1$ be fixed. Then*

$$B(n, \tau) \geq D(n, \tau) := \binom{n+k-2+\varepsilon}{n-1} + \binom{n+k-2}{n-1}, \qquad (2)$$

where $\tau = 2k - 1 + \varepsilon$, $\varepsilon \in \{0, 1\}$.

The bounds (1) and (2) are dual in a sense (as Theorems 2.1 and 2.5 are). The most important relation between them follows.

Theorem 2.7 *The bounds (1) and (2) are related by the equalities*

$$L_{\tau-1-\varepsilon}(n, t_{k-1-\varepsilon}^{1,1-\varepsilon}) = L_{\tau-\varepsilon}(n, t_{k-1-\varepsilon}^{1,1-\varepsilon}) = D(n, \tau - \varepsilon), \; \varepsilon \in \{0, 1\}, \qquad (3)$$

at the ends of the intervals I_τ. In particular, if a spherical τ-design $C \subseteq \mathbb{S}^{n-1}$ attains (2), then it attains also (1) at the left end point of the interval I_τ.

Spherical codes which attain the Levenshtein bound $L_\tau(n, s)$ need to have special structure. First, the set of their distinct inner products coincides with the set of zeros of the Levenshtein polynomials (see below for more) and, second, they are spherical designs of strength τ. Therefore, such codes carry associations schemes of k classes (see [32, 57], where $\tau = 2k - 1 + \varepsilon, \varepsilon \in \{0, 1\}$). Results on the existence of spherical codes which attain the Levenshtein bound $L_\tau(n, s)$ via investigations of the distance distributions of the corresponding codes (a classical approach in Coding theory) were obtained in the MFI group [19]. In particular, it was shown in [19] that the even bounds $L_{2k}(n, s)$ cannot be attained for any s in the interior of the interval I_{2k}.

2.3 Energy Bounds

Recently, the interplay between Levenshtein and Delsarte-Goethals-Seidel universal bounds was extended by the addition a new huge segment—universal lower bounds on energy of codes (cf. [21]). The three problems were proved to be closely related.

Definition 2.8 Given a function $h(t) : [-1, 1] \to [0, +\infty]$, the potential energy (or h-energy) of a spherical code $\mathcal{C} \subset \mathbb{S}^{n-1}$ is defined by

$$\mathcal{E}_h(\mathcal{C}) := \sum_{x,y \in \mathcal{C}, x \neq y} h(\langle x, y \rangle). \tag{4}$$

The class of absolutely monotone potential functions is of great interest.

Definition 2.9 The function $h(t) : [-1, 1] \to (0, +\infty]$ is called *absolutely monotone* (in the interval $[-1, 1]$), if $h^{(k)}(t) \geq 0$ for every $t \in [-1, 1)$ and every integer $k \geq 0$, where $h(1) = \lim_{t \to 1^-} h(t)$.

Among the most prominent absolutely monotone potentials we list

$$h(t) = [2(1 - t)]^{1-n/2}, \text{ Newton potential,}$$

$$h(t) = [2(1 - t)]^{-\alpha/2}, \ \alpha > 0, \text{ Riesz potential,}$$

$$h(t) = e^{-\alpha(1-t)}, \text{ Gaussian potential,}$$

$$h(t) = -\log[2(1 - t)], \text{ Logarithmic potential.}$$

We consider the energy minimization problem, that is, to determine

$$\mathcal{E}_h(n, M) := \inf_{|\mathcal{C}|=M} \{E_h(\mathcal{C})\}, \tag{5}$$

the minimum possible h-energy of a spherical code $\mathcal{C} \subset \mathbb{S}^{n-1}$ of fixed cardinality $|\mathcal{C}| = M$. Yudin [77] was first to prove that LP works well for the energy minimization problem.

Theorem 2.10 (LP bound for minimum energy [77]) *Let $n \geq 2$ and $M \geq 2$ be fixed positive integers and let $h : [-1, 1] \to [0, +\infty]$. Let the polynomial $f(t) \in \mathbb{R}[t]$ be positive definite and such that $f(t) \leq h(t)$ for any $-1 \leq t < 1$. Then $\mathcal{E}_h(n, M) \geq M(f_0 M - f(1))$.*

Definition 2.11 For $(a, b) = (0, 0), (1, 0)$ and $(1, 1)$ we set

$$T_k^{a,b}(u, v) = \sum_{i=0}^{k} r_i^{a,b} P_i^{a,b}(u) P_i^{a,b}(v),$$

where $r_i^{a,b} = 1/c^{a,b} \int_{-1}^{1} \left(P_i^{a,b}(t) \right)^2 (1 - t)^a (1 + t)^b d\mu(t)$, $c^{1,0} = c^{0,0} = \gamma_n$, $c^{1,1} = \gamma_{n+2}$. Let $\alpha_i, i = 0, 1, \ldots, k + \varepsilon$, be the roots of the polynomial used by Levenshtein for obtaining the bound (1), $s = \alpha_{k+\varepsilon}$, $\tau = 2k - 1 + \varepsilon$, $\varepsilon \in \{0, 1\}$. Let

$$\rho_1 = \frac{T_k^{0,0}(s, 1)}{T_k^{0,0}(-1, -1) T_k^{0,0}(s, 1) - T_k^{0,0}(-1, 1) T_k^{0,0}(s, -1)} \quad \text{for } \varepsilon = 1,$$

$$\rho_{i+\varepsilon} = \frac{1}{c^{1,\varepsilon}(1 + \alpha_{i+\varepsilon})^\varepsilon (1 - \alpha_{i+\varepsilon}) T_{k-1}^{1,\varepsilon}(\alpha_{i+\varepsilon}, \alpha_{i+\varepsilon})}, \quad i = 1, 2, \ldots, k,$$

The following universal lower bound (ULB) was derived in 2016 by a team which includes the first author [21].

Theorem 2.12 ([21]) *Let $n \geq 2$ and $\tau = 2k - 1 + \varepsilon \geq 1$, $\varepsilon \in \{0, 1\}$, be fixed positive integers. Let the function h be absolutely monotone in $[-1, 1]$. Then for every fixed $M \in (D(n, \tau), D(n, \tau + 1)]$ the following holds true:*

$$\mathcal{E}_h(n, M) \geq M^2 \sum_{i=1}^{k+\varepsilon} \rho_i h(\alpha_i). \tag{6}$$

If a spherical code $\mathcal{C} \subset \mathbb{S}^{n-1}$ of cardinality $|\mathcal{C}| = M$ attains the bound (6), then $\mathcal{C}$ is a spherical τ-design and its inner products are exactly the roots of the Levenshtein polynomial $\alpha_0, \alpha_1, \ldots, \alpha_{k+\varepsilon}$.

The conditions for attaining the bounds (1) and (6) are the same; i.e., a spherical code attains the Levenshtein bound (1) if and only if it attains the bound (6). In particular, every spherical design attaining the Delsarte-Goethals-Seidel bound (2) also attains (6).

The ULB (6) was preceded (and motivated, in a sense) by the paper by Cohn and Kumar [27] from 2007, where universally optimal spherical codes were defined and investigated.

Definition 2.13 A spherical code $\mathcal{C} \subset \mathbb{S}^{n-1}$ is called *universally optimal* if it (weakly) minimizes the h-energy among all codes on S^{n-1} with cardinality $|\mathcal{C}| = M$ and for every absolutely monotone potential function h.

Definition 2.14 A spherical code which is a spherical $(2k - 1)$-design and has exactly k distinct inner products of distinct points is called *sharp configuration*.

Theorem 2.15 ([27]) *All sharp configurations are universally optimal. The 600-cell is also universally optimal.*

Corollary 2.16 *Every code which attains the bound* (1) *(equivalently,* (6)) *is universally optimal.*

It is still unknown if there are other codes (apart from the 600-cell) which are universally optimal but do not attain the bounds (1) and (6). On the other hand, since the 600-cell attains a bound from Theorem 2.1 (obtained by a degree 17 polynomial), we arrive to the following conjecture.

Conjecture 2.17 Every universally optimal code attains an LP bound from Theorem 2.10.

2.4 Test Functions and Second Level Bounds

The above discussion naturally motivates investigations of the problem of the possibilities for improving the bounds (1) and (6). The Levenshtein polynomials and the polynomials used for (6) are optimal in the following sense—they are the best among the polynomials of the same or lower degree; i.e., the following is true.

Theorem 2.18 ([71] for (1), [21] for (6)) *The bounds* (1) *and* (6) *cannot be improved by using in Theorems 2.1 and 2.10, respectively, polynomials of the same (as the corresponding Levenshtein polynomials) or lower degree.*

On the other hand, improvement of higher degrees polynomials are possible as shown already in 1978 by Odlyzko and Sloane [67] for the kissing numbers problem (see [20] for a recent survey; see also [64] for the last found kissing number). More general investigation of the optimality of the Levenshtein bound was started by Boyvalenkov et al. [18] (see also [17]) and extended for the ULB case by Boyvalenkov et al. [21].

For fixed n and $s \in I_\tau$, define

$$Q_j(n, s) := \frac{1}{L_\tau(n, s)} + \sum_{i=1}^{k} \rho_i P_j^{(n)}(\alpha_i), \quad j \geq \tau = 2k - 1 + \varepsilon, \ \varepsilon \in \{0, 1\}.$$

Similarly, for fixed n and $M \in (D(n, \tau), D(n, \tau + 1)]$, define

$$R_j(n, M) := \frac{1}{M} + \sum_{i=1}^{k} \rho_i P_j^{(n)}(\alpha_i), \quad j \geq \tau = 2k - 1 + \varepsilon, \ \varepsilon \in \{0, 1\},$$

where the parameters (ρ_i, α_i) come from fixing s by $M = L_\tau(n, s)$. Note that the values of $R_j(n, M)$ are in fact particular values of $Q_j(n, s)$.

Theorem 2.19 (a) [18] *Given n and s, the bound* (1) *can be improved if and only if there exists some $j > \tau$ such that $Q_j(n, s) < 0$.*

(b) [21] *Given n and M, the bound* (6) *can be improved if and only if there exists some $j > \tau$ such that $R_j(n, M) < 0$.*

Investigations of the better bounds "promised" by Theorem 2.19 (a) started already in 90s (see [17, 18]) but a major development came very recently with the so-called lifting of the Levenshtein framework [24].

Definition 2.20 [24] Let n and N be positive integers. A space $\Lambda \subset C[-1, 1]$ is called a *ULB-space* for dimension n and cardinality N if the following two conditions hold:

(i) there exists a $1/N$-quadrature rule (see [21, Sect. 2.2]) that is exact for Λ;

(ii) for any absolutely monotone function h there exists some $f \in \Lambda$ that satisfies the conditions of Theorem 2.1 and agrees with h at the nodes of the $1/N$-quadrature rule from (i).

In the context of Definition 2.20, it follows by results from [21, 56] that the spaces $\mathcal{P}_\tau$ of the real polynomials of degree at most τ are ULB-spaces with the Levenshtein quadrature rule. However, the next step requires more involved ULB-spaces. In certain ULB-spaces, called skip-two/add-two,[2] a framework was developed in [24] to provide a second level of the Levenshtein and energy bounds. Moreover, a third level lift was performed in [24] for a new proof of the universal optimality of the 600-cell, and furthermore, for a complete characterization of the optimal LP polynomials of degree at most 17, thus providing a conceptual explanation.

2.5 Some Research Directions

Most of the above results have their counterparts in more general setting of Polynomial metric spaces (PM-spaces). The class of PM-spaces include the q-ary Hamming spaces, the Johnson spaces and certain infinite projective spaces. So it is also a natural field for the research of the group in MFI, both in 1990s and since 2015. For example, the ULB and related results were recently extended to Polynomial metric spaces in [23].

We conclude with some directions for research which are natural in the above context.

---

[2]This name is inspired by the behaviour of the test functions.

The asymptotic behaviour of the ULB (6) is known (see [23, Sect. 6]) only in the asymptotic process where the number τ is fixed and the dimension n and cardinality tend to infinity in certain relation depending on τ. It will be quite interesting to investigate the process when M tends to infinity while n is fixed.

In [22] Levenshtein-type and ULB-type bounds were obtained for spherical codes with inner products in a prescribed subinterval of $[-1, 1]$. The Hamming counterparts of these results will be considered elsewhere, but applications in other PM-spaces are desired as well, in particular in Johnson spaces.

Another recent development (see [23, Sect. 8]) considers upper bounds for the energy of spherical codes of prescribed cardinality and minimum distance. Counterparts of such bounds in other PM-spaces are not claimed yet.

3   Linear Codes over Finite Fields and Arcs in Finite Projective Spaces

Many problems in coding theory are geometric in nature. So, methods and results from finite geometry can be used in the investigation of codes, and vice versa. In the next two sections, we present a major field of research in the Institute of Mathematics and Informatics—the connection between coding theory and finite geometry.

3.1   MDS-Codes

One of the best examples for the relation of coding theory and Galois geometries is the connection between the MDS-codes and the arcs in $\mathrm{PG}(k-1, q)$. The classical monograph by MacWilliams and Sloane [58] calls the MDS-codes the most interesting codes in the whole history of coding theory. Apart from the pure mathematical importance, these codes have many applications in data protection.

Theorem 3.1 (Singleton bound) *For every linear $[n, k, d]_q$-code, it holds $d \leq n - k + 1$.*

Definition 3.2 A linear code whose parameters meet the Singleton bound, i.e. codes with $d = n - k + 1$, are called maximum distance separable, or MDS-codes.

The following theorem gives the fundamental properties of MDS-codes and and allows their interpretation as arcs in suitable Galois geometries.

Theorem 3.3 *Let C be a linear $[n, k, d]_q$-code. The following conditions are equivalent:*

(1) C is a linear $[n, k, n-k+1]_q$-code;
(2) every k columns in any generator matrix of C are linearly independent;

(3) every n − k columns in any parity check matrix of C are linearly independent;
(4) C$^\perp$ is a linear [n, n − k, k + 1]$_q$-code.

Independently from MDS-codes one can introduce n-arcs in the Galois geometries.

Definition 3.4 A set of n-points $\mathcal{K}$ in PG$(k − 1, q)$ is called an n-arc if any k points from $\mathcal{K}$ are in general position (no three on a line, no four in a plane etc.). An n-arc in PG$(k − 1, q)$ is said to be complete if it is not contained in an $(n + 1)$-arc in PG$(k − 1, q)$.

Theorem 3.5 *The pointset $\mathcal{K} = \{\mathbf{g}_1, \ldots, \mathbf{g}_n\}$ is an n-arc in PG$(k − 1, q)$ if and only if the $(k \times n)$-matrix $G = (\mathbf{g}_1 \ldots \mathbf{g}_n)$ is a generator matrix of a code with parameters $[n, k, n − k + 1]_q$.*

Here the vectors $\mathbf{g}_i \in \mathbb{F}_q^k$ represent points in PG$(k − 1, q)$ in homogeneous coordinates. According to Theorems 3.3 and 3.5 the existence of an n-arc in $\mathcal{K}$ in PG$(k − 1, q)$ implies the existence of an n-arc $\widetilde{\mathcal{K}}$ in PG$(n − k − 1, q)$. The standard example for n-arcs in PG$(k − 1, q)$ is given by the normal rational curves.

Definition 3.6 A normal rational curve in PG$(k − 1, q)$ is every set of points in this geometry, which is projectively equivalent to the set $\{(1, t, \ldots, t^{k-1}) \mid t \in \mathbb{F}_q^+\}$, where $\mathbb{F}_q^+ = \mathbb{F}_q \cup \{\infty\}$.

The codes associated with normal rational curves are known as the doubly extended Reed-Solomon codes. Let us denote by $m(k − 1, q)$ the maximal number of points in an n-arc in PG$(k − 1, q)$. The problem of determining the exact value of $m(k − 1, q)$, together with the problem of the characterization of the arcs of this cardinality, is central in coding theory and in finite geometry. In the terms of linear codes it can be formulated as the problem of determining the maximal length of a MDS-code of fixed dimension k. It is easily checked that $m(k − 1, q) \le q + k − 1$.
It is well-known from [16] that

$$m(2, q) = \begin{cases} q + 2 & \text{for } q \text{ even,} \\ q + 1 & \text{for } q \text{ odd.} \end{cases}$$

An arc of cardinality $m(2, q)$ is called an oval for q odd and hyperoval for q even. The next remarkable theorem by Segre [69] characterizes the ovals in PG$(2, q)$ for odd q. It states that every set of $q + 1$ points in a projective plane of odd order, no three of which are colinear, satisfies an algebraic equation of second degree.

Theorem 3.7 *Every oval in PG$(2, q)$ for q odd consists of the rational points of a conic.*

For planes of even order $q = 2^h$ examples of hyperovals are obtained by taking the points of a conic plus the nucleus (defined as the common points of all tangents). A hyperoval of this type is called regular. Segre [70] proved that for $q = 2, 4, 8$ every hyperoval is regular. For $h \geq 4$ there exist hyperovals that are not regular. Several classes of non-regular hyperovals are known, but the problem of their classification seems to be one of the toughest in finite geometry. The following partial result has been proved in [70].

Theorem 3.8 *Every hyperoval in* $PG(2, q), q = 2^h, h > 1$, *is projectively equivalent to*

$$\mathcal{D}(F) = \{(1, t, F(t)) \mid t \in \mathbb{F}_q\} \cup \{(0, 1, 0), (0, 0, 1)\},$$

where F is a permutation polynomial over $\mathbb{F}_q$ of degree at most $q - 2$, with $F(0) = 0$, $F(1) = 1$ and

$$F_s(X) = \frac{F(X + s) + f(X)}{X}$$

is a permutation polynomial for every s, satisfying $F_s(0) = 0$.

Let us return to the problem of determining the exact value of $m(k - 1, q)$. The conjecture that $m(k - 1, q) = q + 1$ for all k and q, with the exception for q even, $k = 3, q - 1$, in which case $m(k - 1, q) = q + 2$, is known as the MDS-conjecture. It is stated by Segre in [69]. Shortly before that Bush [25] proved that in $PG(k - 1, q), k \geq q$, it holds $m(k - 1, q) = k + 1$. This value is achieved iff the pointset is projectively equivalent to $\{e_1, \ldots, e_k, e_1 + \ldots + e_k\}$, where $e_1, \ldots, e_k$ is a basis in $\mathbb{F}_q^k$.

During the years it was proved that the conjecture is true for all $q \leq 27$ and all $k \leq 5$ and $k \geq q - 3$, as well as for $k = 6, 7, q - 4, q - 5$ with a few exceptions [45]. The maximal value of $q + 1$ (for $k \neq 3, q - 1$, q even) is achieved only for normal rational curves (doubly extended Redd-Solomon codes) with two exceptions constructed by Glynn [43] and by Hirschfeld [44].

A remarkable progress towards proving the MDS-hypothesis was made by S. Ball in [7], where he proved the following theorem.

Theorem 3.9 *Let S be a set of vectors S in the vector space $\mathbb{F}_q^k$, $q = p^h$, where $3 \leq q - p + 1 \leq k \leq q - 2$, for which every subset fo cardinality k is a basis. Then S has cardinality at most $q + 1$. In case of equality S is projectively equivalent to a normal rational curve.*

An important corollary from this theorem is that the MDS-conjecture is true in the case of prime fields. In the language of coding theory this means that the longest MDS-codes of dimension at most p are doubly extended Reed-Solomon codes. A significant step towards proving the MDS-conjecture for non-prime fields was made in [8], where it is proved for "small" dimensions.

Theorem 3.10 *Let S be a set of vectors in the vector space $\mathbb{F}_q^k$, $q = p^h, h > 1$, where $k \leq q - 2$, such that every subset of cardinality k is a basis. Then $|S| \leq q + 1$.*

3.2 Near-MDS Codes

One of the obvious consequences of the MDS-conjecture in coding theory is that the length of the MDS-codes is bounded by the size of the field plus one. One class of codes that preserves all significant properties of the MDS-codes and in the same time contains codes that are twice as long was introduced by Dodunekov and Landjev in [35]. This class contains remarkable codes as the ternary Golay code, the quadratic residue $[11, 6, 5]$-code over $\mathbb{F}_4$, the extended quadratic residue $[12, 6, 6]$-code over the same field [39], as well as many good algebraic geometric codes obtained from elliptic curves [1, 2, 42].

The most natural definition for the near-MDS codes is in terms of the generalized Hamming weights.

Definition 3.11 Let C be a linear $[n, k]_q$-code. The r-th generalized Hamming weight $d_r(C)$ is defined as the minimal cardinality of the support[3] of an $[n, r]$-subcode of C:

$$d_r(C) = \min\{|\mathrm{supp}D| \mid D \text{ is an } [n, r]\text{-subcode of } C\}.$$

For the generalized Hamming weights we have the generalized Singleton bound:

$$d_r(C) \leq n - k + r,$$

for every $r = 1, \ldots, k$.

Definition 3.12 The linear $[n, k]_q$-code C is a near-MDS code if

$$d_i(C) = n - k + i, \text{ for } i = 2, \ldots, k,$$

and

$$d_1(C) = n - k.$$

An equivalent definition is the following: a linear $[n, k]_q$-code is near-MDS iff $d(C) + d(C^\perp) = n$. Other classes related to near-MDS codes were proposed in [13, 41], but these were not that beautiful mathematically. Not every $[n, k, n - k]_q$-code is a near-MDS code, but this is true for lengths $n > k + q$. Similarly to the MDS-codes, the orthogonal to a near-MDS codes is again a near-MDS code. The spectrum of every near-MDS cpde can be computed up to a parameter [35].

---

[3] The support of a code is the set of all coordinate positions which are not identically zero.

Theorem 3.13 *Let C be an $[n, k]_q$ near-MDS code and let (A_i) and A'_i be the spectra of C and $C^\perp$, respectively. Then for every $s \in \{1, \ldots, k\}$ it holds:*

$$A_{n-k+s} = \binom{n}{k-s} \sum_{j=0}^{s-1} (-1)^j \binom{n-k+s}{j}(q^{s-j} - 1) + (-1)^s \binom{k}{s} A_{n-k},$$

$$A'_{k+s} = \binom{n}{k+s} \sum_{j=0}^{s-1} (-1)^j \binom{k+s}{j}(q^{s-j} - 1) + (-1)^s \binom{n-k}{s} A'_k.$$

Similarly to MDS-codes, we denote by $m'(k - 1, q)$ the maximal length of a near-MDS code of fixed dimension k over a fixed field $\mathbb{F}_q$. A major problem in coding theory is to determine the exact value of $m'(k - 1, q)$. There exists an upper bound on the length of a near-MDS $[n, k]_q$-code: $n \leq 2q + k$. By the classical geometric result for the non-existence of maximal arcs in geometries over fields of odd characteristic [9, 10] this bound can be improved to

$$n \leq 2q + k - 2.$$

This bound is achieved for small dimensions and small fields. Good near-MDS codes are obtained from geometric constructions. For example, the intersection points of the 10 lines of the Desarguesian configuration in PG(2, 7) are associated with an $[15, 3, 12]_7$ near-MDS code which meets the upper bound. An $[n, k]_q$-code, $q = p^h$, is called an elliptic code if it is associated with an elliptic curve with n rational points. Elliptic codes turn out to be NMDS-codes. If $N_q(1)$ denotes the maximal number of $\mathbb{F}_q$-rational points on an elliptic curve, defined over $\mathbb{F}_q$, then from a classical result by [76], one gets that

$$N_q(1) = \begin{cases} q + \lfloor 2\sqrt{q} \rfloor & \text{if } p \text{ divides } \lfloor 2\sqrt{q} \rfloor \text{ and } h \geq 3 \text{ is odd,} \\ q + \lfloor 2\sqrt{q} \rfloor + 1 & \text{otherwise.} \end{cases}$$

Hence near-MDS $[n, k]_q$-codes exist for all lengths $n \leq N_q(1)$ and all dimensions $k = 2, \ldots, n - 2$.

Due to extensive computational work by Bartoli et al. [12, 36, 37, 60, 63], the exact values of $m'(k, q)$ were determined for all fields of order $q \leq 9$, as well as lower and upper bounds on the size of the longest near-MDS code for some larger fields. Even more results were obtained for the case of dimension three, which corresponds to the problem of the maximal size of an $(n, 3)$-arc in $PG(2, q)$ (cf. [59, 61, 62]). These results are summarized in the table below.

q/k	2	3	4	5	7	8	9	11	13	16
2	6	8	10	12	16	18	20	24	28	34
3	7	9	9	11	15	15	17	21	23	28
4	8	10	10	12	14	16	16	20–21	21–24	
5		11	11	11	13	15	16	18–22	21–25	
6		12	12	12	13	14	16	18–23	21–36	
7			9	11	14	15	17	18–24	21–27	
8			10	12	13	16	18	18–25	21–28	
9				11	13	14	19	19–26	21–29	
10				12	14	15	20	20–27	21–30	
11					14	15	16	18–28	21–31	
12					15	16	16	18–29	21–32	
13					15	15	16	18–30	21–33	
14					16	16	17	18–31	21–34	
15						17	17	18–32	21–35	
16						18	18	18–33	21–36	

It is obvious that there exists a huge gap between the lower bound of $N_q(1)$ and the upper bound of $2q + k - 2$. There exists a conjecture for the near-MDS codes, analogous to the MDS-conjecture, according to which the best near-MDS codes do not differ too much from the elliptic curves:

Near-MDS conjecture: For every prime power $q = p^h$ there exists a constant c (not depending on q) such that $m'(k - 1, q) \leq N_1(q) + c$.

This problem seems to be extremely difficult. It has not been resolved even in the plane case, i.e. for codes of dimension 3. We end up with a problem posed by A. Blokhuis (cf. [11]) and is worth 10,000 Hungarian Florins:

Open Problem by A. Blokhuis. For an $(n, 3)$-arc in $PG(2, q)$, determine a constant c such that $n/q < c < 2$ for q large enough, or else a construction where $n/q > c > 1$.

4 Linear Codes over Finite Chain Rings and Arcs in Projective Hjelmslev Spaces

A third direction of research in coding theory is on codes over algebraic structures that are weaker than finite fields, notably finite chain rings. During the years it became clear that codes over rings can outperform the classical linear codes over finite fields. Groundbreaking in this direction is the research by Nechaev [65, 66] and by Calderbank et al. [26] who gave a representation of two notorious families of nonlinear codes, the Kerdock and Preparata codes [51, 68], as images of codes that are linear over $\mathbb{Z}_4$. In what follows we present the general theory of linear codes over finite chain rings, their connection with arcs in projective Hjelmslev spaces, and a characterization result on codes of constant homogeneous weight.

4.1   Linear Codes over Finite Chain Rings

Let R be a chain ring with $|R| = q^m$, $R/\text{rad } R \cong \mathbb{F}_q$, let θ be a generator of rad R, and consider the set R^n of all n-tuples over R. The set R^n has the structure of an $(R\text{-}R)$-bimodule with respect to component-wise addition and left/right multiplication by elements from R. We say that θ^i is the *period* of the vector $\mathbf{x} \in R^n$ if i is the smallest nonnegative integer with $\theta^i \mathbf{x} = \mathbf{0}$ (equivalently, $\mathbf{x} \in R^n \theta^{m-i}$). We denote this by $\theta^{m-i} \parallel \mathbf{x}$. The set of vectors in R^n of period θ^m is denoted by $(R^n)^*$. Since $R\theta^i = \theta^i R$ for all $i \geq 0$, the concept of period is left-right symmetric even for noncommutative chain rings.

Definition 4.1   A *code C of length n over R* is a nonempty subset of R^n. The vectors of C are called *codewords*. The code C is *left* (resp., *right*) *linear* if it is an R-submodule of $_R R^n$ (resp., of R_R^n). A *linear* code is one which is either left or right linear.

Definitions and results in the sequel will be stated for left linear codes, most of them having obvious right counterparts.

A *partition* $\lambda \vdash n$ of an integer n is a sequence of nonnegative integers $\lambda_0 \geq \lambda_1 \geq \lambda_2 \geq \ldots$ with $\sum_{i \geq 0} \lambda_i = n$. The trailing zeros of this sequence will be suppressed. The following theorem generalizes the structure theorem for finite abelian p-groups (see e. g. [53, Ch. 15, § 2]):

Theorem 4.2   ([46])   *Every linear code C over a chain ring R with radical $N = \text{rad } R$ is a direct sum of cyclic R-modules. The partition $\lambda = (\lambda_1, \ldots, \lambda_k) \vdash \log_q |C|$ satisfying*

$$_R C \cong R/N^{\lambda_1} \oplus \cdots \oplus R/N^{\lambda_k} \tag{7}$$

is uniquely determined by $_R C$. Moreover, the partition $\mu = \lambda' \vdash \log_q |C|$ conjugate to λ has components $\mu_i = \dim \theta^{i-1} C/\theta^i C$.

Definition 4.3   The *shape* of a linear code C over R is the partition

$$\lambda = (\lambda_1, \ldots, \lambda_k) \vdash \log_q |C|,$$

which satisfies $_R C \cong R/N^{\lambda_1} \oplus \cdots \oplus R/N^{\lambda_k}$. The partition λ' conjugate to λ is called the *conjugate shape* of C. The integer $k = \lambda_1' = \dim_{R/N}(C/\theta C)$ is called the *rank* of C and is denoted by rk C. A subset $\{\mathbf{x}_1, \ldots, \mathbf{x}_k\} \subseteq C \setminus \{\mathbf{0}\}$ is called a *basis* of C if $_R C = R\mathbf{x}_1 \oplus \cdots \oplus R\mathbf{x}_k$.

Definition 4.4   Let $C \leq {}_R R^n$ be a linear code of rank rk $C = k$. A *generator matrix* of C is a $k \times n$-matrix having as its rows a basis of C, so that, in particular, $C = \{\mathbf{x}G; \mathbf{x} \in R^k\}$.

For two vectors $\mathbf{u} = (u_1, \ldots, u_n) \in R^n$ and $\mathbf{v} = (v_1, \ldots, v_n) \in R^n$ we define their inner product $\mathbf{u} \cdot \mathbf{v}$ by

$$\mathbf{u} \cdot \mathbf{v} := u_1 v_1 + \cdots + u_n v_n. \tag{8}$$

Given a code $C \subseteq R^n$, we define

$$C^{\perp} = \{\mathbf{y} \in R^n \mid \mathbf{x} \cdot \mathbf{y} = 0 \text{ for every } \mathbf{x} \in C\},$$

$$^{\perp}C = \{\mathbf{y} \in R^n \mid \mathbf{y} \cdot \mathbf{x} = 0 \text{ for every } \mathbf{x} \in C\}.$$

The linear code $C^{\perp} \leq R_R^n$ (resp., $^{\perp}C \leq {}_R R^n$) is called the *right* (resp., *left*) *orthogonal code* of C.

Theorem 4.5 ([46]) *Let $C \leq {}_R R^n$ be a linear code of shape $\lambda = (\lambda_1, \ldots, \lambda_n)$ and rank $\lambda_1' = k$.*

1. *The orthogonal code $C^{\perp}$ has shape $(m - \lambda_n, m - \lambda_{n-1}, \ldots, m - \lambda_1)$ and conjugate shape $(n - \lambda_m', n - \lambda_{m-1}', \ldots, n - \lambda_1')$. In particular, C is free as an R-module if and only if $C^{\perp}$ is free if and only if $\mathrm{rk}(C^{\perp}) = n - k$;*
2. $^{\perp}(C^{\perp}) = C$;
3. *if in addition $C' \leq {}_R R^n$ then $(C \cap C')^{\perp} = C^{\perp} + C'^{\perp}$, $(C + C')^{\perp} = C^{\perp} \cap C'^{\perp}$.*

Corollary 4.6 *Let $\mathbf{G} \in M_{m,n}(R)$ be any matrix. The linear codes $C \leq {}_R R^n$ and $D \leq R_R^m$ generated by the rows and columns of $\mathbf{G}$, respectively, have the same shape.*

Definition 4.7 A *parity check matrix* of a linear code $C \leq {}_R R^n$ is an $(n - \lambda_m') \times n$-matrix whose rows form a basis of the orthogonal code $C^{\perp}$ (so that in particular $C^{\perp} = \{\mathbf{H}^T \mathbf{y}; \mathbf{y} \in R^{n-\lambda_m'}\}$).

Note that if $\mathbf{H}$ is a parity check matrix of C, then by Part 2 of Theorem 4.5 we have $\mathbf{x} \in C$ if and only if $\mathbf{x} \cdot \mathbf{H}^T = \mathbf{0}$. The number of (and periods of the) columns of $\mathbf{H}$ are determined by Part 1 of Theorem 4.5. For $\mathbf{x} = (x_1, \ldots, x_n) \in R^n$ we set

$$a_i(\mathbf{x}) = |\{j \mid 1 \leq j \leq n \text{ and } \theta^i \| x_j\}|.$$

Definition 4.8 The sequence $(a_0(\mathbf{x}), \ldots, a_m(\mathbf{x}))$ is called the *type* of the word $\mathbf{x} \in R^n$.

Definition 4.9 An *automorphism of the code R^n* is a bijective mapping $\phi \colon R^n \to R^n$ which satisfies $a_i(\mathbf{x} - \mathbf{y}) = a_i(\phi(\mathbf{x}) - \phi(\mathbf{y}))$ for all $\mathbf{x}, \mathbf{y} \in R^n$ and all $i \in \{0, 1, \ldots, m\}$.

Definition 4.10 Two codes $C_1, C_2 \subseteq R^n$ are said to be *isomorphic* (resp., *semilinearly isomorphic*) if there exists a code automorphism (resp., semilinear code automorphism) ϕ of R^n with $\phi(C_1) = C_2$.

4.2   The Equivalence of Arcs in Projective Hjelmslev Spaces And Linear Codes over Finite Chain Rings

We start by introducing the projective Hjelmslev geometries $\text{PHG}(_R R^n)$. This is done in the same way one defines the coordinate (Desarguesian) projective geometries over finite fields. Consider a chain ring R with $|R| = q^m$, $R/\operatorname{rad} R \cong \mathbb{F}_q$, $q = p^r$. Set $M = {}_R R^n$ and $M^* = M \setminus \theta M$, where θ is a fixed generator of $\operatorname{rad} R$. Let $\mathcal{P} = \{Rx \mid x \in M^*\}$ be the set of all free rank 1 submodules of M and let $\mathcal{L} = \{Rx + Ry \mid x, y \text{ linearly independent }\}$ be the set of all free rank 2 submodules of M. We call the elements of $\mathcal{P}$ and $\mathcal{L}$ points and lines, respectively, and define incidence by set-theoretical inclusion. The obtained structure is called the $(n-1)$-dimensional (left) projective Hjelmslev geometry over R and is denoted by $\text{PHG}(_R R^n)$, or $\text{PHG}(n-1, R)$, if we decide to consider by default left modules. This is no restriction since every right module over R can be considered as a left module over the opposite ring R^{opp} with multiplication $a \circ b = ba$.

A set of points S in $\text{PHG}(_R R^n)$ is called an Hjelmslev subspace if for every two points $X, Y \in S$ there exists at least one line incident with both of them which is entirely contained in S. Equivalently, the pointset S is an Hjelmslev subspace if it consists of all free rank 1 submodules of a *free* submodule of $_R R^n$. It should be noted that the intersection of Hjelmslev subspaces is not necessarily an Hjelmslev subspace. We say that the pointset S is a subspace in $\text{PHG}(_R R^n)$ if it is the intersection of Hjelmslev subspaces, or, equivalently, if it contains all free rank 1 submodules of some submodule of $_R R^n$. The shape of this submodule is called *the shape of the subspace S*.

Two points $X = Rx$ and $Y = Ry$ are called i-neighbors if the module $Rx + Ry$ has shape $(m, m - i)$, $i \in \{0, 1, \dots, m\}$. This is denoted by $X \frown_i Y$. A point X is an i-neighbor to a subspace S if there exists a point Y on S with $X \frown_i Y$. A subspace S is an i-neighbor of a subspace T if every point from S has an i-neighbor on T. The relation $\frown_i$ is generally not symmetric, but it is an equivalence relation if one restricts it to subspaces of the same shape.

Let $\Pi = \text{PHG}(_R R^n) = (\mathcal{P}, \mathcal{L})$, where $\mathcal{P}$ and $\mathcal{L}$ are the set of all points and all lines of Π, respectively. For a subspace S of Π we denote by $[S]^{(i)}$ the set of subspaces of Π of the same shape as S that are i-neighbors to S. Let us denote by $\mathcal{P}^{(i)}$ (resp. $\mathcal{L}^{(i)}$) the set of all i-neighbor classes of points (resp lines) of Π. The structure obtained from $\text{PHG}(_R R^n)$ by deleting a hyperplane and all of its 1-neighbors is called the affine Hjelmslev geometry $\text{AHG}(_R R^n)$.

Definition 4.11 A linear code $C \leq {}_R R^n$ is said to be *of full length* or *non-degenerate* if for every $i \in \{1, \dots, n\}$ there exists a codeword $\mathbf{c} = (c_1, c_2, \dots, c_n) \in C$ with $c_i \in R^*$.

Let $C \leq {}_R R^n$ be a linear code of full length. Let $S = (\mathbf{c}_1, \dots, \mathbf{c}_k)$ be a sequence of (not necessarily independent) generators for $_R C$ and $\mathbf{G} \in M_{k,n}(R)$ be the $k \times n$-matrix with rows $\mathbf{c}_1, \dots, \mathbf{c}_k$. Denote the columns of $\mathbf{G}$ by $\mathbf{g}_1, \dots, \mathbf{g}_n$. Since C is fat and $\mathbf{c}_1, \dots, \mathbf{c}_k$ generate C, the vectors $\mathbf{g}_j$ have period θ^m and thus define points $\mathbf{g}_j R$

in the projective (right) Hjelmslev geometry $(\mathcal{P}, \mathcal{L}, \mathcal{I}) = \mathrm{PHG}(R_R^k)$. We define the multiset $\mathcal{K}_S$ *induced by the generating sequence S of C* as

$$\mathcal{K}_S : \begin{cases} \mathcal{P} \to \mathbb{N}_0 \\ P \mapsto |\{j \mid P = \mathbf{g}_j R\}|. \end{cases} \tag{9}$$

We say that the *multiset $\mathcal{K}_S$ and the code $C = \sum_{c \in S} R\mathbf{c}$ are associated.* By definition of $\mathcal{K}_S$ we have $|\mathcal{K}_S| = n$. Furthermore, the modules $\langle \mathcal{K}_S \rangle$ and $_R C$ have the same shape and, in particular, the same cardinality; see [46].

The following theorem from [46] is a generalization of a similar result by Dodunekov and Simonis [38] about linear codes over finite fields.

Theorem 4.12 *For every multiset $\mathcal{K}$ of size n in $\mathrm{PHG}(R_R^k)$ there exists a linear code of full length $C \leq {}_R R^n$ and a generating sequence $s = (\mathbf{c}_1, \ldots, \mathbf{c}_n)$ of codewords from C which induces $\mathcal{K}$. Two multisets $\mathcal{K}_1$ in $\mathrm{PHG}(R_R^{k_1})$ and $\mathcal{K}_2$ in $\mathrm{PHG}(R_R^{k_2})$ associated with the linear codes C_1 and C_2 over R, respectively, are equivalent if and only if the codes C_1 and C_2 are semilinearly isomorphic.*

4.3 Linear Codes over Finite Chain Rings of Constant Homogeneous Weight

In [14] A. Bonisoli proved that a linear code in which all non-zero words assume the same weight is a direct sum of simplex codes. For arcs in $\mathrm{PG}(k-1, q)$ such that all hyperplanes have the same multiplicity are the sum of several copies of the whole projective space. In the next theorem, we prove an analogue of Bonisoli's theorem for the homogeneous weight introduced by Constantinescu and Heise [28]. Our approach is a geometric one, so first we introduce so called homogeneous arcs, an analogue of arcs for the homogeneous weight.

For the sake of simplicity we consider chain rings R of nilpotency index $m = 2$. Denote by Π_{k-1} the $(k-1)$-dimensional Hjelmslev geometry over R. The homogeneous weight $\omega_{\mathcal{K}}$ of a subspace S in Π_{k-1} is defined by

$$\omega_{\mathcal{K}}(S) = \mathcal{K}(S) - \frac{1}{q-1}\mathcal{K}([S] \setminus S), \tag{10}$$

where by $[S]$ is the set of all points that are neighbours to the points in S.

A multiset $\mathcal{K}$ is called a homogeneous (N, W)-arc if $\mathcal{K}(\mathcal{P}) = N$, $\omega(H) \leq W$ for every hyperplane H, and there exists a hyperplane H_0 with $\omega(H_0) = W$. A linear code over R is said to be of constant homogeneous weight if every word which has at least one coordinate in R^* has the same homogeneous weight. It is clear from the results in the previous section that arcs of constant homogeneous weight are associated with linear codes of constant homogeneous weight.

Lemma 4.13 *Let K be an (N, W)-homogeneous arc in Π_k. For an arbitrary hyperplane H_0, it holds*

$$\sum_{H \in [H_0]} w(H) = 0,$$

where the sum is over all hyperplanes of Π_{k-1} that are neighbours to H.

Proof Every point in $[H]$ lies on q^{k-1} hyperplanes from $[H]$. On the other hand, there are q^k hyperplanes that are neighbours to a given hyperplane. Hence the contribution of each point to the sum $\sum_{H \in [H]} w(H)$ is $q^{k-1} - \dfrac{1}{q-1}\left(q^k - q^{k-1}\right) = 0$, whence the result follows. □

By this lemma we easily obtain that the sum of the homogeneous weights of all hyperplanes in Π_{k-1} is also 0.

Corollary 4.14 *Let K be an arc in $\Pi_{k-1}(R)$ for which all hyperplanes have constant homogeneous weight W. Then $W = 0$.*

Theorem 4.15 *Every $(N, 0)$-homogeneous arc is a sum of neighbour classes of points.*

Proof Order linearly the points x_i and the hyperplanes H_i, in such way that $x_i \approx x_j$ (resp. $H_i \approx H_j$) iff $\lfloor i/q \rfloor = \lfloor j/q \rfloor$. Here i runs the integers $0, 1, \ldots, q^{k-1}\frac{q^k-1}{q-1} - 1$. For this linear ordering of points and hyperplanes, define the square matrix $A = (a_{ij})$ of size $q^{k-1}\frac{q^k-1}{q-1}$ by

$$a_{ij} = \begin{cases} 1 & \text{if } x_i \in H_j, \\ -\dfrac{1}{q-1} & \text{if } x_i \notin H_j \text{ but } x_i \approx H_j, \\ 0 & \text{otherwise.} \end{cases} \tag{11}$$

With every homogeneous arc K we associate a vector

$$\mathbf{x}_K = \left(K(x_0), K(x_1) \ldots, K(x_{q^{k-1}\frac{q^k-1}{q-1}-1}) \right).$$

If K is a $(N, 0)$ arc then $\mathbf{x}_K A = \mathbf{0}$, where $\mathbf{0} = (0, 0, \ldots, 0)$ is of length $q^{k-1}\frac{q^k-1}{q-1}$.

Now we are going to prove that $\text{rk} A = (q^{k-1} - 1)\frac{q^k-1}{q-1}$. Consider the matrix $A' = (a|B)$, where B is a $q^{k-1}\frac{q^k-1}{q-1}$ by $\frac{q^k-1}{q-1}$ matrix whose columns are the incidence vectors of the neighbour classes of points:

$$b_{ij} = \begin{cases} 1 & \text{if } x_i \in [x_{jq^{k-1}}] \\ 0 & \text{if } x_i \notin [x_{jq^{k-1}}]. \end{cases}$$

The characteristic vectors $\chi(H_i)$ of the hyperplanes of Π_k belong to the vector space spanned by the columns of A'. Since the incidence matrix of all s-dimensional versus all t-dimensional Hjelmslev subspaces is of full rank, the matrix A' is of full rank [52], i.e. $\mathrm{rk}A' = q^{k-1}\dfrac{q^k-1}{q-1}$. This implies that

$$\mathrm{rk}A \geq q^{k-1}\frac{q^k-1}{q-1} - \frac{q^k-1}{q-1} = (q^{k-1}-1)\frac{q^k-1}{q-1}.$$

On the other hand, we have

$$\sum_{j:\lfloor j/q^{k-1}=a\rfloor} A^{(j)} = \mathbf{0},$$

for all $a \in \{0, 1, \ldots, \frac{q^k-1}{q-1} - 1\}$. Here $A^{(j)}$ are the columns of A. This implies that $\mathrm{rk}A = (q^{k-1}-1)\dfrac{q^k-1}{q-1}$, and the space of all solutions is spanned by the vectors

$$\mathbf{b}_a = (\underbrace{0,0,\ldots,0}_{aq^{k-1}}, \underbrace{1,1,\ldots,1}_{q^{k-1}}, \underbrace{0,0,\ldots,0}_{q^{k-1}\frac{q^k-1}{q-1}-q-1}),$$

where $a = 0, 1, \ldots, \frac{q^k-1}{q-1} - 1$.   □

5 Other Security Research in BAS

We describe briefly some other areas and results of research related to different aspects to the security.

Cyclic redundancy check (CRC) codes were investigated in MFI at Institute of Mathematics and Informatics of BAS and found several applications worldwide.

A CRC code is an error-detecting code commonly used in digital networks and storage devices to detect accidental changes to raw data. Each block of data gets a short check value attached, based on its content. At the receiver, the calculation is repeated and, in the event the check value does not match, corrective action can be taken against data corruption. Best CRC codes were proposed in the works [4, 5] (see also [6, 50]). These codes have different lengths of the check value and can be used to protect data in various modern communication systems. Applications are known in rail and railway transport control systems in Italy (Ansaldo STS S.p.A.), Germany (Thales Rail Signalling Solutions GesmbH), India (Safety Critical Railway Project), also in some communication systems in Russia and Turkey, in embedded wireless systems developed in TU Berlin for pseudonyms protection in professional biobanking. Some of the results were used in two US patents—8341510 B2 and 8327251 B2 from 2012.

In another research area, S-boxes were considered. The S-boxes are one of the most important primitives widely used in modern block ciphers in the last 30 years. Among all different S-box generation techniques three main groups can be distinguished—the pseudo-random generation, algebraic constructions and the heuristic methods. In MFI, researchers were focused on applications of genetic algorithms for construction of cryptographically strong S-boxes aimed at obtaining the best known up to now randomly generated S-boxes [33, 47, 48]. In addition many nonequivalent such S-boxes are generated. In comparison with the AES S-box these results are still worse but have the advantages of the random structures.

An accompanying chapter [73] of this book by Tagarev, a researcher from the Institute of Information and Communication Technologies (IICT) of BAS, describes research in the field of cybersecurity. Joint research of institutes and laboratories of BAS is presented with examples on supporting the development of information security policies [74]. Given that IICT operates the "Avitohol" supercomputer and plays a lead role in the development of the national capacity for high-performance computing, of particular interest are the studies of using supercomputer systems in the design and operation of systems aiming to provide cybersecurity and resilience [75].

References

1. Abatangelo, V., Larato, B.: Near-MDS codes arising from algebraic curves. Discret. Math. **301**(1), 5–19 (2005)
2. Abatangelo, V., Larato, B.: Elliptic near-MDS codes. Des. Codes Cryptogr. **46**, 167–174 (2008)
3. Abramowitz, M., Stegun, I.A.: Handbook of Mathematical Functions. Dover, New York (1965)
4. Baicheva, T., Dodunekov, S., Kazakov, P.: On the cyclic redundancy-check codes with 8-bit redundancy. Comput. Commun. **21**, 1030–1033 (1998)
5. Baicheva, T., Dodunekov, S., Kazakov, P.: On the undetected error probability performance of cyclic redundancy-check codes of 16-bit redundancy. IEE Proc. Commun. **147**, 253–256 (2000)
6. Baicheva, T.: Determination of the best CRC codes with up to 10-bit redundancy. IEEE Trans. Commun. **56**, 1214–1220 (2008)
7. Ball, S.: On sets of vectors of a finite vector space in which every subset of a basis is a basis. J. Eur. Math. Soc. **14**, 733–748 (2012)
8. Ball, S., De Beule, J.: On sets of vectors of a finite vector space in which every subset of a basis is a basis II. Des. Codes Cryptogr. **65**, 323–329 (2012)
9. Ball, S., Blokhuis, A.: An easier proof of the maximal arcs conjecture. Proc. Am. Math. Soc. **126**, 3377–3380 (1998)
10. Ball, S., Blokhuis, A., Mazzocca, F.: Maximal arcs in Desarguesian planes of odd order do not exist. Combinatorica **17**, 31–41 (1997)
11. Ball, S., Hirschfeld, J.: Bounds on (n, r)-arcs and their application to linear codes. Finite Fields Appl. **11**, 326–336 (2005)
12. Bartoli, D., Marcugini, S., Pambianco, F.: The non-existence of some NMDS codes and the extremal sizes of complete $(n, 3)$-arcs in $PG(2, 16)$. Des. Codes Cryptogr. **72**(1), 129–134 (2014)
13. de Boer, M.: Almost MDS codes. Des. Codes Cryptogr. **9**(2), 143–155 (1996)
14. Bonisoli, A.: Every equidistant linear code is a sequence of dual Hamming codes. Ars Comb. **18**, 181–186 (1984)

15. Borodachov, S., Hardin, D., Saff, E.: Discrete Energy on Rectifiable Sets. Springer (2019) (to appear)
16. Bose, R.C.: Mathematical theory of the symmetric factorial design. Sankhya **8**, 107–166 (1947)
17. Boyvalenkov, P.: Extremal polynomials for obtaining bounds for spherical codes and designs. Discr. Comp. Geom. **14**, 167–183 (1995)
18. Boyvalenkov, P., Danev, D., Bumova, S.: Upper bounds on the minimum distance of spherical codes. IEEE Trans. Inf. Theory **41**, 1576–1581 (1996)
19. Boyvalenkov, P., Danev, D., Landgev, I.: On maximal spherical codes II. J. Combin. Des. **7**, 316–326 (1999)
20. Boyvalenkov, P., Dodunekov, S., Musin, O.: A survey on the kissing numbers. Serdica Math. J. **38**, 507–522 (2012)
21. Boyvalenkov, P., Dragnev, P., Hardin, D., Saff, E., Stoyanova, M.: Universal lower bounds for potential energy of spherical codes. Constr. Approx. **44**, 385–415 (2016)
22. Boyvalenkov, P., Dragnev, P., Hardin, D., Saff, E., Stoyanova, M.: On spherical codes with inner products in prescribed interval. Des. Codes Cryptogr. **87**, 299–315 (2019)
23. Boyvalenkov, P., Dragnev, P., Hardin, D., Saff, E., Stoyanova, M.: Energy bounds for codes in polynomial metric spaces. Anal. Math. Phys. (2019) (to appear). arXiv:1804.07462. https://link.springer.com/article/10.1007/s13324-019-00313-x
24. Boyvalenkov, P., Dragnev, P., Hardin, D., Saff, E., Stoyanova, M.: Next levels universal bounds for spherical codes: the Levenshtein framework lifted. Submitted (2019). arXiv:1906.03062. https://urldefense.proofpoint.com/v2/url?u=-3A__www.ams.org_journals_mcom_2021-2D90-2D329_S0025-2D5718-2D2021-2D03621-2D2_&d=DwIGaQ&c=vh6FgFnduejNhPPD0fl_yRaSfZy8CWbWnIf4XJhSqx8&r=dRWH_06vC_UztnDis8AvFdkc_1wCChKvYhvz_oR6rpc&m=QyCZkpKw4bcscq5XSWjG_Hh6_NsoGWGip5fB_vrj_yw&s=97-lTiWnzjHmcjEvKcjYouDKSq2Mkimx4fXUQSyZtN4&e=
25. Bush, K.A.: Orthogonal arrays of index unity. Ann. Math. Statist. **23**, 426–434 (1952)
26. Calderbank, A., Hammons, A., Vijay Kumar, P., Sloane, N., Solé, P.: A linear construction for certain Kerdock and Preparata codes. Bull. AMS **29**, 218–222 (1993)
27. Cohn, H., Kumar, A.: Universally optimal distribution of points on spheres. J. Am. Math. Soc. **20**, 99–148 (2007)
28. Constantinescu, I., Heise, W.: A metric for codes over residue class rings. Probl. Inf. Transm. **33**(3), 208–213 (1997)
29. Conway, J.H., Sloane, N.J.A.: Sphere Packings, Lattices and Groups. Springer, New York (1988)
30. Delsarte, P.: An algebraic approach to the association schemes in coding theory. Philips Res. Rep. Suppl. **10** (1973)
31. Delsarte, P., Goethals, J.-M., Seidel, J.J.: Spherical codes and designs. Geom. Dedicata **6**, 363–388 (1977)
32. Delsarte, P., Levenshtein, V.I.: Association schemes and coding theory. Trans. Inf. Theory **44**, 2477–2504 (1998)
33. Dimitrov, M., Baicheva, T., Esslinger, B.: Efficient generation of cryptographically strong S-boxes with high nonlinearity. Submitted
34. Dodunekov, S.: Optimal linear codes. DrSci Dissertation, IMI-BAS (1986)
35. Dodunekov, S., Landjev, I.: On near-MDS codes. J. Geom. **54**, 30–43 (1995)
36. Dodunekov, S., Landjev, I.: On the quaternary [11, 6, 5] and [12, 6, 6] codes. In: Gollmann, D. (ed.) Applications of Finite Fields. IMA Conference Series 59, pp. 75–84. Clarendon Press, Oxford (1996)
37. Dodunekov, S., Landjev, I.: Near-MDS codes over some small fields. Discr. Math. **213**, 55–65 (2000)
38. Dodunekov, S., Simonis, J.: Codes and projective multisets. Electron. J. Comb. **5**, R37 (1998)
39. Dumer, I.I., Zinoviev, V.A.: Some new maximal codes over $\mathbb{GF}(4)$. Problemi Peredachi Informacii **14**, 24–34 (1978). (in Russian)
40. Ericson, T., Zinoviev, V.: Codes on Euclidean spheres. Elsevier Science B. V. (2001)
41. Faldum, A., Willems, W.: Codes of small defect. Des. Codes Cryptogr. **10**, 341–350 (1997)

42. Giulietti, M.: On the extendability of near-MDS elliptic codes. AAECC **15**(1), 1–11 (2004)
43. Glynn, D.G.: The non-classical 10-arc of PG(4, 9). Discret. Math. **59**, 43–51 (1986)
44. Hirschfeld, J.W.P.: Rational curves on quadrics over finite fields of characteristic two. Rend. Mat. **3**, 772–795 (1971)
45. Hirschfeld, J. W. P., Storme, L.: The packing problem in statistics, coding theory and finite projective spaces: update 2001. In: Proceedings of the Fourth Isle of Thorns Conference Developments in Mathematics, vol. 3. Kluwer, pp. 201–246 (2001)
46. Honold, T., Landjev, I.: Linear codes over finite chain rings. Electron. J. Comb. **7**(11) (2000)
47. Ivanov, G., Nikolov, N., Nikova, S.: Reversed genetic algorithms for generation of bijective s-boxes with good cryptographic properties. Cryptogr. Commun. Discret. Struct. Boolean Funct. Seq. **8**, 247–276 (2015)
48. Ivanov, G., Nikolov, N., Nikova, S.: Cryptographically strong S-boxes generated by modified immune algorithm. In: Pasalic, E., Knudsen, L.R. (eds.) International Conference on Cryptography and Information Security "BalkanCryptSec 2015". Lecture Notes in Computer Sciences, vol. 9540, pp. 31–42 (2016)
49. Kabatiansky, G.A., Levenshtein, V.I.: Bounds for packings on a sphere and in space. Probl. Inf. Transm. **14**, 1–17 (1978)
50. Kazakov, P.: Fast calculation of the number of minimum-weight words of CRC codes. IEEE Trans. Inf. Theory **47**, 1190–1195 (2001)
51. Kerdock, A.M.: A class of low-rate nonlinear binary codes. Inf. Control **20**, 182–187 (1972)
52. Landjev, I., Vanderdriesche, P.: On the rank of incidence matrices in projective Hjelmslev spaces. Des. Codes Cryptogr. **73**, 615–623 (2014)
53. Lang, S.: Algebra, 2nd edn. Addison-Wesley Publ, Company (1984)
54. Levenshtein, V.I.: On bounds for packings in n-dimensional Euclidean space, Dokl. Akad. Nauk SSSR **245**, 1299–1303. in Russian. English translation in Soviet Math. Dokl. **20**, 417–421 (1979)
55. Levenshtein, V.I.: Bounds for packings in metric spaces and certain applications. Probl. Kibernetiki **40**, 44–110 (1983). (in Russian)
56. Levenshtein, V.I.: Designs as maximum codes in polynomial metric spaces. Acta Appl. Math. **25**, 1–82 (1992)
57. Levenshtein, V.I.: Universal bounds for codes and designs (Ch. 6). In: Pless, V.S., Huffman, W.C. (eds.) Handbook of Coding Theory. Elsevier, Amsterdam, pp. 499–648 (1998)
58. MacWilliams, F.J., Sloane, N.J.A.: The Theory of Error-correcting Codes. North Holland, North Holland Math. Library, vol. 16. Amsterdam (1977)
59. Marcugini, S., Milani, A., Pambianco, F.: Maximal $(n, 3)$-arcs in PG(2, 11). Discret. Math. **208/209**, 421–426 (1999)
60. Marcugini, S., Milani, A., Pambianco, F.: NMDS codes of maximal length over F_q, $8 \leq q \leq 11$. IEEE Trans. Inf. Theory **48**(4), 963–966 (2002)
61. Marcugini, S., Milani, A., Pambianco, F.: Classification of the $(n, 3)$-arcs in PG(2, 7). J. Geom. **80**, 179–184 (2004)
62. Marcugini, S., Milani, A., Pambianco, F.: Maximal $(n, 3)$-arcs in PG(2, 13). Discret. Math. **294**, 139–145 (1999)
63. Marcugini, S., Milani, A., Pambianco, F.: Classification of linear codes exploting an invariant. Contrib. Discret. Math. **1**(1), 1–7 (2006)
64. Musin, O.: The kissing number in four dimensions. Ann. Math. **168**, 1–32 (2008)
65. Nechaev, A.A.: Kerdock code in cyclic form. Disk. matematika **1**(4), 123–139 (1989) (in Russian). English version: Discret. Math. Appl. **1**(4), 365–384 (1991)
66. Nechaev, A.A., Kuzmin, A.S.: Linearly presentable codes. In: Proceedings of the IEEE International Symposium on Information Theory and Its Applications (Victoria B. C., Canada), pp. 31–34 (1996)
67. Odlyzko, A.M., Sloane, N.J.A.: New bounds on the number of unit spheres that can touch a unit sphere in n dimensions. J. Comb. Theory A **26**, 210–214 (1979)
68. Preparata, F.P.: A class of optimum non-linear double-error-correcting codes. Inf. Control **13**(4), 378–400 (1968)

69. Segre, B.: Ovals in a finite projective plane. Can. J. Math. **7**, 414–416 (1955)
70. Segre, B.: Sui k-archi nei piani finiti di caratteristica due. Rev. Math. Pures Appl. **2**, 289–300 (1957)
71. Sidelnikov, V.M.: On extremal polynomials used to estimate the size of codes. Probl. Inf. Transm. **16**, 174–186 (1980)
72. Szegő, G.: Orthogonal Polynomials, vol. 23. AMS Col. Publ., Providence, RI (1939)
73. Tagarev, T.: ICT research for security and defence (this is a chapter in this volume)
74. Tagarev, T., Polimirova, D.: Main considerations in elaborating organizational information security policies. In: Proceedings of 20th International Conference on Computer Systems and Technologies (CompSysTech'19), 21-22 June 2019, Ruse, Bulgaria
75. Tagarev, T., Sharkov, G.: Computationally intensive functions in designing and operating distributed cyber secure and resilient systems. In: Proceedings of 20th International Conference on Computer Systems and Technologies (CompSysTech'19), 21-22 June 2019, Ruse, Bulgaria
76. Waterhouse, W.C.: Abelian varieties over finite fields. Ann. Sci. École. Norm. Sup. 2(4), 521–560 (1969)
77. Yudin, V.A.: Minimal potential energy of a point system of charges. Discret. Mat. **4**, 115–121 (1992) (in Russian). English translation: Discr. Math. Appl. **3**, 75–81 (1993)
78. Zong, C.: Sphere Packings. Springer, New York (1999)

Research on Artificial Neural Networks in Bulgarian Academy of Sciences

Petia Koprinkova-Hristova(ⓘ)

Abstract The paper presents a review of research on Artificial Neural Networks (ANNs) done by scientists from the Bulgarian Academy of Sciences (BAS). It summarizes mainly papers refereed in world famous data bases with scientific literature such as Scopus that are considered as representative collection of the work on ANNs done by teams of researchers including at least one scientist form an institute at BAS. The check up in Scopus revealed more than 200 research items published during last 30 years having h factor of 18 without self citations.

Keywords Artificial neural networks · Bulgarian Academy of Sciences

1 Introduction

This review is based on a list of papers dealing with ANNs extracted form the world famous scientific literature data bases (mainly Scopus) and having at least one author whose affiliation is Bulgarian Academy of Sciences. It has not pretensions to be exhaustive rather than to include representative works from BAS researchers that are visible worldwide.

The performed search in Scopus revealed that first works on ANNs in BAS date from 1991 thus fulfilling 30 years of research in the area up to now. Figure 1 represents number of referred papers per year. It reveals significant increase in publication activity during last few years.

The obtained list of more than 200 sources is sorted according to their content into the following groups:

– Theoretical advances and analyses
– Hardware and parallel and distributed computing implementations
– Applications

P. Koprinkova-Hristova (✉)
Institute of Information and Communication Technologies, Bulgarian Academy of Sciences,
Sofia, Bulgaria
e-mail: pkoprinkova@bas.bg

© The Author(s), under exclusive license to Springer Nature Switzerland AG 2021   287
K. T. Atanassov (ed.), *Research in Computer Science in the Bulgarian
Academy of Sciences*, Studies in Computational Intelligence 934,
https://doi.org/10.1007/978-3-030-72284-5_14

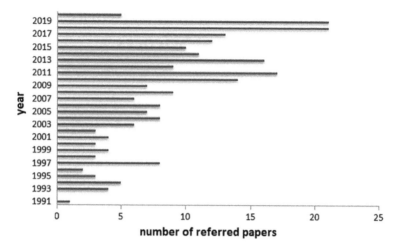

Fig. 1 Number of refereed publications on ANN having at least one author from BAS research institute (numbers are calculated at the beginning of 2020)

Next sections summarize the research on ANNs in BAS belonging to each one of these three main areas.

2 Theoretical Advances and Analyses

2.1 Theoretical Analyses

The monograph [180] summarized huge work of a group from BAS working intensively on Cellular Neural Networks (CNNs) theory and applications. Their basic directions of research are related to CNN stability analyses [47, 179], control of hysteresis and oscillations [7–9], solution of various partial differential equations [136, 139, 152–155, 163, 178, 183, 186–189, 194, 195, 197–199, 213] and applications to modelling of disease spreading [181], tsunami waves [200], management risk [182] and image processing [46, 190, 191].

BAS has long history of research on theoretical analyses of stability and equilibrium states properties of various Recurrent Neural Network (RNN) architectures such as Hopfield ANNs [49, 51], Cohen-Grossberg [50], resonance of a binary attractor ANNs [52], quadratic ANNs [56]. In [12–19, 40] theoretical stability analyses of various types of RNNs (CNNs, Hopfield, Cohen-Grossberg) with different type of delays were reported. In [109] effect of intrinsic plasticity of Echo state networks on their stability properties was investigated. In [121–125, 230] attractor neural network behaviour was investigated. Sensitivity analyses and effect of randomness on ANNs parameters were reported in [75, 76].

2.2   Training Approaches and Novel Designs

Another direction of theoretical research is related to various training approaches of ANNs. In [89, 90, 93] Hebbian vs. gradient approaches for training Echo state networks (ESN) within actor-critic designs were proposed and compared.

A novel approach for features extraction aimed at clustering and visualization of multidimensional data using a member of reservoir computing family of RNNs (Echo State Network) based on Intrinsic Plasticity (IP) rule was developed in [91, 92, 94, 115]. It was tested on numerous data sets and practical tasks [44, 45, 95, 98, 99, 101, 102, 113].

In [106, 107] approach for on-line fast Adaptive Critic Design (ACD) approach using ESNs was proposed. It was implemented in numerous optimization tasks [105, 108, 113, 114, 145].

In [224, 226] novel fast training algorithms for neo-fuzzy and fuzzy-neural ANNs were proposed. In [240] training of Kohonen maps using alternative heighborhood functions was investigated.

In [67] a simplified neural cell model similar to the RBF ANNs was proposed and analysed.

In [167–169] a novel research on presentation of ANNs as networks of flow capacities was initiated recently.

Following contemporary advance in biologically plausible neural network designs BAS researcher started also working in this direction creating spike timing models of visual information processing and decision making [100, 119, 140] based on recent neuro-biological findings about related human brain structures and their connectivity.

2.3   Neuro-Hybrid Approaches

These include various combinations of ANNs with other intelligent techniques such as fuzzy logic, generalized nets, heuristic optimization techniques and evolutionary algorithms aimed at improved structural representation, faster training and incorporation of expert knowledge.

Two monographs [129, 202] gather all developed generalized net models of numerous ANNs (multilayer perceptron; recurrent neural network; Elman neural network; time delay neural network; distributed time delay neural network; layered digital dynamic network; nonlinear autoregressive network with exogenous inputs; brain-state-in-a-box neural network; hierarchical neural networks) as well as parallel ANN models (parallel optimization of multilayer perceptron; parallel optimization of multilayer perceptron with variable learning rate backpropagation algorithm; parallel optimization of multilayer perceptron with variable learning rate backpropagation algorithm with time limit; parallel optimization of multilayer perceptron with momentum backpropagation algorithm; parallel optimization of multilayer perceptron with conjugate gradient backpropagation algorithm). Applications of these theoretical developments were reported in [20, 21, 24, 26, 31, 137, 203–205, 208, 209, 211].

In [27–29] an interpretation of neural networks (MLP as well as RNN) using index matrices was proposed.

Another highly exploited hybrid approach combines ANNs with fuzzy sets. In [23, 25, 30, 203, 206, 207, 210, 211, 233] Intuitionistic Fuzzy (IF) Logic was incorporated within ANNs in the form of IF weights, IF interval valued estimates and IF Inter-Criteria Analysis for structure optimization. In [133, 138] so called fuzzy neuron was proposed. Recently this direction of research continued to development of various combinations between fuzzy sets, IF sets and ANNs named neo-fuzzy [222, 224], semi-fuzzy [214], fuzzy-neural (or neuro-fuzzy) [215, 218, 223, 225], intuitionistic-neuro-fuzzy [216, 217, 219], intuitionistic-RBF [221] and intuitionistic semi-fuzzy [220] approaches that were applied to various control schemes as well as to non-linear process modelling.

Combination between backpropagation training algorithm of ANN and fuzzy control systems was developed in [88].

In [72, 120] approaches combining ANNs with decision threes were proposed.

Genetic algorithms were also very often combined with ANNs in attempt to solve their structural optimization [73, 77, 78, 134, 238, 239].

3   Hardware and Parallel and Distributed Implementations

In BAS hardware implementations of ANNs started in early nineties [38, 126]. A contemporary research on development of hardware implementations of CNNs using memristor devices were also reported [184, 185, 192, 193, 196]. A new material that resembles ANNs was reported recently in [212].

In light of current developments in parallel and distributed computing environments and need of big processing number of works on parallel implementations of of various ANNs increases. In [22] parallel implementation of MLP was proposed. In [58] distributed deep learning approach based on gossip communication was proposed. In [78] neuro-genetic distributed approach was reported.

4   Applications of ANNs

Applications of ANNs encompass wide range of areas such as engineering, medicine, chemistry, biology, environmental and space research, linguistic etc.

From the very beginning of their emergence ANNs were considered as universal approximates of nonlinear dependencies and were widely applied for modelling of various nonlinear systems such as: biotechnological processes [70, 79–81, 85, 86, 110, 149–151, 159, 174, 175, 177, 231, 232]; vibrations modelling aimed at predictive maintenance of industrial plant [32, 103]; software sensors [111, 161,

234]; chemical processes [36, 37, 141, 142]; in medicine [143, 148]; time series forecasting [33, 173]; groundwater flow simulation [135]; pollution monitoring [144, 176, 235]; sea level prediction [147]; solar [164] and stellar [165] activity; quantum mechanics [166]; financial markets [170–172, 227]; welding process [229]. In [1–6] results of nuclear physics experimental research of a big international team (including researchers from the Institute of Nuclear Research and Nuclear Energy at BAS) that applied among other techniques ANNs for the experimental data processing were reported. In [39] an MLP model of visual information processing in an attempt to mimic human brain was proposed.

ANNs were also widely applied within various control schemes such as: orbit correction in synchrotrons [55]; RNNs within adaptive control schemes of fermentation process and mechanical systems [34, 35]; nonlinear Model Predictive Control schemes were developed in [59–61]; robot control [82]; in [114] ACD with ESN was applied for adaptive control of distillation column; mobile robot control using ACD with ESN [145].

Another direction of applications is related to solving of various optimization tasks. In [68, 69, 71, 87, 105] neuro-dynamic programming approaches were applied to optimization of fermentation processes; in [83, 84] work on optimization of welding process using ANNs was reported; in [96, 97] Adaptive Critic Design (ACD) was implemented within Kalman filter tuning scheme; backpropagation of utilty was applied to optimization of initial conditions of fermentation process [104], steel alloys composition [116–118]; optimization of static memory of embedded devices [236]; recommendation systems [237].

Another intensive area of application of ANNs is in various signal processing tasks such as: bio-signal processing and classification using EEG data for emotion recognition [44, 45, 101], ECG data for heartbeat analyses [48, 74]; voice analyses [43, 65], speakers recognition [62–64, 146].

Another direction of work aims at data analyses and classification of: behavioral data [54, 57, 66, 120]; genome sequencing data sets [127, 128].

There are also some applications in medical diagnostic systems [130–132].

ANNs also had numerous applications in image processing and pattern recognition such as satellite images analyses [41, 53, 98, 99, 228], minimization of the chromatic error [46], face recognition [42], yeasts cells classification [160, 162]; face recognition [201].

Recent developments in Natural Language Processing (NLP) also attracted a lot of attention to ANN exploitation in various NLP tasks. Long-short term memory (LSTM) units were successfully applied [156, 158] and compared with reservoir computing approach [112, 157] on the task of Word Sense Disambiguation (WSD) while in [10, 11] various machine learning approaches among which ANN classifiers were tested on the same WSD task.

5 Conclusions

Analyses of the refereed works of BAS researchers on ANNs revealed divers topics of works having not only application character but also theoretical purposefulness.

The positive trend towards increasing of number of referred research was observed. Even for the first three months of 2020 five papers were already published so we could expect almost the same or even bigger publication activity in this fast developing and modern area of research.

Citation overview in Scopus revealed also that these works were mentioned by about 1500 papers without self-citation having h-factor 18. Almost half of papers included in this review were mentioned at least once by others.

References

1. Aaboud, M.E.A.: Study of hard double-parton scattering in four-jet events in pp collisions at $\sqrt{s} = 7$ tev with the atlas experiment. J. High Energy Phys. **2016**(11) (2016). https://doi.org/10.1007/JHEP11(2016)110
2. Aaboud, M.E.A.: Measurement of the inclusive cross-sections of single top-quark and top-antiquark t-channel production in pp collisions at $\sqrt{s} = 13$ tev with the atlas detector. J. High Energy Phys. **2017**(4) (2017). https://doi.org/10.1007/JHEP04(2017)086
3. Aaboud, M.E.A.: Measurement of the production cross-section of a single top quark in association with a z boson in proton-proton collisions at 13 tev with the atlas detector. Phys. Lett. Sect. B: Nucl. Elem. Particle High-Energy Phys. **780**, 557–577 (2018). https://doi.org/10.1016/j.physletb.2018.03.023
4. Aad, G.E.A.: Search for flavour-changing neutral currents in processes with one top quark and a photon using 81 fb-1 of pp collisions at s=13 tev with the atlas experiment. Phys. Lett. Sect. B: Nucl. Elem. Particle High-Energy Phys. **800** (2020). https://doi.org/10.1016/j.physletb.2019.135082
5. Acciarri, M.E.A.: Measurement of exclusive branching fractions of hadronic one-prong tau decays. Phys. Lett. B **345**(1), 93–102 (1995)
6. Adrian, O.E.A.: Measurement of γbb γhad from hadronic decays of the z. Phys. Lett. B **307**(1-2), 237–246 (1993)
7. Agranovich, G., Litsyn, E., Slavova, A.: Discrete-continuous control of bifurcations and oscillatory behaviour in a class of cellular neural networks. Neural, Parallel Sci. Comput. **13**(3–4), 393–410 (2005)
8. Agranovich, G., Litsyn, E., Slavova, A.: Impulsive control of a hysteresis cellular neural network model. Nonlinear Anal. Hybrid Syst. **3**(1), 65–73 (2009)
9. Agranovich, G., Litsyn, E., Slavova, A.: Stabilizing control of a hysteresis cellular neural network model. Comptes Rendus de L'Academie Bulgare des Sciences **63**(1), 47–54 (2010)
10. Agre, G., Petrov, D., Keskinova, S.: A new approach to the supervised word sense disambiguation. Lecture Notes in Computer Science (including subseries Lecture Notes in Artificial Intelligence and Lecture Notes in Bioinformatics), vol. 11089 LNAI (2018)
11. Agre, G., Petrov, D., Keskinova, S.: Word sense disambiguation studio: A flexible system for wsd feature extraction. Information (Switzerland) **10**(3) (2019)
12. Akca, H., Al-Zahrani, E., Covachev, V., Covacheva, Z.: Existence of periodic solutions for the discrete-time counterpart of a neutral-type cellular neural network with time-varying delays and impulses. In: AIP Conference Proceedings. vol. 1863 (2017)
13. Akca, H., Alassar, R., Covachev, V., Covacheva, Z.: Discrete counterparts of continuous-time additive hopfield-type neural networks with impulses. Dyn. Syst. Appl. **13**(1), 77–92 (2004)

14. Akca, H., Alassar, R., Covachev, V., Covacheva, Z., Al-Zahrani, E.: Continuous-time additive hopfield-type neural networks with impulses. J. Math. Anal. Appl. **290**(2), 436–451 (2004)
15. Akca, H., Alassar, R., Covachev, V., Yurtsever, H.A.: Discrete-time impulsive hopfield neural networks with finite distributed delays. Comput. Assist. Mech. Eng. Sci. **14**(2), 145–158 (2007)
16. Akca, H., Covachev, V.: Impulsive cohen-grossberg neural networks with s-type distributed delays. Tatra Mountains Math. Publ. **48**(1), 1–13 (2011)
17. Akca, H., Covachev, V., Altmayer, K.S.: Exponential stability of neural networks with time-varying delays and impulses. Adv. Intell. Soft Comput. **56**, (2009)
18. Akca, H., Covachev, V., Covacheva, Z.: Discrete-time counterparts of impulsive hopfield neural networks with leakage delays. Springer Proc. Math. Stat. **47**, 351–358 (2013)
19. Akca, H., Covachev, V., Covacheva, Z.: Global asymptotic stability of cohen-grossberg neural networks of neutral type. J. Math. Sci. (United States) **205**(6), 719–732 (2015)
20. Aladjov, H.T., Atanassov, K.T., Shannon, A.G.: Generalized net model of temporal learning algorithm for artificial neural networks. In: 2002 1st International IEEE Symposium. vol. 1, pp. 190–193 (2002)
21. Antonov, A.: Generalized net model for parallel optimization of hidden units in neural networks with radial basis functions. Comptes Rendus de L'Academie Bulgare des Sciences **66**(9), 1239–1246 (2013)
22. Antonov, A., Hadjitodorov, S.: Concurrent algorithm for learning of neural networks. In: IS'2012—2012 6th IEEE International Conference Intelligent Systems, Proceedings, pp. 225–228 (2012)
23. Atanassov, K.: Intuitionistic fuzzy logics as tools for evaluation of data mining processes. Knowl. Based Syst. **80**, 122–130 (2015)
24. Atanassov, K., Krawczak, M., Sotirov, S.: Generalized net model for parallel optimization of feed-forward neural network with variable learning rate backpropagation algorithm. In: 2008 4th International IEEE Conference Intelligent Systems, IS 2008. vol. 3, pp. 1616–1619 (2008)
25. Atanassov, K., Pasi, G., Yager, R.: Intuitionistic fuzzy interpretations of multi-criteria multi-person and multi-measurement tool decision making. Int. J. Syst. Sci. **36**(14), 859–868 (2005)
26. Atanassov, K., Sotirov, S.: Optimization of a neural network of self-organizing maps type with time-limits by a generalized net. Adv. Stud. Contemp. Math. **13**(2), 213–220 (2006)
27. Atanassov, K., Sotirov, S.: Representation of the neural networks by the game method for modelling. Adv. Stud. Contemp. Math. (Kyungshang) **22**(3), 347–354 (2012)
28. Atanassov, K., Sotirov, S.: Index matrix interpretation of the multilayer perceptron. In: 2013 IEEE International Symposium on Innovations in Intelligent Systems and Applications. IEEE INISTA 2013 (2013)
29. Atanassov, K., Sotirov, S.: Index matrix interpretation of one type of extended neural networks. Int. J. Reasoning-based Intell. Syst. **6**(3–4), (2014)
30. Atanassov, K., Sotirov, S., Angelova, N.: Intuitionistic fuzzy neural networks with interval valued intuitionistic fuzzy conditions. Stud. Comput. Intell. **862**, (2020)
31. Atanassov, K., Sotirov, S., Antonov, A.: Generalized net model for parallel optimization of feed-forward neural network. Adv. Stud. Contemp. Math. **15**(1), 109–119 (2007)
32. Balabanov, T., Hadjiski, M., Koprinkova-Hristova, P., Beloreshki, S., Doukovska, L.: Neural network model of mill-fan system elements vibration for predictive maintenance. In: INISTA 2011-2011 International Symposium on Innovations in Intelligent Systems and Applications, pp. 410–414 (2011)
33. Balabanov, T.D., Blagoev, I.I., Dineva, K.I.: Self rising tri layers MLP for time series forecasting. Commun. Comput. Inform. Sci. **919**, (2018)
34. Baruch, I.S., Cortes, J.B., Medina, J.P., Hernandez, L.A.P.: An adaptive neural control of a fed-batch fermentation processes. IEEE Conf. Control Appl. Proc. **2**, 808–812 (2003)
35. Baruch, I.S., Martinez, A.D.C., Thomas, F., Garrido, R.: An integral-plus-state adaptive neural control of mechanical system. IEEE Conf. Control Appl. Proc. **2**, 813–818 (2003)

36. Binev, Y., Corvo, M., Aires-de Sousa, J.: The impact of available experimental data on the prediction of 1 h nmr chemical shifts by neural networks. J. Chem. Inform. Comput. Sci. **44**(3), 946–949 (2004)

37. Binev, Y., Aires-de Sousa, J.: Structure-based predictions of 1h nmr chemical shifts using feed-forward neural networks. J. Chem. Inform. Comput. Sci. **44**(3), 940–945 (2004)

38. Bochev, V.: Distributed arithmetic implementation of artificial neural networks. IEEE Transactions on Signal Processing **41**(5), 2010–2013 (1993)

39. Bojilov, L., Bocheva, N.: Neural network model for visual discrimination of complex motions. Comptes Rendus de L'Academie Bulgare des Sciences **65**(10), 1379–1386 (2012)

40. Bolle, D., Dominguez, D.R.C., Erichsen Jr., R., Korutcheva, E., Theumann, W.K.: Time evolution of the extremely diluted blume-emery-griffiths neural network. Phys. Rev. E Stat. Nonlinear, Soft Matter Phys. **68**(6 1), 629011–629014 (2003)

41. Borisova, D., Jelev, G., Atanassov, V., Koprinkova-Hristova, P., Alexiev, K.: Algorithms for lineaments detection in processing of multispectral images. In: Proceedings of SPIE: The International Society for Optical Engineering, vol. 9245 (2014)

42. Boumbarov, O., Sokolov, S., Gluhchev, G.: Combined face recognition using wavelet packets and radial basis function neural network. In: ACM International Conference Proceeding Series. vol. 285 (2007)

43. Boyanov, B., Hadjitodorov, S.: Acoustic analysis of pathological voices: a voice analysis system for the screening and laryngeal diseases. IEEE Eng. Med. Biol. Mag. **16**(4), 74–82 (1997)

44. Bozhkov, L., Koprinkova-Hristova, P., Georgieva, P.: Learning to decode human emotions with echo state networks. Neural Netw. **78**, 112–119 (2016)

45. Bozhkov, L., Koprinkova-Hristova, P., Georgieva, P.: Reservoir computing for emotion valence discrimination from eeg signals. Neurocomputing **231**, 28–40 (2017)

46. Cancelliere, R., Gai, M., Slavova, A.: Application of polynomial cellular neural networks in diagnosis of astrometric chromaticity. Appl. Math. Model. **34**(12), 4243–4252 (2010)

47. Cancelliere, R., Slavova, A.: Dynamics and stability of generalized cellular nonlinear network model. Appl. Math. Comput. **165**(1), 127–136 (2005)

48. Christov, I., Bortolan, G.: Ranking of pattern recognition parameters for premature ventricular contractions classification by neural networks. Physiol. Measur. **25**(5), 1281–1290 (2004)

49. Condon, M., Grahovski, G.G.: On stability and model order reduction of perturbed nonlinear neural networks. In: Proceedings—22nd European Conference on Modelling and Simulation, pp. 292–298. ECMS 2008 (2008)

50. Covachev, V., Akca, H., Sarr, M.: Discrete-time counterparts of impulsive cohen-grossberg neural networks of neutral type. Neural, Parallel Sci. Comput. **19**(3–4), 345–360 (2011)

51. Covachev, V., Covacheva, Z.: Existence of periodic solutions for the discrete-time counterpart of a complex-valued hopfield neural network with time-varying delays and impulses. In: Proceedings of the International Joint Conference on Neural Networks, vol. 2018-July (2018)

52. de la Casa, M.A., Korutcheva, E., Parrondo, J.M.R., de la Rubia, F.J.: System-size resonance in a binary attractor neural network. Phys. Rev. E Stat. Nonlinear Soft Matter Phys. **72**(3) (2005)

53. Dimitrov, P., Dong, Q., Eerens, H., Gikov, A., Filchev, L., Roumenina, E., Jelev, G.: Subpixel crop type classification using proba-v 100 m ndvi time series and reference data from sentinel-2 classifications. Remote Sensing **11**(11), (2019)

54. Dimitrova, M., Boyadjiev, D., Butorin, N.: Interface adaptation to style of user-computer interaction. Lecture Notes in Computer Science (including subseries Lecture Notes in Artificial Intelligence and Lecture Notes in Bioinformatics), vol. 1892, (2000)

55. Dinev, D.: Closed-orbit correction in synchrotrons. Phys. Particles Nuclei **28**(4), 398–417 (1997)

56. Dominguez, D.R.C., Korutcheva, E., Theumann, W.K., Erichsen Jr., R.: Flow diagrams of the quadratic neural network, Lecture Notes in Computer Science (including subseries Lecture Notes in Artificial Intelligence and Lecture Notes in Bioinformatics), vol. 2415 LNCS (2002)

57. Fijalkowski, J., Ganzha, M., Paprzycki, M., Fidanova, S., Lirkov, I., Badica, C., Ivanovic, M.: Mining smartphone generated data for user action recognition—preliminary assessment. In: AIP Conference Proceedings, vol. 2025 (2018)

58. Georgiev, D., Gurov, T.: Distributed Deep Learning on Heterogeneous Computing Resources Using Gossip Communication, Lecture Notes in Computer Science (including subseries Lecture Notes in Artificial Intelligence and Lecture Notes in Bioinformatics), vol. 11958 LNCS (2020)

59. Grancharova, A., Johansen, T.A.: Explicit NMPC based on neural network models. Lecture Notes in Control and Information Sciences, vol. 429, (2012)

60. Grancharova, A., Kocijan, J., Johansen, T.A.: Dual-mode explicit output-feedback predictive control based on neural network models. In: IFAC Proceedings Volumes (IFAC-PapersOnline), vol. 43, pp. 545–550 (2010)

61. Grancharova, A., Kocijanb, J., Johansend, T.A.: Explicit output-feedback nonlinear predictive control based on black-box models. Eng. Appl. Artif. Intell. **24**(2), 388–397 (2011)

62. Hadjitodorov, S., Boyanov, B.: Pc-based system for robust speaker recognition. J. Comput. Inform. Technol. **6**(4), 415–423 (1998)

63. Hadjitodorov, S., Boyanov, B., Dalakchieva, N.: A two-level classifier for text-independent speaker identification. Speech Commun. **21**(3), 209–217 (1997)

64. Hadjitodorov, S., Boyanov, B., Ivanov, T., Dalakchieva, N.: Text-independent speaker identification using neural nets and ar-vector models. Electron. Lett. **30**(11), 838–840 (1994)

65. Hadjitodorov, S., Boyanov, B., Teston, B.: Laryngeal pathology detection by means of class-specific neural maps. IEEE Trans. Inform. Technol. Biomed. **4**(1), 68–73 (2000)

66. Heinrich, H., Moll, G.H., Dickhaus, H., Kolev, V., Yordanova, J., Rothenberger, A.: Time-on-task analysis using wavelet networks in an event-related potential study on attention-deficit hyperactivity disorder. Clin. Neurophysiol. **112**(7), 1280–1287 (2001)

67. Ilchev, V., Ilchev, S.: Simplified information neural cell model and its basic properties. In: 2016 IEEE 8th International Conference on Intelligent Systems, IS 2016–Proceedings, pp. 81–89 (2016)

68. Iliev, V., Kostov, G., Stoycheva, J., Koprinkova-Hristova, P., Angelov, M., Popova, S.: Bioethanol production optimization using acd with esn critic. In: INISTA 2011: 2011 International Symposium on Innovations in Intelligent Systems and Applications, pp. 606–610 (2011)

69. Ilkova, T., Petrov, M.: Dynamic and neuro-dynamic optimization of a fed-batch fermentation process, Lecture Notes in Computer Science (including subseries Lecture Notes in Artificial Intelligence and Lecture Notes in Bioinformatics), vol. 5253 LNAI (2008)

70. Ilkova, T., Petrov, M.: Neuro-fuzzy based model of batch fermentation of kluyveromyces marxianus var. lactis mc5. Biotechnol. Biotechnol. Equip. **28**(5), 975–979 (2014)

71. Ilkova, T., Petrov, M., Roeva, O.: Optimization of a whey bioprocess using neuro-dynamic programming strategy. Biotechnol. Biotechnol. Equip. **26**(5), 3249–3253 (2012)

72. Ivanova, I., Kubat, M.: Initialization of neural networks by means of decision trees. Knowl. Based Syst. **8**(6), 333–344 (1995)

73. Ivanova, P.I., Tagarev, T.D.: Indicator space configuration for early warning of violent political conflicts by genetic algorithms. Ann. Oper. Res. **97**(1–4), 287–311 (2000)

74. Jekova, I., Bortolan, G., Christov, I.: Assessment and comparison of different methods for heartbeat classification. Med. Eng. Phys. **30**(2), 248–257 (2008)

75. Kapanova, K.G., Dimov, I., Sellier, J.M.: A neural network sensitivity analysis in the presence of random fluctuations. Neurocomputing **224**, 177–183 (2017)

76. Kapanova, K.G., Dimov, I., Sellier, J.M.: On randomization of neural networks as a form of post-learning strategy. Soft Comput. **21**(9), 2385–2393 (2017)

77. Kapanova, K.G., Dimov, I., Sellier, J.M.: A genetic approach to automatic neural network architecture optimization. Neural Comput. Appl. **29**(5), 1481–1492 (2018)

78. Ketipov, R., Kostadinov, G., Petrov, P., Zankinski, I., Balabanov, T.: Human-computer mobile distributed computing for time series forecasting. In: Communications in Computer and Information Science, vol. 1141 CCIS (2019)

79. Kirilova, E., Vaklieva-Bancheva, N., Vladova, R.: Prediction of temperature conditions of autothermal thermophilic aerobic digestion bioreactors at wastewater treatment plants. Int. J. Bioautom. **20**(2), 289–300 (2016)

80. Kirilova, E., Yankova, S., Ilieva, B., Vaklieva-Bancheva, N.: A new approach for modeling the biotransformation of crude glycerol by using narx ann. J. Chem. Technol. Metallurgy **49**(5), 473–478 (2014)

81. Kirilova, E.G., Vaklieva-Bancheva, N.G.: Ann modeling of a two-stage industrial atad system for the needs of energy integration. Bulgarian Chem. Commun. **50**, 90–99 (2018)

82. Kiryazov, K., Kiriazov, P.: Efficient learning approach for optimal control of human and robot motion. In: Emerging Trends in Mobile Robotics. Proceedings of the 13th International Conference on Climbing and Walking Robots and the Support Technologies for Mobile Machines, CLAWAR 2010, pp. 1219–1226 (2010)

83. Koleva, E., Christova, N., Velev, K.: Neural network based approach for quality improvement of orbital arc welding joints. In: 2010 IEEE International Conference on Intelligent Systems, IS 2010—Proceedings, pp. 290–295 (2010)

84. Koleva, E., Mladenov, G.: Process parameter optimization and quality improvement at electron beam welding, pp. 101–166. Welding: Processes, Quality, and Applications (2011)

85. Koprinkova, P., Petrova, M., Patarinska, T., Bliznakova, M.: Neural network modeling of fermentation processes: specific kinetic rate models. Cybern. Syst. **29**(3), 303–317 (1998)

86. Koprinkova, P.D., Patarinska, T.D., Petrova, M.N.: Memory effects description by neural networks with delayed feedback connections. Int. J. Intell. Syst. **19**(4), 341–351 (2004)

87. Koprinkova-Hristova, P.: Acd approach to optimal control of mixed culture cultivation for phb production process—sugar's time profile synthesis. In: 2008 4th International IEEE Conference Intelligent Systems, IS 2008, vol. 3, pp. 1229–1232 (2008)

88. Koprinkova-Hristova, P.: Backpropagation through time training of a neuro-fuzzy controller. Int. J. Neural Syst. **20**(5), 421–428 (2010)

89. Koprinkova-Hristova, P.: Adaptive critic design and heuristic search for optimization. Lecture Notes in Computer Science (including subseries Lecture Notes in Artificial Intelligence and Lecture Notes in Bioinformatics), vol. 8353. LNCS (2014)

90. Koprinkova-Hristova, P.: On-line training of ESN and IP tuning effect. Lecture Notes in Computer Science (including subseries Lecture Notes in Artificial Intelligence and Lecture Notes in Bioinformatics), vol. 8681 LNCS (2014)

91. Koprinkova-Hristova, P.: On effects of ip improvement of esn reservoirs for reflecting of data structure, vol. 2015-September (2015). https://doi.org/10.1109/IJCNN.2015.7280703

92. Koprinkova-Hristova, P.: Multi-dimensional data clustering and visualization via echo state networks. Intell. Syst. Ref. Library **108**, 93–122 (2016)

93. Koprinkova-Hristova, P.: Three approaches to train echo state network actors of adaptive critic design. Lecture Notes in Computer Science (including subseries Lecture Notes in Artificial Intelligence and Lecture Notes in Bioinformatics), vol. 9886 LNCS (2016)

94. Koprinkova-Hristova, P., Alexiev, K.: Echo state networks in dynamic data clustering. Lecture Notes in Computer Science (including subseries Lecture Notes in Artificial Intelligence and Lecture Notes in Bioinformatics), vol. 8131 LNCS (2013)

95. Koprinkova-Hristova, P., Alexiev, K.: Sound fields clusterization via neural networks. In: INISTA 2014—IEEE International Symposium on Innovations in Intelligent Systems and Applications, Proceedings, pp. 368–374 (2014)

96. Koprinkova-Hristova, P., Alexiev, K.: ACD with ESN for tuning of MEMS kalman filter. Lecture Notes in Computer Science (including subseries Lecture Notes in Artificial Intelligence and Lecture Notes in Bioinformatics), vol. 9374 (2015)

97. Koprinkova-Hristova, P., Alexiev, K.: Neuro-fuzzy tuning of kalman filter. In: 2016 IEEE 8th International Conference on Intelligent Systems, IS 2016—Proceedings, pp. 651–657 (2016)

98. Koprinkova-Hristova, P., Alexiev, K., Borisova, D., Jelev, G., Atanassov, V.: Recurrent neural networks for automatic clustering of multispectral satellite images. In: Proceedings of SPIE—The International Society for Optical Engineering, vol. 8892 (2013)

99. Koprinkova-Hristova, P., Angelova, D., Borisova, D., Jelev, G.: Clustering of spectral images using echo state networks. In: 2013 IEEE International Symposium on Innovations in Intelligent Systems and Applications, IEEE INISTA 2013 (2013)
100. Koprinkova-Hristova, P., Bocheva, N., Nedelcheva, S.: Investigation of feedback connections effect of a spike timing neural network model of early visual system. In: 2018 IEEE (SMC) International Conference on Innovations in Intelligent Systems and Applications, INISTA 2018 (2018)
101. Koprinkova-Hristova, P., Bozhkov, L., Georgieva, P.: Echo state networks for feature selection in affective computing. In: Lecture Notes in Artificial Intelligence (Subseries of Lecture Notes in Computer Science), vol. 9086, pp. 131–141 (2015)
102. Koprinkova-Hristova, P., Doukovska, L., Kostov, P.: Working regimes classification for predictive maintenance of mill fan systems. In: 2013 IEEE International Symposium on Innovations in Intelligent Systems and Applications, IEEE INISTA 2013 (2013)
103. Koprinkova-Hristova, P., Hadjiski, M., Doukovska, L., Beloreshki, S.: Recurrent neural networks for predictive maintenance of mill fan systems. Int. J. Electron. Telecommun. 57(3), 401–406 (2011)
104. Koprinkova-Hristova, P., Kostov, G., Angelov, M., Pandzharov, P.: Intelligent optimisation of batch fermentations initial conditions. Int. J. Reasoning-based Intell. Syst. 2(3–4), 285–292 (2010)
105. Koprinkova-Hristova, P., Kostov, G., Popova, S.: Intelligent optimization of a mixed culture cultivation process. Int. J. Bioautom. 19, S113–S124 (2015)
106. Koprinkova-Hristova, P., Oubbati, M., Palm, G.: Adaptive critic design with echo state network. In: Conference Proceedings: IEEE International Conference on Systems, Man and Cybernetics, pp. 1010–1015 (2010)
107. Koprinkova-Hristova, P., Oubbati, M., Palm, G.: Heuristic dynamic programming using echo state network as online trainable adaptive critic. Int. J. Adaptive Control Signal Process. 27(10), 902–914 (2013)
108. Koprinkova-Hristova, P., Palm, G.: Adaptive critic design with ESN critic for bioprocess optimization. In: Lecture Notes in Computer Science (including subseries Lecture Notes in Artificial Intelligence and Lecture Notes in Bioinformatics), vol. 6353 LNCS (2010)
109. Koprinkova-Hristova, P., Palm, G.: ESN intrinsic plasticity versus reservoir stability. In: Lecture Notes in Computer Science (including subseries Lecture Notes in Artificial Intelligence and Lecture Notes in Bioinformatics), vol. 6791 LNCS (2011)
110. Koprinkova-Hristova, P., Patarinska, T.: Neural network modelling of continuous microbial cultivation accounting for the memory effects. Int. J. Syst. Sci. 37(5), 271–277 (2006)
111. Koprinkova-Hristova, P., Patarinska, T.: Neural network software sensors design for lysine fermentation process. Appl. Artif. Intell. 22(3), 235–253 (2008)
112. Koprinkova-Hristova, P., Popov, A., Simov, K., Osenova, P.: Echo state network for word sense disambiguation. In: Lecture Notes in Computer Science (including subseries Lecture Notes in Artificial Intelligence and Lecture Notes in Bioinformatics), vol. 11089 LNAI (2018)
113. Koprinkova-Hristova, P., Stefanova, M., Genova, B., Bocheva, N.: Echo state network for classification of human eye movements during decision making. In: IFIP Advances in Information and Communication Technology, vol. 19, (2018)
114. Koprinkova-Hristova, P., Todorov, Y., Paraschiv, N., Olteanu, M., Terziyska, M.: Adaptive control of distillation column using adaptive critic design. In: Proceedings of the 2017 21st International Conference on Process Control, PC 2017, pp. 434–439 (2017)
115. Koprinkova-Hristova, P., Tontchev, N.: Echo state networks for multi-dimensional data clustering. In: Lecture Notes in Computer Science (including subseries Lecture Notes in Artificial Intelligence and Lecture Notes in Bioinformatics), vol. 7552 LNCS (2012)
116. Koprinkova-Hristova, P., Tontchev, N., Popova, S.: Neural networks approach to optimization of steel alloys composition. In: IFIP Advances in Information and Communication Technology, vol. 363 AICT (2011)
117. Koprinkova-Hristova, P., Tontchev, N., Popova, S.: Multi-criteria optimization of steel alloys for crankshafts production. In: INISTA 2012—International Symposium on Innovations in Intelligent Systems and Applications (2012)

118. Koprinkova-Hristova, P., Tontchev, N., Popova, S.: Two approaches to multi-criteria optimi- sation of steel alloys for crankshafts production. Int. J. Reasoning-based Intell. Syst. **5**(2), 96–103 (2013)
119. Koprinkova-Hristova, P.D., Bocheva, N., Nedelcheva, S., Stefanova, M.: Spike timing neural model of motion perception and decision making. Front. Comput. Neurosci. **13**, (2019)
120. Koprinska, I., Pfurtscheller, G., Flotzinger, D.: Sleep classification in infants by decision tree-based neural networks. Artif. Intell. Med. **8**(4), 387–401 (1996)
121. Koroutchev, K., Korutcheva, E.: Conditions for the emergence of spatially asymmetric retrieval states in an attractor neural network. Central Eur. J. Phys. **3**(3), 409–419 (2005)
122. Koroutchev, K., Korutcheva, E.: Spatial asymmetric retrieval states in binary attractor neural network. AIP Conf. Proc. **780**, 603–606 (2005)
123. Koroutchev, K., Korutcheva, E.: Bump formation in a binary attractor neural network. Phys. Rev. E Stat. Nonlinear, Soft Matter Phys. **73**(2) (2006)
124. Korutcheva, E., Del Prete, V., Nadal, J.: A perturbative approach to nonlinearities in the information carried by a two layer neural network. Int. J. Modern Phys. B **15**(3), 281–295 (2001)
125. Korutcheva, E., Koroutchev, K.: On the local-field distribution in attractor neural networks. Int. J. Modern Phys. C **7**(4), 463–483 (1996)
126. Kovatchev, M., Hieva, R.: Neural networks and computers based on inphase optics. In: Pro- ceedings of SPIE—The International Society for Optical Engineering, vol. 1621, pp. 259–267 (1991)
127. Krachunov, M., Nisheva, M., Vassilev, D.: Machine learning-driven noise separation in high variation genomics sequencing datasets. In: Lecture Notes in Computer Science (including subseries Lecture Notes in Artificial Intelligence and Lecture Notes in Bioinformatics), vol. 11089 LNAI (2018)
128. Krachunov, M., Nisheva, M., Vassilev, D.: Machine learning models for error detection in metagenomics and polyploid sequencing data. Information (Switzerland) **10**(3) (2019)
129. Krawczak, M., Sotirov, S., Atanassov, K.: Multilayer Neural Network Modellig by General- ized Nets. Warsaw School of Information Technologies (2010)
130. Kuncheva, L.: An aggregation of pro and con evidence for medical decision support systems. Comput. Biol. Medicine **23**(6), 417–424 (1993)
131. Kuncheva, L.: Two-level classification schemes in medical diagnostics. Int. J. Bio-med. Com- put. **32**(3–4), 197–210 (1993)
132. Kuncheva, L.I.: Fuzzy two-level classifier for high-g analysis: medical diagnosis of acceler- ation effects. IEEE Eng. Med. Biol. Mag. **13**(5), 717–722 (1994)
133. Kuncheva, L.I.: Pattern recognition with a model of fuzzy neuron using degree of consensus. Fuzzy Sets Syst. **66**(2), 241–250 (1994)
134. Kuncheva, L.I.: Initializing of an rbf network by a genetic algorithm. Neurocomputing **14**(3), 273–288 (1997)
135. Liolios, K., Tsihrintzis, V., Angelidis, P., Georgiev, K., Georgiev, I.: Numerical simulation for horizontal subsurface flow constructed wetlands: A short review including geothermal effects and solution bounding in biodegradation procedures. In: AIP Conference Proceedings, vol. 1773 (2016)
136. Melton, T., Slavova, A.: Travelling wave solutions of fitzhugh-nagumo cnn model with hys- teresis. Comptes Rendus de L'Academie Bulgare des Sciences **64**(5), (2011)
137. Mengov, G., Georgiev, K., Pulov, S., Trifonov, T., Atanassov, K.: Fast computation of a gated dipole field. Neural Netw. **19**(10), 1636–1647 (2006)
138. Mitra, S., Kuncheva, L.I.: Improving classification performance using fuzzy mlp and two-level selective partitioning of the feature space. Fuzzy Sets Syst. **70**(1), 1–13 (1995)
139. Mladenov, V., Slavova, A.: On the periodic solutions in one dimensional cellular nonlin- ear networks based on josephson junctions (jj's). In: Proceedings of the IEEE International Workshop on Cellular Neural Networks and their Applications (2006)
140. Nedelcheva, S., Koprinkova-Hristova, P.: Orientation selectivity tuning of a spike timing neural network model of the first layer of the human visual cortex. Stud. Comput. Intell. **793**, (2019)

141. Nestorov, I., Rowland, M., Hadjitodorov, S.T., Petrov, I.: Empirical versus mechanistic modelling: comparison of an artificial neural network to a mechanistically based model for quantitative structure pharmacokinetic relationships of a homologous series of barbiturates. AAPS J. **1**(4), xiii–xiv (1999)
142. Nestorov, I., Rowland, M., Hadjitodorov, S.T., Petrov, I.: Empirical versus mechanistic modelling: comparison of an artificial neural network to a mechanistically based model for quantitative structure pharmacokinetic relationships of a homologous series of barbiturates. AAPS Pharm Sci **1**(4), 1–9 (1999)
143. Nikolov, S., Nenov, M.: Modelling vaccine quantity in mathematical models of melanoma treatment. Series Biomech. **32**(4), 19–25 (2018)
144. Nikolova, N., Lavrova-Popova, S., Petkova, P., Tsakovski, S., Pribylova, P.: Passive air sampling monitoring of pops in southeastern Europe at high mountain station beo-Moussala, Bulgaria. J. Chem. Technol. Metallurgy **53**(2), 267–274 (2018)
145. Oubbati, M., Kschele, M., Koprinkova-Hristova, P., Palm, G.: Anticipating rewards in continuous time and space with echo state networks and actor-critic design. In: ESANN 2011 proceedings, 19th European Symposium on Artificial Neural Networks, Computational Intelligence and Machine Learning, pp. 117–122 (2011)
146. Ouzounov, A.: Text-independent speaker identification using a hybrid neural network and conformity approach. In: IEEE International Conference on Neural Networks—Conference Proceedings, vol. 4, pp. 2098–2102 (1997)
147. Pashova, L., Popova, S.: Daily sea level forecast at tide gauge Burgas, Bulgaria using artificial neural networks. J. Sea Res. **66**(2), 154–161 (2011)
148. Pavlova, P.E., Sliakev, N.G., Borisova, E.G.: Comparative analysis of methods for ascertainment the similarity between reflected spectra obtained from skin lesions. IFAC-Papers online, vol. 52, pp. 365–369 (2019)
149. Petrova, M., Koprinkova, P., Patarinska, T.: Neural model taking into account culture memory. Biotechnol. Biotechnol. Equip. **8**(1), 88–92 (1994)
150. Petrova, M., Koprinkova, P., Patarinska, T.: Neural network modelling of fermentation processes. microorganisms cultivation model. Bioprocess Eng. **16**(3), 145–149 (1997)
151. Petrova, M., Koprinkova, P., Patarinska, T., Bliznakova, M.: Neural network modelling of fermentation processes: specific growth rate model. Bioprocess Eng. **18**(4), 281–287 (1998)
152. Popivanov, P., Slavova, A.: Cellular neural network model for nonlinear waves in medium with exponential memory. In: Lecture Notes in Computer Science (including subseries Lecture Notes in Artificial Intelligence and Lecture Notes in Bioinformatics), vol. 1988 (2001)
153. Popivanov, P., Slavova, A.: Smooth and nonsmooth solutions of several equations of mathematical physics and their cellular neural network realization. In: Lecture Notes in Computer Science (including subseries Lecture Notes in Artificial Intelligence and Lecture Notes in Bioinformatics), vol. 5434 LNCS (2009)
154. Popivanov, P., Slavova, A., Zecca, P.: Periodic solutions of the burgers-hopf equation with small parameter and its cellular neural network model. Mediterranean J. Math. **5**(1), 1–19 (2008)
155. Popivanov, P., Slavova, A., Zecca, P.: Compact travelling waves and peakon type solutions of several equations of mathematical physics and their cellular neural network realization. Nonlinear Anal. Real World Appl. **10**(3), 1453–1465 (2009)
156. Popov, A.: Neural network models for word sense disambiguation: an overview. Cybern. Inform. Technol. **18**(1), 139–151 (2018)
157. Popov, A., Koprinkova-Hristova, P., Simov, K., Osenova, P.: Echo State vs. LSTM Networks for Word Sense Disambiguation, Lecture Notes in Computer Science (including subseries Lecture Notes in Artificial Intelligence and Lecture Notes in Bioinformatics), vol. 11731 LNCS (2019)
158. Popov, A., Sikos, J.: Graph embeddings for frame identification. In: International Conference Recent Advances in Natural Language Processing, RANLP, vol. 2019-September, pp. 939–948 (2019)

159. Popova, S.: Parameter identification of a model of yeast cultivation process with neural network. Bioprocess Eng. **16**(4), 243–245 (1997)
160. Popova, S., Chaker, N., Wagenknecht, M., Kostova, S.: Yeast cells classification by kohonen neural network. In: IFAC Proceedings Volumes (IFAC-PapersOnline), vol. 37, pp. 213–216 (2004)
161. Popova, S., Koprinkova, P., Patarinska, T.: Neural network based biomass and growth rate estimation aimed to control of a chemostat microbial cultivation. Appl. Artif. Intell. **17**(4), 345–360 (2003)
162. Popova, S., Mitev, V.: Application of artificial neural networks for yeast cells classification. Bioprocess Eng. **17**(2), 111–113 (1997)
163. Rangelov, T., Slavova, A.: Dynamic behaviour of piezoelectric solid via cnn approach. Comptes Rendus de L'Academie Bulgare des Sciences **66**(6), 801–808 (2013)
164. Saiz, E., Cerrato, Y., Cid, C., Dobrica, V., Hejda, P., Nenovski, P., Stauning, P., Bochnicek, J., Danov, D., Demetrescu, C., Gonzalez, W.D., Maris, G., Teodosiev, D., Valach, F.: Geomagnetic response to solar and interplanetary disturbances. J. Space Weather Space Clim. **3**, (2013)
165. Schierscher, F., Paunzen, E.: An artificial neural network approach to classify sdss stellar spectra. Astronomische Nachrichten **332**(6), 597–601 (2011)
166. Sellier, J.M., Kapanova, K.G., Leygonie, J., Caron, G.M.: Machine learning and signed particles, an alternative and efficient way to simulate quantum systems. Int. J. Quant. Chem. **119**(23), (2019)
167. Sgurev, V.: Artificial neural networks as a network flow with capacities. Comptes Rendus de L'Academie Bulgare des Sciences **71**(9), 1245–1252 (2018)
168. Sgurev, V., Drangajov, S., Jotsov, V.: Network flow interpretation of artificial neural networks. In: 9th International Conference on Intelligent Systems 2018: Theory, Research and Innovation in Applications, IS 2018—Proceedings, pp. 494–498 (2018)
169. Sgurev, V., Drangajov, S., Jotsov, V.: A new network flow platform for building artificial neural networks. Stud. Comput. Intell. **864**, (2020)
170. Shahpazov, V.L., Doukovska, L.A., Karastoyanov, D.N.: Artificial intelligence neural networks applications in forecasting financial markets and stock prices. In: BMSD 2014—Proceedings of the 4th International Symposium on Business Modeling and Software Design, pp. 282–288 (2014)
171. Shahpazov, V.L., Velev, V.B., Doukovska, L.A.: Design and application of artificial neural networks for predicting the values of indexes on the bulgarian stock market. In: 2013 Signal Processing Symposium, SPS 2013 (2013)
172. Shahpazov, V.L., Velev, V.B., Doukovska, L.A.: Forecasting price movement of sofix index on the bulgarian stock exchange—sofia using an artificial neural network model. In: BMSD 2013: Proceedings of the 3rd International Symposium on Business Modeling and Software Design, pp. 298–303 (2013)
173. Shopov, V., Markova, V.: Identification of non-linear dynamic system. In: 2019 International Conference on Information Technologies, InfoTech 2019—Proceedings (2019)
174. Simeonov, I., Chorukova, E.: Neural networks modelling of two biotechnological processes. In: 2004 2nd International IEEE Conference 'Intelligent Systems'—Proceedings, vol. 1, pp. 331–336 (2004)
175. Simeonov, I., Chorukova, E.: Anaerobic digestion modelling with artificial neural networks. Comptes Rendus de L'Academie Bulgare des Sciences **61**(4), 505–512 (2008)
176. Simeonova, P., Lovchinov, V., Dimitrov, D., Radulov, I.: Environmetric approaches for lake pollution assessment. Environ. Monitor. Assessment **164**(1–4), 233–248 (2010)
177. Simeonova, V., Tasheva, K., Kosturkova, G., Vasilev, D.: A soft computing qsar adapted model for improvement of golden root in vitro culture growth. Biotechnol. Biotechnol. Equip. **27**(3), 3877–3884 (2013)
178. Slavova, A.: Modeling nonlinear waves and pdes via cellular neural networks. Annali dell'Universita di Ferrara **45**(1), 311–326 (1999)

179. Slavova, A.: Stability analysis of cellular neural networks with nonlinear dynamics. Nonlinear Anal. Real World Appl. **2**(1), 93–103 (2001)
180. Slavova, A.: Cellular neural networks: dynamics and modelling. Math. Modelling: Theory Appl. **16**, (2003)
181. Slavova, A.: Dynamics and traveling waves in cnn vector disease model. IEEE Trans. Circ. Syst. II: Express Briefs **53**(11), 1304–1307 (2006)
182. Slavova, A.: Cellular neural networks model of risk management. In: Proceedings of the IEEE International Workshop on Cellular Neural Networks and their Applications, pp. 181–185 (2008)
183. Slavova, A.: New wave profiles in viscoelastic burgers' rtd-based cellular neural networks model. In: ECCTD 2009— European Conference on Circuit Theory and Design Conference Program, pp. 81–84 (2009)
184. Slavova, A.: Memristor cnn model for image denoising. In: 2019 26th IEEE International Conference on Electronics, Circuits and Systems (ICECS), pp. 221–224 (2019)
185. Slavova, A., Bobeva, G.: Determination of edge of chaos in hysteresis cnn model with memristor synapses. In: 2017 European Conference on Circuit Theory and Design (ECCTD), pp. 1–4 (2017)
186. Slavova, A., Kyurkchiev, N.: On cnn model of black scholes equation with leland correction. Comptes Rendus de L'Academie Bulgare des Sciences **71**(2), 169–175 (2018)
187. Slavova, A., Markova, M.: Receptor-based cellular neural network models. WSEAS Trans. Math. **4**(3), 212–217 (2005)
188. Slavova, A., Markova, M.: Receptor-based cnn model with hysteresis for pattern formation. In: Proceedings of the IEEE International Workshop on Cellular Neural Networks and their Applications (2006)
189. Slavova, A., Markova, M.: Polynomial lotka-volterra cnn model. dynamics and complexity. Comptes Rendus de L'Academie Bulgare des Sciences **60**(12), 1271–1276 (2007)
190. Slavova, A., Rashkova, V.: Convection diffusion model for image processing. Comptes Rendus de L'Academie Bulgare des Sciences **64**(3), 339–344 (2011)
191. Slavova, A., Rashkova, V.: A novel cnn based image denoising model. In: 2011 20th European Conference on Circuit Theory and Design, ECCTD 2011, pp. 226–229 (2011)
192. Slavova, A., Tetzlaff, R.: Math. Anal. Memristor CNN (2019). https://doi.org/10.5772/intechopen.86446
193. Slavova, A., Tetzlaff, R.: Memristor cnns with hysteresis **793**, (2019)
194. Slavova, A., Tetzlaff, R., Markova, M.: Cnn computing of the interaction of fluxons. In: 2011 30th URSI General Assembly and Scientific Symposium, URSIGASS 2011 (2011)
195. Slavova, A., Zafirova, Z.: Dynamics of viscoelastic burgers' cellular neural networks model. In: AIP Conference Proceedings (2019)
196. Slavova, A., Zafirova, Z., Tetzlaff, R.: Edge of chaos in nanoscale memristor cnn. In: 2019 IEEE International Symposium on Circuits and Systems (ISCAS), pp. 1–4 (2019)
197. Slavova, A., Zecc, P.: Travelling wave solution of polynomial cellular neural network model for burgers-huxley equation. Comptes Rendus de L'Academie Bulgare des Sciences **65**(10), 1335–1342 (2012)
198. Slavova, A., Zecca, P.: Cnn model for studying dynamics and travelling wave solutions of fitzhugh-nagumo equation. J. Comput. Appl. Math. **151**(1), 13–24 (2003)
199. Slavova, A., Zecca, P.: Complex behavior of polynomial fitzhugh-nagumo cellular neural network model. Nonlinear Anal. Real World Appl. **8**(4), 1331–1340 (2007)
200. Slavova, A., Zecca, P.: Cellular neural networks modeling of tsunami waves. In: International Workshop on Cellular Nanoscale Networks and their Applications (2012)
201. Sokolov, S., Boumbarov, O., Gluhchev, G.: Face recognition using combination of wavelet packets, pca and lda. In: ISSPIT 2007: 2007 IEEE International Symposium on Signal Processing and Information Technology, pp. 257–262 (2007)
202. Sotirov, S., Atanassov, K.: Generalized Nets and Neural Networks, Generalized Nets in Artificial Intelligence, vol. 6. Prof. M. Drinov Academic Publishing House, Sofia

203. Sotirov, S., Atanassov, K.: Intuitionistic fuzzy feed forward neural network. Cybern. Inform. Technol. **9**(2), 62–68 (2009)

204. Sotirov, S., Atanassov, K., Krawczak, M.: Generalized net model for parallel optimization of feed-forward neural network with variable learning rate backpropagation algorithm with time limit. Stud. Comput. Intell. **299**, (2010)

205. Sotirov, S., Atanassov, K., Krawczak, M.: Generalized net model for parallel optimization of multilayer perceptron with momentum backpropagation algorithm. In: 2010 IEEE International Conference on Intelligent Systems, IS 2010: Proceedings, pp. 281–285 (2010)

206. Sotirov, S., Atanassova, V., Sotirova, E., Bureva, V., Mavrov, D.: Application of the intuitionistic fuzzy intercriteria analysis method to a neural network preprocessing procedure. In: 16th World Congress of the IFSA, 9th Conference of the EUSFLAT, pp. 1559–1564. Atlantis Press (2015)

207. Sotirov, S., Atanassova, V., Sotirova, E., Doukovska, L., Bureva, V., Mavrov, D., Tomov, J.: Application of the intuitionistic fuzzy intercriteria analysis method with triples to a neural network preprocessing procedure. Comput. Intell. Neurosci. **2017**, (2017). https://doi.org/10.1155/2017/2157852

208. Sotirov, S., Krawczak, M., Atanassov, K.: Modelling the brain-state-in-a-box neural network with a generalized net. In: New trend in Fuzzy Sets, Intuitionistic Fuzzy Sets, Generalized Nets and related topics. Applications, vol. II. System Research Institute, Polish Academy of Science, Warsaw (2013)

209. Sotirov, S., Orozova, D., Sotirova, E.: Generalized net model of the process of the prognosis with feedforward neural network. In: XVIth International Symposium on Electrical Apparatus and Technologies, SIELA 2009, Proceedings, vol. 1, pp. 272–278 (2009)

210. Sotirov, S., Sotirova, E., Atanassova, V., Atanassov, K., Castillo, O., Melin, P., Petkov, T., Surchev, S.: A hybrid approach for modular neural network design using intercriteria analysis and intuitionistic fuzzy logic. Complexity **2018**, (2018)

211. Sotirov, S., Sotirova, E., Melin, P., Castilo, O., Atanassov, K.: Modular neural network preprocessing procedure with intuitionistic fuzzy InterCriteria analysis method. Adv. Intell. Syst. Comput. **400**, (2016)

212. Spasova, M., Stoilova, O., Manolova, N., Rashkov, I.: Electrospun plla/peg scaffolds: materials resemble neural network. Mater. Today **28**, 114–115 (2019)

213. Stoynov, P.: Cellular neural networks and their applications. In: AIP Conference Proceedings, vol. 2159 (2019)

214. Terziyska, M., Doukovska, L., Petrov, M.: Implicit GPC based on semi fuzzy neural network model. Adv. Intell. Syst. Comput. **322**, (2015)

215. Terziyska, M., Todorov, Y.: Fuzzy-neural predictive control using fast optimisation polices. Int. J. Reason.-based Intell. Syst. **6**(3–4), 136–144 (2014)

216. Terziyska, M., Todorov, Y.: Intuitionistic neo-fuzzy network for modeling of nonlinear systems dynamics. In: 2016 IEEE 8th International Conference on Intelligent Systems, IS 2016: Proceedings, pp. 616–621 (2016)

217. Terziyska, M., Todorov, Y.: Intuitionistic neo-fuzzy predictive control. In: 2016 IEEE 8th International Conference on Intelligent Systems, IS 2016: Proceedings, pp. 635–640 (2016)

218. Terziyska, M., Todorov, Y.: Reduced rule-base fuzzy-neural networks. Stud. Comput. Intell. **681**, (2017)

219. Terziyska, M., Todorov, Y., Doiieva, M., Metodieva, P.: Distributed adaptive neuro intuitionistic fuzzy architecture for prediction of the dose in gamma irradiated milk products. IFAC-PapersOnLine. **52**, 75–80 (2019)

220. Terziyska, M., Todorov, Y., Olteanu, M.: Input space selective fuzzification in intuitionistic semi fuzzy-neural network. In: Proceedings of the 8th International Conference on Electronics, Computers and Artificial Intelligence, ECAI 2016 (2017)

221. Todorov, Y., Koprinkova-Hristova, P., Terziyska, M.: Intuitionistic fuzzy radial basis functions network for modeling of nonlinear dynamics. In: Proceedings of the 2017 21st International Conference on Process Control, PC 2017, pp. 410–415 (2017)

222. Todorov, Y., Terziyska, M.: Modeling of chaotic time series by interval type-2 NEO-fuzzy neural network. In: Lecture Notes in Computer Science (including subseries Lecture Notes in Artificial Intelligence and Lecture Notes in Bioinformatics), vol. 8681 LNCS (2014)

223. Todorov, Y., Terziyska, M.: State-space fuzzy-neural network for modeling of nonlinear dynamics. In: INISTA 2014: IEEE International Symposium on Innovations in Intelligent Systems and Applications, Proceedings, pp. 212–217 (2014)

224. Todorov, Y., Terziyska, M.: Simple heuristic approach for training of type-2 neo-fuzzy neural network. In: Proceedings of the 2015 20th International Conference on Process Control, PC 2015, vol. 2015-July, pp. 278–283 (2015)

225. Todorov, Y., Terziyska, M., Doukovska, L.: Distributed fuzzy-neural state-space predictive control. In: Proceedings of the 2015 20th International Conference on Process Control, PC 2015, vol. 2015-July, pp. 31–36 (2015)

226. Todorov, Y., Terzyiska, M., Petrov, M.: Recurrent fuzzy-neural network with fast learning algorithm for predictive control. In: Lecture Notes in Computer Science (including subseries Lecture Notes in Artificial Intelligence and Lecture Notes in Bioinformatics), vol. 8131 LNCS (2013)

227. Trifonov, R., Yoshinov, R., Pavlova, G., Tsochev, G.: Artificial neural network intelligent method for prediction. In: AIP Conference Proceedings, vol. 1872 (2017)

228. Tsaneva, M.G., Krezhova, D.D., Yanev, T.K.: Development and testing of a statistical texture model for land cover classification of the black sea region with modis imagery. Adv. Space Res. **46**(7), 872–878 (2010)

229. Tsonevska, T.S., Koleva, E.G., Koleva, L.S., Mladenov, G.M.: Modelling the shape of electron beam welding joints by neural networks. In: Journal of Physics: Conference Series, vol. 1089 (2018)

230. Turiel, A., Korutcheva, E., Parga, N.: The mutual information of a stochastic binary channel: Validity of the replica symmetry ansatz. J. Phys. A: Math. General **32**(10), 1875–1894 (1999)

231. Tzonkov, S., Koprinkova, P.: Neural network models of S. carlsbergensis batch cultivation. Biotechnol. Biotechnol. Equip. **8**(2), 64–67 (1994)

232. Vaklieva-Bancheva, N.G., Vladova, R.K., Kirilova, E.G.: Simulation of heat-integrated autothermal thermophilic aerobic digestion system operating under uncertainties through artificial neural network. Chem. Eng. Trans. **76**, 325–330 (2019)

233. Vankova, D., Sotirov, S., Doukovska, L.: An application of neural network to health-related quality of life process with intuitionistic fuzzy estimation. Adv. Intell. Syst. Comput. **559**, (2018)

234. Vassileva, S., Wang, X.Z.: Neural network systems and their applications in software sensor systems for chemical and biotechnological processes. Intelligent Systems: Technology and Applications, Six Volume Set, pp. I–291–I–335 (2002)

235. Videnova, I., Nedialkov, D., Dimitrova, M., Popova, S.: Neural networks for air pollution nowcasting. Appl. Artif. Intell. **20**(6), 493–506 (2006)

236. Yanakiev, V., Paunova, E.: Static memory optimization by clustering and neural networks in embedded devices. In: ACM International Conference Proceeding Series, vol. 578 (2011)

237. Zaluski, A., Ganzha, M., Paprzycki, M., Badica, C., Badica, A., Ivanovic, M., Fidanova, S., Lirkov, I.: Experimenting with facilitating collaborative travel recommendations. In: 2019 23rd International Conference on System Theory, Control and Computing, ICSTCC 2019: Proceedings, pp. 260–265 (2019)

238. Zankinski, I.: Effects of the neuron permutation problem on training artificial neural networks with genetic algorithms. In: Lecture Notes in Computer Science (including subseries Lecture Notes in Artificial Intelligence and Lecture Notes in Bioinformatics), vol. 10187 LNCS (2017)

239. Zankinski, I., Barova, M., Tomov, P.: Hybrid approach based on combination of backprop-agation and evolutionary algorithms for artificial neural networks training by using mobile devices in distributed computing environment. In: Lecture Notes in Computer Science (including subseries Lecture Notes in Artificial Intelligence and Lecture Notes in Bioinformatics), vol. 10665 LNCS (2018)

240. Zankinski, I., Kolev, K., Balabanov, T.: Alternatives for Neighborhood Function in Kohonen Maps, Lecture Notes in Computer Science (including subseries Lecture Notes in Artificial Intelligence and Lecture Notes in Bioinformatics), vol. 11958 LNCS (2020)

An Overview of Multi-criteria Decision Making Models and Software Systems

Daniela Borissova ⓘ

Abstract The article describes an overview of proposed multi-criteria decision making models and their applications. Several research directions are discussed including multi-criteria models with a priori and exact articulation of DM preferences; models with fuzzy preferences; models with interactive participation of DM; models based on bi-level optimization, group decision making models and software systems for decision making. This overview concerns the latest achievements in multi-criteria decision making of the scientists from Bulgarian Academy of Sciences.

Keywords Decision making · Multi-criteria methods · Bi-level optimization · Group decision making · Software systems

1 Introduction

Due the globalization and growing economy today is difficult to make successful business decisions without assistance of experts at different levels. Such decisions often involve business intelligence to gain the maximum information from available data in order to make effective business decisions [1]. Business intelligence relies on analytics experts with capabilities to provide believable information to help making effective and high quality business decisions [2]. The core of any decision support system is to help the higher management in different decision problems. Increasing of competition and developing information and communications technology make the problem of business decisions more significant and more complex. The complexity is related with the essence of the selection process that involves various quantitative and qualitative criteria. For such problems the multi-criteria decision making (MCDM) methods can be used to tackle with different and conflicting criteria.

The general formulation of the multi-criteria optimization problem is as follows:

D. Borissova (✉)
Institute of Information and Communication Technologies at the Bulgarian Academy of Sciences, 1113 Sofia, Bulgaria
e-mail: dborissova@iit.bas.bg

© The Author(s), under exclusive license to Springer Nature Switzerland AG 2021
K. T. Atanassov (ed.), *Research in Computer Science in the Bulgarian Academy of Sciences*, Studies in Computational Intelligence 934,
https://doi.org/10.1007/978-3-030-72284-5_15

$$\min \mathbf{F}(\mathbf{x}) = [F_1(x), F_2(x), \ldots, F_k(x)]^T \tag{1}$$

subject to:

$$g_j(\mathbf{x}) \le 0, j = 1, 2, \ldots, m \tag{2}$$

$$h_l(\mathbf{x}) = 0, l = 1, 2, \ldots, e \tag{3}$$

where $\mathbf{F}(x)$ is a vector of criteria, k is number of criteria, m is number of restrictions represented as inequalities, e is number of restrictions represented as equality, $\mathbf{x} = (x_1, x_2, \ldots, x_n)^T \in E^n$ is the vector of the variables and n is number of independent variables x_i.

In multi-criteria optimization tasks, several criteria need to be optimized simultaneously that is why there is no single alternative that optimizes all criteria in general. Any alternative that improves a particular criterion can be seen as a solution of the multicriteria task, but for the final task's solution one preferred alternative has to choose. In order to choose the most preferred alternative, additional information has to be provided by the decision maker (DM). This information from the DM will reflect his preferences regarding the quality of the searched alternative.

2   Multi-criteria Decision Making Models

There exists huge variety of different real life problems that can be formulated as multi-criteria optimization problems and they are constantly increased. This motivated researches to proposed different methods to cope with multidimensional nature of such problems. Depending on the way of DM preferences expression, the MCDM models can be divided on: (1) models with a priori and exact articulation of DM preferences, (2) models with fuzzy DM preferences and (3) models with interactive participation of DM.

2.1   Models with a Priori and Exact Articulation of DM Preferences

The well-known and widely used scalarization techniques are based on weighted sum method, and lexicographic. Both of these methods use can be used both for a priori and a posteriori articulation of DM preferences. The weighted sum method transforms the original multi-criteria problem into weighted linear sum of their normalization:

$$\max \sum_{i=1}^{k} w_i F_i(\mathbf{x}) \tag{4}$$

where w_i are weighted coefficients for criteria importance from DM point of view and $\sum_i w_i = 1$; $F_i(\mathbf{x})$ represent non-dimensional objective functions.

Four objective functions about working range, weight, price and operational time duration are formulated for design of night vision devices [3]. Another MCDM model is formulated for selection of k-best devices taking into account the DM preferences toward devices parameters [4, 5]. These models allow formulation of mixed-integer combinatorial optimization task that is solved by single run and the solution simultaneously determines of k-best devices considering the DM preferences toward given evaluation criteria.

For the goal of design the systems as whole based on standardized modules MCDM models are proposed. The one of them deals with personal computers configuration design [6] while the other concerns the night vision devices design [7]. The essence of these models is in the formulation of restrictions about modules compatibility and taking into account the external surveillance conditions that influence to the parameters of designed device [8]. Multicriteria optimization is used in formulation of problem which solutions determine possible combinations of surveillance conditions that conform to the given value for detecting range of night vision goggles by a standing man [9]. The weighted sum method is used also to determine the proper power supply for night vision device by choice of battery type and capacity [10].

It is shown that MCDM could be helpful in ranking of given set of alternatives taking into account the DM preferences [11]. The proposed ranking algorithm is based on repetitive multi-criteria optimization problem solving using a priori aggregation of preference information and the final result is a ranking list where the devices are ordered accordingly to DM preferences.

The lexicographic method is another way to handle multi-objective optimization problems by a priori articulation of DM preferences. This method requires ranking of objectives by the DM and sequence solving the single-objective optimization problem. The solution of each single objective problem could serve as a limiting measure for that objective which is used to define a restriction on the next step when the next objective is optimized and so on. The mathematical formulation of this method is as follows [12]:

$$\min F_j(x), x \in X, j = 1, 2, \ldots, k \tag{5}$$

subject to

$$F_j(x_j) \le \varepsilon_j F_j(x_j^*), j = 1, 2, \ldots, i - 1, i > 1 \tag{6}$$

where the ranking of objective function position is denoted by index j, $F_j\left(x_j^*\right)$ expresses the optimum of j-th objective function on j-th iteration, and ε_j represents the acceptable tolerance.

Other essential real-life problem that can benefit of multi-criteria optimization is the predictive maintenance based on dynamic response information obtained by sensors in structural health monitoring. The formulated multi-criteria model aims to minimize the number of sensors while keeping the structure health monitoring information as close to the maximal like in the situation when all sensors are present [12]. Using the lexicographic ordering, the first task determines the maximum of the information from structure health monitoring provided by several vibration mode shapes. On the second step another single-criterion problem for sensors number is solved using a restriction about the acceptable deviation from the optimum value of the previous task. Thus the described approach defines sensors on particular locations that can be dropped out in order to provide best fit to the given acceptable accuracy.

A two-stage algorithm combining multi-objective optimization and group decision making is proposed [13]. The formulated bi-objective optimization model aims to maximize annual energy production while minimizing the costs. The weighted sum and lexicographic method are used to generate different alternatives representing different turbines placement and respectively different design of wind farm. Numerical testing shows adequate implementation of both methods for generating of wind farm layout design alternatives. Once the farm layout design alternatives are known, the selection of the most suitable alternative can be done by group decision making.

It should be noted, that the limitations weighted sum approach is related with representation of compromises that DM is willing to make among objectives and requirement for normalization of the objectives. The lexicographic approach is a lite more flexible in tuning of these trade-offs by considering the desirable degree of proximity to the "best" values of objectives. An algorithm that implements a lexicographic approach is suggested for multi-criteria flow network problem when the criteria are ranked by priority [14].

In both of approaches (weighted sum and lexicographic) the DM preferences are expressed by priori articulation of information, i.e. before the actual optimization is conducted by using aggregation single figure of merit.

2.2 Models with Fuzzy Preferences

There exist different ways to express the DM preferences—a priory, posteriorly, interactive or preferences with some probabilistic feature. In the last case, the expressions of preferences can be done by fuzzy relations. Fuzzy sets and fuzzy logic are powerful mathematical tools for modeling and controlling uncertain systems that facilitate approximation reasoning in decision making in the absence of complete and precise information [15]. The overall fuzzy MCDM problem can be formulated as follow:

$$\max F_{fuzzy}(\mathbf{x}) = \min\{\mu_1 f_1(x), \mu_2 f_{12}(x), \ldots, \mu_k f_k(x)\} \tag{7}$$

The fuzziness of $f_i(x)$ is given by μ_i which quantifies how well a solution satisfies the requirements. The fuzzy MCDM models contribute for approximately simulation the human decision making by means of applying of the fuzzy sets theory [16, 17]. The aggregation of fuzzy relations on alternatives with the help of aggregation operators can be considered in two cases: (1) when the weighted coefficients of the criteria are not present in the mathematical formula of the aggregation operators and (2) when fuzzy preference relation between the criteria importance is given. The obtained results show that the properties of the aggregated relation depend with the individual relations' properties [18]. These properties give a possibility to decide the ranking, choice or clustering problems by fuzzy multicriteria decision making [19]. The multicriteria decision making models with fuzzy criteria given by fuzzy relations by each criterion are proposed. In aggregation process different weights for criteria could be used: weighting coefficients, weighting functions, fuzzy numbers and fuzzy preference relation between the criteria importance [20]. It is shown that aggregation procedures depend on the kind of the weights for criteria [21, 22]. In addition, to solve different fuzzy MCDM problems several algorithms are proposed like: algorithms by crisp criteria with real numbers as weighted coefficients; algorithm by crisp criteria with weighted coefficients—weighting functions; algorithm for aggregation of fuzzy relations between alternatives and a fuzzy relation between the weights (importance) of the criteria; algorithms for fuzzy numbers as alternatives' evaluations; and algorithms for testing and obtaining of the fuzzy relations with defined properties [23]. The alternatives' evaluations toward criteria can be real numbers by different scales, fuzzy preference relations by each criterion or trapezoidal fuzzy numbers by the criteria. The weights expressing the criteria importance can be real numbers, weighting functions or fuzzy relation between the couples of criteria according to their importance. Different combinations between the initial information may be used to solve the problems of choice or ordering of the alternatives.

It is shown that some problem for selection of enterprises—potential participants in economic cluster could be done by using of fuzzy sets theory methods [24]. This problem is approached by fuzzy algorithms and results are compared to PROMETHEE II multi-criteria outranking method. The obtained results demonstrate that the usage of fuzzy algorithms gives good quality solution. A decision support method for investment preference evaluation that enables classification of companies under conditions of incomplete data is presented [25]. The proposed investment preference concept used as qualitative criterion aims to estimate the minimal probability of bankruptcy. In some investments problems, the provided fuzzy logic tool is quite adequate to solving a multicriteria investment problem. A fuzzy approach for assessing the quality of an asset and making an investment decision based on this assessment is proposed [26]. The proposed model considers three criteria (return, risk and return-to-risk ratio) to support the investment decision-making process with measure of the quality of financial asset. The results of this model can be used for asset allocation in investment portfolio management.

2.3 Models with Interactive Participation of DM

The main idea of methods with interactive DM participation is the fact that DM takes part at each iteration by specifying reference points or direction for generation of new solutions. A typical way in interactive expressing of DM preferences is the reference direction approach where DM can manage the direction for finding Pareto-optimal solution. There exists variety of such methods and algorithms involving exact or approximation approaches to solve MCDM problems. A review of known reference point scalarizing problems and of the classification based scalarizing problems is given in [27]. The scalarizing problems are an important part of the interactive methods solving multicriteria optimization problems. The reference point method allows DM to express his/her preferences in regard to the given reference levels. It is shown that modification of reference point method is useful for analysis of multiple objective linear programming problems and can be realized by using modified Tchebycheff distance. The obtained results give Pareto solutions only. The formulated theorem gives a rule for improving the obtained solution with respect to one criterion [28]. The multiple objective-optimization problems represented by concave functions that have to be maximized over a feasible set can be solved by two auxiliary scalar optimization problems based on reference points [29]. The first one contains only continuous variables, while the second problem is formulated as mixed integer programming problem. The solutions of these two scalar problems determine nondominated points.

In case of multiple objective linear programming problems [30] and nonlinear programming under multiple objectives optimization problems a scalarization technique using a reference point can be useful also to solve such kind of problems [31]. It is shown that improving the achieved value of one criterion, only the corresponding component of the reference point has to be changed. The property of strict quasiconvexity allows successfully using the reference point method for the analysis of MOLFP problems to get weakly efficient points and then to improve the obtained value of chosen criterion [32].

Similar to reference point method is the reference direction approach, which takes into account determined some aspiration levels for the criteria in accordance DM point of view [33]. A reference direction approach can be used for solving multiple objective integer linear programming problems [34]. At each iteration of the proposed algorithm only one mixed integer linear programming problem is solved to find an efficient solution. The DM has to provide only the reference point at each iteration and obtained intermediate solution is integer. To provide more flexible setting of DM preferences toward objective weights, aspiration levels, aspiration directions, aspiration intervals, an original generalized interactive scalarizing method that incorporates thirteen interactive scalarizing approaches is presented [35]. Based on the reference direction a mixed-integer scalarizing problem can be formulated. By solving such problem approximately it is possible to find one or more integer solutions located close to the efficient surface and at some iteration DM could decide to solve the scalarizing problem to obtain an exact (weak) efficient solution [36]. A variant of reference direction approach with lexicographic formulation is presented for nonlinear multi-objective problems [37].

The scalarizing problems of the reference-neighborhood are presented for solving multi-objective linear integer programming problems [38]. These problems are formulated on the basis of inexplicit classification of the criteria in accordance with DM's preferences. DM provides information about his/her preferences for choice of new Pareto optimal solution with respect to the criteria values at the current solution.

For multicriteria network flow problems, a classification approach for searching of Pareto optimal solutions is proposed [39]. It is shown, that the network could be modified by using the current solution and DM's preferences about improvement of the value only for one criterion while preserving the values for others.

The approximation approaches for solving MCDM problems are based on evolutionary algorithms. An interactive evolutionary method FIEM for solving integer multiple objective problems is proposed [40]. At each iteration DM gives the preference information in terms of reference point and this information is used in an evolutionary algorithm to generate a new population. This algorithm does not depend on the number of objectives, which make possible to use it for convex integer optimization problems solving with great number of objectives. Another interactive evolutionary population-based algorithm is developed to solve multi-objective convex integer optimization problems [41, 42]. The used heuristic procedure contributes to acceleration of the search process and the algorithm performs considerably faster than the usual population-based algorithms [43]. A method based on the strength Pareto evolutionary algorithm designed to solve multi-objective convex integer optimization problems is proposed. It aims to improving the interaction phase with DM where the search process can be quickly directed to the part of the search space, where the location of a desired non-dominated solution is expected [44]. A detailed survey of evolutionary algorithms used in multi-objective optimization is presented in [45]. It is shown that evolutionary multi-objective optimization procedures are designed to achieve the convergence to as close to the Pareto-optimal front as possible and to maintenance of a well distributed set of trade-off solutions.

It should be noted that all interactive algorithms emphasis on the participation of DM during the decision-making process. Their basic advantage is the generation of restricted subset of Pareto optimal solutions that DM have to evaluate and thus the final solution highly depend on the evaluation of current solution and given preferences.

3   Models Based on Bi-level Optimization

Bi-level optimization problems are common for different real-world problems such as domain of economics, decision science, transportation, engineering, etc. The bi-level optimization problem can be seen as multi-objective bi-level optimization problem with multiple objectives at one or both levels. A general multi-objective bi-level optimization problem formulation is as follows [46, 47]:

$$\min_{x \in X, y \in Y} \{F_1(x, y), F_2(x, y), , \ldots, F_p(x, y)\} \tag{8}$$

subject to:

$$G_i(x, y) \le 0, i = \{1, 2, \ldots, I\} \tag{9}$$

$$y \in arg \min_{z \in Y} \{f(x, z) : g_j(x, z) \le 0, j = \{1, 2, \ldots, J\}\} \tag{10}$$

where the upper-level objective function is expressed by F while f represents the lower level objective function; the upper-level decision vector is denoted by x and y represents the lower level decision vector; G_i and g_j are the inequality constraint functions at the upper and lower levels respectively.

In bi-level optimization, the problem of low-level is modified according to the solutions of the upper level optimization one. According to the manner of modifications, it has been derived two general coordination procedures: goal coordination and predictive coordination [47]. The *goal coordination* formalizes the process of modification of the goal function of the low level optimization problem by the solutions of the upper level problem. For this case the interpretation of this coordination is the establishment of costs parameters by the upper problem to the goal functions of the lower problem/s. The *predictive coordination* defines that the solutions of the upper level problem modifies parameters in the constraints of the low level problem. For this case an interpretation of such coordination strategy is given as allocation of resources from the upper problem to the lower one. Nevertheless of the type of the coordination strategies, the solution of such bi-level optimization problem is performed with multiple data transfer of intermediate solutions between the two hierarchically ordered optimization problems. Thus, the solution of the bi-level problem is found with a delay.

The noniterative coordination is formally based on explicit analytical approximations of inexplicit relations of the goal function of the upper level optimization problem and the solutions of the low level sub-problems towards the coordination parameters [48–51]. Having explicit analytical relations, the bi-level optimization problem is solved in a faster way without iterative calculations. The noniterative coordination strategy is very useful when the low level problem can be decomposed to a set of low level sub-problems which share common resources or they have common interconnections among their parameters [49, 52–54].

The bi-level optimization formalism has been applied for solving practical problems in portfolio optimization of financial investments [55]. The used functions $Risk(x)$ and $Return(x)$ have argument $0 \le x_i \le 1$ that express the relative part of the investment allocated to asset i and $\sum_i x_i = 1$. The portfolio lower level problem supposes that investor preference for undertaking risk is given, while the upper level problem targets the evaluation of λ by maximizing the Sharpe ratio:

$$\max_{0 \leq x \leq \infty} \frac{Return(x(\lambda))}{Risk(x(\lambda))} \tag{11}$$

The coefficient λ describes the investor preference for undertaking risk. When $\lambda = 0$ the investor tries to minimize the risk without concerning the portfolio return, while $\lambda = \infty$ corresponds to maximizing the portfolio return without taking into consideration the risk.

The bi-level formalism can be used for solving different problems including control of traffic in urban area [56–59]. The low level problem in a discrete form describes the dynamics of the queue of vehicles in front of the traffic lights at the cross-road sections. These optimization problems for a traffic network give the duration of the green light, assuming fixed duration of the traffic light cycle. The upper level optimization targets the maximization of the traffic flows, which pass through the network. The bi-level optimization formalism has been also applied for the design of state transportation scheme in Bulgaria aiming to intensify the public railway transportation and minimizing the bus transportation [60–62]. The low level problem estimates the maximal passenger flows which can be transported between two main points, taking into all transportation links supported by trains and busses. The transport network capacity is determine by travelling time of each network link. The upper level problem performs optimization of the transport flows, considering low costs on rail transportation and high costs on bus transportation.

The predictive noniterative coordination can be used for some problems of resource allocation. If the hierarchical system operates on steady state with available resources and a request arises to allocate additional resources, the problem how to distribute this additional resource could be done by using of predictive noniterative coordination formulation [49, 51, 54].

Some models of optimal maintenance strategy defining can be formulated as bi-level optimization too. For example, optimal maintenance strategy for replace or repair is expressed as optimization problem based on maximization of the cost–benefit estimations. The first stage answers the question to repair or replace the machine as a whole by optimization task solving. If the solution is to repair the machine, the second stage use other optimization task which solution answer the question to repair or replace of each particular component [63].

4 Group Decision-Making Models

A special case of MCDM is the multi-criteria group decision making (MCGDM) where group of authorized experts have to determine the final decision (alternative). To cope with different competency of the group members' modifications of simple additive weighting model and weighted product model are proposed [64]. These modifications concern the usage of weighted coefficients to distinguish the particular expertise of each expert. Beside this the classical formulation of these models are changed by new formulation as single combinatorial optimization problem with binary integer variables for selection of the preferable alternative. The modified SAW is written as follows:

$$\max\left(\sum_{i=1}^{M} x_i \sum_{k=1}^{K} \lambda^k A_i^k\right) \tag{12}$$

subject to:

$$\forall i = 1, 2, .., M : (\forall k = 1, 2, \ldots, K : A_i^k = \sum_{j=1}^{N} w_j^k e_{i,j}^k) \tag{13}$$

$$\sum_{j=1}^{N} w_j^k = 1, \forall k = 1, 2, \ldots, K \tag{14}$$

$$\sum_{i=1}^{M} x_i = 1, x_i \in \{0, 1\} \tag{15}$$

$$\sum_{k=1}^{K} \lambda^k = 1, \lambda^k \in [0, 1] \tag{16}$$

where the aggregate assessment of the i-th alternative against all criteria is express by A_i^k according to the k-th expert point of view. The relative importance between criteria is expressed by corresponding weighted coefficients w_j^k for each DM. The evaluation of the i-th alternative versus the j-th criterion by k-th DM is denoted by $e_{i,j}^k$. The binary integer variables x_i are used as decision variables to select a single alternative. The introduced coefficients λ^k express the expertise of the members in group.

The modification of WPM takes the following formulation [65]:

$$\max\left(\sum_{i=1}^{M} x_i \sum_{k=1}^{K} \lambda^k R(A_i)^k\right) \tag{17}$$

subject to the same restrictions about w_j^k, x_i, λ^k and additional restriction:

$$\forall i = 1, 2, .., M : (\forall k = 1, 2, \ldots, K : R(A_i)^k = \prod_{j=1}^{N} (e_{ij}^k)^{w_j^k} e_{i,j}^k) \tag{18}$$

where the aggregate assessment of the i-th alternative against all criteria, according to the point of view of the k-th DM in the group is denoted by $R(A_i)^k$.

These modifications are used for real problems concerning the selection of supplier under public procurement [58] and software engineering problem [66].

Other modification of well-known SMART model (simple multi-attribute rating techniques) is proposed to evaluation and ranking the students for different purposes [67]. By using of additional weights for theoretical and practical outcomes it is possible to obtain different ranking of the students. Such a ranking could by suitable to dividing the total number of students into small groups in the relevant specialty.

In case of when more than one alternative are to be determined the described model for k-best alternatives selection via group decision making could be used [68]. The aim is to reduce a predefined set of alternatives to k-best from which the executive managers could make the final choice. This combinatorial optimization model for multi-attribute group decision making for selection of k-best alternatives allow formulation of single criterion task that can be solved by a single run.

Different human aspects can be involved when configure products of modular type taking into account the importance of supply and demand. To tackle with such problem a new utility function is proposed to aggregate the estimations of individual DM, weights for relative importance between evaluation criteria and weights for DM evaluations in forming the final group decision [69]. The proposed modelling approach for product configuration design via group decision making and combinatorial optimization relies on DMs estimations toward the given evaluation criteria concerning the different aspects of product design. The estimations about modules are used in an optimization 0–1 programming model to determine the combination of compatible modules. The compatibilities between particular modules are expressed by restrictions to compose basic configuration for product platform or product family design. The proposed group decision making model in design of product configuration is applied for PC configuration design in a medium-sized enterprise.

The comparison between individual and group decision making approach in MCDM problems show that individual point of view of each expert significantly influence to the final decision [1]. In this respect, to express more transparently the expertise of each group member a two component function of objective and subjective part seems to be useful [70]. The objective part could be expressed by using of component for the experience of each expert in the particular field. For example, previous experience involving acquired knowledge, availability of publications, responsibilities related with different projects, etc. The subjective part could be expressed by component that takes into account the advisability of relevance area of competence to the particular decision making problem. These two parts are represented by aggregation function as follows [70]:

$$\lambda^k = \beta^k + \gamma^k \tag{19}$$

where the coefficient expressing the experience of k-th DM in years is denoted by β^k and γ^k is the coefficient for advisability of expertise field competency to the particular problem for the k-th DM.

In more complicated situations, it should consider that the knowledge reliability of experts can vary over different criteria. In such cases the intuitionistic fuzzy approach of a multi person multi criteria decision making activity can be used. In a series of papers [71–79], new procedures for MCDM are described. They are based on the apparatus of Intuitionistic Fuzzy Sets (IFS) see [80, 81]. In [71] and [72], IF-interpretations of multi-measurement tool and multi-person MCDM procedures are given. In them, the evaluations of the multi-measurement tools and of multi-persons are in the form $\langle m, n \rangle$, where m and n are the positive and negative evaluation degrees, $m, n \in [0, 1]$ and $m, n \leq 1$. In [77, 78], these procedures are described by the tools of generalized nets (see [82, 83]) and they are extended in [74] for the case in which the experts can use not only the previous given criteria, but also own ones. Now, the experts can order the used criteria in their own way and can ignore some of them. The next extension of the MCDM procedure is described in [73]. Now, the experts have their own scores in the form $\langle p, q \rangle$, where p and q correspond to the successfully and non-successfully finished procedures form the respective expert, $p, q \in [0, 1]$ and $p, q \leq 1$. The number of the refused in the middle expert's procedures corresponds to the number $1 - p - q$. Procedures for self-evaluation of the experts with the aim to determine their scores on the base of subjective opinions are discussed in [76, 79]. In [75] is described a method for evaluation of the 'proximity' of the criteria used by the experts. It is based on the intercriteria analysis [84]. In the case that there are two "close" criteria, one of them—the more expensive/difficult for measurement can be omitted and in the next MCDM procedure it will not be given to the experts.

Regarding group decision making, two new algorithms of the EDAS method with fuzzy sets are proposed [85]. The first of them proposed EDAS extension for distance measure between two interval Type-2 fuzzy numbers is applied graded mean integration representation, while the second takes into account the proximity between the fuzzy alternatives and its similarity measure is map distance operator. To deal with inexact expert evaluation in both beneficial and non-beneficial criteria, a fuzzy DEMATEL-VIKOR combination is proposed [86].

5 Software Systems for Decision-Making

Due the specific nature of real-life MCDM problems, a variety of software systems are proposed to solve these problems. The proposed software package MULCRE can use quantitative or qualitative criteria for various methods [87]. The available methods in this software can cope with problems of ranking, obtaining of subset of elements or obtaining of subset and further ranking. A multi-criteria decision support system able to solve linear problems is proposed [88]. It is based on an interactive classification-based algorithm that allows DM to set the local preferences by some acceptable levels, directions and intervals of change in the values of a part or of all the criteria. Other software system called MultiDecision-1 is designed to support DM in solving different multicriteria analysis and multicriteria optimization problems [89]. The next developed software system called MultiDecision-2 consists of two independent parts—MKA-2 subsystem and MKO-2 subsystem [90]. This system helps DM in the solving of different problems of multicriteria analysis and linear (continues and integer) problems of multicriteria optimization.

The developed software system MultiOptima consists of three independent parts—MKA-2 system, MKO-2 system and LIOP-1 system and it is intended to support the DM in modelling and solving different problems of multicriteria analysis and linear and linear integer problems of single and multicriteria optimization [91]. In contrast to the previous systems, the software system Optima-Plus system is designed to support DM in modeling and solving of linear and linear integer single-criterion and multicriteria optimization problems [92]. This system relies on three methods for single-criterion optimization and an innovative generalized interactive method for multicriteria optimization with variable scalarization and parameterization. The interactive method can apply most of the well-known scalarizing problems and is applicable in different expressing the DM preferences. The developed multi-criteria decision support system MultiChoice is designed to support the DM in modeling and solving of multicriteria ranking and multicriteria choice problems [93]. Three different types of methods aiming to expands DM's possibilities to set the preferences about the quality of the most preferred solution are implemented, namely AHP method, PROMETHEE II method and CBIM method.

The next software systems to support decision-making process are also iterative, but involved evolutionary approaches [94, 95]. These web-based systems could be used for solving multiple-objective problems with continuous and/or integer decision variables. It involves fifteen interactive methods incorporated into original generalized scalarizing interactive method and an evolutionary method is also included. The choice of a method is organized in an implicit way on the base of DM preferences.

The core of any software system is the data structures and algorithms. When talking about decision support system, it should be able to react in real-time to changing problem situations, propose alternative actions, and evaluate the merits of such proposals. Such prototype of intelligent Web-based system for modular design that can assist and advise the DM during the decision-making process is described in [96]. For the goal of night vision goggles design, a web-based software system "NVGpro" is proposed [97]. DM can select different elements for designed device and based on this selection the device parameters are calculated. The software system allows DM to set up some limits about designed device parameters and using sorting algorithm the system make the most suitable modules selection conforming the given limits. An integrated framework of designing a decision support system for engineering predictive maintenance is proposed [98]. The advantage of such system is the possibly to integrate traditional decision support system with effective utilization of expert system capability for condition based maintenance policy [99]. There exist many real problems are combinatorial by nature and for this purpose a software system for visual processing of graph data is developed [100]. Such system can be used for solving of practical graph problems and as an e-learning tool for visualization of algorithms performance step by step [101].

To cope with different competency of the group members additional weighted coefficients can be involved. It is shown that group decision-making can overcome the subjective opinion of the individual DMs and contributes to the selection of the most appropriate supplier while satisfying point of view of all group members [102]. A web-based framework of group decision support application is proposed that integrates combinatorial optimization approach [103]. It relies on formulation of mixed integer optimization task whose solution defines an optimal choice of alternative by aggregation of group member opinions.

6 Conclusion

The presented overview of different multi-criteria models and applications emphasize on the latest achievements of the scientists from Bulgarian Academy of sciences. The content cover different models for MCDM concerning models based on scalarization techniques, models with interactive participation of DM, models with fuzzy preferences, models based on bi-level optimization, group decision-making models and some software system based on these models and algorithms.

Due the nature of the real-life problems, the researchers try to find the most suitable model to describe in details the specifics of the investigated multi-criteria decision making problem. This is why large variety of models and algorithms are proposed. Some of them are capable to find exactly compromise solutions while the rest try to find approximately solutions. For the convenience of users, some of these models are implemented in software systems to support decision-making. Some case-study with proposed software systems could be viewed as future investigations to prove the applicability of the implemented algorithms and models.

References

1. Borissova, D., Korsemov, D., Mustakerov, I.: Multi-criteria decision making problem for doing business: comparison between approaches of individual and group decision making. In: Saeed, Kh., Chaki, R., Janev, V. (eds.) Computer Information Systems and Industrial Management. LNCS, vol. 11703, pp. 385–396. Springer (2019)
2. Borissova, D., Mustakerov, I., Korsemov, D.: Business intelligence system via group decision making. Cybern. Inf. Technol. **16**(3), 219–229 (2016)
3. Mustakerov, I., Borissova, D.: Technical systems design by combinatorial optimization choice of elements on the example of night vision devices design. Comptes Rendus de l'Academie Bulgare des Sciences **60**(4), 373–380 (2007)
4. Borissova, D., Mustakerov, I.: K-best night vision devices by multi-criteria mixed-integer optimization modeling. Int. J. Inf. Sci. Eng. **7**(10), 205–210 (2013)
5. Mustakerov, I., Borissova, D., Bantutov, E.: Multiple-choice decision making by multicriteria combinatorial optimization. Int. J. Adv. Model. Optim. **14**(3), 729–737 (2012)
6. Mustakerov, I., Borissova, D.: Modular systems design via multi-objective optimization. Int. J. Adv. Model. Optim. **15**(2), 421–430 (2013)
7. Borissova, D.: Night Vision Devices—Modeling and Optimal Design. Prof. M. Drinov Publishing House of Bulgarian Academy of Sciences, Sofia (2015)

8. Borissova, D., Mustakerov, I.: A generalized optimization method for night vision devices design considering stochastic external surveillance conditions. Appl. Math. Model. **33**(11), 4078–4085 (2009)
9. Borissova, D., Mustakerov, I.: A multicriteria approach to exploring combinations of external surveillance conditions defining a given NVD working range value. Cybern. Inf. Technol. **9**(4), 102–109 (2009)
10. Borissova, D.: Using weighted sum method for the choice of the night vision goggles battery power supply. Eng. Sci. **1**, 16–26 (2007)
11. Mustakerov, I., Borissova, D.: A combinatorial optimization ranking algorithm for reasonable decision making. Comptes Rendus de l'Academie Bulgare des Sciences **66**(1), 101–110 (2013)
12. Mustakerov, I., Borissova, D.: Multi-criteria model for optimal number and placement of sensors for structural health monitoring: lexicographic method implementation. Int. J. Adv. Model. Optim. **16**(1), 103–112 (2014)
13. Borissova, D., Mustakerov, I.: A two-stage placement algorithm with multi-objective optimization and group decision making. Cybern. Inf. Technol. **17**(1), 87–103 (2017)
14. Djelatova, M.: A lexicographic algorithm solving a problem of a multiobjective flow in a network. Probl. Eng. Cybern. Robot. **51**, 85–89 (2001)
15. Peneva, V., Popchev, I.: Fuzzy multicriteria decision-making. Cybern. Inf. Technol. **2**(1), 16–26 (2002)
16. Peneva, V., Popchev, I.: Models of decision making support by multicriteria problems in fuzzy environment. Adv. Stud. Contemp. Math. **12**(2), 291–308 (2006)
17. Peneva, V., Popchev, I.: Models for decision making by fuzzy relations and fuzzy numbers for criteria evaluations. Comptes Rendus de l'Academie Bulgare des Sciences **62**(10), 1217–1222 (2009)
18. Peneva, V., Popchev, I.: Models for weighted aggregation of fuzzy relations to multicriteria decision making problems. Cybern. Inf. Technol. **6**(3), 3–18 (2006)
19. Peneva, V., Popchev, I.: Aggregation of fuzzy preference relations to multicriteria decision making. Fuzzy Optim. Decis. Making **6**(4), 351–365 (2007)
20. Peneva, V., Popchev, I.: Multicriteria decision making based on fuzzy relations. Cybern. Inf. Technol. **8**(4), 3–12 (2008)
21. Peneva, V., Popchev, I.: Models for fuzzy multicriteria decision making based on fuzzy relations. Comptes Rendus de l'Academie Bulgare des Sciences **62**(5), 551–558 (2009)
22. Peneva, V., Popchev, I.: Multicriteria decision making by fuzzy relations and weighting functions for the criteria. Cybern. Inf. Technol. **9**(4), 58–71 (2009)
23. Peneva, V., Popchev, I.: Fuzzy multi-criteria decision making algorithms. Comptes Rendus de l'Academie Bulgare des Sciences **63**(7), 979–992 (2010)
24. Radeva, I.: Multicriteria fuzzy sets application in economic clustering problems. Cybern. Inf. Technol. **17**(3), 29–46 (2017)
25. Popchev, I., Radeva, I.: A decision support method for investment preference evaluation. Cybern. Inf. Technol. **6**(1), 3–16 (2006)
26. Popchev, I., Georgieva, P.: A fuzzy approach to solving multicriteria investment problems. In: Iskander, M. (eds.) Innovative Techniques in Instruction Technology, E-learning, E-assessment, and Education, pp. 427–431. Springer, Dordrecht (2008)
27. Vassilev, V.: A generalized scalarizing problem of multicriteria optimization. Cybern. Inf. Technol. **2**(2), 88–99 (2002)
28. Popchev, I., Metev, B., Yordanova, I.: A Realization of reference point method using the Tchebycheff distance. In: Lewandowski, A., Volkovich, V. (eds.) Multiobjective Problems of Mathematical Programming, 1988. Lecture Notes in Economics and Mathematical Systems, vol. 351, pp. 76–82. Springer, Berlin (1991)
29. Metev, B., Yordanova-Markova, I.: Multi-objective optimization over convex disjunctive feasible sets using reference points. Eur. J. Oper. Res. **98**(1), 124–137 (1997)
30. Metev, B., Yordanova, I.: Use of reference points for MOLP problems analysis. Eur. J. Oper. Res. **68**(3), 374–378 (1993)

31. Metev, B.: Use of reference points for solving MONLP problems. Eur. J. Oper. Res. **80**(1), 193–203 (1995)
32. Metev, B., Gueorguieva, D.: A simple method for obtaining weakly efficient points in multiobjective linear fractional programming problems. Eur. J. Oper. Res. **126**(2), 386–390 (2000)
33. Narula, S.C., Kirilov, L., Vassilev, V.: Reference direction approach for solving multiple objective nonlinear programming problems. IEEE Trans. Syst. Man Cybern. **24**(5), 804–806 (1994)
34. Vassilev, V., Narula, S.: A reference direction algorithm for solving multiple objective integer linear programming problems. J. Oper. Res. Soc. **44**(12), 1201–1209 (1993)
35. Kirilov, L., Guliashki, V., Genova, K., Vassileva, M., Staykov, B.: Generalized scalarizing model GENS in DSS WebOptim. Int. J. Decis. Support. Syst. Technol. (IJDSST) **5**(3), 1–11 (2013)
36. Vassilev, V., Narula, S., Gouljashki, V.: An interactive reference direction algorithm for solving convex nonlinear integer multiobjective programming problems. Int. Trans. Oper. Res. **8**(4), 367–380 (2001)
37. Miettinen, K., Kirilov, L.: Interactive reference direction approach using implicit parametrization for nonlinear multiobjective optimization. J Multi-Criteria Decis Anal **13**, 115–123 (2005)
38. Genova, K., Kirilov, L., Guljashki, V.: New reference-neighbourhood scalarization problem for multiobjective integer programming. Cybern. Inf. Technol. **13**(1), 104–114 (2013)
39. Nikolova, M.: A classification based approach for finding Pareto optimal solutions of the multicriteria network flow. Cybern. Inf. Technol. **3**(1), 11–17 (2003)
40. Kirilov, L., Guliashki, V.: Interactive evolutionary method FIEM for solving integer multiple objective problems. Comptes Rendus de l'Academie Bulgare des Sciences **64**(2), 201–210 (2011)
41. Genova, K., Guliashki, V.: Evolutionary algorithm for multiple objective convex integer problems, In: Mitrovski, C. (ed.) XLV International Scientific Conference on Information, Communication and Energy Systems and Technologies ICEST2010, vol. I, pp. 285–289 (2010)
42. Guliashki, V., Kirilov, L., Genova, K.: An evolutionary algorithm for integer multicriteria optimization (EVALIMCO). In: Kahraman, C., Kerre, E., Bozbura, F. (eds.) 10-th Int. FLINS Conference, pp. 118–123. World Scientific (2012)
43. Guliashki, V., Kirilov, L.: An interactive evolutionary algorithm for multiple objective convex integer problems. In: 12th International Conference on Computer Systems and Technologies, pp. 82–87. Wien, Austria (2011)
44. Guliashki, V., Kirilov, L.: SPEA-based method for MCDM convex integer problems. Cybern. Inf. Technol. **9**(4), 93–101 (2009)
45. Guliashki, V., Toshev, H., Korsemov, C.: Survey of evolutionary algorithms used in multiobjective optimization. Probl. Eng. Cybern. Robot. **60**, 42–54 (2009)
46. Stoilov, T., Stoilova, K.: Noniterative Coordination in Multilevel Systems, 1st edn. Springer (1999)
47. Stoilova, K., Stoilov, T.: Goal and predictive coordination in two level hierarchical systems. Int. J. Gen. Syst. **37**(2), 181–213 (2008)
48. Stoilov, T., Stoilova, K.: Bilevel noniterative interconnected optimization. In: 4th International IEEE Conference Intelligent Systems. IEEE, Bulgaria (2008). https://doi.org/10.1109/IS.2008.4670532
49. Stoilova, K.: Predictive noniterative coordination implemented for fast resource allocation. Comptes Rendus de l'Academie Bulgare des Sciences **61**(8), 1055–1064 (2008)
50. Stoilov, T.: Noniterative Coordination in Hierarchical Systems. Prof. M. Drinov Publishing House of Bulgarian Academy of Sciences, Sofia (1998)
51. Stoilova, K.: Noniterative Predictive Coordination. Prof. M. Drinov Publishing House of Bulgarian Academy of Sciences, Sofia (2010)

52. Stoilov, T., Stoilova, K.: A self-optimization traffic model by multilevel formalism. Auton. Road Transp. Support. Syst. 87–111 (2016)
53. Stoilova, K., Stoilov, T., Nikolov, K.: Autonomic properties in traffic control. Cybern. Inf. Technol. **4**, 18–32 (2013)
54. Stoilova, K., Stoilov, T.: Hierarchical optimization for fast resource allocation. In: Stoilov T. (ed.) Time Management, pp. 31–46. INTECH Publisher (2012)
55. Stoilov, T., Stoilova, K.: Portfolio risk management modelling by bi-level optimization. Handbook on Decision Making. Intelligent Systems Reference Library, vol. 33, pp. 91–110. Springer, Heidelberg (2012)
56. Stoilova, K., Stoilov, T., Abouaissa, H.: Traffic lights optimization with measurements of noise levels. Control and Automation Theory for Transportation Applications, pp. 31–36 (2013)
57. Stoilov, T., Stoilova, K., Stoilova, V.: Bi-level formalization of urban area traffic lights control. Innovative Approaches and Solutions in Advanced Intelligent Systems, pp. 303–318 (2016)
58. Stoilov, T., Stoilova, K., Papageorgiou, M., Papamichail, I.: Bi-level optimization in a transport network. Cybern. Inf. Technol. **15**(5), 37–49 (2015)
59. Stoilova, K., Stoilov, T., Ivanov, V.: Bi-level optimization as a tool for implementation of intelligent transportation systems. Cybern. Inf. Technol. **17**(2), 97–105 (2017)
60. Pavlova, K., Stoilov, T., Stoilova, K.: Bi-level model for public rail transportation under incomplete data. Cybern. Inf. Technol. **17**(3), 75–91 (2017)
61. Pavlova, K., Stoilov, T.: Design of state rail and bus transportation scheme with bi-level optimization model. Inf. Technol. Control **15**(4), 2–9 (2017)
62. Pavlova, K., Stoilov, T.: Design of state rail and bus transportation scheme with bi-level optimization model. Inf. Technol. Control **4**, 2–9 (2017)
63. Mustakerov I., Borissova, D.: An intelligent approach for optimum maintenance strategy defining. In: Innovations in Intelligent Systems and Applications (INISTA). IEEE, Bulgaria (2013). https://doi.org/10.1109/INISTA.2013.6577666
64. Korsemov, D., Borissova, D.: Modifications of simple additive weighting and weighted product models for group decision making. Int. J. Adv. Model. Optim. **20**(1), 101–112 (2018)
65. Korsemov, D., Borissova, D., Mustakerov, I.: Group decision making for selection of supplier under public procurement. In: Kalajdziski S., Ackovska N. (eds.) ICT Innovations 2018. Engineering and Life Sciences. CCIS, vol. 940, pp. 51–58. Springer (2018)
66. Korsemov, D., Borissova, D., Mustakerov, I.: Combinatorial optimization model for group decision-making. Cybern. Inf. Technol. **18**(2), 65–73 (2018)
67. Borissova, D., Keremedchiev, D.: Group decision making in evaluation and ranking of students by extended simple multi-attribute rating technique. Cybern. Inf. Technol. **18**(3), 45–56 (2019)
68. Borissova, D.: Group decision making for selection of k-best alternatives. Comptes Rendus de l'Academie Bulgare des Sciences **69**(2), 183–190 (2016)
69. Borissova, D., Keremedchiev, D.: Product configuration design via group decision making and combinatorial optimization. Comptes Rendus de l'Academie Bulgare des Sciences **72**(9), 1251–1261 (2019)
70. Borissova, D.: A group decision making model considering experts competency: an application in personnel selections. Comptes Rendus de l'Academie Bulgare des Sciences **71**(11), 1520–1527 (2018)
71. Atanassov. K., Pasi, G., Yager, R.: Intuitionistic fuzzy interpretations of multi-measurement tool multi-criteria decision making. In: Kacprzyk, J., Atanassov, K. (eds.) International Conference on Intuitionistic Fuzzy Sets. Notes Intuit. Fuzzy Sets **8**(3), 66–74 (2002)
72. Atanassov, K., Pasi, G., Yager, R.: Intuitionistic fuzzy interpretations of multi-person multi-criteria decision making. In: First International IEEE Symposium Intelligent Systems, Varna, Bulgaria (2002). https://doi.org/10.1109/IS.2002.1044238
73. Atanassov, K., Pasi, G., Yager, R.: Intuitionistic fuzzy interpretations of multi-criteria multi-person and multi-measurement tool decision making. J. Int. J. Syst. Sci. **36**(14), 859–868 (2005)
74. Atanassov, K., Pasi, G., Yager, R., Atanassova, V.: Intuitionistic fuzzy graph interpretations of multi-person multi-criteria decision making. In: Third Conference of the European Society for Fuzzy Logic and Technology EUSFLAT'2003, pp. 177–182. Zittau (2003)

75. Atanassov, K., Szmidt, E., Kacprzyk, J., Atanassova, V.: An approach to a constructive simplification of multiagent multicriteria decision making problems via intercriteria analysis. Comptes Rendus de l'Academie Bulgare des Sciences **70**(8), 1147–1156 (2017)
76. Atanassov, K., Szmidt, E., Kacprzyk, J., Atanassova, V.: Intuitionistic fuzzy approach to the preference degree estimations. Comptes Rendus de l'Academie Bulgare des Sciences **68**(1), 25–32 (2015)
77. Pasi, G., Atanassov, K., Melo Pinto, P., Yager, R., Atanassova, V.: Multi-person multi-criteria decision making: intuitionistic fuzzy approach and generalized net model. In: 10th ISPE Int. Conference on Concurrent Engineering—Advanced Design, Production and Management Systems, pp. 26–30. Madeira (2003)
78. Pasi, G., Yager, R., Atanassov, K.: Intuitionistic fuzzy graph interpretations of multi-person multi-criteria decision making: generalized net approach. In: 2nd International IEEE Conference on Intelligent Systems, Varna, Bulgaria, pp. 434–439 (2004)
79. Atanassov, K., Gluhchev, G., Hadjitodorov, S., Kacprzyk, J., Shannon, A., Szmidt, E., Vassilev, V.: Generalized Nets Decision Making and Pattern Recognition. Warsaw School of Information Technology, Warszawa (2006)
80. Atanassov, K.: Intuitionistic Fuzzy Sets. Springer, Heidelberg (1999)
81. Atanassov, K.: On Intuitionistic Fuzzy Sets Theory. Springer, Berlin (2012)
82. Atanassov, K.: Generalized Nets. World Scientific, Singapore (1991)
83. Atanassov, K.: On Generalized Nets Theory. Prof. M. Drinov Academic Publ. House, Sofia (2007)
84. Atanassov, K., Mavrov, D., Atanassova, V.: Intercriteria decision making: A new approach for multicriteria decision making, based on index matrices and intuitionistic fuzzy sets. Issues Intuit. Fuzzy Sets Gen. Nets **11**, 1–8 (2014)
85. Ilieva, G.: Group decision analysis algorithms with EDAS for interval fuzzy sets. Cybern. Inf. Technol. **18**(2), 51–64 (2018)
86. Ilieva, G.: Group decision analysis with interval type-2 fuzzy numbers. Cybern. Inf. Technol. **17**(1), 31–44 (2017)
87. Danev, B., Slavov, G., Popchev, I., Metev, B.: Multicriteria evaluation—a new practice in BIA activity. In: Sendov, B., Popchev, I., Ivanov, Y. (eds.) Bilateral Bulgarian and American Seminar Between the Bulgarian Academy of Sciences and the National Academy of Sciences of the USA on the Introducing Research Results in to Practice, pp. 115–122. Bulgarian Academy of Sciences (1987)
88. Genova, K., Vassileva, M., Vassilev, V., Andonov, F.: Linear multicriteria decision support system. In: Rachev, B., Smirikarov, A. (eds.) Proceedings of CompSysTech'2003, pp. III.17-1–III.17-7 (2003)
89. Vassilev, V., Genova, K., Vassileva, M.: A multicriteria decision support system MultiDecision-1. Int. J. Inf. Theor. Appl. **13**, 103–111 (2006)
90. Vassilev, V., Vassileva, M., Staykov, B., Genova, K., Andonov, F., Chongova, P.: MultiDecision-2: a multicriteria decision support system. Inf. Technol. Knowl. **2**(1), 203–211 (2008)
91. Vassilev, V., Vassileva, M., Staykov, B., Genova, K., Andonov, F., Dochev, D.: An integrated software system for optimization and decision support MultiOptima. Cybern. Inf. Technol. **2**, 83–101 (2008)
92. Vassilev, V., Vassileva, M., Genova, K., Staykov, B.: A general-purpose software system for linear optimization and decision support "Optima-Plus". International Book Series "Information Science & Computing", vol. 10, no. 3, pp. 17–24 (2009)
93. Vassilev, V., Genova, K., Andonov, F., Staykov, B.: Multicriteria decision support system MultiChoice. Cybern. Inf. Technol. **4**(1), 65–75 (2004)
94. Kirilov, L., Guliashki, V., Genova, K., Zhivkov, P., Staykov, B., Vatov, D.: Interactive environment WebOptim for solving multiple-objective problems using scalarising and evolutionary approaches. Int. J. Reason. Based Intell. Syst. **7**(1/2), 4–15 (2015)
95. Kirilov, L., Guliashki, V., Staykov, B.: Chapter 7 Web-based decision support system for solving multiple-objective decision-making problems. Technological Innovations in Knowledge Management and Decision Support, pp. 150–175 (2019)

96. Mustakerov, I., Borissova, D.: Data structures and algorithms of intelligent web-based system for modular design. Int. J. Comput. Sci. Eng. **7**(7), 87–92 (2013)
97. Borissova, D., Mustakerov, I.: Method of rational choice by sorting in the software system "NVGpro." Cybern. Inf. Technol. **6**(1), 69–75 (2006)
98. Borissova, D., Mustakerov, I.: An integrated framework of designing a decision support system for engineering predictive maintenance. Int. J. Inf. Technol. Knowl. **6**(4), 366–376 (2012)
99. Borissova, D., Mustakerov, I.: A concept of intelligent e-maintenance decision making system. In: Innovations in Intelligent Systems and Applications (INISTA'2013). IEEE, Bulgaria (2013). https://doi.org/10.1109/INISTA.2013.6577668
100. Moustakerov, I.: A software system for visual processing of graph data. Cybern. Inf. Technol. **5**(2), 156–163 (2005)
101. Borissova, D., Mustakerov, I.: E-learning tool for visualization of shortest paths algorithms. Trends J. Sci. Res. **2**(3), 84–89 (2015)
102. Borissova, D., Atanassova, Z.: Multi-criteria decision methodology for supplier selection in building industry. Int. J. 3-D Inf. Model. **7**(4), 49–58 (2018)
103. Mustakerov, I., Borissova, D.: A web application for group decision-making based on combinatorial optimization. In: 4th International Conference on Information Systems and Technologies (ICIST 2014), Valencia, Spain, pp. 46–56 (2014)

Methods for Modelling of Overall Telecommunication Systems

Stoyan Poryazov, Velin Andonov, and Emiliya Saranova

Abstract The aim of the paper is to summerize some of the methods for modeling of overall telecommunication systems developed by the authors at the Institute of Mathematics and Informatics of the Bulgarian Academy of Sciences and to propose new methods using the apparatus of the Generalized Nets (GNs) theory. On the basis of the discussed methods, two basic tasks about overall telecommunication systems are formulated. Analytical models for solving the Quality of Service (QoS) prediction task and the Network Dimensioning/Redimensioning Task (NDT/NRDT) are proposed. A classical model of overall telecommunication system is considered. General teletraffic tasks are formulated on the basis of a proposed conceptual model. Some assumptions for the system are stated which allow for a relatively simple analytical model to be obtained. Analytical expressions for basic teletraffic characteristics of the main tasks about overall telecommunication systems are derived. Graphical representation of the results is included. A comparison with other approaches for network dimensioning is made and is represented graphically

Keywords Overall telecommunication system · Conceptual modeling · Quality of service · Overall network dimensioning/redimensioning

1 Introduction

Despite the long-standing use of the telecommunication systems in practice, their modelling is still an object of current development. The aim of the paper is to summarize some of the methods for conceptual and analytical modeling of overall

S. Poryazov · V. Andonov (✉) · E. Saranova
Institute of Mathematics and Informatics, Bulgarian Academy of Sciences,
Acad. G. Bonchev Str., Block 8, 1113 Sofia, Bulgaria
e-mail: velin_andonov@math.bas.bg

S. Poryazov
e-mail: stoyan@math.bas.bg

E. Saranova
e-mail: emiliya@math.bas.bg

K. T. Atanassov (ed.), *Research in Computer Science in the Bulgarian Academy of Sciences*, Studies in Computational Intelligence 934,
https://doi.org/10.1007/978-3-030-72284-5_16

325

teleocommunication systems (with and without queuing) developed at the Institute of Mathematics and Informatics (IMI) of the Bulgarian Academy of Sciences (BAS). These methods differ from the traditional widely accepted and often used methods in the literature. The motivation behind the development and the application of new methods is based on the fact that the results of the use of the traditional methods are not always presented with satisfying accuracy.

The proposed and used method for conceptual modeling in Sects. 2, 3 and 4 of the present paper is based on an original approach and graphical language, developed at IMI–BAS. The method of the Generalized Nets (GNs) is used in Sect. 5 in the conceptual modeling of overall telecommunication system with queuing.

Two main tasks about overall telecommunication system without queuing are formulated and solved analytically. They are:

1. task for prediction of the Quality of Service (QoS) (Sect. 3);
2. task for dimensioning of networks with QoS guarantees (Sect. 4).

The numerical results of the application of the proposed methods are presented graphically. A comparison is made between the results of the proposed method and other well known and often used in the practice methods.

The analytical model (Sect. 5) for determining the important teletraffic parameters of the overall telecommunication system with queuing in the Switching stage, from a practical point of view, is based on the GNs conceptual model and Queuing theory.

2   Classical Model of Overall Telecommunication System

The classical model of overall telecommunication system is proposed in [1] and developed in more details in [2]. It is a detailed conceptual traffic model of an overall (virtual) circuit switching telecommunication network, like PSTN and GSM, including users' behaviour, with: BPP (Bernoulli-Poisson-Pascal) input flow; repeated calls; limited number of homogeneous terminals; losses due to abandoned and interrupted dialing, blocked and interrupted switching, not available intent terminal, blocked and abandoned ringing and abandoned communication.

The described approach is applicable directly for every (virtual) circuit switching telecommunication system (like GSM and PSTN) and may help considerably for ISDN, BISDN and most of core and access networks traffic modelling. For packet switching systems, like Internet, proposed approach may be used as a comparison basis.

The traffic of the calling (denoted by A) and the called (denoted by B) terminals and user's traffic are considered separately but in their interrelation. Two types of virtual devices are included in the model: base and comprising base devices.

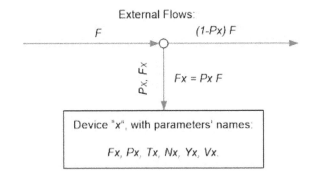

Fig. 1 A graphical representation of a basic virtual device x

2.1 Base Virtual Devices Representation and Their Parameters

At the bottom of the structural model presentation, we consider basic virtual devices that do not contain any other virtual devices. A basic virtual device has a general graphical representation as shown in Fig. 1.

The parameters of the basic virtual device x are the following (see [3] for terms definition):

– Fx—intensity or incoming rate (frequency) of the flow of requests (i.e. the number of requests per time unit) to device x;
– Px—probability of directing the requests towards device x;
– Tx—service time (duration of servicing of a request) in device x;
– Yx—traffic intensity [Erlang];
– Vx—traffic volume [Erlang—time unit];
– Nx—number of lines (service resources, positions, capacity) of device x.

For the better understanding of the model and for a more convenient description of the intensity of the flow, a special notation including qualifiers (see [3]) is used. For example $dem.F$ for demand flow; $inc.Y$ stands for incoming traffic; $ofr.Y$ for offered traffic; $rep.Y$ for repeated traffic.

2.2 Types and Names of the Base Virtual Devices

The graphic representations of the base virtual devices together with their names and types are shown in Fig. 2 (see [1]). The type of each of the basic virtual devices is also shown in Fig. 2. Each basic virtual device belongs to one of the following types: Generator, Terminator, Modifier, Server, Enter Switch, Switch and Graphic connector. With the exception of the Switch, which has one or two entrances and one or two exits, every other virtual device has one entrance and/or one exit.

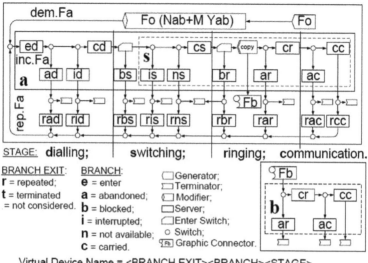

Fig. 2 Classical conceptual model of an overall telecommunication system (see [1])

In the conceptual model each virtual device has a unique name. The names of the devices are constructed according to their position in the model. The model is partitioned into service stages (dialing, switching, ringing and communication). Every service stage has branches (enter, abandoned, blocked, interrupted, not available, carried), corresponding to the modeled possible cases of ends of the calls' service in the branch considered. Every branch has two exits (repeated, terminated) which show what happens with the calls after they leave the telecommunication system. Users may make a new bid (repeated call), or to stop attempts (terminated call). In virtual device name construction, the corresponding bold first letters of the names of stages, branches end exits are used in the order shown below.

Virtual Device Name = <BRANCH EXIT> <BRANCH> <STAGE>

A parameter's name of one virtual device is a concatenation of parameters name letter and virtual device name. For example, "Yid" means "traffic intensity in interrupted dialing case"; "Fid" means "flow (calls) intensity in interrupted dialing case"; "Pid" means "probability for interrupted dialing"; Tid = "mean duration of the interrupted dialing"; "Frid" = "intensity of repeated flow calls, caused by (after) interrupted dialing".

2.3 Comprise Virtual Devices

The following comprising virtual devices denoted by **a, b, s** (see Fig. 2) and **ab** (not shown in Fig. 2) are considered in the model.

- **a** comprises all calling terminals (A-terminals) in the system. It is shown with continuous line box in Fig. 2;
- **b** comprises all called terminals (B-terminals) in the system. It is shown in box with dashed line in the down right corner in Fig. 2;
- **ab** comprises all the terminals (calling and called) in the system. It is not shown in Fig. 2;
- **s** virtual device corresponding to the switching system. It is shown with dashed line box into the a-device in Fig. 2.

The flow of calls (B-calls), with intensity Fb, occupying the B-terminals, is coming from the Copy device. This corresponds to the fact that at the beginning of the ringing a second (B) terminal in the system becomes busy. The second reason for this conceptual modelling trick is that the paths of the A and B-calls are different in the telecommunication system's environment, after releasing the terminals (compare environments of **a** and **b**—devices in Fig. 2). There are two virtual devices of type Enter Switch (see Fig. 2)—before Blocked Switching (**bs**) and Blocked Ringing (**br**) devices. These devices deflect calls if there is no free line in the switching system or the intent B-terminal is busy, respectively. The correspondent transition probabilities depend on the macrostate of the system (*Yab*). The macrostate of a (virtual) device (including the overall network, considered as a device) is defined as the mean number of simultaneously served calls in this device, in the observed time interval (similar to "mean traffic intensity"in (see [3]).

2.4 Paths of the Calls

The network under consideration corresponds to the reference configuration "terminal—subscriber switch—terminal" [4]. We ignore the signalling network. In the present paper, "call" means "call attempt" or "bid" according to [3]. The paths of the calls, generated from (and occupying) the A-terminals in the proposed network traffic model and its environment are shown in Fig. 2. Fo is the intent intensity of calls of one idle terminal; M is a constant, characterizing the BPP (Bernoulli-Pascal-Poisson) flow of demand calls (*dem.Fa*). If $M = -1$, the intensity of demand flow corresponds to Bernoulli (Engset) distribution. If $M = 0$—to the Poisson (Erlang) distributio. If $M = +1$—to the Pascal (Negative Binomial) distribution. In our analytical model every value of M in the interval $[-1, +1]$ is allowed. The BPP-traffic model is very applicable (see [4]) (the handbook is a clear comparison basis for the main ideas, discussed in this paper). In the numerical examples, presented in [1, 2], $M = 0$, because the conclusions made are independent of the input flow model.

2.5   Formulation of General Teletraffic Tasks

System tuples and base tuples. In the conceptual model presented we have at least 35 important virtual devices (31 base and 4 (**a, b, ab** and **s**) comprising). These devices are important because values of their parameters are specific for the characteristics and state of the modeled telecommunication system. Since every device has 5 parameters (P, F, T, Y, N), the total number of parameters is 175.

Definition 1 A system tuple is a finite set of distinguishable (by name and/or position) parameters' values, which fulfills simultaneously the three following requirements:

1. All parameters (parameters' set), evaluated by the system tuple, correspond to one considered (observed, modeled) system;
2. All the values of a system tuple correspondent to the one and the same time interval of measurements or considerations;
3. The instant of beginning and duration of this time interval are elements of the system tuple set.

Every subset of a system tuple is called *subtuple*.

There are many obvious dependencies in a system tuple, corresponding to the Full Parameters' Set of the Conceptual Model. For example, the sum of probabilities of outgoing transitions in every virtual switch devices has value one; in stationary state Little's formula ($Y = FT$) is in force for every virtual device; we assume most of devices with infinite capacity. As a result, there are sets of base parameters (subtuples), with the following property: If we know the values of the base parameters, we may calculate the values of all other parameters of the same system tuple. Several different base parameters' sets may exist. After careful analysis and some assumptions (see below) we have chosen a base parameters' set with 41 parameters. The values of these parameters we call base tuple. The base tuple is a sub-tuple of a system tuple.

Classification of the parameters. The parameters of the chosen base parameters' set may be classified, according origination of their values, in the following fice groups:

1. Human Behaviour Parameters are 21: $Fo, Nab, Prad, Tid, Prid, Pris, Tis,$ $Pns, Tns, Prns, Tbs, Prbs, Tbr, Prbr, Par, Tar, Prar, Tcr, Prac,$ $Tcc, Prcc$;
2. Technical Characteristics Parameters are 4: Pid, Pis, Tcs, Ns;
3. Mix Factors' Parameters are 6: $Ted, Pad, Tad, Tcd, Pac, Tac$;
4. Modeler Chosen Values Parameter (1): M;
5. Derived Parameters from the previous four groups are 9: $Yab, Fa, dem.Fa,$ $rep.Fa, Pbs, Pbr, ofr.Fs, Ts, ofr.Ys$.

In this paper, we propose a short term classification of the chosen base parameters' set with 31 static and 10 dynamic parameters. For the static parameters we assume that their values don't depend on the state of the system and correspondingly on the intensity of the input flow. They may depend on other factors, e.g. the time of the day; seasons, human temperament, Telecom Administration, Gross Domestic Product and so on, but for the observed and modeled time interval we consider them as constants. The 31 static parameters are: $M, Nab, Ns, Ted, Pad, Tad, Prad, Pid, Tid, Prid, Tcd, Tbs, Prbs, Pis, Tis, Pris, Pns, Tns, Tcs, Prns, Tbr, Prbr, Par, Tar, Prar, Tcr, Pac, Tac, Prac, Tcc, Prcc.$

The 10 dynamic parameters, with mutually dependent values are: $Fo, Yab, Fa, dem.Fa, rep.Fa, Pbs, Pbr, ofd.Fs, Ts, ofr.Ys.$

Stationary teletraffic tasks. For a short observation interval, we usually consider processes, in the investigated telecommunication system, as standing in a stationary state, described with a system tuple. Some of the values of the system tuple may be known (measured or stipulated), others—not known. The proposed parameters' classification allows definition of different teletraffic tasks' types. Depending on the task specificity, Mix Factors' Parameters may be considered as belonging to the Human Behaviour or Technical Characteristics Parameters groups. Since, excluding M, we have three groups of parameters' types and corresponding main stationary teletraffic tasks' types:

1. *System State Task* is to find values of the 5th group of parameters, if the values of the rest base parameters, from the same base tuple, are known. Note that Yab is the macrostate of the system and the values of Pbs and Pbr belong to Quality of Service (QoS) parameters;

2. *Technical characteristics task* is to find values of the 4th group of parameters, if the values of the rest base parameters, from the same base tuple, are known. Note that Pid and Pis are caused from technical failures usually, Tcs is limited and Ns is a main network dimensioning parameter. The Network Dimensioning Task (NDT) is for finding Ns if the target values of Pbs and Pbr are known;

3. *Human Behaviour Task* is to find values of the 1st group of parameters, if the values of the rest base parameters, from the same base tuple, are known. This task is difficult due the relatively big number of unknown values. There are some results for finding important parameters as the number of active terminals Nab in [5].

Others criteria for parameters' classification and correspondent teletraffic tasks, inside a system tuple, are theoretically and practically interesting, too. For example Task for Inconvenient Measurements is to find values of the difficulty measured parameters, e.g. intensity of repeated attempts flow, if the values of easy measured parameters are known.

Dynamic teletraffic task. The system's dynamic may be presented with a series of system tuples. There is a difference between long and short term dynamics. In long term considerations, all the system parameters may be with variable values. In short term analysis, some of parameters may be considered as having constant values.

2.6   Main Assumptions

Due to the complexity of the model, in order to obtain relatively simple analytical expressions about the parameters, after a careful analysis the following assumptions are formulated:

A-1. (Closed System Structure) We consider a closed telecommunication system with functional structure shown in Fig. 2;

A-2. (Device Capacity) All base virtual devices in the model have unlimited capacity. Comprising devices are limited: **ab**-device contains all the active terminals $Nab \in [2, \infty]$; switching system (**s**) has capacity of Ns calls (every internal switching line may carry only one call); every terminal has capacity of one call, common for both incoming and outgoing calls;

A-3. (A-Terminal Occupation) Every call, from the flow incoming in the telecommunication system $(inc.Fa)$, falls only on a free terminal. This terminal becomes a busy A-terminal;

A-4. (Stationarity) The system is in stationary state. This means that in every virtual device in the model (including comprising devices like switching system), the intensity of input flow $F(0, t)$, call holding time $T(0, t)$ and traffic intensity $Y(0, t)$ in the observed interval $(0, t)$ converge to the correspondent finite numbers F, T and Y, when $t \to \infty$. In this case we may apply the theorem of Little (1961) and for every device: $Y = FT$;

A-5. (Calls' Capacity) Every call occupies one place in a base virtual device, independently from the other devices (e.g. a call may occupy one internal switching line, if it find free one, independently from the state of the intent B-terminal (busy or free));

A-6. (Environment) The calls in the communication systems' environment (outside the blocks a and b in Fig. 2) don't occupy any telecommunication systems' device and therefore they don't create communication systems' load. (For example, unsuccessful calls, waiting for the next attempt, are in "the head" of the user only. The calls and devices in the environment form the intent and repeated calls flows). Calls leave the environment (and the model) in the instance they enter a Terminator virtual device;

A-7. (Parameters' Independability) We consider probabilities for direction of calls to, and holding times in the base virtual devices as independent of each other and from intensity $Fa = inc.Fa$ of incoming flow of calls. Values of these parameters are determined by users' behavior and Ordinaritytechnical characteristics of the communication system. (Obviously, this is not applicable to the devices of type Enter Switch, correspondingly to Pbs and Pbr.);

A-8. (Randomness) All variables in the analytical model may be random and we are working with their mean values, following the Theorem of Little.

A-9. (B-Terminal Occupation) Probabilities of direction of calls to, and duration of occupation of devices **ar, cr, ac** and **cc** are the same for A and B-calls;

A-10. (Channel Switching) Every call occupies simultaneously places in all the base virtual devices in the telecommunication system (comprised of devices **a** or

b) it passed through, including the base device where it is in the moment of observation. Every call releases all its occupied places in all base virtual devices of the communication system, in the instant it leaves comprising devices **a** or **b**.

A-11. (Homogeneity of the terminals) All terminals are homogeneous, i.e., for every terminal all corresponding characteristics are equal.

A-12. (Direction of the A-calls) Every A-terminal generates all call attempts only towards other terminals, not towards itself.

A-13. (Ordinadirity of the B-flow) The flow directed to the B-terminals (Fb) is ordinary. The only exception is when two or more calls reach a free B-terminal simultaneously.

A-14. (Probability of repeated calls blocking) The probability Pbr for finding the B-terminal busy is one and the same for the first and all of the next repeated attempts.

3   QoS Prediction Task

3.1   QoS Prediction Task Formulation

We consider the conceptual model presented in Fig. 2 and described in the previous section. In this chapter, we consider that the overall telecommunication system provides four services: (1) finding B-terminal; (2) connection to B-terminal; (3) finding B-user (with sound, vibration, message, etc.); and (4) transmission and/or record of messages. The quality of these services depends on many subsystems including the users' behaviour and network.

Types of Parameters. There are two types of parameters—static and dynamic. The 10 basic dynamic parameters (with values dependent on the system state) are: Fo, Yab, Fa, $dem.Fa$, $rep.Fa$, Pbs, Pbr, $ofr.Fs$, Ts, and $ofr.Ys$. All other dynamic parameters can be obtained from these.

Note that the traffic Yab of all terminals is accepted as a system macro-state parameter.

Input Parameters. These are mostly static, i.e., related to the network technical characteristics or the users' behaviour. We choose one dynamic parameter—Fo (the intent intensity of calls of one idle terminal) as an independent input variable. The proposed analytical model allows to find all dynamic values, if Fo and all static parameters are known.

The probability of finding the B-user is considered static (i.e. independent of the system state).

Basic QoS parameters. The basic QoS output parameters are:

− Quality of finding the B-terminal service, represented by the probability of call blocking due to lack of resources (equivalent network switching lines)–blocked switching (Pbs);

- Quality of connection to the B-terminal, represented by the probability of call blocking due to busy B-terminal—blocked ringing (Pbr).
- Network call efficiency (Ec);
- Network time efficiency (Et);
- Network traffic efficiency (Ey).

These two parameters allow determination of many other QoS indicators, related to traffic, time, and flow characteristics of users and terminals.

The goal of this section is to find analytically all unknown basic dynamic parameters, including the basic QoS output parameters.

3.2 Derivation of Equations About the Dynamic Parameters

Here, we shortly present the analytical model of overall telecommunication system from [1]. All other parameters in the equations, except the dynamic parameters, are considered known.

Theorem 1 *The traffic intensity of all the terminals (Yab) is a sum of the traffic intensities of the A (Ya) and B-terminals (Yb):*

$$Yab = Ya + Yb. \tag{1}$$

Proof (i) There are no other terminals apart from the included in Fig. 1, because the modeled system is closed (Assumptions A-1); (ii) Every terminal, at a given time moment, may be free or busy. If it is busy, it may be calling (A) or called (B), but not simultaneously both calling and called, because every terminal has capacity of one call (A-2). Obviously, from (i) and (ii) follows (1). □

Since the number of the terminals is limited to Nab (assumption A-2), and there is no negative occupancy, we have the following obvious terminal traffic limitations in the studied system:

$$0 \le Yab \le Nab. \tag{2}$$

In Eq. 2 are referred to as *absolute terminal traffic limitation*

Theorem 2 *The calls flow intensity occupying all terminals (Fab) is a sum of the intensities of calls flows occupying the A-terminals (Fa) and calls flow occupying B-terminals (Fb):*

$$Fa + Fb = Fab. \tag{3}$$

Proof This follows from assumption A-1 and the conceptual model, in which $Fb = inc.Fb$ is a flow different from Fa and corresponds to the real cases when successful calls, after dialing and switching, occupy the B-terminals. From assumption A-3 follows that $Fa = inc.Fa$ (there are no blocked calls, directed to the communication system). □

Theorem 3 *The traffic intensity of the B-terminals (Yb) can be determined from the equation*

$$Yb = Fb\,Tb,\tag{4}$$

where Tb is the mean holding time of calls in a B-terminal, and:

$$Fb = Fa(1 - Pad)(1 - Pid)(1 - Pbs)(1 - Pis)(1 - Pns)(1 - Pbr),\tag{5}$$

$$Tb = Par\,Tar + (1 - Par)[Tcr + Pac\,Tac + (1 - Pac)Tcc].\tag{6}$$

Proof Equation (4) is Little's formula for device b in stationary state (A-4). Equation (5) expresses the fact that the A-calls have to avoid the six modeled losses before occupying the intent B-terminals, with mean intensity of calls Fb. Equation (6) is direct corollary from Fig. 2 (box for b-device, in the right-down), closed system structure (A-1), calls' capacity (A-5), excluding calls in the environment (A-6) parameters independability (A-7), randomness (A-8) and B-terminal occupation assumption (A-9).

We may derive the expression 6 for the B-terminals holding time (Tb) from the next considerations. From Fig. 2, parameters independability (A-7) and channel switching (A-10), it follows that Yb is a sum of the traffics of the base blocks, comprised in it. The assumption for B-terminal occupation (A-9), implies that Yar, Ycr, Yac and Ycc are the same traffic intensities for A and B-terminals, so:

$$Yb = Yar + Ycr + Yac + Ycc.\tag{7}$$

On the other hand, we may express the traffic intensities using the Little's formula and presenting every flow intensity in the base devices as a function of Fb:

$$Yar = Far Tar = FbParTar;\tag{8}$$

$$Ycr = Fcr\,Tcr = Fb(1 - Par)Tcr;\tag{9}$$

$$Yac = Fac\,Tac = Fb(1 - Par)Pac\,Tac;\tag{10}$$

$$Ycc = Fcc\,Tcc = Fb(1 - Par)(1 - Pac)Tcc.\tag{11}$$

After replacing (8), (9), (10) and (11) in (7), taking in consideration A-9 and using (4) we obtain (6).   □

Theorem 4 *The A-terminals' traffic intensity (Ya) is can be determined using the Little's formula*

$$Ya = Fa\,Ta,\tag{12}$$

where Ta is given by:

$$Ta = Ted + Pad\,Tad + (1 - Pad)(Pid\,Tid + (1 - Pid)$$

$$\cdot(Tcd + Pbs\,Tbs + (1 - Pbs)(Pis\,Tis + (1 - Pis)$$

$$\cdot(Pns\,Tns + (1 - Pns)(Tcs + Pbr\,Tbr + (1 - Pbr)Tb))))). \tag{13}$$

The proof of the theorem is very similar to the proof of Theorem 3, but includes more base devices, shown on Fig. 2.

Theorem 5 *Distinguishing static and dynamic parameters, we have:*

$$Ya = Fa[Sa_1 - Sa_2(1 - Pbs)Pbr - Sa_3Pbs], \tag{14}$$

where

$$Sa_1 = Ted + Pad\,Tad + (1 - Pad)[Pid\,Tid + (1 - Pid)[Tcd + Pis\,Tis+$$

$$+(1 - Pis)[Pns\,Tns + (1 - Pns)[Tcs + Tb]]]],$$

$$Sa_2 = (1 - Pad)(1 - Pid)(1 - Pis)(1 - Pns)[Tb - Tbr],$$

$$Sa_3 = (1 - Pad)(1 - Pid)[Pis\,Tis + (1 - Pis)[PnsTns + (1 - Pns)$$

$$\cdot[Tcs + Tb]]] - (1 - Pad)(1 - Pid)Tbs. \tag{15}$$

Proof Equations (14) and (15) are result of simple mathematical transformations of (13), after application of the Little's formula $Ya = FaTa$.   □

Theorem 6 *The traffic of all simultaneously busy terminals (Yab) may be determined from Eq. 16 as a function of Fa and other parameters:*

$$Yab = Fa\{Ted + Pad\,Tad + (1 - Pad)[Pid\,Tid + (1 - Pid)[Tcd + Pbs\,Tbs$$

$$+ (1 - Pbs)[Pis\,Tis + (1 - Pis)[Pns\,Tns + (1 - Pns)[Tcs + Pbr\,Tbr + 2(1 - Pbr)Tb]]]]]\}, \tag{16}$$

or Eq. (17), after separation of static from dynamic parameters in it:

$$Yab = Fa[S_1 - S_2(1 - Pbs)Pbr - S_3Pbs], \tag{17}$$

where:

$$S_1 = Ted + Pad\,Tad + (1 - Pad)[Pid\,Tid + (1 - Pid)[Tcd + PisTis$$

$$+(1 - Pis)[Pns\,Tns + 1 - Pns[Tcs + 2Tb]]]],$$

$$S_2 = (1 - Pad)(1 - Pid)(1 - Pis)(1 - Pns)[2Tb - Tbr],$$

$$S_3 = (1 - Pad)(1 - Pid)[Pis\ Tis + (1 - Pis)[Pns\ Tns + (1 - Pns)[Tcs + 2Tb]]]$$

$$- (1 - Pad)(1 - Pid)Tbs. \tag{18}$$

Proof Adding Eqs. (4) and (13), using (1), and after elementary mathematical transformations, we obtain (16) and from it (17) and (18). $\square$

We have to determinate the mean intensity of the input flow to the telecommunication system. This is the flow occupying the calling (A) terminals Fa. From the ITU E.600 definitions and Fig. 2 it is obvious that the intensity of incoming flow is a sum of the intensities of primary (demand) calls $(dem.Fa)$ and repeated attempts $(rep.Fa)$:

$$Fa = dem.Fa + rep.Fa. \tag{19}$$

From the definition of BBP-flow we have (see Sect. 2.4):

$$dem.Fa = Fo(Nab + MYab). \tag{20}$$

The proof of the following theorem can be found in [1, 2].

Theorem 7 *The intensity of the flow of repeated call attempts $rep.Fa$ may be determined from the expression (21):*

$$rep.Fa = Fa\{Pad\ Prad + (1 - Pad)[Pid\ Prid + (1 - Pid)[Pbs\ Prbs + (1 - Pbs)$$

$$\cdot [Pis\ Pris + (1 - Pis)[Pns\ Prns + (1 - Pns)[Pbr\ Prbr + (1 - Pbr)$$

$$\cdot [Par\ Prar + (1 - Par)[Pac\ Prac + (1 - Pac)Prcc]]]]]]\}. \tag{21}$$

After separation of static and dynamic parameters in the expression for $rep.Fa$ above, we obtain the following representation of $rep.Fa$ as a function of the dynamic parameters Fa, Pbr and Pbs:

Theorem 8

$$rep.Fa = Fa[R_1 - R_2 Pbr(1 - Pbs) - R_3 Pbs], \tag{22}$$

where

$$Q = Par\ Prar + (1 - Par)[Pac\ Prac + (1 - Pac)Prcc],$$

$$R_1 = Pad\ Prad + (1 - Pad)(Pid\ Prid + (1 - Pid)Pis\ Pris + (1 - Pis)(Pns\ Prns$$

$$+(1 - Pns)Q),$$

$$R_2 = (1 - Pad)(1 - Pid)(1 - Pis)(1 - Pns)(Prbr - Q),$$

$$R_3 = (1 - Pad)(1 - Pid)[Pis\,Pris + (1 - Pis)[Pns\,Prns + (1 - Pns)Q] - Prbs]. \quad (23)$$

In the teletraffic engineering of all the telecommunication networks, parameters characterizing the terminal traffic are used. One of the most important of them is the probability of finding the called terminal (B-terminal) busy (Pbr). The following theorem, proved in [1], gives analytical expression for this probability:

Theorem 9 *The probability of finding the B-terminal busy (Pbr) is*

$$Pbr = \begin{cases} \frac{Yab-1}{Nab-1}, & if \ 1 < Yab \le Nab. \\ \\ 0, & if \ 0 \le Yab \le 1. \end{cases} \quad (24)$$

Using the conceptual model and the assumptions made, in [1] the unknown value of the mean blocking probability (Pbs), due to insufficient equivalent switching lines (Ns) is expressed analytically.

Theorem 10 *The mean holding time of the switching system (Ts) is given by the equation*

$$Ts = [Pis\,Tis + (1 - Pis)[Pns\,Tns + (1 - Pns)[Tcs + Pbr\,Tbr + (1 - Pbr)Tb]]]$$

$$= S_{1,z} - S_{2,z}Pbr, \quad (25)$$

where

$$S_{1,z} = Pis\,Tis + (1 - Pis)[Pns\,Tns + (1 - Pns)(Tb + Tcs)]$$

$$S_{2,z} = (1 - Pis)(1 - Pns)(Tb - Tbr). \quad (26)$$

Theorem 11 *The intensity of the offered flow of call attempts (ofr.Fs), may be expressed through Eq. (27):*

$$ofr.Fs = Fa(1 - Pad)(1 - Pid). \quad (27)$$

Theorem 12 *The probability of blockedg switching (Pbs) is determined from Eqs. (28) and (29):*

$$ofr.Ys = ofr.Fs\,Ts, \quad (28)$$

$$Pbs = Erl_b(Ns, ofr.Ys). \quad (29)$$

Equation (29) simply expresses the use of the Erlang-B formula for determination of the blocking probability in the switching system, on the basis of the number of internal switching lines (Ns) and offered traffic $ofr.Ys$. The expression $Erl_b(Ns, ofr.Ys)$ denotes the famous formula of Erlang:

$$Erl_b(Ns, ofr.Ys) = \frac{(ofr.Ys)^{Ns}}{Ns!} / \sum_{j=0}^{Ns} \frac{(ofr.Ys)^j}{j!}. \qquad (30)$$

The following theorem can be easily verified using the graphic representation of the conceptual model.

Theorem 13 *The intensity of the carried call attempts by the switching system satisfies the equation:*

$$crr.Ys = (1 - Pbs)ofr.Ys. \qquad (31)$$

3.3 A System of Equations Characterizing the Overall State of the Telecommunication System

For the 11 basic dynamic parameters, which are mutually dependent: Fo, Yab, Fa, $dem.Fa$, $rep.Fa$, Pbs, Pbr, $ofr.Fs$, Ts, $ofr.Ys$, in the previous subsection we have derived 10 equations with 6 generalized static parameters. These are: Eq. (17) for Yab; (19) for Fa; (20) for $dem.Fa$; (22) for $rep.Fa$; (24) for Pbr; (25) for Ts; (27) for $ofr.Fs$; (28) for $ofr.Ys$; (29) for Pbs and (31) for $crr.Ys$. We have no equation in which Fo is present in the left-hand side. It is in the right-hand side in (20) only. The system becomes:

$$Yab = Fa[S_1 - S_2(1 - Pbs)Pbr - S_3Pbs], \qquad (32)$$

$$Fa = dem.Fa + rep.Fa. \qquad (33)$$

$$dem.Fa = Fo(Nab + MYab), \qquad (34)$$

$$rep.Fa = Fa[R_1 - R_2Pbr(1 - Pbs) - R_3Pbs], \qquad (35)$$

$$Pbr = \begin{cases} \frac{Yab-1}{Nab-1}, & \text{if } 1 < Yab \leq Nab. \\ \\ 0, & \text{if } 0 \leq Yab \leq 1, \end{cases} \qquad (36)$$

$$Ts = S_{1,z} - S_{2,z}Pbr, \qquad (37)$$

$$ofr.Fs = Fa(1 - Pad)(1 - Pid), \tag{38}$$

$$ofr.Ys = ofr.Fs\,Ts, \tag{39}$$

$$Pbs = Erl_b(Ns, ofr.Ys), \tag{40}$$

$$crr.Ys = (1 - Pbs)ofr.Ys, \tag{41}$$

The equations for the static parameters are:

$$S_1 = Ted + Pad\,Tad + (1 - Pad)[Pid\,Tid + (1 - Pid)[Tcd + Pis\,Tis$$
$$+ (1 - Pis)[Pns\,Tns + 1 - Pns[Tcs + 2Tb]]]], \tag{42}$$

$$S_2 = (1 - Pad)(1 - Pid)(1 - Pis)(1 - Pns)[2Tb - Tbr], \tag{43}$$

$$S_3 = (1 - Pad)(1 - Pid)[Pis\,Tis + (1 - Pis)[Pns\,Tns + (1 - Pns)[Tcs + 2Tb]]], \tag{44}$$

$$S_{1,z} = Pis\,Tis + (1 - Pis)[Pns\,Tns + (1 - Pns)(Tb + Tcs)], \tag{45}$$

$$S_{2,z} = (1 - Pis)(1 - Pns)(Tb - Tbr), \tag{46}$$

$$R_1 = Pad\,Prad + (1 - Pad)(Pid\,Prid + (1 - Pid)Pis\,Pris + (1 - Pis)(Pns\,Prns$$
$$+ (1 - Pns)Q), \tag{47}$$

$$R_2 = (1 - Pad)(1 - Pid)(1 - Pis)(1 - Pns)(Prbr - Q), \tag{48}$$

$$R_3 = (1 - Pad)(1 - Pid)[Pis\,Pris + (1 - Pis)[Pns\,Prns$$
$$+ (1 - Pns)Q] - Prbs], \tag{49}$$

$$Q = Par\,Prar + (1 - Par)[Pac\,Prac + (1 - Pac)Prcc]. \tag{50}$$

3.4 Numerical Results

Numerical results related to the QoS prediction task are shown in Fig. 3. The change of the network call efficiency (Ec), the traffic efficiency (Ycc), the traffic of the B-terminals (Yb), the traffic of the A-terminals (Ya), traffic effectiveness (Ey) and the probability of blocked ringing (Pbr) depending on the average traffic of one terminal in the system (Yab/Nab). The traffic values are shown as a ratio to the number of all terminals (Nab). The values belong to the whole theoretically possible interval for telecommunication systems without losses due to lack of resources. The values used for the input parameters (the static parameters) are typical ones for telephone systems.

The network call efficiency (Ec) is the ratio of the number of successive call attempts to the number of all call attempts.

The network traffic effectiveness (Ey) is the absolute of the carried traffic which is the useful target effect of the work of the telecommunication network. This is the traffic corresponding to the base virtual device **cc** (carried communication—successful call attempt). Obviously, it is a part of the traffic of the B-terminals (Yb) which includes other unavoidable load such as the signalisation.These two pieces of traffic have maximums at one and the same point of the system's load—80.35% of the maximum. The values of all variables illustrated in Fig. 3 are shown for this particular point.

Fig. 3 Boundary and extremal values of the terminal traffic and the efficiency of overall telecommunication network with virtual channel switching without losses due to lack of resources

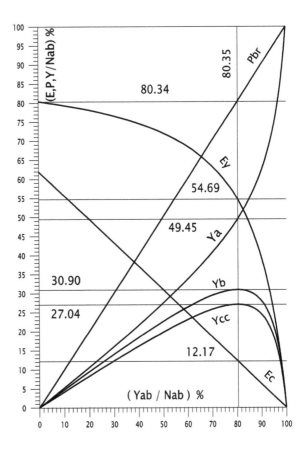

The traffic efficiency (Ey) is the ratio of the successfull traffic to whole traffic of the A-terminals, i.e. $Ey = Ycc/Ya$.

It can be seen that the traffic and call efficiencies have different values, including different maximums—80.34% and approximately 62%, respectively. At the point of maximum theoretical value of the load (100%), there is no useful load and all efficiencies are equal to zero.

An important result is finding the maximum of the B-terminals' traffic (Yb) in networks without technical losses. This means that the required number of switching lines for guaranteeing of connection without blocking, is less than the theoretical— 50% of the number of the terminals (Nab). In our case, the required number of switching lines is 31% of Nab (the maximum in Fig. 3 is 30.90%).

4 Dimensioning and Redimensioning Tasks

4.1 Network Dimensioning Task

One of the main problems which often have to solved by the operators of telecommunication services is to determine the volume of the telecommunication resources in such a way that it is sufficient for the servicing of a given input flow of requests with in advance specified characteristics of the Qualisty of Service (QoS) in the Service Level Agreement (SLA) [6]. The aim is for the user to be satisfied with the offered service at the arranged price [7, 8].

The dimensioning of a telecommunication network is used to determine the capacity of the nodes and switching lines in the network while looking for the balance between the requirements for reaching and sustaining a certain level of QoS and the final value of the offered service.

Formulation of the network dimensioning task—goals, problems, parameters. By dimensioning of a network we mean the process of providing the necessary number of internal switching lines of the switching system (Ns), which are enough to provide the agreed in advance grade of the QoS.

In the process of determining the designed values of the number of internal switching lines $(dsn.Ns)$, the probability of blocked switching (Pbs) is chosen as a target parameter. This parameter represents an important criterion for the Grade of Service (GoS). In a well-dimensioned net, the grade of the blocking due to the lack of resources, should not exceed a certain pre-arranged level of the QoS, i.e., the target value of the blocking $(trg.Pbs)$. The target value $(trg.Pbs)$ of the probability of blocking to the lack of resources is administratively pre-arranged in the Service Level Agreement (SLA) [9].

The problems for network dimensioning with respect to the length of the time interval of the envisaged planning (see [10–13]) can be classified as tasks for:

- *Capacity management* [11], including prediction of the needed capacity, daily or weekly observation of the output of the network and short-term network adjustment.
- *Traffic management*, including dimensioning in short-term plan (up to 2 years). It mainly concerns re-configuration of already existing resources in the network.
- *Redimensioning* of network in which the dimensioning of the network concerns medium-term planning (up to 5 years).
- *Dimensioning* of networks with long-term planning or initial dimensioning of netwokrs (up to 10 years).

In all of the above dimensioning problems, for the purpose of the operative management and planning, the projected resources of the network are determined on the basis of determined, measured, prognostic and target values. This is the purpose of the method for dimensioning developed by us [14]. In our research, two of the total three types of dimensioning problems are studied. These are the Network Dimensioning Task (NDT) in long-term planning and the task for dimensioning of a network in medium-term planning, for repeated dimensioning, referred to as Network Redimensioning Task (NRDT).

The obtained results are applicable to all of the mentioned activities related to the dimensioning of networks. In the long-term dimensioning, methods which allow the prognosis of traffic parameters of the telecommunication system are used. The presented results are used in both of the problems [2].

Classification and notation of the used parameters depending on the way of determining of their values. In many of the problems of Teletraffic Engineering (TE) more than one different values of a certain parameter have to be used [15, 16]. The parameters in our research of system tuples and the sub-collection of the selected base parameters (base tuple) are classified according to way in which their values are determined. The proposed classification is presented in Table 1. The type of the denoted value is denoted by prefix—qualifier before the name of the parameter [5].

The models of the NDT and the NRDT have values of input parameters which are: empirical (obtained through measurement in real functioning networks) [17]; by assumption about the environment of the designed network; administratively determined; designed (for instance, characterizing the QoS). Their output parameters are the designed $(dsn.Pbr, dsn.of r.Ys, dsn.Pbs, dsn.Ns)$.

Base assumptions in the mathematical modeling of the network dimensioning task. On the basis of assumption A-8 and the following observations, we work with the mean values of the teletraffic parameters.

1. The Little's formula, which is the basic formula used in the analytical modeling, does not depend on the distribution of the quantities in the model. This allows us to work with the mean values of the parameters.
2. The formula for determining of the probability of blocking of Poisson, Engset, the B and C formula of Erlang, etc., use only mean values of the input quantities, taking into account their assumed distributions.
3. The number of the designed internal switching lines is searched for as an integer number.

Table 1 Notation of the parameters' values according to their origin

Name according to the origin of the values	Used prefix qualifier	Example
Empirical values (primary parameter, secondary parameter, drived)	*emp.*	*emp.crrs. Ys—empirical values of value carried traffic in the system*
Designed values		
Based on the assumption	*ass*	*ass.Fo—values of input flow, generated by one idle terminal, determined by assumption*
Target	*trg.*	*trg.Pbs—target values of probability blocked watching due to lack of resources*
Threshold	*thr.*	*thr.Fo—threshold values of input flow, generated by one idle terminal*
Administrative	*adm.*	*adm.Nab—administratively determined number of terminals*
Designed	*dsn*	*designed (project)values of :*
		– offered traffic: dsn.ofr.Ys; *– probability of busy terminals: dsn.Pbr* *– required number of switching line in the telecommunication system: dsn.Ns*
Test values		
Test	*tst*	*tst.Pbs—test received values*

A confirmation of the applicability of this approach are ITU recommendations [3, 18–20], etc., and, also, the Cisco instructions for traffic analysis and dimensioning of networks for VoIP [21].

We need to add the following additional assumptions to the 14 stated in the previous section [2]:

A-15. The probabilities $Pad, Pid, Pbs, Pis, Pns, Pbr, Par, Pac$ preserve their values during the repeated call attempts.

A-16. The probabilites of entering of repeated call attempts due to abandoned dialing $Prad$, interrupted dialing $Prid$, interrupted dialing $Prid$, interrupted switching Pis, not available switching $Prns$, abandoned ringing $Prar$ or abandoned communication $Prac$, which characterize the users' behavior also preserve their values.

Parameters characterizing the users' behavior are the mean service time of the separate devices: $Ted, Tad, Tid, Tcd, Tbs, Tis, Tns, Tcs, Tbr, Tar,$ Tcr, Tac and Tcc.

A-17. The mean service time in the various devices, i.e., $Ted, Tad, Tid, Tcd, Tbs,$ $Tis, Tns, Tcs, Tbr, Tar, Tcr, Tac$ and Tcc, remain the same.

As a result of the assumptions A-6 and from A-15 to A-17, the 8 generalized static parameters $S_1, S_2, S_3, R_1, R_2, R_3, S_{1,z}$ and $S_{2,z}$ do not change their values before and after the change of the number of the switching lines.

NDT/NRDT problem statement. The solution of any of the two tasks—NDT and NRDT—requires the achievement of the following goal:

—To determine the required number of switching lines $(dsn.Ns)$ in a working network so that an agreed in advance with the users QoS $(trg.Pbs)$ is achieved or maintained.

The following subproblems have to be solved in the network dimensioning task:

1. Finding of the designed values of parameters $(dsn.Pbr, dsn.ofr.Ys)$ through the least possible number of empiric values of easily measurable variables $(crr.Ys, emp.Fo)$ which describe the designed state of the network (and represent a key for the evaluation of the designed number of internal switching lines, for example $dsn.Yab, dsn.ofr.Ys)$ [22].

2. Finding the designed number of switching lines $(dsn.Ns)$, required for the service of the projected offered traffic, on the basis of the evaluated designed values of the parameters. This should be done in such a way that the designed values for blocking due to lack of resources $(dsn.Pbs)$ does not exceed the target value $trg.Pbs$, i.e. $dsn.Pbs \leq trg.Pbs$ where $dsn.Pbs = Erl_b(dsn.Ns, dsn.ofr.Ys)$.

Various formula exist for the evaluation of the number of internal switching lines in teletraffic theory. We use the Erlang's B-formula recommended by ITU (see Theorem 12, Eq. 30). It represents a functional dependence between the number of switching lines, the offered traffic and the grade of service (the probability of blocking as a measure of quality of traffic service).

The users' behavior in the model used for solving the dimensioning task is described by the following parameters: intent flow intensity of one idle terminal Fo, the probabilities for losses, the probabilities of entering of repeated call attempts due to service refusal and the mean duration of occupying of the separate devices by the requests. Other variables that characterize the users' behavior are the probabilities of losses in the different stages: losses due to abandoned dialing Pad, interrupted dialing by the system Pid, losses due to interrupted switching Pis or not available switching Pns, due to abandoned ringing Par, probability of abandoned communication Pac, i.e., all probabilities for losses with the exception of the probability for blocking due to lack of resource Pbs and the probability of blocked ringing Pbr.

Solving the network dimensioning/redimensioning task. In [2], some dependencies are derived and proven which allow using the least number easily measurable dynamic parameters $(emp.crr.Ys, emp.Pbs)$, values of administratively determined parameters: ($adm.Nab$ and M—modifier for the BPP input flow) and the eight generalized static parameters (S_1, S_2, S_3, R_1, R_2, R_3, $S_{1,z}$, $S_{2,z}$) to evaluate the empirical values of the intensity of the input flow, generated by one idle terminal Fo (Theorem 14) which is needed to determine the designed values in the NDT/NRDT.

Theorem 14 *The empirical value of the intensity of the flow generated by one idle terminal $emp.Fo$ given $Pbr \neq 0$ can be determined by*

$$Fo = \frac{crr.Ys[(1 - R_1 - R_3\, Pbs)S_{2,z} - R_2\Omega]}{(1 - Pad)(1 - Pid)Ts[(1 - Pbs)S_{2,z}(Nab + M) + M(Nab - 1)\Omega]},$$
(51)

where $\Omega = (1 - Pbs)S_{2,z}\, Pbr$.
  When $Pbr = 0$, Fo is given by

$$Fo = \frac{crr.Ys[1 - R_1 - R_3\, Pbs]}{(1 - Pad)(1 - Pid)(1 - Pbs)S_{1,z}Nab + M(S_1 - S_3\, Pbs)crr.Ys},$$
(52)

given that $Pbs = emp.Pbs$ and $crr.Ys = emp.crr.Ys$.

Note: When there are occupied terminals in the system but there are no losses due to occupied terminal, i.e. if $Pbr = 0$ and respectively $0 \leq Yab \leq 1$, then $0 < Fo \leq thr.Fo$. In the case of losses due to finding the B-terminals busy, we have $thr.Fo < Fo$.

Computation of designed parametric values in the NDT/NRDT. In [2] we have proposed analytical method and algorithm for dimensioning (finding the number of switching lines) of overall telecommunication network.

Note: We consider that the following parameters have constant values:

1. The number of active terminals ($adm.emp.Nab = adm.dsn.Nab$) which is denoted by Nab.
2. The designed and empiric values of the activity of one terminal $dsn.Fo = emp.Fo$ which is denoted by Fo.
3. The values of S_1, S_2, S_3, R_1, R_2, R_3, S_{1z} and $S_{2,z}$ are considered constant because they are relatively independent on the system state in the time interval considered.

For brevity, in the derivation of the analytical expressions and when referring to them we will use Pbs instead of $trg.Pbs$.

Theorem 15 *Given BPP type of input flow, in the ND/NRD task the designed value of the probability of blocked ringing ($dsn.Pbr$), when $thr.Fo < Fo$, satisfies the system:*

$$\begin{vmatrix} A(dsn.Pbr)^2 + B\,dsn.Pbr + C = 0 \\ dsn.Pbr \in [0; 1] \end{vmatrix},$$
(53)

where

$$A = (dsn.Fo\,MS_2 - R_2)(1 - Pbs)(Nab - 1),$$

$$B = (1 - Pbs)[dsn.Fo\,S_2(Nab + M) - R_2] + (Nab - 1)[(1 - R_1 - R_3 Pbs)$$

$$-dsn.Fo\,M(S_1 - S_3)],$$

$$C = (1 - R_1 - R_3 Pbs) - dsn.Fo(Nab + M)(S_1 - S_3 Pbs).$$

In [23] on the basis of the assumptions in the NDT/NRDT and the corresponding notation, it is shown that the values of $dsn.Pbr$ can be evaluated and they depend on the measured values of the parameters $dsn.Fo$, Nab and $trg.Pbs$. When $0 < trg.Fo \leq thr.Fo$ then $0 \leq Yab \leq 1$ and based on Eq. 24 from Theorem 9, it follows that $Pbr = 0$ [23].

In [2] it is proved that the equation in (15) has only one real root which satisfies the condition $Pbr \in [0, 1]$.

The following theorem is proven in [24, 25]:

Theorem 16 *The value of $dsn.ofr.Ys$ can be determined under the assumptions and the conditions of the NDT/NRDT using equations the following equations:*

$$dsn.ofr.Ys = \frac{Fo(1 - Pbr)(1 - Pad)(1 - Pid)(S_{1,z} - S_{2,z}Pbr)}{Fo(1 + M\,Pbr)[S_1 - S_2(1 - Pbs)Pbr - S_3\,Pbs]} \tag{54}$$
$$-Pbr[1 - R_1 - R_2(1 - Pbs)Pbr - R_3\,Pbs]$$

when $thr.Fo < Fo$ and

$$dsn.ofr.Ys = \frac{Fo\,Nab(1 - Pad)(1 - Pid)S_{1,z}}{1 - R_1 - R_2 - R_3\,Pbs - Fo\,M(S_1 - S_3\,Pbs)}, \tag{55}$$

when $0 < Fo \leq thr.Fo$.

Derived and proven in [2] are also formula for evaluation of the following designed traffic parameters: $dsn.Fa$, $dsn.dem.Fa$, $dsn.rep.Fa$, $dsn.Yb$, $dsn.Ya$, $dsn.Tb$, $dsn.ofr.Fs$, $dsn.crr.Ys$ etc. Also, their change in the whole theoretical interval is shown graphically.

We will shortly summarize the method for evaluation of the required number of switching lines ($dsn.Ns$) in NDT/NRDT. Based on a method proposed by us, from the expression for $dsn.ofr.Ys$ (Theorem 16) using the inverse B-formula of Erlang ($inv.Erl_b$), the required number of switching lines given the conditions of the NDT/NRDT can be determined directly [2]:

$$dsn.Ns = inv.Erl_b(dsn.ofr.Ys, trg.Pbs). \tag{56}$$

All of the derived equations are valid for the whole theoretically allowed interval of the parameters. The problems for existence and uniqueness of the solutions are

are also studied in [2]. The necessary and sufficient conditions for the evaluation of *dsn.Ns* based on the B-formula of Erlang are derived.

Comparison of the proposed method with some widely accepted and used methods for dimensioning/redimensioning. Studied and quantitatively compared in [26] are some of the widely accepted and used methods for dimensioning/ redimensioning of networks which are based on two recommendations of ITU [3, 19] for offered traffic and those of CISCO [21]. The results obtained using the method proposed by us are compared with them in [27].

Main goal of the method for comparison: To determine the test values of the blocking probabilities due to lack of resources (*test.Pbs*) in the different methods for dimensioning/redimensioning with the desired QoS (*trg.Pbs*).

One and the same methodology is used for all methods, which is based on the equality:

$$test.Pbs(dsn.Ns, test.ofr.Ys) = Erl_b(dsn.Ns, test.ofr.Ys), \qquad (57)$$

where *test.ofr.Ys* is a test offered traffic for the telecommunication system considered and *dsn.Ns* is the designed number of switching lines for servicing of the offered traffic and $Erl_b(ofr.Ys, Ns)$ is the Erlang formula.

In the recommendation [19], two different methods for evaluation of the offered traffic are presented. The first approach is based on the computational procedure for evaluation of the equivalent offered traffic and the assumption that the offered traffic in the telecommunication system is equal to the equivalent traffic. A drawback of this approach is that repeated attempts are not taken directly into account. In the second approach in the recommendation ITU E.501, the impact of the repeated call attempts is taken into account in the evaluation of the offered traffic.

Presented, tested and analysed are methods for dimensioning, based on the recommendations of ITU for offered traffic, those of CISCO and our method. The different way of evaluation of the offered traffic in the recommendations of ITU and CISCO ([21]) is in the base of the differences between the methods for dimensioning/redimensioning.

The main differences between the studied methods are:

1. The model proposed by us considers the telecommunication system as a whole, not in the separate nodes as it is the case with the other methods.
2. The ways of evaluation of the offered traffic according to the recommendations of ITU [3, 19] and ours [28] are different.

For the purpose of testing of the results (the computational procedure) of the dimensioning/redimensioning in the studied methods we introduce the one new notion—indicative point.

Definition 2 (*Indicative point*) The point in which the empirical state of the system coincides with the target state, i.e. *emp.Pbs* = *trg.Pbs*, is called indicative point.

With the prefix *ind* (indicative) we denote the parametric values of the system tuple in the indicative point, for example $ind.Yab, ind.Fa, ind.Ta$ [2].

Note: We have proposed an approach to the verification whether one considered method for dimensioning/redimensioning (through which the designed number of switching lines $dsn.Ns$ is determined) leads to correct results (that is to obtain equality) through comparison of $emp.Pbs$ with $Erl_b(emp.ofr.Ys, dsn.Ns)$ in the indicative point. The verification is based on the fact that in the indicative point the designed state of the system has to coincide with the current (empirical) state of the system, i.e. $dsn.crr.Ys = emp.crr.Ys$ and, respectively, $dsn.ofr.Ys = emp.ofr.Ys$, $dsn.Ns = emp.Ns$.

The whole theoretical interval in which the telecommunication system is considered is "divided" by the indicative point into left and right intervals. For each of the methods for dimensioning/redimensioning considered, investigated and compared are the results from [29] :

1. The impact of the assumption for discreteness of the switching lines on the accuracy of evaluated designed number (the error because of this assumption represents an unavoidable methodical error) [30].
2. The sensibility of the studied methods (the impact of a change in the number of switching lines by one in the whole theoretical interval on the probability for blocking $test.Pbs$) [31].
3. The applicability of each of the methods in the whole theoretical interval on the basis of the introduced criteria for comparison of their applicability.

Numerical results from the comparison (testing) and their analysis Numerical experiments are carried out for the comparison of the results. The numerical results are tested. Universal method for quantitative comparison of the numerical results obtained through our method and the widely accepted methods for dimensioning is proposed. Computer program for quantitative comparison (testing) of the methods is developed.

The numerical results from the dimensioning carried out through ITU-1, ITU-2, Cisco-1, Cisco-2, [32, 33] and the proposed by us method are shown in Fig. 4. The parameters of the methods from E.501 are denoted by indices 1 and 2; the parameters of the Cisco 1 and Cisco 2 are denoted by indices 4 and 5; the parameters from the approach based on E.600 are denoted by index 6; the parameters of our method are denoted by *Proposed Method*.

Advantages of the proposed method are:

1. Dimensioning with accuracy up to one switching line unlike the other methods in which the dimensioning is inaccurate not only because of the assumption for discreteness;
2. Guarantees the desired level of QoS in more than 98% of the whole theoretical interval unlike the studied widely accepted methods, among which the greatest percentage (Cisco-2) is less than 13.76%.

The better results obtained through our method are due to:

1. The considered model is overall traffic model of telecommunication system;
2. The users' behaviour is included in detail.

Fig. 4 The probability of blocking due to lack of resources with respect to the normalized designed load of the system Yab/Nab. Presented are the test values $test.Pbs$ of ITU-1, ITU-2, Cisco-1, Cisco-2, ITU-2 and the proposed method. The normalized designed load of the system for one terminal of the tuple in the indicative point is denoted by $ind.Yab/Nab$. The target probability of blocking due to lack of resources is $trg.Pbs$ while the empiric is denoted by $emp.Pbs$ [2]

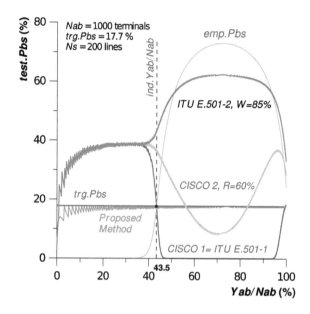

The advantages of the use of our method are:

1. In the NDT, in comparison to other known methods, our method gives better results with respect to the QoS;
2. In the evaluation of the QoS through the proposed method both the technical characteristics and the users'/customers' attitude are taken into account;
3. Because the proposed method considers overall telecommunication system it allows the construction of analytical expressions for evaluation of the QoE.

One *disadvantage* of the proposed method is that additional measurements are needed in comparison to the other studied methods because it is based on detailed account of the users' behaviour.

The *novelties* in our research on the modeling of networks are:

1. In the conceptual modeling of the systems in the formation of QoS both the technical characteristics and the impact of the human behaviour are taken into account. As an example, in the conceptual modeling a telecommunication network is considered because it is the most widely studied through various methods and allows the comparison of the results obtained through the proposed by us method and the results obtained through other widely accepted methods;
2. The analytical model constructed on the basis of the conceptual model is genuine. It is based on:

   – Easily measurable empiric (primary) parameters, for instance the empiric values of the carried by the service system traffic ($emp.crr.Ys$), the duration of service (time occupancy of the called subscriber Tb) etc.;

- The designed (secondary) parameters, evaluated through analytical expressions based on the empirical, target and the administratively determined such as the empirical values of the input flow from one idle terminal ($emp.Fo$), the target value of the blocking due to lack of resources ($trg.Pbs$) and the administratively set number of subscribers in the service system (Nab). Such parameters are, for example, $dsn.Pbr$, $dsn.Ys$ and $dsn.ofr.Ys$.

  For the proposed analytical model:

  (a) assertions for the existence and uniqueness of the solutions of the systems of equations are proved;
  (b) analytical conditions for existence, uniqueness, and boundary conditions are derived.

3. Formulation and evaluation of the $dsn.ofr.Ys$ which is the base for the evaluation of $dsn.Pbr$ which in turns depends on $emp.Fo$. While $emp.Fo$ is a function of $emp.Pbr$, $emp.crr.Ys$, Tb, $emp.Pbs$.

The research can be applied to:

1. Prediction of some parameters (for example the load, the probabilities for blocking etc.), characteristics of the system (efficacy etc.);
2. Management and operational decisions about the work of the system.

5   A Model of Overall Telecommunication System Including a Queuing System in the Switching Stage

5.1   Conceptual Models of Queuing Systems in Service Networks

As a continuation of our work on the modeling of overall telecommunication systems, in [34, 35] we have studied the possible representations of queuing systems in Service Systems Theory. Here, we shall present briefly only two of the models studied there—the classical representation of a queuing system and a Generalized Net (GN, see [36]) model corresponding to it.

The classical representation is shown in Fig. 5. In the classical conceptual model (see [1]), once the Switching system reaches its capacity, the incoming call attempts are blocked and they are redirected to the "blocked switch" branch which begins at the virtual device denoted by **bs** on Fig. 2. With the inclusion of queue in the Switching stage of the model when the Switching system has reached its capacity the incoming call attempts wait in a buffer until a service line in the Switching system becomes available. We consider the buffer size of the queuing system to be of finite length and the number of servers (service lines) also to be finite. In such queuing

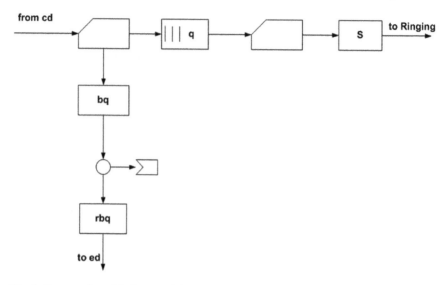

Fig. 5 Conceptual model of a part of the Switching stage of an overall telecommunication system with a queue. **cd** stands for "carried dialing" device, **q** for the Queue device, **s** for "Switching system", **bq** for "blocked queuing", **rbq** for "repeated blocked queuing", **ed** for "enter dialing"

system, the call attempts will be blocked only when both the Switching system and the buffer have reached their capacity.

In comparison to the classical conceptual model in Fig. 2, the branch **bs** is removed because the blocked call attempts from the Enter Switch remain in the queue and they are not redirected to other virtual devices. The Switching system with a queue consists of a device of type Queue denoted by **q**, the Enter Switch before it and all devices of the **bq** branch. The switching system is denoted by **s** in Fig. 5. The Enter Switch device before the **q** device redirects the call attempts when the queue is full. The base device **q** has the same parameters as the other base devices: Fq, Yq, Tq, Pq, Nq. The capacity of the buffer is Nq. The queue discipline considered in the model is FIFO. The Enter switch device between the **q** device and the **s** device has one important parameter—the probability of blocked switching (Pbs) with which the call attempts remain in the **q** device.

The classical representation of the queuing systems does not show explicitly the case when the server has not reached its capacity and the requests are serviced by the buffer device without delay.

A GNs representation of a queuing system is proposed in [34] and it is compared with the conceptual models based on Service Systems Theory. The graphical representation of the GN corresponding to the classical conceptual model of queuing system is shown in Fig. 6.

The places of the net correspond to virtual devices in the following way:

- l_1 and l_2 represent the Generator before Transition 1 in Fig. 5;
- l_3 has no analogue in Fig. 5;

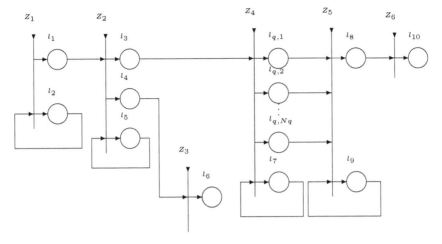

Fig. 6 Generalized net model of a queuing system

- l_4 and l_5 represent the "blocked queuing" (bq) device in Fig. 5;
- l_6 represents the Terminator device after the **bq** device in Fig. 5;
- $l_{q,1}, l_{q,2}, ..., l_{q,Nq}, l_7$ represent the **q** device in Fig. 5;
- l_8 and l_9 represent the s device (comprising device of the Switching system);
- l_{10} corresponds to the Terminator device after the **s** device in Fig. 5.

Each of the six transitions has the following meaning:

- Z_1 represents the function of the Generator before Transition 1 in Fig. 5;
- Z_2 represents the function of Transition 1 in Fig. 5;
- Z_3 represents the function of the Director between the **bq** device and the Terminator device in Fig. 5;
- Z_4 represents the service of the call attempts in the queue;
- Z_5 represents the service of call attempts in the Switching system (the s device in Fig. 5);
- Z_6 represents the function of the Director between the Switching system and the Terminator in Fig. 5.

Four different types of tokens are used in the model. Detailed description of the transitions can be found in [34].

The conceptual model presented in Fig. 5 gives a clearer connection between the functions of the real system and their visual representations. It allows for easy understanding of the connections between the virtual devices and their functions. A downside of this approach is that it uses 5 different virtual devices. It shows the path of the call attempts in the system but they are not shown in the graphical representation of the model. On the other hand, the GN models use less different components to describe the devices and the paths of the calls: places, arcs, transitions. That is why the GN representations are, in a sense, graphically simpler. However, users need special training in order to understand the paths of the calls and the connections between the analogues of the virtual devices in the net. The comparison of the different conceptual models of queuing systems shows that the GN representation is less suitable for the construction of analytical models of overall telecommunication systems.

5.2 Conceptual Model of Overall Telecommunication System Including a Queuing System

The first conceptual model of overall telecommunication system including a queuing system in the switching stage is described in [37]. In this model the classical representation of queuing systems is used. Here we propose a detailization of the model, and in particular a more detailed representation of the queuing system which makes the model easier to understand and more suitable for the derivation of equations for the parameters of the queuing system. The graphical representation of the model is shown in Fig. 7.

In the conceptual model in Fig. 7 there are at least 39 important virtual devices. Of them 34 are base virtual devices and 4 (**a, b, s, ab, w**) are comprising. They are of interest because the values of their parameters characterize the state of the overall telecommunication system. Every device has five parameters: P, F, T, Y and N. Therefore the total number of parameters is 195.

As in the classical model described in Sect. 2, the names of the base virtual device are formed as a concatenation of the first letters of the branch exit, branch and the stage in which the device is situated. The only exception to this are the **bw** and **rbw** devices which stand for "blocked waiting" and "repeated blocked waiting" respectively. The comprise virtual devices with the exception of the **w** device are the same as in the classical model. The inclusion of the **w** device whose name comes from "waiting" allows for a more detailed representation of the queuing system in comparison to the model in Fig. 5. If both the **w** and the **s** device (not shown on the figure) have reached their capacity, the requests are sent to the bw device. The corresponding probability of this to happen is denoted by Pbw. If there are free places in the w device, the requests enter the Enter switch device inside the w device. This happens with probability $1 - Pbw$. From there the requests are serviced in two ways—with waiting or without waiting—depending on the number of requests in the switching system (Ys), as follows.

- If the switching system has reached its capacity, the requests enter the **q** device. The corresponding probability of this to happen is denoted by Pq.
- If the switching system has not reached its capacity, the requests enter the **zq** device. The corresponding probability of this to happen is denoted by Pzq.

We shall consider that the mean service time of the requests in the **zq** device Tzq is equal to 0. In this way the mean service time of the requests in the **w** device can be expressed in the following way

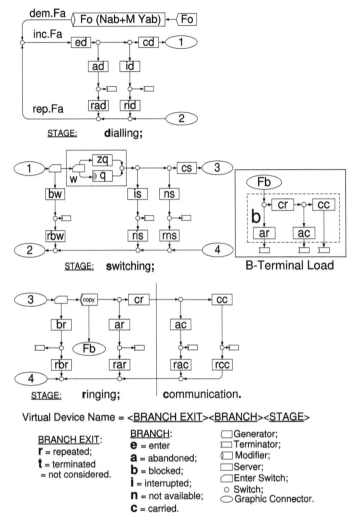

Fig. 7 Conceptual model of overall telecommunication system including a queueing system in the switching stage

$$Tw = Pq\,Tq + (1 - Pq)Tzq = Pq\,Tq. \qquad (58)$$

Similarly to [37], a base tuple of parameters can be determined. The parameters of the base tuple for the present conceptual model may be divided into two groups as follows:

– Static parameters: M', Nab, Ns, Ted, Pad, Tad, $Prad$, Pid, Tid, $Prid$, Ted, Pis, Tis, $Pris$, Pns, Tns, Tes, $Prns$, Tbr, $Prbr$, Par, Tar, $Prar$, Tcr, Pac,

$Tac, Prac, Tcc, Prcc, Nq, Tbw, Trbw, Prbw$. Their values are considered independent of the system state Yab (see [2]) but may depend on other factors. For the model time interval they are considered constants.

- Dynamic parameters: $Yab, Fa, Pbw, Tw, Pbr, Pq, dem.Fa, rep.Fa, ofr.Fw,$ $crr.Fs, T_s, F_s, T_{cs}, Y_s, Yw$. Their values are mutually dependent. Equations expressing their dependencies can be derived with the help of the graphical representation of the conceptual model in Fig. 7.

The parameters can be also classified on the basis of the origin of their values.

- Parameters related to the technical characteristics of the system: $Pid, Pis,$ $Tcs, Ns, Nw = Nq$;
- Parameters describing the human behaviour: $Fo, Nab, Prad, Tid, Prid,$ $Pris, Tis, Pns, Tns, Prns, Tbr, Prbr, Par, Tar, Prar, Tcr, Prac, Tcc,$ $Prcc, Tbw, Trbw, Prbw$;
- Mix factors' parameters: $Ted, Pad, Tad, Tcd, Pac, Tac$. They are dependent on the first two groups;
- Parameters whose value is determined by the modellers: M'. It characterizes a Bernoulli-Poisson-Pascal (BPP) flow;
- Parameters derived from the previous groups: $Yab, Fa, dem.Fa, rep.Fa,$ $Pbs, Pbr, ofr.Fw, crr.Fs, Tw, Pbw$.

The output (dynamic) parameters characterizing the Quality of Service (QoS) are Pbr, Pbs, Tw.

This classification of the parameters allows for different types of teletraffic tasks to be formulated and solved.

In order to be able to derive relatively simple equations for the dynamic parameters, as in the classical model described in Sect. 2, we should make some assumptions. For the present model, all assumptions about the classical model hold. There is a difference only in the following assumptions:

A-1* The telecommunication system considered is represented graphically and functionally in Fig. 7 and it is closed.

A-2* (Device Capacity) All base virtual devices in the model, except the **q** device, have unlimited capacity. The **q** device has capacity Nq. Comprising devices have limited capacities: **ab** device contains all the active terminals; switching system (**s**) has capacity of Ns calls (every internal switching line may carry only one call); the **w** device has finite capacity—Nw; every terminal has capacity of one call, common for both incoming and outgoing calls.

5.3 Derivation of Analytical Expressions for the Parameters of the Queuing System

We consider the conceptual model of overall telecommunication system with queue shown in Fig. 7 and described in Sect. 2. Parameters with known values are: all probabilites for directing the call to a device (the P-parameters), with the exception of Pbw, Pq, Pbr; the holding time parameters of the base virtual devices (T-parameters), except Tq and Tcs. The unknown parameters are all dynamic parameters.

We want to express analytically the unknown parameters' values of the **w** device: waiting time (Tw), probability of blocked waiting (Pbw), length of the queue (Yw).

In order to compactly describe single queuing stations in an unambiguous way, the so called Kendall notation is often used (see [38]). A queuing system is described by 6 identifiers separated by vertical bars in the following way:

Arrivals| Services| Servers | Buffersize | Population | Scheduling

where "Arrivals" characterises the arrival process (arrival distribution), "Service" characterizes the service process (service distribution), "Servers"—the number of servers, "Buffersize"—the total capacity, which includes the customers possibly in the server (infinite if not specified), "Population"—the size of the customer population (infinite if not specified), and finally, "Scheduling"—the employed service discipline.

In our model, the queuing system in the Switching stage of the telecommunication network in Kendall notation is represented as $M|M|Ns|Ns + Nw|Nab|\ FIFO$, where M stands for exponential distribution, Ns is the capacity of the Switching system (number of equivalent internal switching lines) and Nab is the total number of active terminals which can be calling and called. This is related to the derivation of the analytical model of the system.

The queueing system in the switching stage differs from the queuing systems studied in [39, 40] in that it has more exits. The exits are represented in the conceptual model in Fig. 7 with the branches is (interrupted switching), ns (not available switching), br (blocked ringing), ac (abandoned communication), cc (carried communication). In [37], we have derived analytical expressions for the parameters of the queuing system, starting with the simplest queuing system $M|M|1|FIFO$ and gradually advancing to the most complicated system with finite buffer and finite capacity of the server. Here we shall use the results from [37] but adapted to the more detailed conceptual model presented here.

The density functions of the arrival and service times are respectively

$$a(t) = \lambda e^{-\lambda t}, \tag{59}$$

$$b(t) = \mu e^{-\mu t}, \tag{60}$$

where $1/\lambda$ is the mean value of time between two arrivals (interrarival time) and $1/\mu$ is the mean time of service. For our queuing system, $\lambda = ofr.Fw$ and $\mu = Fs$. They are assumed to be statistically independent which results in a birth-death process. Let us denote with p_n the probability that the queuing system is in state n that is

$$p_n = Pr\{\text{there are n call attempts in the queuing system}\}.$$

There are different ways to solve the birth-death equations. The solution is well-known and can be found for example in [39]. First, we notice that the arrival rate λ_n is equal to 0 when $n \geq Ns + Nw$. The probability for the system to be in state n is now given by

$$
p_n = \begin{cases}
\frac{(of r.Fw)^n}{n!(Fs)^n} p_0 & \text{for } 1 \leq n < Ns. \\
\frac{(of r.Fw)^n}{(Ns)^{n-Ns} Ns!(Fs)^n} p_0 & \text{for } Ns \leq n \leq Ns + Nw.
\end{cases}
\tag{61}
$$

Again, the condition that the sum of the probabilities p_n should be equal to 1, gives us the following expression for p_0:

$$
p_0 = \left(\sum_{n=0}^{Ns-1} \frac{(of r.Fw)^n}{n!(Fs)^n} + \sum_{n=Ns}^{Ns+Nw} \frac{(of r.Fw)^n}{(Ns)^{n-Ns} Ns!(Fs)^n} \right)^{-1}.
\tag{62}
$$

In order to simplify the expression we set $r = of r.Fw/Fs$ and $\rho = r/Ns$. After elementary operations above expression for p_0 becomes

$$
p_0^{-1} = \begin{cases}
\sum_{n=0}^{Ns-1} \frac{r^n}{n!} + \frac{r^{Ns}}{Ns!} \frac{1-\rho^{Nw+1}}{1-\rho} & \text{for } \rho \neq 1. \\
\sum_{n=0}^{Ns-1} \frac{r^n}{n!} + \frac{r^{Ns}}{Ns!}(Nw+1) & \text{for } \rho = 1.
\end{cases}
\tag{63}
$$

Using the above, we derive analytical expressions about the important parameters: mean waiting time for all requests (Tw), probability of blocked waiting (Pbw) and expected length of the queue (Yw).

The probability of blocked waiting (Pbw) is equal to the probability that the system is in state $Ns + Nw$ and from (61) we have

$$
Pbw = \frac{(of r.Fw)^{Ns+Nw}}{(Ns)^{Nw} Ns!(Fs)^{Ns+Nw}} p_0.
\tag{64}
$$

For the expected length of the queue in this case which is also equal to Yw we have

$$
Yw = \sum_{n=Ns+1}^{Ns+Nw} (n - Ns)p_n = \frac{p_0 r^{Ns} \rho}{Ns!(1 - \rho)^2}[(\rho - 1)\rho^{Nw}(Nw + 1) + 1 - \rho^{Nw+1}].
\tag{65}
$$

The mean service time of the requests in **w** device for both the waiting and non-waiting requests, given the condition $Tzq = 0$, is

$$
Tw = Pq\, Tq + (1 - Pq)Tzq = Pq\, Tq.
\tag{66}
$$

The mean service time in the **q** device (Tq) is given by

$$Tq = \frac{p_0 r^{Ns} \rho}{Ns!(1 - \rho)^2} \frac{[(\rho - 1)\rho^{Nq}(Nq + 1) + 1 - \rho^{Nq+1}]}{ofr.Fw(1 - Pbw)}. \qquad (67)$$

The probability Pq is the probability that the system is in any of the states $Ns, Ns + 1, ..., Ns + Nw - 1$, i.e.

$$Pq = \sum_{k=Ns}^{Ns+Nw-1} p_k = \sum_{k=Ns}^{Ns+Nw-1} \frac{(ofr.Fw)^k}{(Ns)^{k-Ns} Ns!(Fs)^k} p_0. \qquad (68)$$

After simplification we obtain

$$Pq = \frac{p_0 (Ns)^{Ns} \rho^{Ns}(1 - \rho^{Nw})}{Ns!(1 - \rho)}. \qquad (69)$$

After substitution of (69) and (67) in (66) we obtain

$$Tw = \frac{p_0^2 (Ns\rho r)^{Ns} \rho(1 - \rho^{Nw}).[(\rho - 1)\rho^{Nw}(Nw + 1) + 1 - \rho^{Nw+1}]}{(Ns!)^2(1 - \rho)^3 \quad ofr.Fw(1 - Pbw)}. \qquad (70)$$

6 Directions for Future Research

In the present paper, new methods for conceptual and analytical modeling of overall telecommunication systems are presented including new results regarding overall telecommunication systems with queueing. They have been developed at the Institute of Mathematics and Informatics of the Bulgarian Academy of Sciences. The use of these models gives better accuracy in comparison to the other methods proposed in the scientific literature. Some directions for future research include:

– determining the degree to which the use of GNs makes the analytical modelling of overall telecommunication system easier;
– determining whether the classical teletraffic theory is suitable for the modeling of overall telecommunication systems.
– use of the new results for prediction of the QoE in overall telecommunication systems.

Acknowledgements The work of Stoyan Poryazov is partially supported by the joint research project "Symbolic-Numerical Decision Methods for Algebraic Systems of Equations in Perspective Telecommunication Tasks" of IMI-BAS, Bulgaria and JINR, Dubna, Russia.

The work of Velin Andonov and Emiliya Saranova was supported by the Task 1.2.5. "Prediction and Guaranteeing of the Quality of Service in Human-Cyber-Physical Systems" of National Scientific Program "Information and Communication Technologies for a Single Digital Market in Science, Education and Security (ICT in SES)" financed by the Bulgarian Ministry of Education and Science.

References

1. Poryazov, S.A., Saranova, E.T.: Some general terminal and network teletraffic equations for virtual circuit switching systems. In: Nejat Ince, A., Topuz, E. (eds.) Modeling and Simulation Tools for Emerging Telecommunication Networks, pp. 471–505. Springer, Boston, MA (2006)
2. Poryazov, S., Saranova, E.: Models of Telecommunication Networks with Virtual Channel Switching and Applications. Academic Publishing House "Prof. M. Drinov", Sofia (2012) (in Bulgarian)
3. ITU-T E.600, ITU-T Recommendation E.600: Terms and Definitions of Traffic Engineering, Melbourne (1988); (revised at Helsinki 1993)
4. Iversen, V.B.: Teletraffic Engineering Handbook. ITU-D SG 2/16 & ITC. Draft, December (2003). http://www.tele.dtu.dk/teletraffic/
5. Poryazov, S.A., Saranova, E.T.: On the minimal traffic measurements for determining the number of used terminals in telecommunication systems with channel switching. In: Nejat Ince, A. (ed.) Modeling and Simulation Environment for Satellite and Terrestrial Communication Networks—Proceedings of the European COST Telecommunications Symposium, pp. 135–144. Kluwer Academic Publishers (2002)
6. ITU-T E.860 (2002). Framework of service level agreement
7. Otsetova, A., Saranova, E.: Analysis of service systems characteristics identifying quality of experience. In: Second International Scientific Conference "TIEM 2017", UTP, pp. 91–95 (2017)
8. Otsetova, A., Saranova E.: Quality of a composite service as a function of the qualities of the comprised sub-services. Int. J. Inf. Technol. Knowl. **11**(2), 103–112 (2017)
9. Iversen, V.B.: Teletraffic Engineering & Network Planning. Telenook, 2010/05, DTU, Technical University of Denmark. http://www.fotonik.dtu.dk
10. ITU-T Recommendation E.360.1 (05/2002): Framework for QoS routing and related traffic engineering methods for IP-, ATM-, and TDM-based multiservice networks
11. ITU-T Recommendation E.360.7 (05/2002): QoS routing and related traffic engineering methods—traffic engineering operation requirements
12. ITU-T E.490 (06/92): Traffic measurement and evaluation—general survey
13. ITU-T Recommendation E.734 (10/96): Methods for allocating and dimensioning of Intelligent Network (IN) resources
14. Saranova, E.: Dimensioning of telecommunication network based on quality of services demand and detailed behaviour of users. Int. J. ITK - ITHEA **1**(2), 103–113 (2007)
15. Saranova, E., Poryazov, S.: Some quality characteristics and metrixes in overall telecommunication networks. Int. J. Inf. Models Anal. **6**(2), 131–146 (2017)
16. Saranova, E., Zamiatina, E., Poryazov, S.: Generalized performance characteristics in overall telecommunication networks with QoS guarantees. In: Mathematics of Programming Systems: Interuniversity Collection Scientific Works, vol. 13, pp. 4–11. National Research Institute, Perm, Russia (2016)
17. Saranova, E.: Primary and secondary empirical values in network redimensioning. In: International Book Series Information Science and Computing, ITHEA, vol. 2, pp. 53–59 (2008)
18. ITU-T E.500 (11/98): Traffic intensity measurement principles
19. ITU-T E.501 (05/97): Estimation of traffic offered in the network
20. ITU-T E.800(09/08): Definitions of terms related to quality of service
21. Telephony/Voice over IP (VoIP) Traffic Analysis Cisco Systems. http://www.cisco.com/en/US/docs/ios/solutions_docs/voip_solutions/TA_ISD.html, http://www.cisco.com/cisco/web/psa/default.html?mode=tech&level0=268435881
22. Poryazov, S., Saranova, E.: On the minimal state tuple of VNET with overall QoS guaranties. In: IEEE 19th Telecommunications Forum TELFOR, 22–24 November, 2011. Belgrade, Serbia (2011)
23. Saranova, E., Poryazov, S.: Designed probability determination of finding B-terminal busy. In: DCCN Vishnevsky, V. (ed.) R&D Company Information and Network Technologies, pp. 127–141, Moscow, Russia (2009)

24. Poryazov, S., Saranova, E.: The main perplexities of the ITU-t offered traffic concept. In: National Conference with International Participation TELECOM 2013, pp. 140–145, Sofia, 17–18 October (2013)
25. Saranova, E.: Traffic offered behaviour regarding target QoS parameters in network dimensioning. Int. J. Inf. Technol. Knowl. **2**, 173–180 (2008)
26. Saranova, E., Poryazov, S.: Accuracy analysis of two cisco network re-dimensioning methods. Comptes rendus de l'Académie bulgare des Sci. **65**(11), 1507–1512 (2012)
27. Saranova, E.: ITU network re-dimensioning methods comparison. In: Atanasova, T. (ed.) Proceedings in Modeling and Control of Information Processes, IICT-BAS, IMI-BAS, CTP, pp. 60–74, Sofia (2009)
28. Saranova, E.: Redimensioning of telecommunication network based on ITU definition of quality of services concept. In: DCCN, Vishnevsky, V., Daskalova, H. (eds.), pp. 165–179. Technosphera publisher, Moscow, Russia (2006)
29. Saranova, E.: Influence of some users' behaviour parameters over network redimensioning. In: Proceedings of the International Conference "Information Research, Applications, and Education i.tech", pp. 449–459, Varna, Bulgaria, FOI-COMMERCE-Publisher (2007)
30. Saranova, E., Poryazov, S: Evaluation of the impact of the discreteness of the capacity on the designed probability for blocking in overall network models. In: Proceedings of the National Conference with International Participation TELECOM 2010, pp. 34–41, Sofia, 14–15 October (2010) (in Bulgarian)
31. Saranova, E., Poryazov, S.: Influence of a line on the blocking probability—comparison of three network redimensioning methods. In: Proceedings in DCCN, pp. 190–198, Moscow (2010)
32. Saranova, E., Poryazov, S.: Verification Results for an ITU Network Redimensioning Method. In: DCCN, Vishnevsky, V. (ed.) Technosphera publisher, Moscow, Russia, pp. 137–145 (2008)
33. Saranova, E., Poryazov, S.: Two cisco methods for offered traffic evaluation—analysis and numerical comparison. In: Atanasova, T. (ed.) Proceedings in Modeling and Control of Information Processes, IICT-BAS, IMI-BAS, CTP, pp. 42–50, Sofia (2010)
34. Poryazov, S., Andonov, V., Saranova, E.: Comparison of four conceptual models of a queuing system in service networks. In: Proceedings of the 26th National Conference with International Participation TELECOM 2018, pp. 71–77, Sofia, 25–26 October (2018)
35. Tomov, Zh., Krawczak, M., Andonov, V., Dimitrov, E., Atanassov, K.: Generalized net models of queueing disciplines in finite buffer queueing systems. In: Proceedings of the 16th International Workshop on Generalized Nets, pp. 1–9, Sofia
36. Atanassov, K.: On Generalized Nets Theory. Prof. M. Drinov Academic Publ. House, Sofia, Bulgaria (2007)
37. Andonov, V., Poryazov, S., Otsetova, A., Saranova, E.: A queue in overall telecommunication system with quality of service guarantees. In: Poulkov, V. (eds.) Future Access Enablers for Ubiquitous and Intelligent Infrastructures. FABULOUS 2019. Lecture Notes of the Institute for Computer Sciences, Social Informatics and Telecommunications Engineering, vol. 283, pp. 243–262. Springer, Cham (2019)
38. Haverkort, B.R.: Performance of Computer Communication Systems: A Model-Based Approach. John Wiley & Sons, New York (1998)
39. Schneps, M.: Systems for Distribution of Information. Svyaz Publishing House, Moscow (1979) (in Russian)
40. Vishnevskiy, V.M.: Theoretical Foundations of Computer Networks Planning. Tehnosfera, Moscow (2003) (in Russian)

Results Connected to Time Series Analysis and Machine Learning

Nikolay K. Vitanov

Abstract Machine learning is connected to the scientific study of algorithms and statistical models used by computer systems to perform a specific task without using explicit instructions. In this chapter we describe results of our studies on the methods connected to machine learning and the practical application of these methods to various problems by our team in the last two decades. We discuss in more detail the research on concepts of the nonlinear time series analysis and extreme events theory and their applications to natural, economic and social systems, statistical analysis of flows in channels of networks and the methodology for solving nonlinear differential equations.

Keywords Machine learning · Time series analysis

1 Introduction

About 20 years ago a small research group was formed at the Institute of Mechanics of Bulgarian Academy of Sciences with a goal to start research on the mathematical methodology of data analysis, time series analysis and application of the methods from these scientific fields for analysis of time series from fluid mechanic systems, biomechanical, biological, economic and other systems. In the course of the time the members of the group (the size of the group began slowly to increase) became interested in the new research areas that have been close to areas of studies of the group. One of these research areas was the area of machine learning problems [19, 28, 45]. Our interest was based on the fact that many of the algorithms and statistical models that computer systems use to perform a specific task without using explicit instructions are the same as the algorithms and the models used in our studies, e.g., the methods of computational statistics and time series analysis and especially this part

N. K. Vitanov (✉)
Institute of Mechanics, Bulgarian Academy of Sciences, Akad. G. Bonchev 4, 1113 Sofia, Bulgaria
e-mail: vitanov@imbm.bas.bg

© The Author(s), under exclusive license to Springer Nature Switzerland AG 2021
K. T. Atanassov (ed.), *Research in Computer Science in the Bulgarian Academy of Sciences*, Studies in Computational Intelligence 934,
https://doi.org/10.1007/978-3-030-72284-5_17

of the time series analysis which focuses on making predictions using computers. Another interesting area was the data mining and especially this part of the data mining that focuses on exploratory data analysis.

Below we describe results from our studies on the methodology and application of this methodology to practical problems from the following areas:

1. Nonlinear time series analysis with application to population systems
2. Application of the methods of the time series analysis to time series from natural, biological, technical and social systems
3. Statistical analysis of extreme events
4. Time series analysis of time series from economic systems
5. Data analysis for evaluation of research production
6. Statistical analysis of networks models of social systems
7. Methodology for obtaining of exact solutions of nonlinear partial differential equations.

2 Nonlinear Time Series Analysis and Population Dynamics

Our research on the methodology connected to the machine learning was initiated when we started a project for application of the methodology of the nonlinear time series analysis [21, 23] to the nonlinear dynamics connected to interaction of three populations in presence of possibility for adaptation. In more detail a model was formulated for the evolution of interacting populations in presence of possibility for adaptation. This adaptation can influence the values of the growth rates and the values of the coefficients of interaction among the populations. The model leaded to interesting chaotic attractors mostly connected to the chaos of Shilnikov kind (in addition interesting attractors have been obtained also for fluid dynamics systems—Fig. 1). The characteristics of these attractors have been studied by the methods of the nonlinear time series analysis. The model has been applied not only to the problems of population dynamics but also to some problems from the area of social dynamics (study of competition among ideologies).

The first version of the above-mentioned model is formulated in [13] where the nonlinear dynamics of a system of populations competing for the same limited resource is studied assuming that the populations can adapt their growth rates and competition coefficients with respect to the number of individuals of each population. This adaptation leads to new orbits in the phase space of the system, completely dependent on the adaptation parameters, and in addition the adaptation parameters influence the dynamics of the chaotic attractor connected to the model system of equations. The last fact is demonstrated by calculation of the power spectra of time series corresponding to chaotic motion in the phase space. The study is continued in [14] where regions of parameter space of the model system are considered in which chaotic motion of the Shilnikov kind exists. Transition to chaos by period-doubling bifurcations is observed as well as windows of periodic motion between the regions

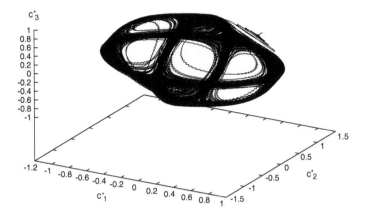

Fig. 1 An example of a chaotic attractor obtained in our studies [10]. The parameters of such attractors are studied by the methodology of the nonlinear time series analysis which is much interesting also for the researchers studying machine learning problems

of chaotic motion and a region of transient chaos after the last window of periodic motion. The obtained Lyapunov dimension for the chaotic attractors is very close to 2 and the Lyapunov spectrum has a structure which allows a topological analysis of the attractors of the investigated system. The research question in [15] is how the adaptation of the competition coefficients of the competing populations for the same limited resource influences the system dynamics in the regions of the parameter space, where chaotic motion of Shilnikov kind exists. The picture of the complex system dynamics is verified by power spectra, histograms and autocorrelations and the Lyapunov exponents and the Kaplan–Yorke dimension are calculated. In [16] a kind of competition possible in a system of at least three populations competing for the same limited resource is investigated for the case where the growth rates and competition coefficients of populations depend on the number of members of all populations. Because of the nonconstant values of the last quantities the system could be repelled from the state of cyclic pairwise competition described by May and Leonard [37]. The competition in a chaotic regime of evolution of the number of members of populations is studied and the result is that the nonconstant competition coefficients can lead to a regularization of the time intervals of domination of each population and the non-constant growth rates can lead to decreasing length of the time intervals of domination as well as to chaos in the occurrence of these intervals. By means of the wavelet transform modulus maxima method the $t(q)$-spectrum and the Hölder exponent are calculated for the time series of the quantity characterizing the time intervals between the successive maxima of the number of the populations individuals. The results of the theory are illustrated by an discussion on qualitative aspects of the dynamics of change of preferences of voters.

In [62] the dynamics of a system of three interacting populations is studied in presence of extinction and substitution: each population whose number of individuals drops under some threshold value becomes extinct, and it is substituted by another population with different fitness and different coefficients of interaction with the other populations. The focus of the study is on the influence of extinction on the system states, which in the absence of extinction can be fixed points, limit cycles or chaotic attractors of Shilnikov kind. The extinction can destabilize each of these states and two possible kinds of evolution are observed in the destabilized system: it can remain forever in the trap of extinction, i.e., the extinctions and substitutions of populations continue for indefinitely long time or it can avoid the trap of extinction by means of the substitution, i.e., the fitness and the coefficients of the interactions between the species move the system attractor away from the zone of the threshold values, the extinction stops, and the system settles on a new attractor. The obtained results are discussed from the point of view of products competing for the preference of buyers that can change their opinion. Another study of the characteristics of the chaos of the Shilnikov kind is presented in [17] where the continuous version of the extended Abramson model from the population dynamics is studied and the fixed points are identified that are responsible for the development of the chaos of Shilnikov kind.

The application of the model and the methodology of the time series analysis continued in [72] where a model for ideological struggles is formulated. In this model several ideologies compete to increase their number of adepts. Such followers can be either converted from one ideology to another or become followers of an ideology though being previously ideologically-free. A reverse process is also allowed. Two kinds of conversions are considered: unitary conversion, e.g. by means of mass communication tools, or binary conversion, e.g. by means of interactions between people. It is obtained that if a the steady state, it depends on the number of ideologies. When the number of ideologies increases some tension arises between them. This tension can change in the course of time. The ideology tensions are measured through an appropriately defined scale index. Finally it is shown that a slight change in the conditions of the environment can prevent the extinction of some ideology; after almost collapsing the ideology can spread again and can affect a significant part of the country's population. Two kinds of such resurrection effects are described as phoenix effects. The study is continued in [75] and discrete time as well as population migration are taken into an account. The complex dynamics of the obtained attracting manifolds was investigated. Conversion from one ideology to another by means of mass media influence and interpersonal relations is considered. Different birth rates are assumed for different ideologies, the rate being assumed to be positive for the reference population, made of initially non-believers. Ideological competition can happen in one or several regions in space. In the latter case, migration of non-believers and adepts is allowed.

In [78] the model of a system of interacting populations is extended to the case when the environment influences growth rates and this influence can be modeled by a Gaussian white noise. The system of model equations for this case is a system of stochastic differential equations with deterministic part in the form of polynomial nonlinearities and state-dependent stochastic part in the form of multiplicative Gaussian white noise. Two cases are discussed where the formal integration of the stochastic differential equations leads to integrals of Ito kind or to integrals of Stratonovich kind. The systems of stochastic differential equations are reduced to the corresponding Fokker–Planck equations. For the Ito case and for the case of 1 population analytic results are obtained for the stationary probability density function of the population density. For the case of more than one population and for both the Ito case and Stratonovich case the detailed balance conditions are not satisfied. As a result the exact analytic solutions of the corresponding Fokker–Planck equations for the stationary probability density functions for the population densities are not known and we obtain approximate solutions for this case by the method of adiabatic elimination.

3 Analysis of Time Series from Natural, Biological, Technical and Social Systems

The methodology connected to data mining and time series analysis has been applied to various systems of interest for the experimental science and for the practice. Below we list several examples.

In [59] experimentally obtained time series are studied from an electric circuit which deterministic part models the signal of the Shinriki oscillator. The stochastic part of the circuit adds δ-correlated Gaussian white noise to the voltage of the circuit. The simplest unstable periodic orbits are extracted for the chaotic attractor of the circuit without influence of the noise. In [6] surface-tension-driven Benard convection at zero Prandtl number is considered. This 3D convection is computed for the smallest possible doubly periodic rectangular domain that is compatible with the hexagonal flow structure at the linear stability threshold of the quiescent state. Upon increasing the Marangoni number beyond this threshold, the initially stationary flow becomes quickly time dependent and a the transition to chaos occurs. A period-doubling scenario is observed for the transition to chaos of the energy attractor. A quantity from the time series analysis: the point correlation dimension PD_2 is used for characterization of nonstationary time series in [60]. PD_2 is calculated for artificially obtained time series and ECG time series for heart activity of *Drosophila melanogaster* with normal or genetically defect heart. The result is that the point correlation dimension in combination with the classical methods of the time series analysis can lead to effective early detection of irregularities of the heart activity of *Drosophila*. This property can be useful for early detection of heart defects of different animals or humans.

An application of the methodology from the area of time series analysis is made in [61] where upper bounds are obtained on the number of degrees of freedom significant for description of oscillations of moving on rough roads machines such as cars, trucks or agricultural machines. The upper bounds can justify or reject the application of low-dimensional mathematical models of the dynamics of the observed machine. For the investigated stationary and nonstationary time series an upper bound is given by the statistical dimension calculated on the basis of the singular spectrum analysis of the time series. For stationary time series the bound can be lowered by calculation of the correlation dimension after a noise reduction by use of the largest principal components of the time series.

An application of the methods for analysis of time series from system with multifractality is presented in [63] where by means of the multifractal detrended fluctuation analysis (MFDFA) long-range correlations in the interbeat time series of heart activity of Drosophila melanogaster are studied. Our main investigation tool are the fractal spectra $f(\alpha)$ and $h(q)$ by means of which the correlation properties of Drosophila heartbeat dynamics are traced for three consequent generations of species. It is observed that opposite to the case of humans the time series of the heartbeat activity of healthy Drosophila do not have scaling properties. The methods for the analysis of fractal properties are applied also in [64] for detection the presence or absence of long-range correlations in pseudo-random sequences obtained by means of computer programs for random number generation. The application of the methodology for study of time series from chaotic regimes in complex system is continued in [44] where two families of Lorenz-like three-dimensional nonlinear dynamical systems are studied. The Lyapunov exponents and Lyapunov dimension for the chaotic attractors are calculated in order to study the influence of the parameters of the Lorenz system on the attractors obtained for various values of the parameters of the model equations.

In [66] the combination of singular spectrum analysis (SSA), principal component analysis (PCA), autocorrelation function analysis (ACFA) and the time delay phase space construction (TDPSC) is studied as a combined tool for analysis of short and nonstationary time series. The opportunities of this bundle of four methods are presented for analysis on the basis of short and nonstationary time series for the piglet prices and production in Japan before and after the Japan government intervention in the agriculture sector aiming at stabilization of the agriculture prices after the oil crisis in 1974. The studied combination of four methods leads us to enough information to make the conclusion that the intervention of the Japan government in agriculture sector was very successful and leaded to stabilization of prices, to a coupling between the prices and production cycles and to decreasing the dimension of the phase space of price and production fluctuations around the year trend thus making their dynamics more forecastable. In [80] another combination of methods is used namely singular spectrum analysis (SSA), principal component analysis (PCA), and multi-fractal detrended fluctuation analysis (MFDFA), for investigating characteristics of vibration time series data from a friction brake. SSA and PCA are used to study the long time-scale characteristics of the time series. MFDFA is applied for investigating all time scales up to the smallest recorded one. It turns out that the majority of the long time-scale dynamics, that is presumably dominated by the structural dynamics of the brake system, is dominated by very few active dimensions only and can well be understood in terms of low dimensional chaotic attractors.

In [2] influence of sex, social dominance, and context on motion-tracked head movements during dyadic conversations is studied. Windowed cross-correlation analyses found high peak correlation between conversants' head movements over short intervals and a high degree of nonstationarity. Nonstationarity in head movements is found to be positively related to the number of men in a conversation. Surrogate data analysis offsetting the conversants' time series by a large lag was unable to reject the null hypothesis that the observed high peak correlations were unrelated to short-term coordination between conversants. The results are consistent with a view that symmetry is formed between conversants over short intervals and that this symmetry is broken at longer, irregular intervals.

Another interesting application of the methodology connected to the machine learning is made in two articles devoted to analysis of data from sport (football) events. In [4] the question is studied whether some implied regularity or structure, as found in the soccer team ranking by the Union of European Football Associations (UEFA), is due to an implicit game result value or score competition conditions. The analysis is based on considerations of complex systems, i.e. finding whether power or other simple law fits are appropriate to describe some internal dynamics. It is observed that the ranking is specifically organized: a major class comprising a few teams emerges after each season. Other classes, which apparently have regular sizes, occur subsequently. Thus, the notion of the Sheppard primacy index is envisaged to describe the findings. Additional primacy indices are discussed for enhancing the features. In [3] we show that the ranking of agents competing with each other in complex systems may lead to paradoxes according to the pre-chosen different measures. A discussion is presented on such rank–rank correlations based on the case of European countries ranked by UEFA and FIFA from different soccer competitions. The conclusion is that the power law form is not the best description contrary to many modern expectations and the stretched exponential law form is much more adequate.

Two classical hypotheses about the population growth in a system of cities are examined in [82]. Hypothesis 1 pertains to Gibrat's and Zipf's theory which states that the city growth–decay process is size independent. Hypothesis 2 pertains to the Yule process which states that the growth of populations in cities happens when (i) the distribution of the city population initial size obeys a log-normal function, (ii) the growth of the settlements follows a stochastic process. The basis for the test is official data on Bulgarian cities at various times. The result is that the population size growth of the Bulgarian cities is size dependent and the population size growth of these cities can be described by a double Pareto log-normal distribution, whence Hypothesis 2 is valid for the Bulgarian city system.

4   Statistical Analysis of Extreme Events

The extreme events in the boundary layer of the atmosphere (e.g., the wind gusts) have ignited our interest in the methodology for prediction of characteristics of time series. This methodology is closely connected to the machine learning. Some of our first results on statistical analysis and prediction of wind gust are presented in [25]. The main methodology for prediction of structures in the turbulent boundary layer of the Earth is presented in [24] where a continuous state Markov chain of suitable order is employed to approximate the dynamics of surface wind speeds recorded at a single site. Using past observations, the model yields probabilistic forecasts of the future. One application of this Markov chain model is for the prediction of turbulent gusts. More results on the model are presented in [26]. In [27] different schemes for the short time (few seconds) prediction of local wind speeds are compared in terms of their performance. Special emphasis is made on the prediction of turbulent gusts, where data driven continuous state Markov chains turn out to be quite successful. Taking into account correlations of several measurement positions in space enhances the predictability. An interesting result is that stronger wind gusts possess a better predictability. Another kind of extreme events (called rogue or freak waves) are studied in [67]. The rogue waves are very large sea waves and because of this they are extremely dangerous events that can lead to loss of cargo and human lives. The very high wave recorded at the North Sea based Draupner oil platforms has increased the scientific interest in the research on the rogue waves. We have applied the methodology of the statistical analysis of extreme events for probability of arising of extremely high waves in some regions of North Sea. Time series are analyzed that contain records of rogue waves and show that the distribution of the maximum height of the waves is well described by the Weibull distribution. On the basis of the obtained distribution the probability of the Draupner wave at corresponding weather conditions is calculated to be $2.3 \cdot 10^{-5}$.

5 Time Series Analysis and Economic Systems

The methodology of time series analysis connected to machine learning and artificial intelligence is much used for analysis of economic system and for forecasting their behavior [5]. We applied such kind of methodology in several articles. In [46] we studied the transition of dynamics observed in an actual real agricultural economic dataset from Japan economy after the intervention of the Japan government to stabilize the agriculture prizes after the oil crises in 1970s. Lyapunov spectrum analysis is conducted on the data to distinguish deterministic chaos and the limit cycle. Chaotic and periodic oscillation are identified before and after the second oil crisis, respectively. The studied time series is stationary and this is shown on the basis of the corresponding recurrence plots. Thus the government intervention might reduce market instability by removing a chaotic market's long-term unpredictability. In [65] an economic system is studied in which one large agent—the Japan government changes the environment of functioning of numerous smaller agents—the Japan agriculture producers by indirect regulation of prices of agriculture goods. By means of analysis of correlations and a combination of singular spectrum analysis (SSA), principal component analysis (PCA), and time delay phase space construction (TDPSC) the influence of the government measures on the domestic piglet prices and production in Japan was studied. The conclusion is that the government regulation politics was successful and lead to a decrease of the nonstationarities and to increase of predictability of the piglet price, to a coupling of the price and production cycles, and to increase of determinism of the dynamics of the fluctuations of piglet price around the year average price.

6 Data Analysis for Evaluation of Research Production

The assessment of research production is a modern research field with large practical significance as the funding organizations try to optimize research funding and in the same time to increase the effectiveness of the research organizations [1, 9, 38, 39, 47]. The models used to study the evolution of research organizations and the location and time behaviors of the research production are very close as concept to the models of population dynamics and especially to the models of epidemics that are very suitable for description of the spreading of research ideas. These models are discussed in [74]. Much larger study of the models, indicators and indexes for research assessment is presented in the book [53] which consists of three parts. The first part is devoted to a brief introduction to the complexity of science and to some of its features, e.g., the triple helix model of a knowledge-based economy is described. The importance of scientometrics and bibliometrics is emphasized, and a mathematical model for quantification of research performance is described. The second part contains a discussion on the indicators and indexes of research production of individual researchers and groups of researchers. If one has to evaluate the research work of collectives of researchers from some department or institute, then one may need additional (to the peer review) methodology, such as a methodology for analysis of citations and publications. The building blocks of such methodology as well as selected indicators and indexes are described, and many examples for the calculation of corresponding indexes are presented. Special attention is devoted to the Lorenz curve and to the definition of sizes of different scientific elites on the basis of this curve. The third part is devoted to the statistical laws and mathematical models connected to research organizations, and the focus is on the models of research production connected to the units of information (such as research publication) and to units of importance of this information (such as citations of research publications). Special attention is devoted to the application of statistical distributions (such as the Yule distribution, Waring distribution, negative binomial distribution, etc.) to modeling features connected to the dynamics of research publications and their citations.

In more detail we discuss in [54] quality of research and research performance as well as measurement of quality and latent variables by sets of indicators. The importance of the non-Gaussianity of many statistical characteristics of social processes is stressed, because non-Gaussianity is connected to important requirements for study of these processes such as the need for multifactor analysis or probabilistic modeling. The sets of quantities that are used in scientometrics are mentioned, and in addition, the importance of understanding the inequality of scientific achievements and the usefulness of knowledge landscapes for understanding and evaluating research performance is stressed. In [55] selected indicators and indexes are discussed (about 45 indexes and indicators constructed on the basis of research publications and/or on the basis of a set of citations of these publications). Among them are the h-index of Hirsch, its variants, and indexes complementary to the h-index; the g-index of Egghe; the i_n-indexes; the m-index; p-index, A-index, R-index, etc. In addition, a short list of indexes for quantitative characterization of research networks and their dynamics is given. In addition to indexes from [55], other indexes useful for assessment of production of groups of researchers may be used. About ninety such indexes are discussed in [56]. The indexes are grouped in the following classes: simple indexes; indexes for deviation from simple tendency; indexes for difference; indexes for concentration, dissimilarity, coherence, and diversity; indexes for advantage and inequality; indexes for stratified data; indexes for imbalance and fragmentation; indexes based on the concept of entropy; Lorenz curve and associated indexes. Several statistical laws that are important for understanding characteristics of research production and for its assessment are presented in [57]. The laws of Lotka, Pareto, Zipf, Zipf–Mandelbrot, and Bradford are discussed from the point of view of their application for description of different aspects of scientific production. In addition to the discussion of statistical laws, we discuss two important effects: the concentration–dispersion effect (which reflects the separation of the researchers into a small group of highly productive ones and a large group of researchers with limited productivity) and the Matthew effect in science (which reflects the larger attention to the research production of the highly ranked researchers). In addition, we mention the invitation paradox (many papers accepted in highly ranked journals are not cited as much as expected) and the Ortega hypothesis (the big discoveries in science are supported by the everyday hard work of ordinary researchers). In [58] selected deterministic and probability models of research dynamics are discussed. From the class of deterministic models we discuss models connected to research publications (SI-model, Goffmann–Newill model, model of Price for growth of knowledge), deterministic model connected to dynamics of citations (nucleation model of growth dynamics of citations), deterministic models connected to research dynamics (logistic curve models, model of competition between systems of ideas, etc. From the class of probability models we discuss a probability model connected to research publications (based on the Yule process), probability models connected to dynamics of citations (Poisson and mixed Poisson models, models of aging of scientific information (death stochastic process model and birth stochastic process model connected to Waring distribution)). The truncated Waring distribution and the multivariate Waring distribution are described. Several probability models of production/citation process (Paretian and Poisson distribution models of the h-index) as well as GIGP model distribution of bibliometric data are presented. A stochastic model of scientific productivity based on a master equation is described, and a probability model for the importance of the human factor in science is discussed.

7 Statistical Analysis of Networks Models of Social Systems

Machine learning methodology is closely connected to the methodology for the study of evolution of networks [8, 48]. Especially interesting for the machine learning is the methodology for the statistical analysis of various kinds of networks. First results from our research on the statistical analysis of networks are presented in [83] where a mathematical model of migration channel based on the truncated Waring distribution is discussed. The model is applied for case of migrants moving through a channel consisting of finite number of countries or cities. It is shown that if the final destination country is very popular then large percentage of migrants may concentrate there. The motion of a substance in a more complicated channel of a network is studied in [85]. This channel has 3 arms and stationary regime of the flow of the substance in the channel is discussed. The obtained statistical distributions for the amount of substance in the nodes of the channel contain as particular case famous long-tail distributions such as Waring distribution, Yule-Simon distribution and Zipf distribution.

A discrete-time model for motion of substance in a channel of a network is discussed in [86]. For the case of stationary motion of the substance and for the case of time-independent values of the parameters of the model we obtain a new class of statistical distributions that describe the distribution of the substance along the nodes of the channel. The case of interaction between a kind of substance specific for a node of the network and another kind of substance that is leaked from the channel is studied in presence of possibility for conversion between the two substances. This study is continued in [87] where different "leakage" is assumed, i.e., some amount of the substance can leave the channel at a node and the rate of leaving can be different for the different nodes of the channel. The main outcome of this study is the distribution of the substance along the nodes of the channel. This distribution has a very long tail. The studied model is applied to the case of migration of humans through a migration channel consisting of chain of countries. In this case the model accounts for the number of migrants entering the channel through the first country of the channel; permeability of the borders between the countries; possible large attractiveness of some countries of the channel; possibility for migrants to obtain permission to reside in a country of the channel. The main outcome of the model is the distribution of migrants along the countries of the channel.

In [89] the problem of the stationary motion of substance is studied for the case of a channel of a network having two arms. Analytical relationships for the distribution of the substance in the nodes of the arms of the channel are obtained. The obtained results are discussed from the point of view of technological applications of the model (e.g., motion of substances such as water in complex technological facilities). The study is continued in [7] for the case of an additional split of the secondary arm of the channel. Analytical relationships for the distribution of the substance in the nodes of the channel are obtained for the case of stationary regime of the motion of the substance in the arms of the channel. The obtained results are discussed from the point of view of application of the model to migration dynamics.

8 Exact Solutions of Nonlinear Partial Differential Equations

Models based on differential equations are much used in many areas of technology. Some of these models, e.g., the models connected with synchronization and predic-

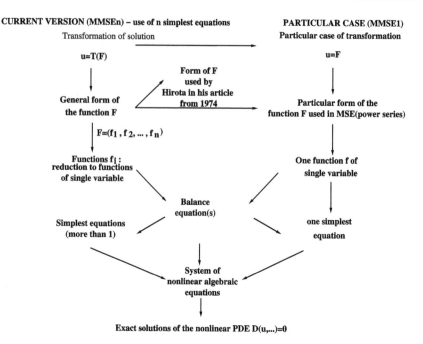

Fig. 2 Schema for application of the modified method of simplest equation based on the use of *n* simplest equations—MMSEn. The steps of the application of the methodology are explained in the text. The use of more than one simplest equations allows for obtaining bisoliton, trisoliton, etc. solutions of the studied nonlinear partial differential equation (if such solutions do exist)

tion of behavior of complex systems are of large interest for the area of machine learning. Since many years we work on methodologies for solving differential equations and especially on the methodology for obtaining exact solutions of nonlinear partial differential equations. The name of this methodology is Modified method of simplest equations and we shall discuss briefly below the last version of the methodology based on the possibility of use of more than one simplest equation and the results of our research over the years. Let us note that the Modified method of simplest equation based on one simplest equation is a version of the Method of simplest equation [31, 32] (the interested reader can obtain more information about this methodology from the numerous articles of Kudryashov after 2008).

The last modification of the modified method of simplest equation is connected to the possibility of use of more than one simplest equation that was applied in [88]. This modification (we shall call it MMSEn Modified method of Simplest Equation with *n* simplest equations—Fig. 2) is as follows.

Let us consider a nonlinear partial differential equation

$$\mathcal{D}(u, \dots) = 0 \tag{1}$$

where $\mathcal{D}(u, \dots)$ depends on the function $u(x, \dots, t)$ and some of its derivatives participate in (u can be a function of more than 1 spatial coordinate). The 7 steps of the methodology of MMSEn follow below.

1. We perform a transformation

$$u(x, \dots, t) = T(F(x, \dots, t)) \tag{2}$$

where $T(F)$ is some function of another function F. In general $F(x, \dots, t)$ is a function of the spatial variables as well as of the time. The transformation $T(F)$ can be the Painleve expansion [20, 29, 30] or another transformation, e.g., $u(x, t) = 4 \tan^{-1}[F(x, t)]$ for the case of the sine–Gordon equation or $u(x, t) = 4 \tanh^{-1}[F(x, t)]$ for the case of sh-Gordon (Poisson-Boltzmann equation) (for applications of the last two transformations see, e.g. [33–36]). In many particular cases one may skip this step (then we have just $u(x, \dots, t) = F(x, \dots, t)$) but in numerous cases we need this step for obtaining a solution of the studied nonlinear PDE. The application of Eqs. (2) to (1) leads to a nonlinear PDE for the function $F(x, \dots, t)$.

2. The function $F(x, \dots, t)$ is constructed as a function of other functions $f_1, \dots, f_N$ that are connected to solutions of some differential equations (these equations can be partial or ordinary differential equations) that are more simple than Eq. (1). We note that the possible values of N are $N = 1, 2, \dots$ (there may be infinite number of functions f too). Numerous forms of the function $F(f_1, \dots, f_N)$ are possible. One example is

$$F = \alpha + \sum_{i_1=1}^{N} \beta_{i_1} f_{i_1} + \sum_{i_1=1}^{N} \sum_{i_2=1}^{N} \gamma_{i_1, i_2} f_{i_1} f_{i_2} + \cdots +$$

$$\sum_{i_1=1}^{N} \cdots \sum_{i_N=1}^{N} \sigma_{i_1, \dots, i_N} f_{i_1} \cdots f_{i_N} \tag{3}$$

where α, β_{i_1}, γ_{i_1, i_2}, $\sigma_{i_1, \dots, i_N}$ $\dots$ are parameters. Note that the relationship (3) contains as particular case the relationship used by Hirota [20]. The power series $\sum_{i=0}^{N} \mu_n f^n$ (where μ is a parameter) used in the previous versions of the methodology of the modified method of simplest equation are a particular case of the relationship (3) too.

3. In general the functions $f_1, \dots, f_N$ are solutions of partial differential equations. By means of appropriate ansätze (e.g., traveling-wave ansätze such as $\xi = \hat{\alpha}x + \hat{\beta}t$; $\zeta = \hat{\gamma}x + \hat{\delta}t$, $\eta = \hat{\mu}y + \hat{\nu}t \dots$) the solved differential equations for $f_1, \dots, f_N$ may be reduced to differential equations H_l, containing derivatives of one or several functions

$$H_l\left[a(\xi), a_\xi, a_{\xi\xi}, \dots, b(\zeta), b_\zeta, b_{\zeta\zeta}, \dots\right] = 0; \quad l = 1, \dots, N \tag{4}$$

In many cases (e.g., if the equations for the functions $f_1, \ldots$ are ordinary differential equations) one may skip this step but the step may be necessary if the equations for $f_1, \ldots$ are partial differential equations.

4. We assume that the functions $a(\xi)$, $b(\zeta)$, etc., are functions of other functions, e.g., $v(\xi)$, $w(\zeta)$, etc., i.e.

$$a(\xi) = A[v(\xi)]; \quad b(\zeta) = B[w(\zeta)]; \ldots \tag{5}$$

Note that the kinds of the functions $A, B, \ldots$ are not prescribed. Often one uses a finite-series relationship, e.g.,

$$a(\xi) = \sum_{\mu_1=-\nu_1}^{\nu_2} q_{\mu_1} [v(\xi)]^{\mu_1}; \quad b(\zeta) = \sum_{\mu_2=-\nu_3}^{\nu_4} r_{\mu_2} [w(\zeta)]^{\mu_2}, \ldots \tag{6}$$

where $q_{\mu_1}, r_{\mu_2}, \ldots$ are coefficients. However other kinds of relationships may be used too.

5. The functions $v(\xi), w(\zeta), \ldots$ are solutions of simpler ordinary differential equations called *simplest equations. For several years the methodology of the modified method of simplest equation was based on use of one simplest equation. The discussed here version of the methodology allows for the use of more than one simplest equation.*

6. The application of the steps 1.–5., to Eq. (1) transforms its left-hand side. Let the result of this transformation be a function that is a sum of terms where each term contains some function multiplied by a coefficient. This coefficient contains some of the parameters of the solved equation and some of the parameters of the solution. In the most cases a balance procedure must be applied in order to ensure that the above-mentioned relationships for the coefficients contain more than one term. This balance procedure may lead to one or more additional relationships among the parameters of the solved equation and parameters of the solution. The last relationships are called *balance equations.*

7. We may obtain a nontrivial solution of Eq. (1) if all coefficients mentioned in Step 6., are set to 0. This condition usually leads to a system of nonlinear algebraic equations for the coefficients of the solved nonlinear PDE and for the coefficients of the solution. Any nontrivial solution of this algebraic system leads to a solution the studied nonlinear partial differential equation. Usually the above system of algebraic equations contains many equations that have to be solved with the help of a computer algebra system.

We have started our work on a methodology for obtaining exact solutions of nonlinear partial differential equations almost 30 years ago. In [33] we have discussed a connection between the structures described by the solutions of the two dimensional Poisson-Boltzmann equation and the solutions of the sine-Gordon equation representing standing waves has been established. In [34] by a substitution that leads to two equations for the Jacobi elliptic functions we obtained traveling-wave solutions of (2+1)-dimensional sine-Gordon equation. This research was continued in [35] and in [36] we studied a class of traveling-wave solutions of the (2+1)-dimensional sine-Gordon equation. The parameters of the obtained waves strongly depend on the wave amplitude and there exist forbidden areas for the wavenumber and frequency. Exact traveling-wave solutions of the (2+1)-dimensional sine–Gordon equation possessing a velocity smaller than the velocity of the linear waves in the correspondent model system are obtained in [49]. The dependence of their dispersion relations and allowed areas for the wave parameters on the wave amplitude are discussed. The obtained waves contain as particular cases static structures consisting of elementary cells with zero topological charge. The obtained structures require minimal spatial system sizes for their existence.

A system of nonlinear partial differential equations is discussed in [68] as a model of the spatial-time interaction among populations which reproduction and intensity of interaction depend on their spatial density. For the particular case of two populations with constant growth rates and competition coefficients analytical solutions of the model system are obtained. These solutions describe nonlinear waves of kink kind and these kinks are coupled. The kinks are connected to propagation of the deviations from the stationary densities corresponding to fixed points in the phase space of the population densities. The study is continued in [69] where of interest are the waves caused by migration of the populations. The assumption is that the migration is a diffusion process influenced by the changing values of the birth rates and coefficients of interaction among the populations. For the particular case of one population and one spatial dimension the general model is reduced to analytically tractable PDE with polynomial nonlinearity up to 4th order. Two kinds of solutions of this equation are obtained: approximate solution for small value of the ratio between the coefficient of diffusion and the wave velocity and exact solutions which describe nonlinear kink and solitary waves.

The basis of the study in [70] is the modified method of simplest equation with use of one simplest equation (MMSE1). The equations of Bernoulli, Riccati and the extended tanh-function equation are used as simplest equations. The obtained methodological results are illustrated by exact solutions of versions of the generalized Kuramoto–Sivashinsky equation, reaction–diffusion equation with density-dependent diffusion, and the reaction-telegraph equation.

In [50] a search for traveling-wave solutions of the class of equations

$$\sum_{p=1}^{N_1} \alpha_p \frac{\partial^p Q}{\partial t^p} + \sum_{q=1}^{N_2} \beta_q \frac{\partial^q Q}{\partial x^q} + \sum_{m=1}^{M} \mu_m Q^m = 0 \tag{7}$$

was made. Above α_p, β_q and μ_m are parameters. The MMSE1 version of the modified method of simplest equation is applied. As simplest equations the equation of Bernoulli or the equation of Riccati are used. This research was continued in [71] where the following two classes of nonlinear partial differential equations are investigated:

I. Class of reaction-diffusion equations

$$\frac{\partial Q}{\partial t} + \frac{dD}{dQ}\left(\frac{\partial Q}{\partial x}\right)^2 + D(Q)\frac{\partial^2 Q}{\partial x^2} + F(Q) = 0, \tag{8}$$

II. Class of reaction-telegraph equations

$$\frac{\partial Q}{\partial t} - \alpha\frac{\partial^2 Q}{\partial t^2} - \beta\frac{\partial^2 Q}{\partial x^2} - \gamma\frac{dF}{qQ}\frac{\partial Q}{\partial t} - F(Q) = 0, \tag{9}$$

where a, b, c are parameters and D and F depend on the population density Q. Exact solutions of these equations are obtained by the version MMSE1 of the modified method of simplest equation for the cases when the simplest equation is the equation of Bernoulli or the equation of Riccati.

In [51] the class of equations

$$\sum_{i,j=0}^{m} A_{ij}(u)\frac{\partial^i u}{\partial t^i}\left(\frac{\partial u}{\partial t}\right)^j + \sum_{k,l=0}^{n} B_{ij}(u)\frac{\partial^k u}{\partial x^k}\left(\frac{\partial u}{\partial x}\right)^l = C(u) \tag{10}$$

is studied where $A_{ij}(u)$, $B_{ij}(u)$ and $C(u)$ are functions of u(x, t) as follows: A_{ij}, B_{ij} and C are polynomials of u or A_{ij}, B_{ij} and C can be reduced to polynomials of u by means of Taylor series for small values of u. The version MMSE1 of the modified method of simplest equation leads to traveling-wave solutions of the investigated class of equations are obtained. The methodology is illustrated by obtaining exact traveling-wave solutions of the Swift–Hohenberg equation and of the generalized Rayleigh equation for the cases when the extended tanh-equation or the equations of Bernoulli and Riccati are used as simplest equations.

In [73] the class of studied equations is extended to

$$\sum_{i_1=0}^{\bar{n}_1}\sum_{i_2=0}^{n_1^*}\sum_{j=0}^{n_2}\sum_{k_1=0}^{\bar{n}_3}\sum_{k_2=0}^{n_3^*}\sum_{l=0}^{n_4}\sum_{p_1=0}^{\bar{n}_5}\sum_{p_2=0}^{n_5^*}\sum_{q=0}^{n_6}\sum_{r_1=0}^{\bar{n}_7}\sum_{r_2=0}^{n_7^*}\sum_{s=0}^{n_8}\left(\frac{\partial^{i_1+i_2}u}{\partial x^{i_1}\partial t^{i_2}}\right)^j \cdot \left(\frac{\partial^{k_1+k_2}u}{\partial x^{k_1}\partial t^{k_2}}\right)^j \cdot$$

$$\left(\frac{\partial^{p_1+p_2}u}{\partial x^{p_1}\partial t^{p_2}}\right)^j \cdot \left(\frac{\partial^{r_1+r_2}u}{\partial x^{r_1}\partial t^{r_2}}\right)^j \cdot A_{i_1,i_2,j,k_1,k_2,l,p_1,p_2,q,r_1,r_2,s}(u) = G(u)$$

$$\tag{11}$$

where it is assumed that

$$\frac{\partial^0 u}{\partial x^0} = \frac{\partial^0 u}{\partial t^0} = \frac{\partial^0 u}{\partial x^0 \partial t^0} = 0$$

and G(u) and A(u) are polynomials

$$G(u) = \sum_{\epsilon=0}^{\kappa} g_{\epsilon} u^{\epsilon};$$

$$A_{i_1,i_2,j,k_1,k_2,l,p_1,p_2,q,r_1,r_2,s}(u) = \sum_{m=0}^{h_{i_1,i_2,j,k_1,k_2,l,p_1,p_2,q,r_1,r_2,s,m}} a_{i_1,i_2,j,k_1,k_2,l,p_1,p_2,q,r_1,r_2,s,m} u^m$$

(12)

This class of equations is treated by the MMSE1 methodology. As examples for application of the methodology exact traveling-wave solutions of the generalized Degasperis–Processi equation and of the b-equation are obtained. As simplest equations we use the equations of Bernoulli and Riccati. Several methodological remarks about the extension of the exp-function method are presented.

In [52] the role of the simplest equation for the application of the modified method of simplest equation MMSE1 is discussed. The main idea is that each function constructed as polynomial of a solution of a simplest equation is a solution of a class of nonlinear PDEs. As examples three simplest equations are discussed: the equations of Bernoulli and Riccati and the equation for the Jacobi elliptic functions. In [11] the MMSE1 is applied for obtaining exact solutions of the model equations for traveling waves in lattices. The relation between the G'/G-expansion method and Modified method of simplest equation MMSE1 is discussed in [12]. A remark on the integrability of some differential equations with fluid mechanics application in connection with MMSE1 is presented in [18]. In [76] the MMSE1 is applied to the extended Korteweg–de Vries equation and to generalized Camassa–Holm equation. Exact traveling wave solutions of these two nonlinear partial differential equations are obtained for the case when the equations of Bernoulli, Riccati and the extended tanh-equation are used as simplest equations. In [77] MMSE1 is applied for obtaining exact solutions of the model nonlinear partial differential equations for density waves for the case or negligible random fluctuations of the population densities. The MMSE1 version of the modified method of simplest equation is applied further in [79] for obtaining exact solitary traveling-wave solutions of nonlinear partial differential equations that contain monomials of odd and even grade with respect to participating derivatives.

Special interest from the methodological point of view presents [81]. There the ordinary differential equation

$$\left(\frac{d^k g}{d\xi^k}\right)^l = \sum_{j=0}^{m} d_j g^j$$

(13)

is proposed as simplest equation. Above $k = 1, \ldots,$ $l = 1, \ldots,$ and m and d_j are parameters. The solution of Eq.(13) defines a special function that contains as particular cases, e.g.,: (i) trigonometric functions; (ii) hyperbolic functions; (iii) elliptic functions of Jacobi; (iv) elliptic function of Weierstrass. The methodology is illustrated by obtaining solitary wave solutions for the generalized Korteweg–de Vries equation and by obtaining solutions of the higher order Korteweg–de Vries equation.

In [41] propagation of traveling waves in a blood- filled hyper-elastic artery with a local dilatation (an aneurysm) is discussed. A long-wave approximation of the model equations is used and as a result a version of the perturbed Korteweg-de Vries kind equation with variable coefficients is obtained. Exact traveling-wave solutions of this equation are obtained by the MMSE1 version of modified method of simplest equation where the differential equation of Abel is used as a simplest equation. This research is continued in [42] and in [40].

In [84] the MMSE1 is applied for obtaining exact solitary traveling-wave solutions of nonlinear partial differential equations that contain monomials of odd and even grade with respect to participating derivatives. The used simplest equation is

$$f_\xi^2 = n^2(f^2 - f^{(2n+2)/n}) \tag{14}$$

which has the solution $f(\xi) = \frac{1}{\cosh^n(\xi)}$ where $\xi = \alpha x + \beta t$. The classes of studied nonlinear partial differential equations that contain: (i) only monomials of odd grade with respect to participating derivatives; (ii) only monomials of even grade with respect to participating derivatives; (iii) monomials of odd and monomials of even grades with respect to participating derivatives. The obtained solitary wave solution for the case (i) contains as particular cases the solitary wave solutions of Korteweg–deVries equation and of a version of the modified Korteweg–deVries equation. One of the obtained solitary wave solutions for the case (ii) is a solitary wave solution of the classic version of the Boussinesq–type equation. In [22] a class of hyperbolic reaction-diffusion equations is discussed and MMSE1 is applied for obtaining an exact solution of an equation that contains polynomial nonlinearity of fourth order. The equation of Bernoulli is used as a simplest equation. Similar results from point of view of the use of the methodology are presented in [43].

In [88] MMSE2 is applied to the nonlinear Schrödinger equation, i.e., we apply an extension of the methodology of the modified method of simplest equation to the case of use of two simplest equations. It is shown that the methodology works also for other equations of the nonlinear Schrödinger kind.

9   Concluding Remarks

We intend to continue our research in all research areas discussed above. The plan is to focus the future research on the methodology of the nonlinear time series analysis that is of interest for the machine learning. In addition the machine learning methods will be studied that are useful for forecasting behavior of time series. This will be made by means of use of the methodology of the theory of networks as well as by the use of the opportunities supplied by the languages R and PYTHON. Special attention will be given to the applications of machine learning to fluid mechanics problems, e.g., problems connected to obtaining solutions of the complicated nonlinear equations used in the models of fluid flows.

Acknowledgements We acknowledge the partial support by the project BG05M2OP001-1.001-0008 "National Center for Mechatronics and Clean Technologies", funded by the Operating Program "Science and Education for Intelligent Growth" of Republic of Bulgaria.

References

1. Andres, A.: Measuring Academic Research. How to Undertake a Bibliometric Study. Chandos, Oxford (2009)
2. Ashenfelter, K.T., Boker, S.M., Waddell, J.R., Vitanov, N.: Spatiotemporal symmetry and multifractal structure of head movements during dyadic conversation. J. Exp. Psychol. Hum. Percept. Perform. **35**, 1072–1091 (2009)
3. Ausloos, M., Cloots, R., Gadomski, A., Vitanov, N.K.: Ranking structures and rank-rank correlations of countries: the FIFA and UEFA cases. Int. J. Modern Phys. C **25**, 1450060 (2014)
4. Ausloos, M., Gadomski, A., Vitanov, N.K.: Primacy and ranking of UEFA soccer teams from biasing organization rules. Physica Scripta **89** (2014)
5. Basuchoudhary, A., Bang, J.T., Sen, T.: Machine-Learning Techniques in Economics. New Tools for Predicting Economic Growth. Springer, Cham (2017)
6. Boeck, T., Vitanov, N.K.: Low-dimensional chaos in zero-Prandtl-number Benard–Marangoni convection. Phys. Rev. E **65** (2002)
7. Borisov, R., Vitanov, N.K.: Human migration: model of a migration channel with a secondary and a tertiary arm. In: AIP Conference Proceedings, vol. 2075, pp. 150001. AIP Publishing (2019)
8. Dehmer, M., Basac, S.C.: Machine Learning Approaches for Network Analysis. Wiley, New York (2012)
9. Ding, Y., Rousseau, R., Wolfram, D. (eds.): Measuring Scholarly Impact. Springer, Cham (2014)
10. Dimitrova, Z.: Fluctuations and dynamics of the chaotic attractor connected to an instability in heating from below fluid layer. Comptes Rendus de l'Academie bulgare des Sciences **60**(10), 1065–1071 (2007)
11. Dimitrova, Z.: On traveling waves in lattices: the case of Riccati lattices. J. Theor. Appl. Mech. **42**(3), 3–22 (2012)
12. Dimitrova, Z.I.: Relation between the G'/G-expansion method and modified method of simplest equation. Comptes Rendus de l'Academie bulgare des Sciences **65**(11), 1513–1520 (2012)
13. Dimitrova, Z.I., Vitanov, N.K.: Influence of adaptation on the nonlinear dynamics of a system of competing populations. Phys. Lett. A **272**, 368–380 (2000)
14. Dimitrova, Z.I., Vitanov, N.K.: Dynamical consequences of adaptation of the growth rates in a system of three competing populations. J. Phys. A Math. General **34**, 7459–7473 (2001)
15. Dimitrova, Z.I., Vitanov, N.K.: Adaptation and its impact on the dynamics of a system of three competing populations. Physica A **300**, 91–115 (2001)
16. Dimitrova, Z.I., Vitanov, N.K.: Chaotic pairwise competition. Theor. Popul. Biol. **66**, 1–12 (2004)
17. Dimitrova, Z.I., Vitanov, N.K.: Shilnikov chaos in a generalized system for modelling dynamics of competing populations. Comptes Rendus de l'Academie Bulgare des Sciences **58**, 257–264 (2005)
18. Dimitrova, Z.I., Vitanov, K.N.: Integrability of differential equations with fluid mechanics application: from Painleve property to the method of simplest equation. J. Theor. Appl. Mech. **43**(2), 31–42 (2013)
19. Goyal, P., Pandey, S., Jain, K.: Deep Learning for Natural Language Processing. Creating Neural Networks with Python. Apress, Springer, New York (2018)
20. Hirota, R.: Exact solution of Korteweg-de Vries equation for multiple collisions of solitons. Phys. Rev. Lett. **27**, 1192–1194 (1971)

21. Hegger, R., Kantz, H., Schreiber, T.: Practical implementation of nonlinear time series methods: the TISEAN package. CHAOS **9**, 413–435 (1999)
22. Jordanov, I.P., Vitanov N.K.: On the exact travelling wave solutions of a hyperbolic reaction-diffusion equation. In: Georgiev, K., Todorov, M., Georgiev, I. (eds.) Advanced Computing in Industrial Mathematics. BGSIAM 2017, pp. 199–201. Springer, Cham (2019)
23. Kantz, H., Schreiber, T.: Nonlinear Time Series Analysis. Cambridge University Press, Cambridge (2004)
24. Kantz, H., Holstein, D., Ragwitz, M., Vitanov, N.K.: Markov chain model for turbulent wind speed data. Physica A **342**, 315–321 (2004)
25. Kantz, H., Holstein, D., Ragwitz, M., Vitanov, N.K.: Extreme events in surface wind: predicting turbulent gusts. In: AIP Conference Proceedings, vol. 742, pp. 315–324 (2004)
26. Kantz, H., Holstein, D., Ragwitz, M., Vitanov, N.K.: Predicting probability for stochastic processes with local Markov property. In: Peinke, J., Kittel, A., Brath, S., Oberlack, M. (eds.) Progress in Turbulence, pp. 95–98. Springer, Berlin (2005)
27. Kantz, H., Holstein, D., Ragwitz, M., Vitanov, N.K.: Short time prediction of wind speeds from local measurements. In: Peinke, J., Schaumann, P., Barth, S. Wind Energy (eds.) Proceedings of the Euromech Colloqium, pp. 93–98. Springer, Berlin (2007)
28. Konar, A., Bhattacharaya, D.: Time-Series Predictiona and Applications. A Machine Intelligence Approach. Springer, Cham (2017)
29. Kudryashov, N.A.: Exact solutions of the generalized Kuramoto-Sivashinsky equation. Phys. Lett. A **147**, 287–291 (1990)
30. Kudryashov, N.A.: On types of nonlinear nonitegrable equations with exact solutions. Phys. Lett. A **155**, 269–275 (1991)
31. Kudryshov, N.A.: Simplest equation method to look for exact solutions of nonlinear differential equations. Chaos, Solitons Fractals **24**, 1217–1231 (2005)
32. Kudryashov, N.A., Loguinova, N.B.: Extended simplest equation method for nonlinear differential equations. Appl. Math. Comput. **205**, 361–365 (2008)
33. Martinov, N., Vitanov, N.: On the correspondence between the self-consistent 2D Poisson-Boltzmann structures and the sine-Gordon waves. J. Phys. A Math. General **25**, L51–L56 (1992)
34. Martinov, N., Vitanov, N.: On some solutions of the two-dimensional sine-Gordon equation. J. Phys. A Math. General **25**, L419–L426 (1992)
35. Martinov, N., Vitanov, N.: Running wave solutions of the two-dimensional sine-Gordon equation. J. Phys. A Math. General **25**, 3609–3613 (1992)
36. Martinov, N.K., Vitanov, N.K.: New class of running-wave solutions of the (2+1)-dimensional sine-Gordon equation. J. Phys. A Math. General **27**, 4611–4617 (1994)
37. May, R.M., Leonard, W.J.: Nonlinear aspects of competition between three species. SIAM J. Appl. Math. **29**, 243–253 (1975)
38. Moed, H.F., Glänzel, W., Schmoch, U. (eds.): Handbook of Quantitative Science and Technology Research. Springer, Netherlands (2005)
39. Moed, H.: Citation Analysis in Research Evaluation. Springer, Netherlands (2005)
40. Nikolova, E.V.: On nonlinear waves in a blood-filled artery with an aneurysm. In: AIP Conference Proceedings, vol. 1978 (2018)
41. Nikolova, E.V., Jordanov, I.P., Dimitrova, Z.I., Vitanov, N.K.: Evolution of nonlinear waves in a blood-filled artery with an aneurysm. In: AIP Conference Proceedings, vol. 1895 (2017)
42. Nikolova, E.V., Jordanov, I.P., Dimitrova, Z.I., Vitanov, N.K.: Nonlinear evolution equation for propagation of waves in an artery with an aneurysm: an exact solution obtained by the modified method of simplest equation. In: Georgiev, K., Todorov, M., Georgiev, I. (eds.) Advanced Computing in Industrial Mathematics. Studies in Computational Intelligence, vol. 728, pp. 131–144. Springer, Cham (2018)
43. Nikolova, E.V., Serbezov, D.Z., Jordanov, I.P.: On the spatio-temporal dynamics of interacting economic agents: application of the modified method of simplest equation. In: AIP Conference Proceedings, vol. 2075 (2019)

44. Panchev, S., Spassova, T., Vitanov, N.K.: Analytical and numerical investigation of two families of Lorenz-like dynamical systems. Chaos, Solitons Fractals **33**, 1658–1671 (2007)
45. Ramasubramanian, K., Singh, A.: Machine Learning Using R. Springer, New York (2017)
46. Sakai, K., Managi, S., Vitanov, N.K., Demura, K.: Transition of chaotic motion to a limit cycle by intervention of economic policy: an empirical analysis in agriculture. Nonlinear Dyn. Psychol. Life Sci. **11**, 253–265 (2007)
47. Scharnhorst, A., Börner, K., Van den Besselaar, P. (eds.): Models for Science Dynamics. Springer, Berlin (2012)
48. Silva, T.C., Zhao, L.: Machine Learning in Complex Networks. Springer, Cham (2016)
49. Vitanov, N.K.: On travelling waves and double-periodic structures in two-dimensional sine-Gordon systems. J. Phys. A Math. General **29**, 5195–5207 (1996)
50. Vitanov, N.K.: Application of simplest equations of Bernoulli and Riccati kind for obtaining exact traveling-wave solutions for a class of PDEs with polynomial nonlinearity. Commun. Nonlinear Sci. Numer. Simul. **15**, 2050–2060 (2010)
51. Vitanov, N.K.: Modified method of simplest equation: powerful tool for obtaining exact and approximate traveling-wave solutions of nonlinear PDEs. Commun. Nonlinear Sci. Numer. Simul. **16**, 1176–1185 (2011)
52. Vitanov, N.K.: On modified method of simplest equation for obtaining exact and approximate solutions of nonlinear PDEs: the role of the simplest equation. Commun. Nonlinear Sci. Numer. Simul. **16**, 4215–4231 (2011)
53. Vitanov, N.K.: Science Dynamics and Research Production. Springer, Cham (2016)
54. Vitanov, N.K.: Science and society. Assessment of research. In: Vitanov, N.K. (eds.) Science Dynamics and Research Production, pp. 3–52. Cham, Springer (2016)
55. Vitanov, N.K.: Commonly used indexes for assessment of research production. In: Vitanov, N.K. (ed.) Science Dynamics and Research Production, pp. 55–99. Springer, Cham (2016)
56. Vitanov, N.K: Additional indexes and indicators for assessment of research production. In: Vitanov, N.K. (eds.) Science Dynamics and Research Production, pp. 101–154. Cham, Springer (2016)
57. Vitanov, N.K: Frequency and rank approaches to research production. Classical statistical laws. In: Vitanov, N.K.: Science Dynamics and Research Production, pp. 157–193. Cham, Springer (2016)
58. Vitanov, N.K: Selected models for dynamics of research organizations and research production. In: Vitanov, N.K. (eds.) Science Dynamics and Research Production, pp. 195–268. Cham, Springer (2016)
59. Vitanov, N.K., Siefert, M., Peinke, J.: Topological analysis of the chaotic behaviour of Shinriki oscillator. Comptes Rendus de l'Academie Bulgare des Sciences **55**, 31–36 (2002)
60. Vitanov, N., Yankulova, E.: On some properties of the point correlation dimension. Comptes Rendus de l'Academie Bulgare des Sciences **56**, 25–30 (2003)
61. Vitanov, N.K., Sakai, K.: Upper bounds on the number of significant degrees of freedom of noise influenced oscillations of moving machines. Syst. Anal. Model. Simul. **43**, 815–828 (2003)
62. Vitanov, N.K., Dimitrova, Z.I., Kantz, H.: On the trap of extinction and its elimination. Phys. Lett. A **349**, 350–355 (2006)
63. Vitanov, N.K., Yankulova, E.D.: Multifractal analysis of the long-range correlations in the cardiac dynamics of Drosophila melanogaster. Chaos, Solitons Fractals **28**, 768–775 (2006)
64. Vitanov, N.K., Tarnev, K., Kantz, H.: 2006. Hölder-exponent-MFDFA-based test for long-range correlations in pseudorandom sequences. J. Theor. Appl. Mech. **36**(2), 47–64 (2006)
65. Vitanov, N.K., Sakai, K., Jordanov, I.P., Managi, S., Demura, K.: Analysis of a Japan government intervention on the domestic agriculture market. Physica A **382**, 330–335 (2007)
66. Vitanov, N.K., Sakai, K., Dimitrova, Z.I.: SSA, PCA, TDPSC, ACFA: Useful combination of methods for analysis of short and nonstationary time series. Chaos, Solitons Fractals **37**, 187–202 (2008)
67. Vitanov, N.K., Hoffmann, N.: On probability for rogue waves in the North sea. Comptes Rendus de l'Academie Bulgare des Sciences **62**, 187–194 (2009)

68. Vitanov, N.K., Jordanov, I.P., Dimitrova, Z.I.: On nonlinear dynamics of interacting populations: coupled kink waves in a system of two populations. Commun. Nonlinear Sci. Numer. Simul. **14**, 2379–2388 (2009)
69. Vitanov, N.K., Jordanov, I.P., Dimitrova, Z.I.: On nonlinear population waves. Appl. Math. Comput. **215**, 2950–2964 (2009)
70. Vitanov, N.K., Dimitrova, Z.I., Kantz, H.: Modified method of simplest equation and its application to nonlinear PDEs. Appl. Math. Comput. **216**, 2587–2595 (2010)
71. Vitanov, N.K., Dimitrova, Z.I.: Application of the method of simplest equation for obtaining exact traveling-wave solutions for two classes of model PDEs from ecology and population dynamics. Commun. Nonlinear Sci. Numer. Simul. **15**, 2836–2845 (2010)
72. Vitanov, N.K., Dimitrova, Z.I., Ausloos, M.: Verhulst-Lotka-Volterra (VLV) model of ideological struggle. Physica A **389**, 4970–4980 (2010)
73. Vitanov, N.K., Dimitrova, Z.I., Vitanov, K.N.: On the class of nonlinear PDEs that can be treated by the modified method of simplest equation. Application to generalized Degasperis–Processi equation and b–equation. Commun. Nonlinear Sci. Numer. Simul. **16**, 3033–3044 (2011)
74. Vitanov, N.K., Ausloos, M.R.: 2012. Knowledge epidemics and population dynamics models for describing idea diffusion. In: Scharnhorst, A., Börner, K., van den Besselaar, P. (eds.) Models of Science Dynamics, pp. 69–125. Springer, Berlin (2012)
75. Vitanov, N.K., Ausloos, M., Rotundo, G.: Discrete model of ideological struggle accounting for migration. Adv. Complex Syst. **15**(supp01), 1250049 (2012)
76. Vitanov, N.K., Dimitrova, Z.I., Kantz, H.: Application of the method of simplest equation for obtaining exact traveling-wave solutions for the extended Korteweg-de Vries equation and generalized Camassa-Holm equation. Appl. Math. Comput. **219**, 7480–7492 (2013)
77. Vitanov, N.K., Dimitrova, Z.I., Vitanov, K.N.: Traveling waves and statistical distributions connected to systems of interacting populations. Comput. Math. Appl. **66**, 1666–1684 (2013)
78. Vitanov, N.K., Vitanov, K.N.: Population dynamics in presence of state dependent fluctuations. Comput. Math. Appl. **68**, 962–971 (2014)
79. Vitanov, N.K., Dimitrova, Z.I.: 2014. Solitary wave solutions for nonlinear partial differential equations that contain monomials of odd and even grades with respect to participating derivatives. Appl. Math. Comput. **247**, 213–217 (2014)
80. Vitanov, N.K., Hoffmann, N.P., Wernitz, B.: Nonlinear time series analysis of vibration data from a friction brake: SSA, PCA, and MFDFA. Chaos, Solitons Fractals **69**, 90–99 (2014)
81. Vitanov, N.K., Dimitrova, Z.I., Vitanov, K.N.: Modified method of simplest equation for obtaining exact analytical solutions of nonlinear partial differential equations: further development of the methodology with applications. Appl. Math. Comput. **269**, 363–378 (2015)
82. Vitanov, N.K., Ausloos, M.: Test of two hypotheses explaining the size of populations in a system of cities. J. Appl. Stat. **42**, 2686–2693 (2015)
83. Vitanov, N.K., Vitanov, K.N.: Box model of migration channels. Math. Soc. Sci. **80**, 108–114 (2016)
84. Vitanov, N.K., Dimitrova, Z.I., Ivanova, T.I.: On solitary wave solutions of a class of nonlinear partial differential equations based on the function $1/cosh^n(x + t)$. Appl. Math. Comput. **315**, 372–380 (2017)
85. Vitanov, N.K., Borisov, R.: Statistical characteristics of a flow of substance in a channel of network that contains three arms. In: Georgiev, K., Todorov, M., Georgiev, I. (eds.) Advanced Computing in Industrial Mathematics. BGSIAM 2017, Studies in Computational Intelligence, vol. 793, pp. 421–432. Springer, Cham (2019)
86. Vitanov, N.K., Vitanov, K.N.: Discrete-time model for a motion of substance in a channel of a network with application to channels of human migration. Physica A **509**, 635–650 (2018)
87. Vitanov, N.K., Vitanov, K.N.: On the motion of substance in a channel of a network and human migration. Physica A **490**, 1277–1294 (2018)
88. Vitanov, N.K., Dimitrova, Z.I.: Modified method of simplest equation applied to the nonlinear Schrödinger equation. J. Theor. Appl. Mech. **48**(1), 59–68 (2018)
89. Vitanov, N.K., Borisov, R.: A model of a motion of substance in a channel of a network. J. Theor. Appl. Mech. **48**(3), 74–84 (2018)

Metaheuristic Algorithms: Theory and Applications

Simeon Ribagin and Velislava Lyubenova

Abstract Metaheuristic is a collective concept of a series of intelligent strategies to enhance the efficiency of heuristic procedures. Metaheuristic algorithms are becoming an important part of modern optimization. A wide range of metaheuristic algorithms have emerged over the last two decades, and are becoming increasingly popular. This article presents a brief overview of the scientific research on new metaheuristic algorithms, as well as their modifications and hybridizations and its various fields of application. The results presented are limited to those proposed by scientists from the Bulgarian Academy of Sciences for the last 20 years.

Keywords Metaheuristic algorithms · Hybrid algorithms · Modifications · Parameter identification

1 Introduction

One of the contemporary fields of Artificial Intelligence is the field of metaheuristic algorithms—a scientific method to problem solving that extends the idea of heuristic algorithms, where "meta" denotes "beyond" or "on a higher level" [101].

According to [227], a metaheuristic is "an iterative generation process which guides a subordinate heuristic by combining intelligently different concepts for exploring and exploiting the search space, learning strategies are used to structure information in order to find efficiently near-optimal solutions."

S. Ribagin (✉)
Institute of Biophysics and Biomedical Engineering Bulgarian Academy of Sciences, Sofia, Bulgaria
e-mail: sim_ribagin@mail.bg

V. Lyubenova
Institute of Robotics "St. Ap. and Gospeller Matthew" Bulgarian Academy of Sciences, Sofia, Bulgaria
e-mail: v_lyubenova@ir.bas.bg

© The Author(s), under exclusive license to Springer Nature Switzerland AG 2021
K. T. Atanassov (ed.), *Research in Computer Science in the Bulgarian Academy of Sciences*, Studies in Computational Intelligence 934,
https://doi.org/10.1007/978-3-030-72284-5_18

Metaheuristic algorithms can be divided into three main groups, as follows [88, 221]:

- Evolutionary algorithms, which simulate the natural selection as one of the basic principles of the evolution. An initial population of individuals is generated at the beginning, where each individual is a possible solution to the problem that is being solved. Combined characteristics of selected members of the current population are passed to the new generation. The least fit members of the population are eliminated, while the fittest ones survive until better solutions are produced. Among the evolutionary algorithms are the following families of algorithms: Genetic algorithms (GAs), Genetic programming, Differential evolution, Evolutionary strategies, Evolutionary programming, Harmony search;
- Swarm-based algorithms, which simulate the collective behaviour of a group of insects or animals as they try to survive. The experience of each member of the swarm is constantly communicated with all or part of the swarm. Thus, the movement of the swarm as a whole is directed towards the most promising areas of the search space, based on the information accumulated by previous experience. Some swarm-based algorithms are Particle swarm optimization, Firefly algorithm (FA), Cuckoo search algorithm (CS), Ant colony optimization (ACO), Artificial Bee Colony (ABC), Bat algorithm (BA);
- Trajectory-based algorithms, which improve a single solution. At each iteration, the current solution is moved to another neighbouring solution in the search space based on a certain neighbourhood structure. The most common trajectory-based algorithms are Tabu search (TS), Simulated annealing (SA), Threshold acceptance (TA), Hill climbing.

Metaheuristics are more and more popular in different research areas and industries. In the last 20 years at the Bulgarian Academy of Sciences several scientific groups have been working on the development of a new theory, i.e. new metaheuristic algorithms, as well as their modifications and hybridizations. In parallel, much work is being done on the applications of metaheuristics in various problems and fields. Among the most active groups of scientists are those from the Institute of Biophysics and Biomedical Engineering, Institute of Information and Communication Technologies, Institute of Chemical Engineering and Institute of Robotics "St. Ap. and Gospeller Matthew". This chapter presents and discusses the major contributions to the field of metaheuristic algorithms of the scientists from Bulgarian Academy of Sciences.

2   Genetic Algorithms

GA originated from the studies of cellular automata, conducted by John Holland and his colleagues at the University of Michigan. Holland's book [91], published in 1975, is generally acknowledged as the beginning of the research of genetic algorithms. The GA is a model of machine learning which derives its behaviour from a metaphor of the processes of evolution in nature [90]. This is done by the creation within a machine of a population of individuals represented by chromosomes. A chromosome could be an array of real numbers, a binary string, a list of components in a database, all depending on the specific problem.

2.1 Modifications of Genetic Algorithms

Aiming to improve GA performance many modifications of single-population GA (SGA) and multi-population GA (MpGA) with standard implementation order of the main genetic operators, namely selection, crossover and mutation (here denoted as SCM), have been developed. Following the idea that GA imitate processes occurred in the nature it could be said that the probability mutation to come first followed by crossover is comparable to that both processes to occur in a reverse order, or selection can occur before or after crossover and mutation, no matter of their order. As a result, two new modifications, named SGA-MSC (mutation, selection, crossover) and SGA-CSM (crossover, selection, mutation) are proposed in [19] and applied to parameter identification of *S. cerevisiae* fed-batch process model. Another three modifications, namely SGA-CMS (crossover, mutation, selection), SGA-SMC (selection, mutation, crossover) and SGA-MCS (mutation, crossover, selection) have been proposed in [27] and thoroughly investigated in [26], again for the purposes of parameter identification of *S. cerevisiae* fed-batch process model. Two modifications that omit mutation operator—SGA-SC (selection, crossover) and SGA-CS (crossover, selection) have been also developed and applied [23] for all aforementioned process.

The basic GA operators are described in a series of researches using the theory of Generalised Nets (GNs) [133, 134, 147, 202, 208]. In [28, 178] descriptions of SGA modifications based on GNs are presented.

A modified GA for a parameter identification of an *E. coli* fed-batch fermentation model is proposed in [162, 163]. In the simple GA the best chromosome does not always keep on improving in each generation. Obtained good solutions could be destroyed by either crossover or mutation or both operations. The aim of the modified genetic algorithm is to prevent this disadvantage. In view of the fact that there is no general theory about tuning the genetic algorithm parameters, some modifications and adjustments of genetic parameters, according to the regarded problem, are done to improve the conventional genetic algorithm. The simulation results illustrate that the use of proposed modified genetic algorithm for a parameter identification of fermentation processes is highly efficient and effective. The functions and parameter adjustments, proposed here, enhance the performance of the algorithm. Still more, the implementation of the modified GA leads to noticeable decreasing of the solution time.

In [183] a modified MpGA without the performance of the mutation operator is proposed. The idea is to reduce the convergence time and therefore to increase the identification procedure effectiveness for on-line application of the algorithm. Modified MpGA, classical multi-population GA and two other modifications are tested for parameter identification problem of an non-linear model. The obtained results show that the highest accuracy for parameter identification of the considered model is achieved with the multi-population GA with Modification 1. The best calculation time is shown by the multi-population GA without mutation.

When the focus is pointed to MpGA, a modified MpGA named MpGA-CMS is elaborated in [14], provoked by the idea of SGA-CMS. Another two new modifications of the standard MpGA-SCM, namely MpGA-CSM and MpGA-MSC, in which the operator selection is executed between the operators crossover and mutation, have been presented in [138]. Altogether six realizations of MpGA have been investigated in [15], among them MpGA-MCS, MpGA-SMC, proposed in [26] and another two without mutation—MpGA-SC and MpGA-CS, elaborated for the first time. All aforementioned MpGA modifications have been developed and firstly applied to the parameter identification of S. cerevisiae fed-batch process model.

Three of the newly proposed SGA (SGA-MCS, SGA-MSC and SGA-CSM) as well as three modifications of MpGA (MpGA-CSM, MpGA-MSC and MpGA-SMC) reduce the algorithms convergence time without affecting the model accuracy when have been applied to parameter identification of S. cerevisiae fed-batch process model.

Further investigations in tuning genetic algorithms parameters, such as generation gap (GGAP), crossover (XOVR), mutation rates (MUTR), insertion (INSR) and migration rates (MIGR), improve the convergence time of the modified algorithms with almost 40%, while saving the model accuracy when GA have been applied to parameter identification of S. cerevisiae fed-batch process model [15, 23, 25, 164].

In a series of papers the relation and influence of the main GA parameters are investigated based on InterCriteria Analysis (ICrA) [7, 142, 146, 185, 186]. ICrA has assisted in the assessment of the performance of some of aforementioned modifications of SGA [135], MpGA [12], as well as altogether SGA and MpGA [11], for the purposes of parameter identification of yeast fed-batch cultivation process model.

A procedure for purposeful genesis concerning intervals of variations of model parameters is proposed, aiming to improve the SGA-SCM effectiveness [22]. In the beginning GA search for solutions in wide, but reasonably chosen boundaries of model parameters. Further, after assembling of obtained values of model parameters, a shrinkage of the preliminary defined boundaries might be proposed, which is the main idea of the procedure of purposeful model parameters genesis. The procedure is further validated to SGA-MCS and MpGA-SCM [20, 137]. Application of the proposed procedure for parameter identification of S. cerevisiae fed-batch process model leads to significant decrease of algorithms convergence time without affecting the model accuracy, as follow: over than 10% for MpGA-SCM, approximately 30% for SGA-MCS and 40% for SGA-SCM.

A procedure based on the intuitionistic fuzzy logic (IFL) is proposed in [21] and further proved as successful alternative for assessing the quality of different algorithms. Effectiveness of the procedure has been tested in assessment of the performance of different kinds of SGA [16, 18], different kinds of MpGA [8, 17, 140], standard and modified SGA towards standard and modified MpGA [13, 139], and GA with different values of the most sensitive parameter of GA in regard to algorithms convergence time, namely generation gap [139], when all aforementioned GA have been applied to parameter identification of *S. cerevisiae* fed-batch process model. In addition, the procedure has been implemented to assess the genetic operators significance in SGA [9] and MpGA [10]. Thus, the most significant genetic operator has been distinguished and its influence for finding the global optimal solution for the both algorithm has been evaluated. IFL has been also implemented to assess the aforementioned purposeful model parameters genesis [136].

It is worth to note, that both proposed procedures are universal tools and could be appropriately and successfully implemented to other stochastic optimization algorithms, as well as to different objects of model parameter identification.

BASIC genetic algorithm is developed in Bulgarian Academy of Sciences—Institute of Chemical Engineering [222]. It proposes a morphogenesis function for representation of all types of variables (continuous, binary, integer) in the continuous search space. In order to be applied for different types of optimization problems (LP, NLP, MINLP, etc.), BASIC GA includes a number of schemes of genetic operators for selection (roulette-wheel, rank-based and tournament), recombination (one-point, two-point, uniform, arithmetic, blend, etc.) and mutation (uniform, non-uniform, breeder, etc.). Thus, the user can choose the appropriate one for a given optimization problem. Moreover, it proposes new scheme for selection for replacement. BASIC GA provides opportunity to be adjusted to the given optimization problems by set up of its global and local parameters. BASIC GA is designed to deal with constrained optimization problems by using static and dynamic penalty functions.

Hierarchical genetic algorithm (HGA) is proposed in [156] and implemented in the MotCo software package (http://motco.dir.bg). A muscle consists of many different motor units (MUs) which are the smallest muscle part controlled individually by nervous system. MUs develop forces as a result of neural impulsation—"twitch" if the pulse is only one and "tetanus" if many pulses are applied. The control of nervous system on the muscle forces is realized through impulsation of each MUs with individual frequency but with non-constant interpulse intervals. A HGA was developed to simulate the MUs activity. The moments of neural impulsations of all MUs are design variables coding the problem in the terms of HGA. The main idea is using different genetic operations to find these time moments, so that the sum of MUs twitches (so, the muscle force) satisfies imposed goals (like minimal deviation from the given joint moment, minimal total muscle force, minimal MUs activation, etc.) [156].

2.2 Hybridization of Genetic Algorithms

In the paper [172] a hybrid scheme using GA and Sequential Quadratic Programming (SQP) method is proposed. In the hybrid GA-SQP the role of the GA is to explore the search place in order to either isolate the most promising region of the search space. The role of the SQP is to exploit the information gathered by the GA. The results show that the GA-SQP takes the advantages of both GA's global search ability and SQP's local search ability.

In [175] two hybrid schemes using FA and GA are introduced. Based on the numerical and simulation result, it is shown that the model obtained by the proposed hybrid algorithms are highly competitive with standard FA and GA. The hybrid algorithms obtain similar objective function values compared to pure GA and FA, but using four times less population size and seven times less computation time. Thus, the hybrids have two advantages—take much less running time and required much less memory compared to standard GA and FA.

A hybrid scheme FA-GA applied to PID controller parameter tuning is introduced in [218]. The hybrid FA-GA adjustments are done based on several pre-tests. The observed results are compared to the ones obtained by applying the pure FA and pure GA. The comparison shows that the proposed hybrid algorithm is highly competitive with standard FA and GA for considered here optimization problem.

A hybrid scheme using GA and ACO is introduced in [87]. In the hybrid GA-ACO, the GA is used to find feasible solutions to the considered optimization problem. Further ACO exploits the information gathered by GA. This process obtains a solution, which is at least as good as—but usually better than—the best solution devised by GA. The results show that the hybrid GA-ACO enhancing the overall search ability and computational efficiency. In [195] a hybrid algorithm ACO-GA is discussed. The proposed ACO-GA hybrid performs better than both ACO and GAs—almost twice less computational time and approximately five times smaller populations needed.

Hybrids of three metaheuristics, namely, ACO, FA and GA are proposed in [191]. The basic idea is to generate initial solutions by the ACO method, and then serve these solutions to the GA and FA as their initial population of individuals. Thus, the GA or FA will start with a population, which is not randomly generated, as in the general case, but one rather closer to an optimal solution. The hybrid algorithms obtain similar objective function values compared to pure ACO, GA and FA, but using smaller population size and in almost twice as little computational time.

In [114] a hybridization of GA and genetic programming is proposed. The hybrid method was tested with use of typical non-linear modelling benchmarks.

2.3 Genetic Algorithms Applications

Model Parameter Identification Problems

Genetic algorithms are one of the most promising metaheuristic techniques. For the first time the application of GA in the area of modeling of fermentation processes is proposed in [89] and the priority of GA are discussed. Since then GA have been successfully applied and proved as an appropriate tool for solving such a difficult problem as parameter identification of *S. cerevisiae* [4, 24, 145, 148, 228] and *E. coli* [160, 161, 166, 199, 200, 204] batch and fed-batch process models. Also, GA have been shown as able to reach a satisfactory solution when functional state modeling approach has been applied to fermentation processes of yeast [143, 144, 205] and bacteria [111, 181, 201, 203, 206], or both [209, 237], in order to overcome the main disadvantages of global process model, namely complex structure and a big number of model parameters.

In [167] a system of six ordinary differential equations is proposed to model the dynamics of an non-linear fermentation process. Parameter estimation is carried out using real experimental data set from an *E. coli* MC4110 fed-batch fermentation process. In order to study and evaluate the links and magnitudes existing between the model parameters and variables sensitivity analysis is carried out. A procedure for consecutive estimation of four definite groups of model parameters based on sensitivity analysis is proposed. The application of that procedure and GA leads to a successful parameter identification.

In [132] the application of genetic algorithm technique to optimize the parameters of an industrial process model was investigated.

In [168] some adjustments of GA, according to the identification problem of fermentation process models, are done to improve the GA performance. Based on the proposed procedure for genetic parameters tuning and series of experiments a successful adjustment of the algorithm is done. The results show that the identification procedure leads to higher order of accuracy of the obtained parameter estimations and to 2.5 times decrease of the computation time. In order to validate the proposed genetic algorithm two experimental data sets of a recombinant *E. coli* BL21(DE3)pPhyt109 fermentation process are used. Based on the regarded GA successful identification is done. The fulfilled model verification shows the model adequateness.

An investigation of the influence of the population size on the GA performance for a model parameter identification problem, is considered in [197]. Population sizes between 5 and 200 chromosomes in the population are tested with constant number of generations. In order to obtain meaningful information about the influence of the population size a considerable number of independent runs of the GA are preformed. The observed results show that the optimal population size is 100 chromosomes for 200 generations. In this case accurate model parameters values are obtained in reasonable computational time. Further increase of the population size, above 100 chromosomes, does not improve the solution accuracy. Moreover, the computational time is increased significantly.

In [85] the Ant Colony Optimization algorithm and Genetic algorithm are proposed for parameter identification of an *E. coli* fed-batch fermentation process model. The objective function was formulated as a distance between the model predicted and the experimental data. Two different distances were used and compared—the Least Square Regression and the Hausdorff Distance. The investigations showed that better results concerning model accuracy are obtained using the objective function with a Hausdorff Distance between the modelled and the measured data. Although the Hausdorff Distance is more time consuming than the Least Square Distance, this metric is more realistic for the considered problem.

Results from GA and ACO for parameter identification of a system of non-linear differential equations are discussed in [190]. The parameter estimation is performed based upon the use of Hausdorff metric, in place of most common used metric—

Least Squares regression. A modified Hausdorff Distance is proposed to evaluate the mismatch between experimental and model predicted data. Parameters of the two algorithms (GA and ACO) were tuned based on several pre-tests according considered here optimization problem. Based on the obtained result it is shown that the best value of the objective function J is achieved by the ACO algorithm. Compared the worst results the results achieved by ACO algorithm are worst than these achieved by the GA.

In [165] a MpGA is applied for identification of complex fermentation process models. Several conventional algorithms for parameter identification (Gauss-Newton, Simplex Search and Steepest Descent) are compared to the MpGA. A general comment on this study is that traditional optimization methods are generally not the universal and most successful ones on any particular domain, especially for the parameter optimization considered here. They have been fairly successful at solving problems of type which exhibit bad behaviour like multimodal or non-differentiable for more conventional based techniques.

Proposed in [172] hybrid scheme using GA and SQP method is applied for modeling of *E. coli* MC4110 fed-batch fermentation process. To demonstrate the usefulness of the presented approach, two cases for parameter identification of different complexity are considered.

In [175] GA-FA hybrid schemes are applied to parameter identification problem of a non-linear mathematical model. Parameter optimization is performed using a real experimental data set. In the considered mathematical model five parameters are estimated, namely maximum specific growth rate, two saturation constants and two yield coefficients. The proposed in [195] ACO-GA hybrid algorithm is applied to the estimation of the parameters of the same fermentation process model. The presented results show a better performance of the hybrid algorithm. Another hybrid scheme is used [87] in order to estimated the model parameters. The results show that the hybrid GA-ACO takes the advantages of both GA and ACO.

In [191] the estimation of the parameters of a real *E. coli* fed-batch fermentation process model is done using hybrids of three metaheuristics, e.g. ACO, FA and GA. Based on the numerical and simulation results, it is shown that the models obtained by the proposed hybrid algorithms are competitive with standard metaheuristic algorithms.

The GA performance, solving the model parameter identification problem, is investigated based on ICrA in [141, 187, 194, 207]. The ICrA approach is applied to establish relations and dependencies between the main GA parameters, as well as the convergence time, model accuracy and model parameters. For this purpose several GA with different parameter values are applied to various model parameter identification problems.

PID Controller Parameter Tuning

In [216] GA controller tuning to achieve good closed-loop system performance is proposed. Using four objective functions reflecting the performance of the proportional-integral-derivative (PID) controller, the significance of the tuning procedure is evaluated. As a result, the optimal PID controller settings are obtained. The presented results indicate high quality and better performance of the designed control system. For a short time the controller sets the control variable and maintains it at the desired set point during the fermentation process. It is demonstrated that the GA provide a simple, efficient and accurate approach to PID controllers tuning. Moreover, GA tuning can be regarded as an effective methodology for attaining improved performance of a process.

In [226] the work is focused on the design of a glucose concentration control system based on non-linear model plant. Due to significant time delay in real time glucose concentration measurement, a correction is proposed in glucose concentration measurement and a Smith predictor (SP) control structure based on universal PID controller is designed. To achieve good closed-loop system performance GA based optimal controller tuning procedure is applied. A standard binary encoding GA is used. The GA parameters and operators are specified for the considered here problem. As a result the optimal PID controller settings are obtained. The results show that the control system based on SP with Extended Kalman Filter (EKF) has a better performance than the one without EKF.

Further, in [225] a feedforward feedback PID controller designed for control of glucose concentration during the *E. coli* fed-batch fermentation process is presented. The controller is used to control the feed rate and to maintain glucose concentration at a desired set point. For the PID controller tuning a GA is used. Tuning of the controller on the basis of a GA leads to higher level of accuracy and efficiency of the system performance.

In [213] an optimal tuning of a universal digital PID controller using metaheuristics as GA, SA and TS is proposed. Objective function values and CPU time were used as criteria to compare the performance of the three metaheuristic algorithms—GA, SA and TS. A series of procedures for PID controller tuning were performed using competing techniques and criteria. As a result the set of optimal PID controller settings was obtained. The simulation results indicate that the proposed metaheuristic algorithms are effective and efficient, and demonstrate that the applied techniques exhibit a significant performance improvement over classical optimization methods.

Further, authors in [217] present a comparison of population-based and single point search metaheuristic methods applied to an optimal tuning of a universal digital PID controller. In the group of population-based metaheuristics GA, FA and ACO are considered. In the group of single point search metaheuristic methods SA, TA and TS are considered. The simulation results indicate that the population-based metaheuristics performs better than the single point search methods considered here. Moreover, GA and ACO show better performance than FA. Regarding the applied single point search methods it should be noted that although TS is much simpler than the population-based and SA and TA methods, obtained TS results are comparable to those of FA. The results show that SA and TA algorithms failed to solve the considered here PID controller tuning problem, compared to the other four metaheuristics.

The hybrid FA-GA proposed in [218] is applied to PID controller parameter tuning in Smith Predictor for a non-linear control system. Good closed-loop system performance is achieved on the basis of the considered PID controllers tuning procedures.

Synthesis of Feed Rate Profiles

In [170] GA are used for synthesis of feed rate profiles during an *E. coli* MC4110 fermentation process. Development of a suitable feeding strategy is critical in fed-batch operation modes. During the fed-batch fermentation the system states change considerably, from a low initial to a very high biomass and product concentration. This dynamic behaviour motivates the development of optimization methods to find the optimal input feeding trajectories in order to improve the process. A GA using different chromosome lengths is proposed in order to optimize the feeding trajectory. This algorithm is found to be an effective and efficient method for solving the optimal feed rate profile problem. The GA is capable of simultaneously optimizing feed rate profile for a given objective function. However, the results seem to indicate that the feed profile formed using chromosome with 60 genes is superior to the rest feeding trajectories. Based on obtained feed rate profile cell concentration has an ideal increase for the complete fermentation period, achieving final cell concentration of $5.26 \, gl^{-1}$ using 1.38 l feeding solution. This is a satisfactory result for the fermentation system due to the economical effect and process effectiveness.

A feed rate profiles design based on GA is also proposed in [169, 182]. An *E. coli* MC4110 fed-batch fermentation process is considered. The feed rate profiles based on three different lengths of chromosomes are synthesized. The ration of the substrate concentration and the difference between actual cell concentration and theoretical maximum cell concentration is used as an objective function. As a result the GA synthesized optimal feed profiles fulfilling the defined criterion.

The parameters of a GA significantly affect the speed of convergence to the near optimal solution, and the accuracy of the solution itself. An investigation of the effects of the different GA parameters on the outcome of the GA enhanced simulation is discussed in [171]. Authors were performed a lot of tests to choose the appropriate GA parameters for the considered here problem. As a result a GA was proposed in order to optimize the feeding trajectory for two *E. coli* fermentation processes— strain *E. coli* MC4110 and strain *E. coli* BL21(DE3)pPhyt109 fed-batch fermentation processes.

Further, in [211] based on GA a feed rate profile that hold defined functional state is synthesized. The obtained results have presented the ability of the GA to generate an optimal feed rate profile based on defined rules. The GA generates feed rate profile that for a short time leads the process to the desired functional state, determined as optimal, and maintains this state during the process. As a result a high final biomass concentration is achieved under optimal conditions for amount and assimilation of the substrate. In the systems there is not accumulation of the residual substrate and any inhibition effects.

In [131] an approach using GA for development of optimal time profiles of key control variable of Poly-HydroxyButyrate (PHB) production process in combination with fuzzy logic controllers is proposed. The achieved results are compared with a knowledge-based approach to the same optimal control task.

Other Applications

BASIC genetic algorithm [222] has been implemented successfully for various optimization problems as follows: minimizing the environmental impact of multipurpose batch chemical or biochemical plants with application in dairy industry [1, 234–236]; solution of supply chain optimization model for products' portfolio design of chemical production complexes [2, 232, 233]; learning of artificial neural networks models applied for description of the process of Autothermal Thermophilic Aerobic Digestion (ATAD) of sludge in bioreactors wastewater treatment system [104, 106, 108, 229], the process of bioconversion of crude glycerol—a byproduct of biodiesel production [103, 105]. BASIC GA is also implemented for solution of stochastic optimization problem for capturing of uncertainties in heat-integrated ATAD system for wastewater treatment [230, 231].

It is used also for optimization of operating conditions of smart patch/layer lightweight hygro-thermo-piezoelectric structures which have a variety of applications in high technology industries such as aeronautics, aerospace, electronics, automotive etc. [100, 107, 152, 153].

In [31, 114] a new approach for non-linear systems modelling combining Takagi-Sugeno neuro-fuzzy systems and population based algorithms was developed.

Using the software MotCo [156] human elbow flexion and extension movements with different velocities are simulated. The used planar model consists of two extensor and three flexor muscles having different nearly real numbers and types (slow, fast fatigable, fast resistant to fatigue) MUs. The results of simulation with MotCo software were compared with the results obtained through using static optimization, where each muscle is modeled by one force, with an optimization criterion "minimum of a weighted sum of the muscle forces raised to the power of n". The results showed that HGA was a suitable means for precise investigation of motor control. Many experimentally observed phenomena (such as antagonistic co-contraction, three-phasic behavior of the muscles during fast movements) can find their explanation by the properties of the MUs twitches.

Since in dynamic conditions many authors use rheological muscle models the comparison was also made between these three approaches—HGA controlling the activity of the MUs, static optimization using paradigm one muscle-one force-one control signal and rheological muscle models where the optimization parameter is muscle "activation" [157, 159]. The conclusion was that the using of rheological models could not help in predicting suitable muscle synergism and muscle antagonistic co-contraction and were suitable only for calculation of the current maximal muscle forces that could be used as weight factors in objective functions.

The influence of the weight coefficients of one optimization criterion (this one that reflects the importance of the movement accuracy on the predicted results), was investigated. Two variants concerning the muscle MUs structure were also compared: each muscle was composed of four distinct types MUs and the MUs twitch parameters were uniformly distributed [158].

An investigation using MotCo software was performed of learning fast elbow flexion movements of the human upper limb in horizontal plane. The input parameters were experimentally measured elbow joint moments. The first flexion and the fastest one, after training, were simulated [154, 155]. Three hypotheses were tested and confirmed: (1) peak muscle forces and antagonist co-contraction increased during training; (2) there was an increase in the firing frequency and the synchronization between MUs; and (3) the recruitment of fast MUs dominated the action. Comparing the simulation results of the very first trial of a particular subject with those of the last trail (at the end of the learning process), it can be concluded that the speed of limb motion and muscle forces increase initially as a result of the more synchronous MUs activation and the increase of firing rate of active MUs.

3   Ant Colony Optimization Algorithm

First idea to use ant behaviour for solving optimization problems is proposed by Marco Dorigo [32]. Later some modifications are proposed, mainly in pheromone updating rules [34]. The basic in ACO methodology is the simulation of ants behaviour. The problem is represented by graph. The solutions are represented by paths in a graph and we look for shorter path corresponding to given constraints.

3.1   Ant Colony Optimization Modifications and Improvements

The success of ACO is based on exploration and exploitation of the search space. In traditional ant algorithm the ants start to create their solutions from random node of the graph in every iteration. It is kind of diversification of the search in a search space. The random start helps to avoid trapping in local minimums. Pheromone information concentrates the search close to the best so far solutions, because is expected that close to them can be found better.

The main contribution to ACO theory is applying start strategies instead of completely random start and imitating flying ant [56, 57, 64].

The random start of the "artificial" ants is very important. It is a kind of diversification of the search and partially avoiding local minimums. In some problems the solution consists of all nodes of the graph of the problem and important is their order. In another problems the solution consists of part of the nodes of the graph and their order is not important, they are so called subset problems. For subset problems is important from which node the ants start to construct their solution. I an ant starts from node which do not belong to the optimal solution it is impossible to construct it. The main idea is to divide the set of the nodes of the graph of the problem to subsets. After every iteration the subsets are estimated according how many good solutions start from the subset and how many bad solutions started from the subset. Than the ants chose the subset with higher estimation with higher probability. The subset with lower estimation become forbidden for several iterations. Thus the start continue to be random, but we control the probability. This method is called semi-random start [56, 57, 64]. The idea is tested on Multiple Knapsack Problem (MKP) and shows improvement of the algorithm performance [56, 57]. Several ideas of set of nodes estimation are explore. Sensitivity of the algorithm according the parameters is investigated [58, 69, 70, 74].

In traditional ACO algorithm the ants can take in to account only the information for neighbour nodes of the graph of the problem and includes the node with a higher probability. In [59] is imitated flying ant. The main difference with traditional ACO algorithm is that an ant can make decision about the direction to move, taking in to account the information from more than one step ahead. The idea is to improve the traditional ant algorithm, giving to the ants a possibility to receive more information about the search space imitating flying ants. The information coming directly from neighbour nodes have more influence than this from other nodes. The algorithm is tested on two very different optimization problems, Global Positioning System (GPS) surveying problem and Multiple Knapsack Problem (MKP) and shows improved performance.

$$
p_{i_p}^k(t) = \begin{cases} \dfrac{\tau_{i_j i_p} \eta_{i_p}(\tilde{S}_k(t))}{\sum_{i_q \in allowed_k(t)} \tau_{i_j i_q} \eta_{i_q}(\tilde{S}_k(t))} & \text{if } i_p \in allowed_k(t) \\ 0 & \text{otherwise} \end{cases} ,
\tag{1}
$$

where $\tau_{i_j i_p}$ is a pheromone level on the arc (i_j, i_p), $\eta_{i_p}(\tilde{S}_k(t))$ is the heuristic and $allowed_k(t)$ is the set of remaining feasible items. Thus the higher the value of $\tau_{i_j i_p}$ and $\eta_{i_p}(\tilde{S}_k(t))$, the more profitable it is to include item i_p in the partial solution.

Proposed different modifications of ACO are investigated applying ICrA in [77, 84, 86, 189, 193].

3.2   Application of Ant Colony Optimization

ACO algorithm is among the most powerful methods for solving combinatorial optimization problems. It is constructive method and achieves unfeasible solutions more rare than other metaheuristic methods. Therefore it is very suitable for solving discrete optimization problems with strong constraints. Bellow, some of the ACO applications are discussed.

Multiple Knapsack Problem
The Multiple Knapsack problem is resource allocation problem and is representative of so called subset problems. The solutions of this problem can have different length and the order of the elements of the solution is not important. In [38] ACO algorithm for solving MKP with additional reinforcement of the pheromone of non used elements of the graph of the problem is proposed. It is a kind of the diversification of the search in a search space. In [39, 40, 44] various types of heuristic information including in a different way the parameters of the objective function

and constraints are proposed. Part of this heuristic information is static and does not depend to the current state of the partial solution. Other proposed heuristics are dynamic and depends to current state of the problem. A variants of calculation of transition probability are described and analysed in [41, 49]. In [43] the deposit of the pheromone, on the node or on the arcs of the graph of the problem is discussed. It looks more natural the pheromone to be on the nodes, but the experiments and detailed analyses show that it is more appropriate the pheromone to be on the arcs [43].

Cutting Stock Problem

Cutting stock problem is NP-hard problem which arise in many industries. There are two variants of the problem, 1D variant and 2D variant. Most of the authors solve simplified 2D problem when the cutting shapes are rectangular. This variant arise in glass industry and in paper industry. The 1D variant is easier, but still is NP. It raises in wooden industry and in building construction industry. In [37] ACO algorithm for solving 1D cutting stock problem is proposed. The algorithm is tested on real data coming from building construction and is compared with commercial products for cutting used by professionals.

Parameter Identification of *E. coli* fermentation Process

In [83, 188] ACO algorithm for parameter identification of *E. coli* fermentation process is proposed. Influence of the ACO parameters on algorithm performance is analyzed in [81]. The hybrid algorithms combine advantages of the used algorithms. In [198] ACO and GA are combined. First is apply ACO for several iterations and achieved solutions are used as initial population for GA. Thus the needed computational resources like time and memory are decreased.

ACO algorithm performance is compared to the GA performance is a case of modelling of fermentation process dynamics in [85, 190]. Hybrid ACO-GA algorithms are proposed in [87, 195].

GPS Surveying Problem

Global Positioning System (GPS) Surveying Problems is an NP-hard problem. For designing GPS surveying network, a given set of earth points must be observed consecutively. The survey cost is the sum of the distances or time to go from one point to another one. This kind of problems is hard to be solved with traditional numerical methods. In [48, 50, 52] different variants of ACO are apply for solving the problem. Several local search procedures are developed and used after every ACO iteration for improvement of achieved solutions [46, 55].

Partition Graph Coloring Problem

Partition Graph Coloring Problem belongs to a class of combinatorial optimization problems commonly referred to as generalized network design problems. It belongs to the class of NP-complete problems. This problem has important real-world applications in telecommunication, network design, scheduling problems, resource allocation, transportation problems, software engineering, etc. Given an undirected graph $G = (V, E)$ and a partition of its nodes into p node sets $V_1, \ldots, V_p$, the *partition graph coloring problem* consists in finding a subset $V^* \subseteq V$ containing exactly one node from each node set V_i, $i \in \{1, ...p\}$ and such that the chromatic number of the graph induced in G by V^* is minimum. In [75] ACO algorithm for solving partition graph coloring problem is proposed. In [76] local search procedure, which is combined with the ACO algorithm is developed. In a result improved algorithm performance and results are achieved.

GRID Scheduling Problem

GRID computing is a form of distributed computing that involves coordinating and sharing computing, application, data storage or network resources across dynamic and geographically dispersed organizations. The goal of grid task scheduling is to achieve high system throughput and to match the application needed with the available computing resources. This is matching of resources in a non-deterministically shared heterogeneous environment. The complexity of scheduling problem increases with the size of the grid and becomes highly difficult to solve effectively. In [45] an ACO algorithm for GRID scheduling problem is proposed. The algorithm can work in dynamic way, when new tasks arrive and need to be scheduled. If some of the nodes of the grid stops to work, their queues are returned and rescheduled without stopping the GRID work. Thus the GRID can work in efficient way.

Image Edge Detection

The aim of the image edge detection is to find the points, in a digital image, at which the brightness level changes sharply. Normally they are curved lines called edges. Edge detection is a fundamental tool in image processing, machine vision and computer vision, particularly in the areas of feature detection and feature extraction. Edge detection may lead to finding the boundaries of objects. It is one of the fundamental steps in image analysis. Edge detection is a hard computational problem. In [78] is proposed special algorithm based on ACO for image edges detection. With algorithm parameters can be controlled how detailed to be the edges.

Sensor Layout Problem

Wireless sensor networks are formed by spatially distributed sensors, which communicate in a wireless way. This network can monitor various kinds of environment and physical conditions like movement, noise, light, humidity, images, chemical substances etc. A given area needs to be fully covered with minimal number of sensors and the energy consumption of the network needs to be minimal too. A special device called high energy communication node collects data from all the sensors and sends them to be processed. The problem is multi-objective with two objective function. In [66] mono-objective ACO algorithm to solve the problem is proposed. The algorithm is converted to mono-objective by multiplying the two objective functions. In [61, 68, 224] the problem is converted to mono-objective using weighted sum and is investigated on optimal weights. Multi-objective ACO algorithm is proposed in [60, 62, 63, 65]. The mono-objective variants of the ACO are compared with multi-objective and is measured the distance between the Pareto front of multi-objective ACO algorithm and the solutions found by the mono-objective variants. In [67, 71–73] the influence of the algorithm parameters on algorithm performance and optimal parameters tuning are investigated.

Transport Modeling

In [51] is proposed ACO algorithm for transport modeling and the solution is a set of non-dominated solutions. The model can be used for optimization of the transportation when different types of transportation is used. It shows the preference of passengers of transportation time and price.

Workforce Planning Problem

The workforce planning is a difficult optimization problem. It is important real life problem which helps organizations to determine workforce which they need. The problem consists to select set of employers from a set of available workers and to assign this staff to the tasks to be performed. The objective is to minimize the costs associated to the human resources needed to fulfil the work requirements. A good workforce planing is important for an organization to accomplish its objectives. The complexity of this problem does not allow the utilization of exact methods for instances of realistic size. The problem has very strong constraints and even for constructive methods like ACO sometimes is difficult to find feasible solutions. In [79, 80] are proposed ACO algorithms for solving workforce planning problem. In [196] the ACO algorithm is combined with a local search procedure. The procedure is applied only on unfeasible solutions and only ones disregarding if the new solution is feasible or not. Thus this procedure is not time consuming. The new hybrid algorithm needs less number of iterations (computational time) to find close to optimal solutions. In [82] is explored algorithm performance according number of ants and number of iterations.

4 Other Metaheuristic Algorithms

4.1 Cuckoo Search

Another metaheuristic algorithm inspired by animal behavior phenomena is CS. It takes as a metaphor the reproduction strategy of cuckoo species in the nature. CS is based on the interesting breeding behaviour called brood parasitism that certain cuckoo species exhibit. The CS algorithm was proposed by Xin-She Yang and Suash Deb [239].

In [219] CS algorithm is adapted and applied for a model parameter identification of an *E. coli* fed-batch fermentation process. The dynamics of bacteria growth and substrate (glucose) utilization is described by a system of ordinary non-linear differential equations. Using real experimental data set a parameter optimization is performed. As a result, a model with high degree of accuracy is obtained applying the CS. The simulation results and comparison with GA and ACO algorithm confirm the effectiveness of the applied CS algorithm in solving a fermentation model parameter identification problem.

Further, in [5] CS is applied to parameter identification of *S. cerevisiae* fed-batch fermentation process model. Aiming to improve the model accuracy and the algorithm convergence time, a detailed pre-tests adjustments of CS parameters have been done in order to reflect the optimization object requirements. Obtained results confirm the effectiveness of the CS algorithm for solving such a complex problem.

In [220] a GN-model of the optimization process based on cuckoos' behaviour is proposed. Developed model executes the algorithm procedure performing basic steps and realizes an optimal search. Further, in [179], the functioning and the results of the work of CS and FA, are described using the apparatus of GNs.

4.2   Firefly Algorithm

The FA, proposed in [238], is a metaheuristic algorithm which is inspired from flashing light behaviour of fireflies in nature. The pattern of flashes is often unique for a particular species of fireflies. The two basic functions of such flashes are to attract mating partners or communicate with them, and to attract potential victim. Additionally, flashing may also serve as a protective warning mechanism.

In [173] FA is proposed and tested for application to the parameter identification of a non-linear dynamical model of a fermentation process. The identification procedure is formulated as an optimization problem. Numerical and simulation results reveal that correct and consistent results can be obtained using the FA. As a result, an adequate, high-quality mathematical model is obtained. The results confirm that the FA is powerful and efficient tool for identification of the parameters in the non-linear dynamic model of fermentation processes.

Further, in [212], the performance of FA and GA are compared for an model identification problem. Parameters of both meta-heuristics are tuned on the basis of several pre-tests. Based on the numerical and simulation result, it is shown that the model obtained by the FA is more accurate and adequate than the one obtained using the GA. Presented results prove FA superiority and powerfulness in solving considered problem.

An application of FA to PID controller parameter tuning in SP is presented in [214]. The FA adjustments are done based on several pre-tests. Simulation results indicate that the applied FA is effective and efficient. Good closed-loop system performance is achieved on the basis of the considered PID controllers tuning procedures. Moreover, the observed results are compared to the ones obtained by applying GA. The comparison of both metaheuristics shows superior performance for FA PID controller tuning of the considered non-linear control system than GA tuned controller.

In [184] the apparatus of generalized nets is applied to describe the FA. Although the FA has many similarities with other swarm intelligence based algorithms, such as Particle Swarm Optimization, ABC algorithm, and Bacterial Foraging Algorithm, it is indeed much simpler both in concept and implementation. The proposed generalized net model provides the opportunity to describe the logic of FA.

In a series of papers different FA hybrid algorithms are proposed [175, 191, 218].

4.3 Artificial Bee Colony Algorithms

Artificial bee colony algorithm is a swarm algorithm introduced by Karaboga [102] that simulates the foraging behaviour of the honey bees. ABC is one of the competitive optimization algorithms that is highly effective for the purpose of multi-dimensional numerical problems.

In [177] ABC algorithm, based on the foraging behaviour of honey bees, is introduced for a numerical optimization problem. To demonstrate the usefulness of the presented approach, the ABC algorithm is applied to parameter identification of an *E. coli* MC4110 fed-batch fermentation process model. Eight ABC algorithms are applied. The best ABC algorithms are compared with four population-based biological-inspired algorithms, namely GA, ACO, FA and CS algorithm. The results clearly show that the ABC algorithm outperforms the biological-inspired algorithms under consideration, taking into account the overall search ability and computational efficiency.

Further, in [6] the ABC algorithms, have been applied to parameter identification of *S. cerevisiae* fed-batch fermentation process model. Obtained results show the efficacy of the ABC algorithm for solving considered parameter identification problem.

4.4 Evolutionary Algorithms

Among the optimization methods, the EA which are generally known as general purpose optimization algorithms are known to be capable of finding the near-optimum solution to the numerical real-valued test problems.

The EA are applied for parameter identification of the following bioprocess models: gluconic acid fermentation by *Aspergillus Nniger* [92, 93, 98, 99, 115, 116], simultaneous saccharification—fermentation process from starch to ethanol [95, 117, 119, 130], ethanol and beer production with free and immobilized cells [3, 112, 113, 151, 223], α-amylase production by *Bacillus subtilis* [93, 97, 115, 117, 122], proteins production by *E. coli* [115, 123, 241–244], poly-β-hydroxybutyrate (PHB) production by mixed culture [93, 95, 96, 115, 117, 119, 126], continuous alcohol fermentation with immobilized yeasts *S. cerevisiae* BO 213 [117, 118], fermentation of yoghurt starter culture by strains *Streptococcus thermophilus* 13a and *Lactobacillus bulgaricus* 2-11 [94, 117].

The EA are applied at optimal parameter tuning as part of software sensor's synthesis: in [124] for monitoring of activated sludge wastewater treatment processes, in [125] for monitoring of dynamics of fed-batch bioprocesses in stirred tank reactors.

In [120, 121], the EA are built in an interactive teaching system for kinetics modelling of biotechnological processes.

In [29, 30] a method for adaptation of the equivalent linearization technique of the non-linear state equation with the coefficients obtained by the fuzzy rules automatically generated using evolutionary strategy is proposed.

4.5 Bat Algorithm

The BA is a metaheuristic optimization method proposed by Yang in [240] based on the behaviour of micro bats which use echolocation pulses with different emission and sound.

In [149] authors describe the (B) and propose an enhancement using fuzzy and intuitionistic fuzzy systems to dynamically adapt BA parameters. New methods for dynamic parameter adaptation in the BA using Type-1, interval Type-2 fuzzy logic and intuitionistic fuzzy logic are proposed. The presented methods perform dynamic adaptation of considered parameters, during the algorithm run, trying to improve the BA performance.

Further, in [180] the BA is applied to different optimization problems—five benchmark functions. The modification of the BA integrating Type-1 and Interval Type-2 fuzzy logic systems was successfully applied. The both fuzzy systems perform well even the increase of the benchmark functions complexity. Obtained results show that the fuzzy systems, Type-1 and Interval Type-2, have similar performance in dynamic BA parameter adaptation.

One more successful modification of the BA integrating Type-1 and Type-2 fuzzy logic systems is presented in [150]. A deeper integration of a fuzzy system that works in conjunction with the algorithm and its parameters is proposed. The aim is to reach optimal results leveraging the convergence speed of the algorithm and avoiding premature convergence to a local minimum. The both applied fuzzy systems perform well even increasing the benchmark functions complexity (dimensions).

In [192] a hybrid scheme using BA and SQP method is introduced. In the hybrid BA-SQP, the role of the BA is to generate feasible solutions to the problem. The role of the SQP is to exploit the information gathered by the BA. This process obtains a solution which is at least as good as—but usually better than—the best solution devised by BA. To demonstrate the usefulness of the presented approach, the hybrid scheme is applied for parameter identification of E. coli MC4110 fed-batch fermentation process model. This case study has not been solved previously in the literature, applying BA or BA hybrids. Moreover, a comparison with both standard BA and SQP method is presented. The results show that the hybrid BA-SQP takes the advantages of both BA's global search ability and SQP's local search ability, hence enhances the overall search ability and computational efficiency.

In [176] the GNs theory is used to model BA. The proposed generalized net model provides the opportunity to describe the logic of BA, executing the algorithm procedure basic steps and realizing an optimal search.

4.6 Water Cycle Algorithm

The idea of the Water Cycle Algorithm (WCA) is inspired from nature and based on
the observation of water cycle and how rivers and streams flow downhill towards the
sea in the real world [36].

In [33] authors describe the enhancement of the WCA using a fuzzy inference
system to dynamically adapt its parameters. The idea of intuitionistic fuzzy systems
for WCA parameter adaptation is discussed, too. The original WCA is compared in
terms of performance with the proposed method called Water Cycle Algorithm with
Dynamic Parameter Adaptation (WCA-DPA). Simulation results on a set of well-
known test functions show that the WCA can be improved with a fuzzy dynamic
adaptation of the parameters.

4.7 Simulated Annealing Algorithm

SA is a stochastic relaxation technique [109]. SA is named by analogy to the annealing
of solids, in which a crystalline solid is heated to its melting point and then allowed to
cool gradually until it is again in the solid phase at some nominal temperature. At the
absolute zero final temperature, the resulting solid achieves its most regular crystal
configuration—corresponding to a (global) minimal value of the system's energy.

In [210] the metaheuristic SA is applied for parameter identification of non-linear
model of cultivation process. The obtained criteria values show that the developed
model is adequate and has a high degree of accuracy. The presented results are a
confirmation of successful application of the SA algorithm and of the choice of SA
algorithm parameters.

Further, in [174] the GA and SA are compared. Real *E. coli* MC4110 fed-batch
fermentation process is considered. Numerical and simulation results reveal that
correct and consistent results can be obtained using both metaheuristic searchers.
Resulting mathematical models predict adequate and accurate the process dynamics.
Effectiveness and efficiency comparisons of both optimization algorithms are presented. The results show that GA obtains identical to or better than SA values of the
objective function. Considering the execution time SA has the advantage over GA.
SA obtains good solution in a short time compared to GA.

In [215] two local search trajectory methods—SA and TA are applied for a PID
controller parameter tuning. Simulation results indicate that the applied trajectory
methods are effective and efficient. Good closed-loop system performance is achieved
on the basis of the considered PID controllers tuning procedures.

SA algorithm is successfully applied for tuning of PID controllers [213] and the
results are compared to the some population-based metaheuristic algorithms and
single point search algorithms in [217].

In [54] is prepared SA algorithm for solving GPS surveying problem. Three local
search trajectories are applied. In [42] Memetic SA is proposed for the same problem.

The difference in this variant is using a set of neighbour solutions instead of only one.

SA algorithm for grid scheduling problem is applied in [47].

Metal nanoparticle simulation is important for metallurgical industry. Several variants of SA algorithm are applied in [127–129] for mono- and bi-metal nanoparticle simulation.

4.8   Tabu Search Algorithm

The TA method, suggested by Dueck and Scheuer [35] is a deterministic analogue of SA. The main difference between these two methods is the level of acceptance of a lower quality solution at each step. In a TA method, this type of decision is taken in a deterministic manner, without having to use the principles of thermodynamic annealing.

In [110] the problem of parameter estimation using TS algorithm is examined. A case study considering the estimation of the parameters of a non-linear dynamic model of *E. coli* fermentation is taken as a test problem. The parameter estimation problem is stated as a non-linear programming problem. This problem is known to be frequently ill-conditioned and multimodal. Thus, traditional (gradient-based) local optimization methods fail to arrive at satisfactory solutions. To overcome this limitation, the use of TS algorithm is explored. Parameters of the TS algorithm are tuned based on several pre-tests. The model parameter estimates with high accuracy are obtained applying proposed tuned TS.

Results from TS application for tuning of PID controllers and a comparison of the TS outcomes with the GA, ACO, FA, SA, and TA performance are presented in [213, 217].

Efficient TS algorithm is proposed in [53] for solving GPS surveying problem.

5   Conclusion

Presented here is a brief overview of the intensive research performed by the scientists from the Bulgarian Academy of Sciences for the last 20 years on metaheuristic algorithms and their modifications and hybridizations.

In the article are considered typical representatives of the three main groups of metaheuristic algorithms, namely:

- Evolutionary algorithms:

  - Genetic Algorithms,
  - Genetic Programming;

- Swarm-based algorithms:

- Ant Colony Optimization,
- Artificial Bee Colony,
- Firefly Algorithm,
- Cuckoo Search Algorithm,
- Bat Algorithm;

- Trajectory-based algorithms:

  - Tabu Search,
  - Simulated Annealing,
  - Threshold Acceptance.

A number of metaheuristic algorithms modifications have been proposed, as well as various hybrid algorithms. New application-specific metaheuristics and software packages for their implementation are also discussed.

Metaheuristic algorithms have been successfully applied to a wide range of problems, including:

- model parameter identification problems considering bio-processes such as fermentations of *E. coli, S. cerevisiae, Aspergillus Nniger, Bacillus subtilis, Streptococcus thermophilus* 13a, *Lactobacillus bulgaricus*, etc.;
- solution of stochastic optimization problem for capturing of uncertainties in heat-integrated ATAD system for wastewater treatment;
- learning of ANN applied for description of the process of ATAD of sludge in bioreactors wastewater treatment system or the process of bioconversion of crude glycerol;
- minimizing the environmental impact of multipurpose batch chemical or biochemical plants with application in dairy industry;
- PID controller parameter tuning;
- synthesis of feed rate profiles;
- simulation of the motor units activity;
- multiple Knapsack problem;
- cutting stock problem;
- GPS survey problem;
- sensor layout problem;
- transport modelling;
- workforce planning problem;
- image edge detection;
- GRID scheduling problem;
- partition graph colouring problem;
- etc.

Considered here applications of this collective concept of a series of intelligent strategies in diverse fields of application confirm its efficiency and demonstrates a part of its increasing popularity. As such, metaheuristic algorithms and their modifications and hybridizations are proven again as an important part of modern optimization.

Acknowledgements Work presented here is partially supported by the National Science Fund of Bulgaria under grants DN02/10 "New Instruments for Knowledge Discovery from Data, and their Modelling" and KP-06-H3/23 "Interactive System for Education in Modelling and Control of Bioprocesses (InSEMCoBio)".

References

1. Adonyi, R., Shopova, E.G., Vaklieva-Bancheva, N.G.: Optimal schedule of the dairy manufacture. Chem. Biochem. Eng. Quart. **23**(2), 231–237 (2009)
2. Adonyi, R., Kirilova, E.G., Vaklieva-Bancheva, N.G.: Systematic approach for designing and activities scheduling of supply chain network. Bulgar. Chem. Commun. **45**(3), 288–295 (2013)
3. Angelov, M., Kostov, G., Ignatova, M., Koprinkova-Hristova, P., Lyubenova, V., Popova, S.: Kinetics and Control of Bioprocesses, 1st edn. Agency 7D, Plovdiv (2012) (in Bulgarian)
4. Angelova, M., Tzonkov, S., Pencheva, T.: Parameter identification of a fed-batch cultivation of *S. cerevisiae* using genetic algorithms. Serdica J. Comput. **4**(1), 11–18 (2010)
5. Angelova, M., Roeva, O., Pencheva, T.: Cuckoo search algorithm for parameter identification of fermentation process model. In: Lecture Notes in Computer Science, vol. 11189, pp. 39–47 (2019)
6. Angelova, M., Roeva, O., Pencheva, T.: Artificial bee colony algorithm for parameter identification of fermentation process model. In: Lecture Notes in Electrical Engineering, vol. 574, pp. 317–323 (2019)
7. Angelova, M., Roeva, O., Pencheva, T.: InterCriteria analysis of crossover and mutation rates relations in simple genetic algorithm. In: Proceedings of the Federated Conference on Computer Science and Information Systems, vol. 5, pp. 419–424 (2015)
8. Angelova, M., Pencheva, T.: How to assess multi-population genetic algorithms performance using intuitionistic fuzzy logic. In: Advanced Computing in Industrial Mathematics, Studies in Computational Intelligence, vol. 793, pp. 23–35 (2018)
9. Angelova, M., Pencheva, T.: Genetic operators significance assessment in simple genetic algorithm. In: Lecture Notes in Computer Science, vol. 8353, pp. 223–231 (2014)
10. Angelova, M., Pencheva, T.: Genetic operators' significance assessment in multi-population genetic algorithms. Int. J. Metaheurist. **3**(2), 162–173 (2014)
11. Angelova, M., Pencheva, T.: Intercriteria analysis approach for comparison of simple and multi-population genetic algorithms performance. In: Studies in Computational Intelligence, vol. 795, pp. 117–130 (2019)
12. Angelova, M., Pencheva, T.: Intercriteria analysis of multi-population genetic algorithms performance. In: Annals of Computer Science and Information Systems, vol. 13, pp. 77–82 (2017)
13. Angelova, M., Atanassov, K., Pencheva, T.: Intuitionistic fuzzy logic as a tool for quality assessment of genetic algorithms performances. In: Studies in Computational Intelligence, vol. 470, pp. 1–13 (2013)
14. Angelova, M., Tzonkov, S., Pencheva, T.: Modified multi-population genetic algorithm for yeast fed-batch cultivation parameter identification. Int. J. Bioautomat. **13**(4), 163–172 (2009)
15. Angelova, M., Pencheva, T.: Improvement of multi-population genetic algorithm convergence time. In: Monte Carlo Methods and Applications, pp. 1–9 (2013)
16. Angelova, M., Atanassov, K., Pencheva, T.: Intuitionistic fuzzy logic based quality assessment of simple genetic algorithm. In: Proceedings of the 16th International Conference on System Theory, Control and Computing (ICSTCC), vol. 2, Sinaia, Romania, October 12–14, 2012
17. Angelova, M., Atanassov, K., Pencheva, T.: Multi-population genetic algorithm quality assessment implementing intuitionistic fuzzy logic. In: Proceedings of the Federated Conference on Computer Sciences and Information Systems—FEDCSIS 2012, Workshop on Computational Optimization—WCO'2012, Wrocław, Poland, pp. 365–370 (2012)

18. Angelova, M., Atanassov, K., Pencheva, T.: Intuitionistic fuzzy estimations of purposeful model parameters genesis. In: IEEE 6th International Conference "Intelligent Systems", Sofia, Bulgaria, pp. 206–211 (2012)
19. Angelova, M., Melo-Pinto, P., Pencheva, T.: Modified simple genetic algorithms improving convergence time for the purposes of fermentation process parameter identification. WSEAS Trans. Syst. **11**(7), 256–267 (2012)
20. Angelova, M., Pencheva, T.: Purposeful model parameter genesis in multi-population genetic algorithm. In: 10-th National Young Scientific-Practical Session, Sofia, Bulgaria, pp. 250–254 (2012) (in Bulgarian)
21. Angelova, M., Pencheva, T.: Quality assessment procedure for genetic algorithms performance using intuitionistic fuzzy logics. In: 10-th National Young Scientific-Practical Session, Sofia, Bulgaria, pp. 244–249 (2012) (in Bulgarian)
22. Angelova, M., Atanassov, K., Pencheva, T.: Purposeful model parameters genesis in simple genetic algorithms. Comput. Math. Appl. **64**, 221–228 (2012)
23. Angelova, M., Pencheva, T.: Algorithms improving convergence time in parameter identification of fed-batch cultivation. Comptes rendus de l'Académie bulgare des Sciences **65**(3), 299–306 (2012)
24. Angelova, M., Pencheva, T.: Sensitivity analysis for the purposes of parameter identification of a S. cerevisiae fed-batch cultivation. In: Lecture Notes in Computer Science, vol. 7116, pp. 165–172 (2012)
25. Angelova, M., Pencheva, T.: Tuning genetic algorithm parameters to improve convergence time. Int. J. Chem. Eng. Article ID 646917 (2011)
26. Angelova, M., Tzonkov, S., Pencheva, T.: Genetic algorithms based parameter identification of yeast fed-batch cultivation. In: Lecture Notes in Computer Science, vol. 6046, pp. 224–231 (2011)
27. Angelova, M., Pencheva, T.: An investigation of the effect of the genetic operators execution sequence on the fermentation process parameter identification. In: 8th National Young Scientific-Practical Session, Sofia, Bulgaria, pp. 50–55 (2010) (in Bulgarian)
28. Atanassov, K., Pencheva, T.: Generalized net model of simple genetic algorithm modifications. Issues in Intuitionistic Fuzzy Sets and Generalized Nets **10**, 97–106 (2013)
29. Bartczuk, L., Przybyl, A., Koprinkova-Hristova, P.: New method for non-linear correction modelling of dynamic objects with genetic programming. In: Lecture Notes in Artificial Intelligence (Subseries of Lecture Notes in Computer Science), vol. 9120, pp. 318–329 (2015)
30. Bartczuk, L., Przybyl, A., Koprinkova-Hristova, P.: New method for nonlinear fuzzy correction modelling of dynamic objects. In: Lecture Notes in Computer Science (including subseries Lecture Notes in Artificial Intelligence and Lecture Notes in Bioinformatics), vol. 8467 LNAI (PART 1), pp. 169–180 (2014)
31. Bartczuk, L., Lapa, K., Koprinkova-Hristova, P.: A new method for generating of fuzzy rules for the nonlinear modelling based on semantic genetic programming. In: Lecture Notes in Computer Science (including subseries Lecture Notes in Artificial Intelligence and Lecture Notes in Bioinformatics), vol. 9693, pp. 262–278 (2016)
32. Bonabeau, E., Dorigo, M., Theraulaz, G.: Swarm Intelligence: From Natural to Artificial Systems. Oxford University Press, New York (1999)
33. Castillo, O., Ramirez, E., Roeva, O.: Water cycle algorithm augmentation with fuzzy and intuitionistic fuzzy dynamic adaptation of parameters. Notes on Intuitionistic Fuzzy Sets **23**(1), 79–94 (2017)
34. Dorigo, M., Stutzle, T.: Ant Colony Optimization. MIT Press (2004)
35. Dueck, G., Scheuer, T.: Threshold accepting: a general purpose algorithm appearing superior to simulated annealing. J. Comput. Phys. **90**, 161–175 (1990)
36. Eskandar, H., Sadollah, A., Bahreininejad, A., Hamdi, M.: Water cycle algorithm–a novel metaheuristic optimization method for solving constrained engineering optimization problems. Comput. Struct. **110–111**, 151–166 (2012)
37. Evtimov, G., Fidanova, S.: Ant Colony optimization algorithm for 1D cutting stock problem. In: Gueorguiev, K., Gueorguiev, I. (eds.) Advanced Computing in Industrial Mathematics, Studies of Computational Intelligence, Springer, vol. 728, pp. 25–31 (2018)

38. Fidanova, S.: Evolutionary algorithm for multiple knapsack problem. In: International Conference Parallel Problems Solving from Nature, Real World Optimization Using Evolutionary Computing, Granada, Spain (2002)

39. Fidanova, S.: ACO algorithm for MKP using various heuristic information. In: 5th International Conference of Numerical Methods and Applications, Lecture Notes in Computer Science, vol. 2542, pp. 434–440 (2003)

40. Fidanova, S.: Monte Carlo method for multiple knapsack problem. In: International Conference of Large-Scale Scientific Computing, Lecture Notes in Computer Science, vol. 2907, pp. 136–143 (2004)

41. Fidanova, S.: Ant colony optimization for multiple knapsack problem and model bias. In: NAA'04, Lecture Notes in Computer Sciences, Springer, vol. 3401, pp. 282–289 (2005)

42. Fidanova, S.: Simulated annealing: A Monte Carlo method for GPS surveying. In: Computational Science–2006, Lecture Notes in Computer Science, vol. 3991, pp. 1009–1012 (2006)

43. Fidanova, S.: Ant colony optimization and multiple knapsack problem. In: Renard, J.-P. (ed.) Handbook of Research on Nature Inspired Computing for Economics and Management, Chapter 33, Idea Group Inc., pp. 498–509 (2006)

44. Fidanova, S.: Heuristics for multiple knapsack problem. In: IADIS Applied Computing 2005 Conference. Algavre, Portugal, pp. 255–260 (2005)

45. Fidanova, S., Durchova, M.: Ant algorithm for grid scheduling problem. In: Large Scale Computing, Lecture Notes in Computer Science, vol. 3743, pp. 405–412 (2006)

46. Fidanova, S.: Hybrid heuristics algorithms for GPS surveying problem. In: Numerical Methods and Applications, Lecture Notes in Computer Science, vol. 4310, pp. 239–248 (2007)

47. Fidanova, S.: Simulated annealing for grid scheduling problem. In: IEEE Computer Society, Proceedings of IEEE JVA International Symposium on Modern Computing, Sofia, Bulgaria, pp. 41–45 (2006)

48. Fidanova, S.: An heuristic method for GPS surveying problem. In: Computational Science, Lecture Notes in Computer Science, vol. 4490, pp. 1084–1090 (2007)

49. Fidanova, S.: Probabilistic model of an colony optimization for multiple knapsack problem. In: Large Scale Scientific Computing, Lecture Notes in Computer Science, vol. 4818, pp. 545–552 (2008)

50. Fidanova, S.: MMAS and ACS for GPS surveying problem. In: Proceeding of International Conference on Evolutionary Computing, Sofia, Bulgaria, pp. 87–91 (2008)

51. Fidanova, S.: Metaheuristic method for transport modelling and optimization. In: Margenov, S., Angelova, G., Agre, G. (eds.) Innovative Approaches and Solutions in Advanced Intelligent Systems, Studies in Computational Intelligence, Chapter 19, Springer, Germany, vol. 648, pp. 295–302 (2016)

52. Fidanova, S., Saleh, H.A.: Ant colony optimization for scheduling the surveying activities of satellite positioning networks. In: International Conference on Information Systems and Data Grids, Sofia, Bulgaria, pp. 43–54

53. Fidanova, S., Saleh, H.A.: Efficient Tabu search procedures for the GPS surveying. In: Proceedings of Metaheuristic International Conference, Vienna, pp. 342–347 (2005)

54. Fidanova, S., Alba, E., Molina, G.: Memetic simulated annealing for GPS surveying problem numerical analysis and applications. In: Lecture Notes in Computer Science, vol. 5434, pp. 281–288 (2009)

55. Fidanova, S., Alba, E., Molina, G.: Hybrid ACO algorithm for the GPS surveying problem. In: Large Scale Scientific Computing. Lecture Notes in Computer Science, vol. 5910, pp. 318–325 (2010)

56. Fidanova, S., Atanassov, K., Marinov, P., Parvathi, R.: Ant colony optimization for multiple knapsack problem with controlled start. Int. J. Bioautomat. **13**(4), 271–280 (2009)

57. Fidanova, S., Atanassov, K., Marinov, P.: Start strategies of ACO applied on subset problems. In: Numerical methods and applications, Lecture Notes in Computer Science No. 6046, Springer, Germany, pp. 248–255 (2011)

58. Fidanova, S., Atanassov, K., Marinov, P.: Intuitionistic fuzzy estimation of the ant colony optimization starting points. In: Large Scale Scientific Computing, Lecture Notes in Computer Science, vol. 7116, pp. 219–226 (2012)

59. Fidanova, S., Atanassov, K.: Flaying ant colony optimization algorithm for combinatorial optimization. Studia Informatica, J. Polish Inf. Soc. **38**(4), 31–40 (2018)

60. Fidanova, S., Shindarov, M., Marinov, P.: Wireless sensor positioning ACO algorithm. In: Atanassov, K., Kasparzik, J. (eds.) Studies of Computational Intelligence, Recent Contributions in Intelligent Systems, Studies of Computational Intelligence, vol. 657, book Chapter 3, pp. 33–44 (2017)

61. Fidanova, S., Shindarov, M., Marinov, P.: Mono-objective algorithm for wireless sensor layout. In: Proceedings of OMCO-NET Conference, Southempton, UK, pp. 57–63 (2012)

62. Fidanova, S., Shindarov, M., Marinov, P.: Multi-objective ant algorithm for wireless sensor network positioning. Proc. Bulgar. Acad. Sci. **66**(3), 353–360 (2013)

63. Fidanova, S., Shindarov, M., Marinov, P.: Optimal sensor layout using multi-objective meta-heuristic. In: Proceedings of International Conference of Information Systems and Grid Technologies, Sofia, Bulgaria, St. Kliment Ohridski University Press, pp. 114–122 (2011)

64. Fidanova, S., Marinov, P., Atanassov, K.: Sensitivity analysis of ACO start strategies for subset problems. In: Numerical Methods and Applications, Lecture Notes in Computer Science, vol. 6046, pp. 256–263 (2011)

65. Fidanova, S., Marinov, P., Alba, E.: ACO for optimal sensor layout. In: Filipe, J., Kacprzyk, J. (eds.) Proceedings of International Conference on Evolutionary Computing, Valencia, Spain, SciTePress—Science and Technology Publications Portugal, pp. 5–9 (2010)

66. Fidanova, S., Marinov, P., Alba, E.: Ant algorithm for optimal sensor deployment. In: Madani, K., Correia, A.-D., Rosa, A., Filipe, J. (eds.) Computational Intelligence, Studies in Computational Intelligence, vol. 399, pp. 21–29 (2012)

67. Fidanova, S., Marinov, P.: Influence of the number of ants on mono-objective ant colony optimization algorithm for wireless sensor network layout. In: Proceedings of BGSIAM'12, Sofia, Bulgaria, pp. 59–66 (2012)

68. Fidanova, S., Marinov, P., Alba, E.: Wireless sensor network layout. In: Sabelfeld, K.K., Dimov, I. (eds.) Monte Carlo methods and applications, Chapter 10, De Gruyter, Berlin, Germany, pp. 87–96 (2012)

69. Fidanova, S., Marinov, P.: Influence of the parameter R on ACO start strategies. In: Proceedings of BGSIAM '11, Sofia, Bulgaria, pp. 38–43

70. Fidanova, S., Marinov, P.: Ant colony optimization start strategies performance according some of the parameters. In: Dimov, I., Farago, I., Vulkov, L. (eds.) Numerical Analysis and Applications, Lecture Notes in Computer Sciences, Springer, Germany, vol. 8236, pp. 287–294 (2013)

71. Fidanova, S., Marinov, P.: Number of ants versus number of iterations on ant colony optimization algorithm for wireless sensor layout. In: Proceedings of International Conference on Robotics Automation and Mechatronics, RAM 2013, Bankya, Bulgaria, pp. 90–93 (2013)

72. Fidanova, S., Marinov, P., Paprzycki, M.: Influence of the number of ants on multy-objective ant colony optimization algorithm for wireless sensor network layout. In: Large-Scale Scientific Computing, Lecture Notes in Computer Science, vol. 8353, pp. 208–215 (2014)

73. Fidanova, S., Marinov, P., Paprzycki, M.: Multi-objective ACO algorithm for WSN layout: performance according number of ants. J. Metaheurist. **3**(2), 149–161 (2014)

74. Fidanova, S., Marinov, P., Atanassov, K.: New estimations of ant colony optimization start nodes. Int. J. Control Cybernet. **43**, 471–486 (2014)

75. Fidanova, S., Pop, P.: An ant algorithm for the partitioned graph coloring problem. In: Numerical Methods and Applications, Lecture Notes in Computer Science, vol. 8962, pp. 78–84 (2015)

76. Fidanova, S., Pop, P.: An improved hybrid ant-local search algorithm for the partition graph coloring problem. J. Comput. Appl. Math. **293**, 55–61 (2016)

77. Fidanova, S., Atanassova, V., Roeva, O.: Ant colony optimization application to GPS surveying problems: InterCriteria analysis. In: Uncertainty and Imprecision in Decision Making and Decision Support: Cross Fertilization, New Models and Applications, Advances in Intelligent Systems and Computing, Springer, vol. 559, pp. 251–264 (2018)

78. Fidanova, S., Ilcheva, Z.: Application of ants ideas on image edge detection. In: Large Scale Scientific Computing, Lecture Notes in Computer Science, pp. 200–207 (2016)

79. Fidanova, S., Luque, G., Roeva, O., Paprzycki, M., Gepner, P.: Ant colony optimization algorithm for workforce planning. In: FedCSIS '2017, IEEE Xplorer, IEEE catalog number CFP1585N-ART, pp. 415–419 (2017)

80. Fidanova, S., Luque, G., Roeva, O., Ganzha, M.: Ant colony optimization algorithm for workforce planning: influence of the evaporation parameter. In: Proceedings of the 2019 Federated Conference on Computer Science and Information Systems, vol. 8859747, pp. 177–181

81. Fidanova, S., Roeva, O.: Influence of ant colony optimization parameters on the algorithm performance. In: Large Scale Scientific Computing, Lecture Notes in Computer Science, vol. 10665, pp. 358–365 (2018)

82. Fidanova, S., Roeva, O., Luque, G.: Ant colony optimization algorithm for workforce planning: influence of the algorithm parameters. In: Studies of Computational Intelligence, vol. 793, pp. 119–128 (2019)

83. Fidanova, S., Roeva, O., Ganzha, M.: ACO for parameter settings of *E. coli* fed-batch cultivation model. In: Proceedings of FedCSIS 2012, IEEE Xplorer, pp. 407–414

84. Fidanova, S., Roeva, O., Mucherino, A., Kapanova, K.: InterCriteria analysis of ant algorithm with environment change for GPS surveying problem. In: Dichev, C., Agre, G. (eds.) AIMSA 2016, LNAI, vol. 9883, pp. 271–278 (2016)

85. Fidanova, S., Roeva, O., Ganzha, M.: ACO and GA for parameter settings of *E. coli* fed-batch cultivation model. In: Fidanova, S. (ed.) Recent Advances in Computational Optimization. Studies of Computational Intelligence, Springer International Publishing Switzerland, vol. 470, pp. 51–71

86. Fidanova, S., Roeva, O., Paprzycki, M., Gepner, P.: InterCriteria analysis of ACO start strategies. In: Proceedings of the 2016 Federated Conference on Computer Science and Information Systems, pp. 547–550 (2016)

87. Fidanova, S., Paprzycki, M., Roeva, O.: Hybrid GA-ACO algorithm for a model parameters identification problem. In: Proceedings of the Federated Conference on Computer Science and Information Systems (FedCSIS), WCO 2014, IEEE, Poland, pp. 413–420. https://doi.org/10.15439/2014F373

88. Fister, I., Fister, Jr., I., Yang, X.-S., Brest, J.: A comprehensive review of firefly algorithms. Swarm Evol. Comput. **13**, 34–46 (2013)

89. Georgieva, O., Tzonkov, S.: An analysis of genetic algorithms as parametric identification tool. In: International Symposium "Bioprocess Systems '2002—BioPS'02", Sofia, Bulgaria, pp. I.44–I.47

90. Goldberg, D.E.: Genetic algorithms in search, optimization and machine learning. Addison Wesley Longman, London (2006)

91. Holland, J.H.: Adaptation in Natural and Artificial Systems, 2nd edn. MIT Press, Cambridge (1992)

92. Ignatova, M., Lyubenova, V., Fernandes, C., Garsia, M., Alonso, A.: Model for control of gluconic acid fermentation by Aspergillus niger and its application for observers design. In: Proceedings of the International Conference "Agricultural and Food Sciences, Processes and Technologies", 12–13 May Sibiu. Romania, vol. 2, pp. 182–189 (2005)

93. Ignatova, M., Lyubenova, V.: Control of Biotechnological Processes—New Formalization of Kinetics: Theoretical Aspects and Applications, LAP LAMBERT Academic Publishing, GmbH & Co., Saarbrücken Germany, p. 120 (2011)

94. Ignatova, M., Lyubenova, V., Angelov, M., Kostov, G.: pH control during continuous prefermentation of yogurt starter culture by strains streptococcus thermophilus 13a and lactobacillus bulgaricus 2–11. Comptes rendus de l'Académie bulgare des Sciences **62**(12), 1587–1594 (2009)

95. Ignatova, M., Lyubenova, V.: Control of class bioprocesses using on-line information of intermediate metabolite production and consumption rates. Acta Universitatis Cibiniensis, Series E: Food Technology **11**(1), 3–16 (2007)

96. Ignatova, M., Lyubenova, V.: Adaptive control of fed-batch process for poly-beta-hydroxybutyrate production by mixed culture. Comptes rendus de l'Académie bulgare des Sciences **60**(5), 517 (2007)

97. Ignatova, M., Lyubenova, V., Eerikäinen, T., Salonen, K., Kiviharju, K.: Software sensor of substrate kinetics and its application for control of alpha-amylase production by Bacillus subtilis. Comptes rendus de l'Académie bulgare des Sciences **61**, 1449–1458 (2008)

98. Ignatova, M., Lyubenova, V., García, M., Vilas, C.: Adaptive linearizing control of gluconic acid fermentation by *Aspergillus niger*. In: National conference with international participation "Automatics and Informatics", pp. 3–6 (2006)

99. Ignatova, M., Lyubenova, V., García, M., Vilas, C., Alonso, A.: Indirect adaptive linearizing control of a class of bioprocesses–estimator tuning procedure. J. Process Control **18**(1), 27–35 (2008)

100. Ivanova, J., Petrova, P., Kirilova, E., Becker, W.: Optimal parameters of a dynamically loaded patch/layer structure against the elastic-brittle interface debonding. Eng. Trans. **65**(1), 97–103 (2017)

101. Jamil, M., Yang, X.-S.: A literature survey of benchmark functions for global optimization problems. Int. J. Math. Model. Num. Opt. **4**(2), 150–194 (2013)

102. Karaboga, D.: An idea based on honeybee swarm for numerical optimization. Technical Report TR06. Erciyes University, Engineering Faculty, Computer Engineering Department (2005)

103. Kirilova, E.G., Yankova, S., Ilieva, B., Vaklieva-Bancheva, N.G.: Modelling of bioconversion of crude glycerol from biodiesel production by using dynamic artificial neural network. In: Proceedings of the 1st Annual International Conference on Industrial, Systems and Design Engineering, 23th–26th June, 2013, Greece, CD, Athens (2013)

104. Kirilova, E.G., Vaklieva-Bancheva, N.G.: Modelling of Two-Stage ATAD Bioreactor System by Using Artificial Neural Network. In: Book Series on Computer Aided Chemical Engineering (2012). https://pdfs.semanticscholar.org/3837/6dd042e5faa785fac62928d5cdf6e05586e8.pdf

105. Kirilova, E.G., Yankova, S., Ilieva, B., Vaklieva-Bancheva, N.G.: A new approach for modeling the biotransformation of crude glycerol by using NARX ANN. J. Chem. Technol. Metallurgy **49**(5), 473–478 (2014)

106. Kirilova, E.G., Vaklieva-Bancheva, N.G., Vladova, R.: Prediction of temperature conditions of autothermal thermophilic aerobic digestion bioreactors at wastewater treatment plants. Int. J. Bioautomat. **20**(2), 289–300 (2016)

107. Kirilova, E.G., Petrova, P., Becker, W., Ivanova, J.: Influence of the geometry and the frequency range on the interface delamination in smart patch/layer structures under combined dynamic loading. ZAMM–J. Appl. Math. Mech./Zeitschrift für Angewandte Mathematik und Mechanik, First published: 24 April 2017. https://doi.org/10.1002/zamm.201600273

108. Kirilova, E.G., Vaklieva-Bancheva, N.G.: ANN modeling of a two-stage industrial ATAD system for the needs of energy integration. Bulgar. Chem. Commun. Special Issue K **50**, 97–106 (2018)

109. Kirkpatrick, S., Gelatt, C.D., Vecchi, M.P.: Optimization by simulated annealing. Science, New Series **220**(4598), 671–680 (1983)

110. Kosev, K., Trenkova, T., Roeva, O.: Tabu search for parameter identification of an fermentation process model. J. Int. Sci. Publ. Mater. Methods Technol. **6**(2), 457–464 (2012)

111. Kosev, K., Roeva, O.: Application of functional states for modeling of acetate production in *E. coli* fed-batch cultivation process. In: Fifth National Young Scientific-Practical Session, Bulgaria, pp. 62–66 (2007) (in Bulgarian)

112. Kostov, G., Pircheva, D., Naydenova, V., Iliev, V., Lyubenova, V., Ignatova, M.: Kinetics investigation of bio-ethanol production with free and immobilized cells. Comptes rendus de l'Académie bulgare des Sciences **66**(10), 1463–1472 (2013)

113. Kostov, G., Lyubenova, V., Shopska, V., Petelkov, I., Ivanov, K., Iliev, V., Denkova, R., Ignatova, M.: Software sensors for monitoring the biomass concentration and the kinetics of continuous beer fermentation with immobilized cells. Comptes rendus de l'Académie bulgare des Sciences **68**(11), 1439–1448 (2015)

114. Lapa, K., Cpałka, K., Koprinkova-Hristova, P.: New method for fuzzy nonlinear modelling based on genetic programming. In: Lecture Notes in Computer Science (including subseries Lecture Notes in Artificial Intelligence and Lecture Notes in Bioinformatics), vol. 9692, pp. 432–449 (2016)

115. Lyubenova, V.: Monitoring the kinetics of biotechnological processes. Doctor of Science Thesis, Technical Science 5.2 Electronics, Electrical Engineering and Automatics, Bulgarian Academy of Sciences, 189 pp. (2016) (in Bulgarian)

116. Lyubenova, V.: Monitoring the kinetics of measured bioprocess variables–theory and applications. Inf. Technol. Control, The John Atanasoff Society of Automatics and Informatics **14**(1), 2–12 (2017)

117. Lyubenova, V., Kostov, G.: On-line Monitoring and Control of Biotechnological Processes, 187 pp. Academic Publishing House of the University of Food Technology (2016) (in Bulgarian)

118. Lyubenova, V., Kostov, G., Denkova, R., Pircheva, D., Ignatova, M., Angelov, M.: Adaptive control of continuous fermentation with immobilized yeasts Saccharomyces cerevisiae BO 213. In: Proceedings of the 1st WSEAS International Conference on Industrial and Manufacturing Technologies, May 14–16, pp. 35–40. Greece, Athens (2013)

119. Lyubenova, V., Ignatova, M.: On-line estimation of physiological states for monitoring and control of bioprocesses. AIMS Bioeng. **4**(1), 93–112 (2017)

120. Lyubenova, V., Ignatova, M., Kostov, G.: Interactive teaching system for structural and parametric identification of bioprocess models. Comptes rendus de l'Académie bulgare des Sciences **71**(6), 820–828 (2018)

121. Lyubenova, V., Ignatova, M., Kostov, G., Shopska, V., Petre, E., Roman, M.: An interactive teaching system for kinetics modelling of biotechnological processes. In: Proceedings of the IEEE International Conference on System Theory, Control and Computing—ICSTCC 2018, October 10–12, 2018, Sinaia, Romania, pp. 366–371 (2018)

122. Lyubenova, V., Ignatova, M., Salonen, K., Kiviharju, K., Eerikäinen, T.: Control of α-amylase production by Bacillus subtilis. Bioprocess Biosyst. Eng. **34**(3), 367–374 (2011)

123. Lyubenova, V., Junne, S., Ignatova, M., Neubauer, P.: Software sensor design considering oscillating conditions as present in industrial scale fed-batch cultivations. Biotechnol. Bioeng. **110**(7), 1945–1955 (2013)

124. Lyubenova, V., Ignatova, M.: Cascade software sensors for monitoring of activated sludge wastewater treatment processes. Comptes rendus de l'Académie bulgare des Sciences **64**(3), 395–404 (2011)

125. Lyubenova, V., Ignatova, M.: Monitoring of dynamics of fed-batch bioprocesses in stirred tank reactors. Comptes rendus de l'Académie bulgare des Sciences **66**(9), 1323–1330 (2013)

126. Lyubenova, V., Ignatova, M., Novak, M., Patarinska, T.: Reaction rate estimators of fed-batch process for poly-β-hydroxybutyrate (PHB) production by mixed culture. Biotechnol. Biotechnol. Equip. **21**(1), 113–116 (2007)

127. Myasnichenko, V., Kirilov, L., Mikhov, R., Fidanova, S., Sdobnyakov, N.: Simulated annealing method for metal nanoparticle structures optimization. In: Studies of Computational Intelligence, vol. 793, pp. 277–289 (2019)

128. Myasnichenko, V., Sdobnyakov, N., Kirilov, L., Mikhov, R., Fidanova, S.: Structural instability of gold and bimetallic nanowires using Monte Carlo simulation. In: Studies in Computational Intelligence, vol. 838, pp. 133–145 (2020)

129. Myasnichenko, V., Sdobnyakov, N., Kirilov, L., Mikhov, R., Fidanova, S.: Monte-Carlo approach for optimizing of metal nanowires and nanoalloys structure. In: Lecture Notes in Computer Science, vol. 11189, pp. 133–141 (2019)

130. Ochoa, S., Lyubenova, V., Repke, J., Ignatova, M., Wozny, G.: Adaptive control of the simultaneous saccharification-fermentation process from starch to ethanol. Comput. Aided Chem. Eng., Elsevier **25**, 489–494 (2008)

131. Olteanu, M., Paraschiv, N., Koprinkova-Hristova, P.: Genetic Algorithms vs. knowledge-based control of PHB production. Cybernet. Inf. Technol. **19**(2), 104–116 (2019)

132. Olteanu, M., Paraschiv, N., Koprinkova-Hristova, P., Todorov, Y.: Genetic algorithm for system modelling. In: Proceedings of the 9th International Conference on Electronics, Computers and Artificial Intelligence, ECAI 2017, pp. 1–4 (2017)

133. Pencheva, T.: Generalized nets model of crossover technique choice in genetic algorithms. Issues Intuitionistic Fuzzy Sets Generalized Nets **9**, 92–100 (2011)

134. Pencheva, T., Atanassov, K., Shannon, A.: Generalized nets model of offspring reinsertion in genetic algorithms. Ann. "Informatics" Sect. Union Scientists Bulgaria **4**, 29–35 (2011)

135. Pencheva, T., Angelova, M.: InterCriteria analysis of simple genetic algorithms performance. Stud. Comput. Intell. **681**, 147–159 (2017)

136. Pencheva, T., Angelova, M.: Intuitionistic fuzzy logic implementation to assess purposeful model parameters genesis. Stud. Comput. Intell. **657**, 179–203 (2017)

137. Pencheva, T., Angelova, M.: Purposeful model parameters genesis in multi-population genetic algorithm. Global J. Technol. Optimizat. **5**, 164 (2014)

138. Pencheva, T., Angelova, M.: Modified multi-population genetic algorithms for parameter identification of yeast fed-batch cultivation. Bulgar. Chem. Commun. **48**(4), 713–719 (2016)

139. Pencheva, T., Angelova, M., Atanassov, K.: Genetic algorithms quality assessment implementing intuitionistic fuzzy logic, Chapter 11. In: Vasant, P. (ed.) Handbook of Research on Novel Soft Computing Intelligent Algorithms: Theory and Practical Applications, pp. 327–354. Hershey, Pennsylvania (USA), IGI Global (2013)

140. Pencheva, T., Angelova, M., Atanassov, K.: Quality assessment of multi-population genetic algorithms performance. Int. J. Sci. Eng. Res. **4**(12), 1870–1875 (2013)

141. Pencheva, T., Angelova, M., Vassilev, P., Roeva, O.: InterCriteria analysis approach to parameter identification of a fermentation process model. In: Atanassov, K.T., Castillo, O., Kacprzyk, J., Krawczak, M., Melin, P., Sotirov, S., Sotirova, E., Szmidt, E., De Tré, G., Zadrożny, S. (eds.) Novel Developments in Uncertainty Representation and Processing, Part V, vol. 401 of the Series Advances in Intelligent Systems and Computing, pp. 385–397 (2016)

142. Pencheva, T., Angelova, M., Atanassova, V., Roeva, O.: InterCriteria analysis of genetic algorithm parameters in parameter identification. Notes on Intuitionistic Fuzzy Sets **21**(2), 99–110 (2015)

143. Pencheva, T., Tzonkov, S., Hitzmann, B.: Structural and parameter identification of dissolved oxygen limitation state during yeast fed-batch cultivation. In: 2009 Annual Bulletin of the Australian Institute of High Energetic Materials, vol. 1, pp. 68–74 (2010)

144. Pencheva, T., Tzonkov, S., Hitzmann, B.: Identification of dissolved oxygen limitation state during yeast fed-batch cultivation. In: Proceedings of the 2009 Interdisciplinary Conference on Chemical, Mechanical and Materials Engineering (2009 ICCMME)—World Conference Series with Virtual Participation, Australian Institute of High Energetic Materials, Melbourne, Australia, pp. 151–156 (2009)

145. Pencheva, T., Roeva, O., Hristozov, I.: Functional State Approach to Fermentation Processes Modelling, Tzonkov, S., Hitzmann, B. (eds.), Prof. Marin Drinov Academic Publishing House, Sofia (2006)

146. Pencheva, T., Roeva, O., Angelova, M.: Investigations of genetic algorithm performance based on different algorithms for InterCriteria relations calculation. In: Lecture Notes in Computer Science, vol. 10665, pp. 390–398 (2018)

147. Pencheva, T., Roeva, O., Shannon, A.: Generalized net models of basic genetic algorithm operators. In: Studies in Fuzziness and Soft Computing, vol. 332, pp. 305–325 (2016)

148. Pencheva, T., Petrov, M., Ilkova, T., Roeva, O., Vanags, J.: Bioprocess Engineering, Viesturs, U., Tzonkov S. (eds.), Avangard Prima, Sofia, Bulgaria (2006)

149. Perez, J., Valdez, F., Roeva, O., Castillo, O.: Parameter adaptation of the Bat Algorithm, using type-1, interval type-2 fuzzy logic and intuitionistic fuzzy logic. Notes on Intuitionistic Fuzzy Sets **22**(2), 87–98 (2016)

150. Perez, J., Valdez, F., Castillo, O., Roeva, O.: Bat Algorithm with parameter adaptation using interval type-2 fuzzy logic for benchmark mathematical functions. In: 2016 IEEE 8th International Conference on Intelligent Systems, Sofia, pp. 120–127 (2016)

151. Petelkov, I., Lyubenova, V., Zlatkova, A., Shopska, V., Denkova, R., Kaneva, M., Kostov, G.: Encapsulation of brewing yeast in Alginate/Chitosan matrix: kinetic characteristics of the fermentation process at a constant fermentation temperature. Comptes rendus de l'Académie bulgare des Sciences **69**(10), 1355–1364 (2016)
152. Petrova, T., Kirilova, E.G., Becker, W., Vaklieva-Bancheva, N.G., Ivanova, J.: Optimal analysis of adhesive lightweight joints. ZAMM—J. Appl. Math. Mech./Zeitschrift für Angewandte Mathematik und Mechanik **96**(11), 1280–1290 (2016)
153. Petrova, T., Kirilova, E.G., Becker, W., Ivanova, J.: Monitoring of adhesive joint used in lightweight devices. Pliska Studia Matematica **25**, 119–128 (2015)
154. Raikova, R.T., Gabriel, D.A., Aladjov, H.T.: Experimental and modelling investigation of learning a fast elbow flexion in the horizontal plane. J. Biomech. **38**(10), 2070–2077 (2005)
155. Raikova, R.T., Gabriel, D.A., Aladjov, H.T.: Modeling investigation of learning a fast elbow flexion in the horizontal plane–prediction of muscle forces and motor units action. Comput. Methods Biomech. Biomed. Eng. **9**(4), 211–219 (2006)
156. Raikova, R.T., Aladjov, H.T.: Hierarchical genetic algorithm versus static optimization–investigation of elbow flexion and extension movements. J. Biomech. **35**(8), 1123–1135 (2002)
157. Raikova, R.T., Aladjov, H.T.: The influence of the way the muscle force is modeled on the predicted results obtained by solving indeterminate problems for a fast elbow flexion. Comput. Methods Biomech. Biomed. Eng. **6**(3), 181–196 (2003)
158. Raikova, R.T., Aladjov, H.T.: Simulation of the motor units control during a fast elbow flexion in the sagittal plane. J. Electromyogr. Kinesiol. **14**(2), 227–238 (2004)
159. Raikova, R., Aladjov, H.T.: Comparison between two muscle models under dynamic conditions. Comput. Biol. Med. **35**(5), 373–387 (2005)
160. Roeva, O.: Application of genetic algorithms in fermentation process identification. J. Bulgar. Acad. Sci., CXVI **3**, 39–43 (2003)
161. Roeva, O.: Genetic algorithms for a parameter estimation of a fermentation process model: a comparison. Int. J. Bioautomat. **3**, 19–28 (2005)
162. Roeva, O.: A modification of simple genetic algorithms. In: International Symposium "Bioprocess Systems '2005—BioPS '05", Sofia, Bulgaria, October 25, 26, 2005, I.1–I.14 (2005)
163. Roeva, O.: A modified genetic algorithm for a parameter identification of fermentation processes. Biotechnol. Biotechnol. Equip. **20**(1), 202–209 (2006)
164. Roeva, O.: Tuning of genetic algorithm for parameter identification of fed-batch cultivation process. In: Fifth National Young Scientific-Practical Session, Bulgaria, pp. 94–98 (2007) (in Bulgarian)
165. Roeva, O.: Multipopulation genetic algorithm: a tool for parameter optimization of cultivation processes models. In: Lecture Notes on Computer Science, Springer, Berlin, Heidelberg, vol. 4310, pp. 255–262 (2007)
166. Roeva, O.: Functional state modelling of *Escherichia coli* cultivation with application of genetic algorithms, Ph.D. thesis, Technical University, Sofia, Bulgaria (2007) (in Bulgarian)
167. Roeva, O.: Parameter estimation of a monod-type model based on genetic algorithms and sensitivity analysis. In: Lecture Notes on Computer Science, Springer, Berlin Heidelberg, vol. 4818, pp. 601–608 (2008)
168. Roeva, O.: Improvement of genetic algorithm performance for identification of cultivation process models. In: Artificial Intelligence Series—WSEAS, Advanced Topics on Evolutionary Computing, Book Series, pp. 34–39 (2008)
169. Roeva, O.: An application of genetic algorithms for fermentation processes optimization. In: Sixth National Young Scientific-Practical Session, Bulgaria, pp. 93–98 (2008) (in Bulgarian)
170. Roeva, O.: Feed rate profiles synthesis using genetic algorithms, AIMSA 2010. In: Lecture Notes in Artificial Intelligence, Springer, Berlin Heidelberg, vol. 6304, pp. 281–282 (2010)
171. Roeva, O.: Genetic algorithms for static optimisation of fed-batch fermentation processes, Ch. 4. In: Tzonkov, S., Drinov, M. (eds.) Contemporary Approaches to Modelling, Optimisation and Control of Biotechnological Processes, pp. 105–134. Academic Publ. House, Sofia (2010)

172. Roeva, O.: A hybrid genetic algorithm for parameter identification of bioprocess models. In: Lecture Notes on Computer Science, Springer, Germany, vol. 7116, Lecture Notes in Computer Science (including subseries Lecture Notes in Artificial Intelligence and Lecture Notes in Bioinformatics), LNCS, vol. 7116, pp. 247–255 (2012)

173. Roeva, O.: Optimization of *E. coli* cultivation model parameters using firefly algorithm. Int. J. Bioautomat. **16**(1), 23–32 (2012)

174. Roeva, O.: A comparison of simulated annealing and genetic algorithm approaches for cultivation model identification. In: Monte Carlo Methods and Applications, Walter de Gruyter GmbH & Co. KG, Genthiner Strasse 13, D-10785 Berlin/Germany, pp. 193–201 (2013)

175. Roeva, O.: Genetic algorithm and firefly algorithm hybrid schemes for cultivation processes modelling. In: Kowalczyk, R., Fred, A., Joaquim, F. (eds.) Transactions on Computational Collective Intelligence XVII, Series Lecture Notes in Computer Science, Springer, vol. 8790, pp. 196–211 (2014)

176. Roeva, O.: Bat algorithm in terms of generalized net. In: Proceedings of 15th International Workshop on Generalized Nets, Burgas, pp. 1–6 (2014)

177. Roeva, O.: Application of artificial bee colony algorithm for model parameter identification. In: Zelinka, I., Vasant, P., Duy, V., Dao, T. (eds.) Innovative Computing, Optimization and Its Applications, Studies in Computational Intelligence, vol. 741, pp. 285–303 (2018)

178. Roeva, O., Shannon, A., Pencheva, T.: Description of simple genetic algorithm modifications using generalized nets. In: IEEE 6th International Conference on IS 2012, Sofia, Bulgaria, vol. 2, pp. 178–183 (2012)

179. Roeva, O., Zoteva, D., Atanassova, V., Atanassov, K., Castillo, O.: Cuckoo search and firefly algorithms in terms of generalized net theory. Soft. Comput. **24**, 4877–4898 (2020)

180. Roeva, O., Perez, J., Valdez, F., Castillo, O.: InterCriteria analysis of bat algorithm with parameter adaptation using type-1 and interval type-2 fuzzy systems. Notes on Intuitionistic Fuzzy Sets **22**(3), 91–105 (2016)

181. Roeva, O., Kosev, K., Tzonkov, S.: Mathematical description of functional states in *E. coli* fed-batch cultivation processes. Int. J. Bioautomat. **7**, 34–45

182. Roeva, O., Kosev, K.: Genetic algorithms for feed rate profiles design. In: Proceedings of the International Conference on Automatics and informatics, Sofia, pp. I-45–I-48 (2010)

183. Roeva, O., Kosev, K., Trenkova, T.: A modified multi-population genetic algorithm for parameter identification of cultivation process models. In: Proceedings of the ICEC 2010, Valencia, pp. 348–351 (2010)

184. Roeva, O., Melo-Pinto, P.: Generalized net model of Firefly algorithm. In: Proceedings of 14th International Workshop on Generalized Nets, Burgas, pp. 22–27 (2013)

185. Roeva, O., Vassilev, P.: InterCriteria analysis of generation gap influence on genetic algorithms performance. In: Atanassov, K.T., Castillo, O., Kacprzyk, J., Krawczak, M., Melin, P., Sotirov, S., Sotirova, E., Szmidt, E., De Tré, G., Zadrożny, S. (eds.) Novel Developments in Uncertainty Representation and Processing, Part V, vol. 401 of the series Advances in Intelligent Systems and Computing, pp. 301–313 (2016)

186. Roeva, O., Vassilev, P., Angelova, M., Pencheva, T.: InterCriteria analysis of parameters relations in fermentation processes models. In: Lecture Notes in Computer Science, vol. 9330, pp. 171–181 (2015)

187. Roeva, O., Vassilev, P., Angelova, M., Su, J., Pencheva, T.: Comparison of different algorithms for InterCriteria relations calculation. In: 2016 IEEE 8th International Conference on Intelligent Systems, Sofia, September 04, 2016–September 06, pp. 567–572 (2016)

188. Roeva, O., Fidanova, S.: Metaheuristic techniques for optimization of an *E. coli* cultivation model. Biotechnol. Biotechnol. Equip. **27**(3), 3870–3876 (2013)

189. Roeva, O., Fidanova, S.: Comparison of different metaheuristic algorithms based on intercriteria analysis. J. Comput. Appl. Math. **340**, 615–628 (2018)

190. Roeva, O., Fidanova, S.: Chapter 13. Application of genetic algorithms and ant colony optimization for modeling of *E. coli* cultivation process. In: Real-World Application of Genetic Algorithms, In Tech, pp. 261–282 (2012)

191. Roeva, O., Fidanova, S.: Parameter identification of an *E. coli* cultivation process model using hybrid metaheuristics. Int. J. Metaheuristics **3**(2), 133–148 (2014)

192. Roeva, O., Fidanova, S.: Hybrid bat algorithm for parameter identification of an *E. coli* cultivation process model. Biotechnol. Biotechnol. Equip. **27**(6), 4323–4326 (2013)

193. Roeva, O., Fidanova, S., Paprzycki, M.: Comparison of different ACO start strategies based on InterCriteria analysis. In: Fidanova, S. (ed.) Recent Advances in Computational Optimization, Results of the Workshop on Computational Optimization WCO 2016. Studies in Computational Intelligence, vol. 717, pp. 53–72 (2018)

194. Roeva, O., Fidanova, S., Vassilev, P., Gepner, P.: InterCriteria analysis of a model parameters identification using genetic algorithm. In: Proceedings of the Federated Conference on Computer Science and Information Systems, vol. 5, pp. 501–506 (2015)

195. Roeva, O., Fidanova, S., Atanassova, V.: Hybrid ACO-GA for parameter identification of an E. coli cultivation process model. In: Lecture Notes in Computer Science, vol. 8353, pp. 313–320 (2014)

196. Roeva, O., Fidanova, S., Luque, G., Paprzycki, M., Gepner, P.: Hybrid Ant Colony Optimization Algorithm for Workforce Planning, FedCSIS'2018, IEEE Xplorer, pp. 233–236

197. Roeva, O., Fidanova, S., Paprzycki, M.: Influence of the population size on the genetic algorithm performance in case of cultivation process modelling. In: IEEE Proceedings of the Federated Conference on Computer Science and Information Systems (FedCSIS), WCO 2013, Poland, pp. 371–376 (2013)

198. Roeva, O., Fidanova, S., Paprzycki, M.: InterCriteria analysis of ACO and GA hybrid algorithms. In: Studies in Computational Intelligence, vol. 610, pp. 107–126 (2016)

199. Roeva, O., Tzonkov, S.: Parametric identification of fermentation processes using multy-population genetic algorithms. Technical Ideas **XL**(3–4), 18–26 (2003)

200. Roeva, O., Tzonkov, S.: Modelling of escherichia coli cultivations: acetate inhibition in a fed-batch culture. Int. J. Bioautomat. **4**, 1–11 (2006)

201. Roeva, O., Tzonkov, S.: An improvement of functional state local models of escherichia coli MC4110 fed-batch cultivation. Inf. Technol. Control, Year V **4**, 47–52 (2007)

202. Roeva, O., Pencheva, T., Shannon, A., Atanassov, K.: Generalized nets in artificial intelligence. In: Generalized Nets and Genetic Algorithms, vol. 7. Prof. Marin Drinov Academic Publishing House, Sofia (2013)

203. Roeva, O., Pencheva, T., Georgieva, Y., Hitzmann, B., Tzonkov, S.: Implementation of functional state approach for modelling of escherichia coli fed-batch cultivation. Biotechnol. Biotechnol. Equip. **18**(3), 207–214 (2004)

204. Roeva, O., Pencheva, T., Hitzmann, B., Tzonkov, S.: A genetic algorithms based approach for identification of escherichia coli fed-batch fermentation. Int. J. Bioautomat. **1**, 30–41 (2004)

205. Roeva, O., Pencheva, T., Tzonkov, S., Hitzmann, B.: Functional state modelling of cultivation processes: dissolved oxygen limitation state. Int. J. Bioautomat. **19**(1, Suppl. 1), S93–S112 (2015)

206. Roeva, O., Pencheva, T., Tzonkov, S., Arndt, M., Hitzmann, B., Kleist, S., Miksch, G., Friehs, K., Flaschel, E.: Multiple model approach to modelling of escherichia coli fed-batch cultivation extracellular production of bacterial phytase. Electron. J. Biotechnol. **10**(4), 592–603 (2007)

207. Roeva, O., Pencheva, T., Angelova, M., Vassilev, P.: InterCriteria analysis by pairs and triples of genetic algorithms application for models identification. In: Studies in Computational Intelligence, Springer, vol. 655, pp. 193–218 (2016)

208. Roeva, O., Pencheva, T., Atanassov, K., Shannon, A.: Generalized net model of selection operator of genetic algorithms. In: IEEE International Conference on Intelligent Systems, July 7–9, London, UK, pp. 286–289 (2010)

209. Roeva, O., Pencheva, T.: Functional state modelling approach validation for yeast and bacteria cultivations. Biotechnol. Biotechnol. Equip. **28**(5), 968–974 (2014)

210. Roeva, O., Trenkova, T.: Modelling of a fed-batch culture applying simulated annealing. BIOMATH **1**(2), 1211114, 1–6. https://doi.org/10.11145/j.biomath.2012.11.114 (2012)

211. Roeva, O., Trenkova, T.: Cultivation process optimization based on genetic algorithms. J. Int. Sci. Publ. Mater. Methods Technol. **4**(2), 121–136 (2010)

212. Roeva, O., Trenkova, T.: Genetic algorithms and firefly algorithms for non-linear bioprocess model parameters identification. In: Proceedings of the 4th International Joint Conference on Computational Intelligence (ECTA), 5–7 October 2012, Barcelona, Spain, pp. 164–169 (2012)

213. Roeva, O., Slavov, T.: PID controller tuning based on metaheuristic algorithms for bioprocess control. Biotechnol. Biotechnol. Equip. **26**(5), 3267–3277 (2012)

214. Roeva, O., Slavov, T.: Firefly algorithm tuning of PID controller for glucose concentration control during *E. coli* fed-batch cultivation process. In: Proceedings of the Federated Conference on Computer Science and Information Systems, WCO 2012, Poland, pp. 455–462 (2012)

215. Roeva, O., Slavov, T.: Application of simulated annealing and threshold accepting for a parameter tuning problem. In: Proceedings of OMCONET, Southampton, England, pp. 64–71 (2012)

216. Roeva, O., Slavov, T.: Fed-batch cultivation control based on genetic algorithm PID controller tuning. In: Lecture Notes on Computer Science, Springer, Berlin, Heidelberg, vol. 6046, pp. 289–296 (2011)

217. Roeva, O., Slavov, T., Fidanova, S.: Population-based vs. single point search meta-heuristics for a PID controller tuning. In: Vasant, P. (ed.) Handbook of Research on Novel Soft Computing Intelligent Algorithms: Theory and Practical Applications, pp. 200–230. IGI Global (2014)

218. Roeva, O., Slavov, T.: A new hybrid GA-FA tuning of PID controller for glucose concentration control. In: Fidanova, S. (ed.). Recent Advances in Computational Optimization. Studies of Computational Intelligence, Springer International Publishing Switzerland, vol. 470, pp. 155–168 (2013)

219. Roeva, O., Atanassova, V.: Cuckoo search algorithm for model parameter identification. Int. J. Bioautomat. **20**(4), 483–492 (2016)

220. Roeva, O., Atanassova, V.: Generalized net model of cuckoo search algorithm. In: Proceedings of the 2016 IEEE 8th International Conference on Intelligent Systems, Sofia, September 04, 2016–September 06, 2016, pp. 589–592 (2016)

221. Shehab, M., Khader, A.T., Al-Betar, M.A.: A survey on applications and variants of the cuckoo search algorithm. Appl. Soft. Comput. **61**, 1041–1059 (2017)

222. Shopova, E.G., Vaklieva-Bancheva, N.G.: BASIC–a genetic algorithm for engineering problem solution. Comput. Chem. Eng. **30**(8), 1293–1309 (2006)

223. Shopska, V., Denkova, R., Lyubenova, V., Kostov, G.: Kinetic characteristics of alcohol fermentation in brewing: state of art and control of the fermentation process. In: Fermented Beverages. Woodhead Publishing, pp. 529–575 (2019)

224. Shindarov, M., Fidanova, S., Marinov, P.: Wireless sensor positioning algorithm. In: Proceedings of IEEE Conference on Intelligent Systems, Sofia, Bulgaria, pp. 419–424 (2012)

225. Slavov, T., Roeva, O.: Application of genetic algorithm to tuning a PID controller for glucose concentration control. WSEAS Trans. Syst. **11**(7), 223–233 (2012)

226. Slavov, T., Roeva, O.: Genetic algorithm tuning of PID controller in Smith predictor for glucose concentration control. Int. J. Bioautomat. **15**(2), 101–114 (2011)

227. Tangherloni, A., Spolaor, S., Cazzaniga, P., et al.: Biochemical parameter estimation vs. benchmark functions: a comparative study of optimization performance and representation design. Appl. Soft. Comput. **81**, 105494. 2019.105494 (2019)

228. Tzonkov, S., Kostov, Y., Petrov, M., Pencheva, T., Zlateva, P., Ljakova, K., Ilkova, T., Hristozov, I., Iliev, B., Roeva, O., Hristova, Y.: Bioprocess Systems: Modeling, Control and Optimization, East-West, Sofia, Bulgaria (2004) (in Bulgarian)

229. Vaklieva-Bancheva, N.G., Vladova, R., Kirilova, E.G.: Simulation of heat-integrated autothermal thermophilic aerobic digestion system operating under uncertainties through artificial neural network. Chem. Eng. Trans. **76**, 325–330 (2019)

230. Vaklieva-Bancheva, N.G., Vladova, R., Kirilova, E.G.: Genetic algorithm approach for optimization of energy integrated ATAD systems under uncertainties. In: Proceedings of the 17th

International Symposium on Thermal Science and Engineering of Serbia "Energy-Ecology-Efficiency", Sokobanja, Serbia, pp. 851–870 (2015)

231. Vaklieva-Bancheva N.G., Kirilova, E.G., Vladova, R.K.: Capturing uncertainties for sustainable operation of autothermal thermophilic aerobic digestion systems. In: book series on Computer Aided Chemical Engineering, vol. 33B, pp. 1729–1734 (2014)

232. Vaklieva-Bancheva, N.G., Shopova, E.G., Espuña, A., Puigjaner, L.: Product portfolio optimization for dairy industry. In: Proceedings of International Mediterranean Modelling Multiconference (October 4th–6th, 2006), Spain, Barcelona, pp. 101–110 (2006)

233. Vaklieva-Bancheva, N.G., Espuña, A., Shopova, E.G., Puigjaner, L., Ivanov, B.B.: Multiobjective optimization of dairy supply chain. In: book series on Computer Aided Chemical Engineering, vol. 24, pp. 781–786. Elsevier (2007)

234. Vaklieva-Bancheva, N.G., Kirilova, E.G.: Working frame for environmental benign management of multipurpose batch chemical and biochemical plants. Asian Chem. Lett., ANITA Publications 14(2), 157–170 (2010)

235. Vaklieva-Bancheva, N.G., Kirilova, E.G.: Cleaner manufacture of multipurpose batch chemical and biochemical plants. Scheduling and optimal choice of production recipes. J. Clean. Prod. 18(13), 1300–1310 (2010)

236. Vaklieva-Bancheva, N.G., Kirilova, E.G.: Reduction the impact of peaks emissions of pollutants from multipurpose batch chemical and biochemical plants. Bulgar. Chem. Commun. 45(1), 47–54 (2013)

237. Viesturs, U., Simeonov, I., Pencheva, T., Vanags, J., Petrov, M., Pavlov, Y., Roeva, O., Ilkova, T., Vishkins, M., Hristozov, I.: Contemporary Approaches to Modelling, Optimisation and Control of Biotechnological Processes, Tzonkov, S., (ed.). Prof. Marin Drinov Academic Publishing House, Sofia, Bulgaria (2010)

238. Yang, X.-S.: Nature-Inspired Meta-Heuristic Algorithms. Luniver Press, Beckington, UK (2008)

239. Yang, X.-S., Deb, S.: Cuckoo search via Lévy flights. In: Proceedings of World Congress on Nature & Biologically Inspired Computing (NaBIC 2009 India), IEEE Publications, USA, pp. 210–214 (2009)

240. Yang, X.-S.: A New Metaheuristic Bat-Inspired Algorithm. Department of Engineering, University of Cambridge, Trumpington Street, Cambridge CB2 1PZ, UK (2010)

241. Zlatkova, A., Lyubenova, V., Dudin, S., Ignatova, M.: Marker for switching of multiple models describing E. coli cultivation. Comptes rendus de l'Académie bulgare des Sciences, 70(2), 263–272 (2017)

242. Zlatkova, A., Lyubenova, V.: Dynamics monitoring of fed-batch E. coli fermentation. Int. J. Bioautomat. 21(1), 121–132 (2017)

243. Zlatkova, A., Lyubenova, V.: Monitoring the dynamics of bioprocesses using intermediate metabolite. In: International Conference Automatics & Informatics, October 4–6, pp. 231–234 (2017)

244. Zlatkova, A., Lyubenova, V., Ignatova, M.: Monitoring of oxidative-fermentative physiological states in fed-batch E. coli fermentation. In: Proceedings of the International Conference Agriculture and Food for the XXI Century, May 11–13, Sibiu, Romania, pp. 25–36 (2017)

In silico Studies of Biologically Active Molecules

Ilza Pajeva, Ivanka Tsakovska, Tania Pencheva, Petko Alov,
Merilin Al Sharif, Iglika Lessigiarska, Dessislava Jereva,
and Antonia Diukendjieva

Abstract Contemporary *in silico* (computer-aided) approaches to rational drug design and human health and environmental chemical risk assessment combine various ligand- and structure-based methods, including classical and three-dimensional (3D) quantitative structure–activity relationship (QSAR) models, pharmacophore and homology modelling, docking and virtual screening. These approaches are highly interdisciplinary and integrate knowledge from various disciplines in natural sciences. Their ultimate purpose is to quantitatively characterise the relationship between the compounds' chemical structures and their effects, expressed by theoretical models, while the effect can be different—therapeutic, toxic, etc. *In silico* approaches effectively help in understanding and elucidation of the mechanisms by which chemical compounds interact with target biomacro-molecules, thereby explaining fundamental processes in the living organisms. The chapter describes main methods in the computer-aided design and computational toxicology. The methods are classified according to the information available for modelling. Several case studies are described to illustrate the application of various *in silico* methods alone or in a combination. The case studies summarize the most recent results of the authors related to development of *in silico* models, algorithms and software applications to important biologically active molecules, both small drugs and drug-like compounds, and biomacromolecules, including nuclear receptors.

Keywords Computer-aided drug design · Modelling of bioactive molecules · QSAR · Pharmacophore modelling · Docking · Virtual screening

I. Pajeva (✉) · I. Tsakovska · T. Pencheva · P. Alov · M. Al Sharif · I. Lessigiarska · D. Jereva · A. Diukendjieva
Department of QSAR and Molecular Modelling, Institute of Biophysics and Biomedical Engineering, Bulgarian Academy of Sciences, Acad. G. Bonchev Str., Bl. 105, 1113 Sofia, Bulgaria
e-mail: pajeva@biomed.bas.bg

© The Author(s), under exclusive license to Springer Nature Switzerland AG 2021
K. T. Atanassov (ed.), *Research in Computer Science in the Bulgarian Academy of Sciences*, Studies in Computational Intelligence 934,
https://doi.org/10.1007/978-3-030-72284-5_19

421

1   Introduction

Drug discovery and development is a time consuming and costly process: it takes in average 13.5 years, only one of every 10,000–12,000 tested chemical compounds reaches the pharmaceutical market and the costs of developing a new drug range from $150 million to $3 billion in the past years [34]. The computer-aided drug design approaches, referred to as *in silico* approaches[1], applied at the early stage of drug discovery and development, have been proved as effective tools to optimise the process and to help in explanation of the drug action at a molecular level. It is worth noting that none of the modern medicines has been developed without the help of such approaches. New concepts in the field evolve in a tight correlation with the steadily increasing amount of data on potential new targets (4,500 drugable genes out of 25,000 in the human genome [16]), resolved 3D protein structures and their complexes with ligands, as well as with the large number of biologically active compounds derived from natural products or synthesised for testing against various diseases. Similarly, in recent years, it has become increasingly important to use computational approaches in the complex assessment of risk of chemicals on human health and environment thus resulting in a new research field that has acquired citizenship as a "regulatory computational toxicology".

In silico investigation of biologically active substances is an extremely challenging and dynamic research field that requires significant scientific capacity both in academic institutions and in the R&D departments of companies that deal with health problems and biotechnology. The need for synergy between academia and industry has led to the emergence of such organisations as Academic Drug Discovery Consortium, USA (ADDC, https://addconsortium.org), its European analogue European Lead Factory (ELF, https://www.europeanleadfactory.eu/), as well as the world's largest public–private partnership in the field of new drug development, the Innovative Medicines Initiative (IMI, https://www.imi.europa.eu/). Their goal is to develop new generations of drugs and therapies through partnerships between pharmaceutical and biotechnological companies, universities and research institutes. Further, *in silico* approaches are reflected in the current European legislative directives such as REACH (EC 1907/2006, Registration, Evaluation, Authorization and Restriction of Chemicals) and the Regulation of the European Parliament and of the Council on cosmetic products (Regulation (EC) No 1223/2009). They stimulate the increased use of *in silico* models as an alternative to *in vivo* experiments for supporting risk assessment of the use of chemicals. At the same time a number of activities have been undertaken over the years at EFSA (European Food Safety Authority) to support the development of harmonised methodologies for combined exposure to multiple chemicals. Among the key activities, structuring of new data types arising from new methodologies, including validated *in silico* methods are outlined [14].

---

[1]The term "*in silico*" has been introduced relatively recently as a synonym for "computer-aided", the latter dates from 1980s with the massive introduction of personal computers in the research practice.

The *in silico* studies of biologically active molecules represent highly interdisciplinary research field that integrates knowledge from many different branches of science such as chemistry, physics, biology, pharmacology, medicine, toxicology, mathematics, and informatics. According to the "Glossary of terms used in computational drug design" (IUPAC Recommendations 1997) it "involves all computer-assisted techniques used to discover, design and optimise biologically active compounds with a putative use as drugs" [32]. In this chapter we describe some of the main methods used in computer-aided drug design and computational toxicology and illustrate their applicability by case studies conducted in the Department of QSAR and Molecular Modelling at the Institute of Biophysics and Biomedical Engineering of the Bulgarian Academy of Sciences.

2 Methods

2.1 In silico *Approaches—Main Principles*

At the heart of all *in silico* approaches is the understanding for existence of an objective relationship between the structure of the compound and its biological activity (BA). The quantification of this dependence for a particular series of compounds reflects the changes (Δ) in the structures S of the compounds that result in changes in their BA values:

$$\Delta \text{ BA} = f(\Delta S) \tag{1}$$

In general, the *in silico* approaches are based on the thermodynamic relations expressed by Eqs. 2 and 3:

$$\Delta G = \Delta H - T\Delta S \tag{2}$$

$$\Delta G = 2.303 \text{ R T } \log K_D \tag{3}$$

where: ΔG is the change in the standard Gibbs free energy of formation of the complex between the unbound drug and the protein (receptor); ΔG is formed by the enthalpy (ΔH) and entropy (ΔS) component of the drug-receptor interaction; K_D is the equilibrium dissociation constant (drug affinity); T ($^\circ$K) is the absolute temperature; R is the universal gas constant (8.314 J/mol K).

There are two basic principles formulated more than a century ago that underpin the evaluation of interactions between drugs (usually small molecules) and biomacromolecules (enzymes, transport and signal proteins, ion channels, DNA, etc.). In 1894, the German scientist Emil Fischer formulated the principle of complementarity in the drug-enzyme interaction known as a "*key-lock model*": "The fit of a substrate to the catalytic site of an enzyme is like the fit of key and lock". Several years later, in 1913, another German scientist, Paul Ehrlich, in his lecture at 17th International Congress of Medicine "Chemotherapeutic Tools—Scientific Principles, Methods and Results", postulated the principle of "Corpora non agunt nisi fixata", meaning that compounds do not act if not bound. The American Daniel Kuschland evolved the key-lock model of Fisher to the more sophisticated "*induced-fit*" or "*hand-glove*" model defining it as a structural correspondence between molecules that is achieved in the process of their interaction. The most up-to-date interpretation of the drug-protein interactions is given by the "*multiconformational ensemble model*" that considers the possibility for an ensemble of several low energy conformations of drug-protein complexes existing in a dynamic equilibrium.

The interactions between the drug and the protein molecule take place in a special region in the protein structure known as a "binding (recognition, active) site" or "receptor"—a specific 3D arrangement of biomacromolecule monomers (amino acids, nucleotides, sugars) to which the drug molecule binds. The interactions, called "weak interactions", can be electrostatic (ion, dipole, etc.), arene-arene, cation-pi, steric, hydrogen bonding, hydrophobic, etc. [25] and any of them can contribute to the change in the standard Gibbs free energy of formation of the ligand-receptor complex (generally ΔG can vary from -10 to -80 kJ/mol).

The above principles and variety of drug-receptor interactions come to illustrate the complexity of the *in silico* modelling of biologically active molecules and direct to specific methods to be used depending on the available data and particular tasks.

2.2 Classification of the in silico Approaches

The classification depends on the level of information and knowledge available about the studied molecules. Table 1 summarises the classification of modern *in silico* approaches. In the table the term "ligand" is used to denote a molecule (mostly small one) that interacts with a biomacromolecule (receptor). In general, the following three modelling scenarios are possible: (1) unknown 3D structure of the receptor; (2) the 3D structure of the receptor is unknown but there is information about 3D structure of isofunctional proteins; (3) the 3D structure of the receptor is known. Each of these cases uses its own methods that can be applied alone or in combination depending on the particular study.

Modelling under conditions of an unknown 3D structure of the receptor (called "ligand-based drug design") includes the following main groups of methods: classical QSAR analysis; molecular field analysis (most often Comparative Molecular Field Analysis, CoMFA and Comparative Molecular Similarity Indices Analysis, CoMSIA); pharmacophore modelling and virtual screening. The modelling of an unknown 3D structure of the target molecule based on information about 3D structures of isofunctional molecules uses homology modelling techniques. Modelling with a known 3D structure of the receptor (known as "structure-based drug design") applies docking and virtual screening.

Table 1 Classification of *in silico* methods

Data/Knowledge about	*In silico* method/model	Results/Prediction
Unknown 3D structure of the receptor (ligand-based)		
• Ligands: – substrates – inhibitors – agonists – antagonists	• Classical QSAR analysis • Molecular field analysis (3D QSAR) • Pharmacophore modelling • Virtual screening	• Identification of important structural features of ligands • Prediction of activity values of ligands • Pharmacophore models • Design of new molecules
Unknown 3D structure of the receptor, known 3D structures of isofunctional proteins		
• Ligands • Isofunctional proteins	• Homology modelling	• Characterisation of the binding site(s) and subsequent use for docking and virtual screening
Known 3D structure of the receptor (structure-based)		
• Ligands • Ligand-receptor complexes	• Pharmacophore modelling • Docking • Virtual screening	• Identification of important amino acids in the binding site/s • Prediction of activity values of ligands • Design of new molecules

2.3 Classical QSAR Analysis: Hansch Analysis

Classical QSAR models are one of the mostly used *in silico* models in the rational drug design and computational toxicology. The models rely on the following basic assumptions: (a) all ligands belong to a homologous series of compounds; (b) all compounds act by the same mechanism on the receptor; (c) all compounds bind to the receptor active site in a similar manner [26].

The most popular classical QSAR analysis is named after the founder of the modern *in silico* drug design—Corwin Hansch [20]. The Hansch's model correlates the biological activity values of the compounds with their physicochemical properties (denoted as structural or molecular descriptors) by linear and non-linear regression analysis [27]. Equations 2 and 3 (see Sect. 2.1 above) lie at the heart of these models and define some important practical implications: (a) biological activity values are converted to a logarithmic scale to be linearly related to the change in the free energy of binding; (b) logarithms of the reciprocal values of BA are used to relate the higher values to the more active compounds; (c) the use of the logarithmic scale results in a more symmetrical (near-normal) error distribution (given the fact that the experimental error in biological measurements, especially in cell line experiments, may be significant). Linear values can be used exceptionally in cases where the biological data variance is small and the linear and logarithmic values are strongly interrelated.

The Hansch's model [19, 20], also known as "linear free energy relationship", has the following general representation as a multiple linear regression equation:

$$\log (1/C) = k_1 F_1 + k_2 F_2 + \cdots + k_n F_n + k_0 \tag{4}$$

where C is the molar concentration causing a particular biological effect (BA value); F_i are physicochemical properties (molecular descriptors); k_i are the regression coefficients of the model. There are enormous variations in the Hansch's model due to the large number (hundreds to thousands) of available molecular descriptors (experimentally measured or calculated).

In general, the molecular descriptors can be classified into two main groups [27] depending on the level of integrity and the nature of the property they represent:

- According to the level of integrity:

    (i) integral descriptors—characterise the whole molecule, for example, a water–lipid partition coefficient logP, a distribution coefficient logD, molecular weight MW, molecular volume MV, dipole moment μ, etc.;

    (ii) local descriptors—characterise molecular substructures, most often by different substitution constants.

- Depending on the nature of property:

    (i) lipophilic (hydrophobic) descriptors—describe the affinity of the molecules for hydrophobic environments. This group includes: logP, logD, lipophilic contribution π_x of substitute x in the compound's structure, hydrophobic fragment constant f, etc.

    (ii) polarizability descriptors—characterise the ability of molecule to polarise depending on the environment, e.g. molecular refraction MR, parahore PA, etc.;

    (iii) electronic descriptors—characterise the electronic properties of the structure. They can be integral (dipole moment) as well as local descriptors that characterise the influence of a substituent on the distribution of electron density in a molecule, for example, Hammet electron constant σ, the ionization constants pKa, quantum-chemical indices, etc.

    (iv) steric descriptors—characterise the size and shape of the molecule as a whole or parts thereof, such as the Taft constant E_s, Charton constant v_x, van-der-Waals volume V_w, a molecular surface available for the solvent SASA, STERIMOL parameters [27], etc. Other molecular descriptors are topological indices, indicator variables, similarity indices [18].

The obtained QSAR models are evaluated by the following statistical criteria: multiple regression correlation coefficient R^2 (Eq. 5), standard error of estimate SEE (Eq. 6) and F-ratio (Eq. 7):

$$R^2 = 1 - \frac{\sum (y_{obs} - y_{calc})^2}{\sum (y_{obs} - y_{mean})^2} \tag{5}$$

$$SEE = \sqrt{\frac{\sum (y_{obs} - y_{calc})^2}{n - k - 1}} \tag{6}$$

$$F = \frac{R^2(n - k - 1)}{k(1 - R^2)} \tag{7}$$

where: n—the number of objects (compounds) in the set; k—the number of independent variables (molecular descriptors) in the regression equation; y_{obs}—the observed BA values of the compounds; y_{calc}—the calculated BA values; y_{mean}—the average of the observed BA values.

Based on the experience gained so far in the development and application of QSAR models, the so-called "Good QSAR practice" has been formulated [27]. It summarizes a number of rules to build reliable models that include:

- Requirements to the molecular descriptors (parameters in the regression model):

  - Selection of independent molecular descriptors (parameters in the regression model) that are biologically meaningful.
  - Selection of molecular descriptors in the final model that are only significant terms.

- Requirements to the model:

  - High ratio between the number of objects (compounds) and the number of parameters (descriptors) in order to avoid the risk of chance correlations.
  - High R^2, low SEE, and high F-ratio.
  - Validation of the model by using cross-validation procedures.
  - Testing the model on a set of compounds (test-series) that has not be used to derive the model.
  - Application of parsimony (Occam's razor) in the model selection—in the case of close statistical characteristics, the simpler model is recommended.

2.4 3D QSAR Analysis: COMFA and CoMSIA Approaches

The 3D QSAR aims at establishing a correlation between the biological activity and 3D properties (molecular fields) of a series of structurally related molecules. The concept of this analysis was first proposed by Cramer et al. as "a multidimensional statistical procedure for comparing the experimental property of a molecule with one or more of its fields at various points located in a lattice around the molecule" [10]. It is based on the assumption that differences in the target property (BA) of a series of compounds are "sensitive" to spatially-localized differences in their molecular fields, and the size, intensity and localization of these differences determine the orientation and the interactions of the ligand in the receptor active site.

The procedure consists of the following steps:

(i)   creating a database (training set of compounds) containing at least one conformation per molecule (usually a geometrically optimised low energy conformation, called a conformer);

(ii) alignment (superposition) of the conformers in an appropriate space orientation;
(iii) defining a grid box around the superposed structures in the Cartesian space;
(iv) determining the grid parameters and type of the probe atom located at the grid points;
(v) calculating the molecular fields (in COMFA) or similarity indices (in CoMSIA) at each grid point;
(vi) performing cross-validated statistical analysis on BA (dependent variable) and molecular fields (independent variables) by the partial least square, PLS, method [48] for estimating the internal predictivity of the model;
(vii) external validation of the 3D QSAR model on a test set of compounds not used for training.

The estimation of the 3D QSAR model is based on the following statistical criteria: the predictive residual sum of squares PRESS (Eq. 8); the cross-validation correlation coefficient q_{cv}^2, (Eq. 9); the standard error of prediction SEP (Eq. 10), the external predictive correlation coefficient q_{pr}^2 (Eq. 11):

$$PRESS = \sum_i \left(y_{obs_i} - y_{pred_i} \right)^2 \tag{8}$$

$$q_{cv}^2 = 1 - \frac{PRESS}{\sum_i \left(y_{obs_i} - y_{mean} \right)^2} \tag{9}$$

$$SEP = \sqrt{\frac{PRESS}{n - p - 1}} \tag{10}$$

$$q_{pr}^2 = 1 - \frac{\sum \left(y_{obs}^{new} - y_{pred}^{new} \right)}{\sum \left(y_{obs}^{new} - y_{mean}^{train} \right)} \tag{11}$$

where: n—the number of objects (compounds) in the training set; p—the optimal number of the PLS components (corresponds to the model with the lowest SEP and maximal q_{cv}^2); y_{obs}—the observed BA values of the compounds; y_{pred}—the predicted BA values; y_{mean}—the average of the observed BA values; y_{obs}^{new}—the observed BA of the test set compounds; y_{pred}^{new}—the predicted BA of the test set compounds; y_{mean}^{train}—the average BA value in the training set.

The leave-one-out (LOO) cross-validation is mostly used in the model derivation. The final statistical model is obtained with p components using all observations (non-cross-validated model) and estimated by R^2, SEE and F-ratio.

In CoMFA steric (Lennard–Jones potential) and electrostatic (Coulomb potential) fields are calculated [10]. In CoMSIA the molecular fields are evaluated by molecular similarity indices calculated by Gaussian functions [24]. The following indices are calculated: steric, electrostatic, hydrophobic, hydrogen bond (HB)-donor and HB-acceptor abilities. For this purpose, at the intersection points q of the grid, a probe atom is placed with the following characteristics: radius 1 Å, charge +1, hydrophobicity +1, HB-donor ability +1 and HB-acceptor ability +1. A molecular similarity index is calculated between each molecule and the probe atom (Eq. 12) thus appearing as an indirect measure of the similarity between the aligned molecules:

$$A^q_{F,k}(j) = \sum_j w_{probe,k} w_{i,k} e^{-\alpha r^2_{iq}} \tag{12}$$

where: A—a similarity index at point q of the grid, summed for all atoms i in molecule j; $w_{probe,k}$—a property k of the probe atom; $w_{i,k}$—physicochemical property k of atom i; r_{iq}—the distance between the sample atom at point q of the grid and atom i of the molecule; α—attenuation factor with optimal values of 0.2–0.4 (at higher α values, steeper Gaussian functions and faster A-distance reduction are obtained). Thus, these indices are an indirect measure of the similarity between the aligned molecules.

2.5 Pharmacophore Modelling

According to "Glossary of terms used in computational drug design" [32] a pharma-cophore is an ensemble of steric and electronic features (pharmacophoric features) that is necessary to ensure the optimal supramolecular (noncovalent) interactions with a specific biological target and to trigger (or block) its biological response. Compared to the rest of the compound' structure, the pharmacophoric features are those that provide the major contribution to the affinity of drug binding to the receptor active site [38]. The method can be related to either ligand-based or structure-based *in silico* methods (Table 1).

There is a number of computational methods and programs for pharmacophore generation, but, in general, they all derive pharmacophore models of a series of ligands by aligning their 3D structures (conformers) on pharmacophoric features (atoms and atomic groups able to perform similar interactions with the receptor) and relating them to the ligands' biological activity. The pharmacophore model should be in correspondence with all available data: the active molecules should have all features of the pharmacophore; the inactive molecules should miss some of these features. If several different pharmacophore hypotheses for the same active site are generated they can be overlaid and reduced to their shared features so that common interactions are retained that are present in all active molecules. The final model is the largest common denominator shared by a set of active molecules (a consensus pharmacophore model).

The pharmacophoric features can be classified into three main categories according to annotation schemes implemented in MOE software (CCG Inc., https://www.chemcomp.com) [30]:

(i)   located directly on an atom of a molecule and typically indicating a function related to receptor-ligand binding: HB donor (Don), HB acceptor (Acc), cation (Cat), anion (Ani), metal ligator (ML), etc.;

(ii) located along an implicit lone pair or implicit hydrogen directions and used to annotate the location of possible partners, for example a hydrogen bond or a metal ligation;

(iii) located at the geometric center of a subset of the atoms of a molecule: aromatic (Aro), pi-ring (PiR), hydrophobic (Hyd). The features are presented by spheres of a non-zero radius to encode the permissible variation in the pharmacophore geometry. Different sets of features can be selected to constitute the pharmacophores. The compound conformers are aligned on the particular pharmacophore and the correspondence between the structural elements (atoms, centroids) of the conformers and the pharmacophore features are identified. The alignment is done so that to ensure the minimal distance between them. Spatial constraints imposed on particular atoms (e.g. excluded volume) can be additionally introduced to avoid steric clashes with the receptor and to ensure favourable contact regions.

2.6 Docking and Virtual Screening

These methods belong to the structure-based ones (Table 1) presuming a known 3D structure of the receptor of interest or of its complex with a ligand.

Docking is a computational procedure aiming to find favourable binding configurations between a small to medium-sized ligand and the receptor of interest (usually a protein). A ligand configuration into the receptor binding pocket is characterised by its pose and for each ligand a number of poses can be generated. Each pose is scored in an effort to determine favourable binding mode of the ligand. Virtual screening utilizes docking and it is based on the assumption that potential drug candidates (the so-called hits) can be selected through prediction of protein–ligand binding affinities by using a 3D protein structure and a large number of candidate small molecules from databases [39].

The docking procedure involves the following main steps:

(i)   preparation of the protein structure: it assigns ionization states and positions hydrogens in the protein structure given its 3D coordinates (typically from a crystal structure);

(ii) conformational analysis of the ligands: ligand conformations are generated from a single 3D conformer using a collection of preferred torsion angles of their rotatable bonds;

(iii) placement: a collection of poses is generated from the pool of ligand conformations using appropriate placement methods;

(iv) scoring: poses generated by the placement methodology are scored using various scoring functions. Typically, scoring functions emphasise favourable ligand contacts with the protein (hydrophobic, ionic, HB, etc.).

Docking algorithms can be classified into the following main groups depending on:

- the status of the ligand and protein conformations:
    - (i) rigid docking—the structures of the ligand and the protein do not change during the pose search;
    - (ii) semi-flexible docking—the structure of the ligand is only flexible while the protein remains rigid during docking;
    - (iii) flexible docking—the structure of the ligand remains flexible, while the protein structures can be flexible at different levels (e.g. only side chains of the amino acid residues surrounding the active site or all residues at a given distance around the active site or the whole protein including its backbone atoms).

- the placement methodology: triangle method (the poses are generated by superposition of triplets of ligand atoms on triplets of receptor atoms); principle moment of inertia (PMI) method (the poses are generated by alignment of PMI of the ligands to a randomly generated subsets of spheres of the receptor sites); incremental construction method (the poses are generated by incremental fragment growing strategy); pharmacophore placement (the poses are generated using the features of a preliminary developed pharmacophore).

- the scoring: three main types of scoring functions are currently available and used—empirical, force field and knowledge-based (potential of mean force, PMF) [39]. The empirical scoring functions approximate the interaction energy as a sum of independent terms representing important binding properties: HB, ionic, hydrophobic, loss in the ligand flexibility (entropy), etc. They use data on training sets of X-ray protein–ligand complexes and the multiple linear regression method to derive the function. These functions have proven to be surprisingly successful in spite of the fact that they describe complex molecular recognition phenomena with a very simplistic interaction model. The force-field scoring functions describe the interaction energy as a sum of independent terms, such as van der Waals interaction (calculated by the Lennard–Jones potential) and electrostatic (calculated by Coulomb potentials). The knowledge-based scoring functions are built from statistical analyses of the structural information in the X-ray protein–ligand coordinates presuming that interatomic distances occurring more often than some average value should represent favourable contacts, and vice versa. To derive these functions, one defines a set of general ligand and protein atom types and notes frequencies of occurrence of all possible pairs in binned distance intervals [43].

The selection of an appropriate scoring function is a key step in docking and virtual screening as it guides the whole process: from the ligand positioning, through the selection of the probable binding mode, to the discrimination between the ligands that can bind (binders) and those that cannot (non-binders). In docking and virtual screening protocols different scoring functions can perform differently depending on the types of the studied molecules (e.g. the size, shape, polarity of the active site). This implies using different scoring functions during docking and post-docking optimisation allowing refinement of the scored poses in order to select the best drug candidates among the pool of thousands to millions potential ligands.

3 Case Studies

Below some most recent studies employing various ligand- and structure-based *in silico* approaches are presented. Most of them combine two or more *in silico* methods described in Sect. 2. The case studies have been performed by the research team of the QSAR and Molecular Modelling department in the Institute of Biophysics and Biomedical Engineering—Bulgarian Academy of Sciences and in cooperation with colleagues in the frame of national and international research projects. Among the main subjects of interest are nuclear receptors (the estrogen receptor alpha, ERα, and the peroxisome proliferator-activated receptor gamma, PPARγ) and their ligands. In recent years, the activity of the Department has been expanded with regard to *in silico* investigation of compounds of natural origin as potential lead structures in drug design.

3.1 QSAR Models of Antiradical Effects of Polyphenolic Compounds

Considering the increased interest to compounds with antioxidant activity lately we have been developing QSAR models of polyphenol compounds of natural origin to estimate and predict their antiradical capacity measured in stoichiometric assays [3]. Data about the structures and antiradical capacity (measured in ABTS and DPPH assays) of 84 natural phenolic derivatives (phenolic acids, flavonoids, condensed tannins, coumarins, stilbenes, chalcones, anthra- and naphtoquinones) were collected from the literature. 3D-geometries were optimised using semiempirical quantum-mechanics methods. A number of molecular descriptors pertinent to the phenolic groups of these compounds was calculated: (i) primary: O–H bond-dissociation enthalpies, proton dissociation enthalpies, electron affinities, etc.; (ii) derivative: number of phenolic groups, potentially active in antiradical reactions [4] (Fig. 1); and (iii) integral: vector descriptors, based on the array of calculated vibrational frequencies (EVA descriptors [15]).

QSAR models of antiradical capacity measured in assays with stoichiometric endpoints were based on secondary or integral molecular descriptors, and underlined the importance of accounting for solvation effects in calculation of both the secondary and the integral type of descriptors. The models based on the calculated number of phenolic groups potentially active in antiradical reactions possess very good predictive ability but their applicability domain was restricted to the chromene-derivatives

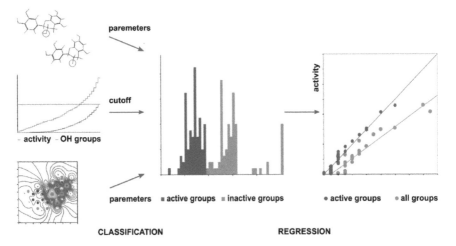

Fig. 1 Schematic representation of the classification approach to obtain the number of potentially active phenolic groups, used as a derivative molecular descriptor for modelling antiradical capacity of polyphenol compounds measured in stoichiometric assays

subset of the studied compounds (flavonoids, condensed tannins, coumarins). EVA-based models [5] possess somewhat lower predictive ability but its applicability domain covers all studied polyphenol classes (Table 2).

Table 2 QSAR models for prediction of TEAC (trolox-equivalent antioxidant, n_T, capacity) of natural polyphenols in the ABTS assay. **A**, by total number of phenolic groups (simple linear regression models); **B**, using integral EVA descriptors (obtained from vibrational frequencies, calculated by PM6 semiempirical Hamiltonian in, n_A, MOPAC [44] with subsequent PLS regression); **C** by number of potentially active phenolic groups (obtained by cutoff classification with subsequent simple linear regression)

Statistical parameters of the models	A. Simple regression QSAR models by total number of phenolic groups		B. PLS regression model using integral EVA descriptors (all compounds)	C. Hybrid regression model by number of active phenolic groups (chromene subset) TEAC = 0.002 + 1.001 n_AOH
	All compounds TEAC = −0.473 + 0.663 n_TOH	Chromene subset TEAC = −0.734 + 0.697 n_TOH		
n	84	40	76	40
R^2	0.802	0.876	0.894	0.963
SEE	0.812	0.840	0.442	0.459
q^2_{pr}	0.723 (67/17)	0.863 (32/8)	0.835 (61/15)	0.962 (32/8)

n—number of compounds used in the building and validation of the models; R^2—regression correlation coefficient; SEE—standard error of estimate; q^2_{pr}—external cross-validation predictive correlation coefficient (the number of compounds in the training/test sets are given in parentheses)

3.2 QSAR Model for Prediction of Membrane Permeability of Bioactive Compounds

Within the SEURAT-1 initiative (https://www.seurat-1.eu/) and in particular the COSMOS project (COSMetics to Optimise Safety, https://www.cosmostox.eu/), *in silico* studies were conducted to evaluate the gastrointestinal absorption (GIA) of bioactive substances. For this purpose, a QSAR model for membrane permeability PAMPA (Parallel Artificial Membrane Permeability Assay) [12] was developed. The model was implemented in the COSMOS free access processing and analysis platform (https://cosmosspace.cosmostox.eu/) and included in the database of the European Union Reference Laboratory for Alternatives to Animal Testing (https://ecvam-dbalm.jrc.ec.europa.eu/).

The QSAR model used Double-Sink PAMPA data of 269 compounds collected from [7] and expressed as effective membrane permeability values (logPe). The following molecular descriptors were included in the model: the logarithm of the apparent octanol/water distribution coefficient, logD, and the ratio of polar to total molecular surface area, PSA/TSA [23]. LogD values were calculated by ACD/Percepta obtained from ChemSpider (https://www.chemspider.com) or by the calculator plugins of ChemAxon Marvin obtained from Chemicalize (https://chemicalize.com). PSA was substituted by TPSA (topological polar surface area [13]). For TSA substitution the polar and total surface areas and their ratio were calculated for all the compounds in the set using MOE software. Sixty two molecular size descriptors were generated and their correlation with TSA and that of TPSA/descriptor ratios to PSA/TSA ratios were assessed (Table 3).

The top-ranked TPSA/descriptor ratios were used in the QSAR models. As a substitute of TSA the molecular weight, MW, was selected as a descriptor showing one of the highest correlations with TSA, also being among the most conveniently calculated ones. The external predictivity of the models was evaluated by splitting the set into training and test sets (4:1 stratified splitting).

Table 3 The CDK- and Indigo-calculated molecular descriptors and TPSA/descriptor ratios with the highest correlation to TSA (A) and to PSA/TSA (B)

A		B	
Descriptors	r	Ratios	r
TSA	1.000	PSA/TSA	1.000
Atomic polarizabilities	0.959	TPSA/VABC volume descriptor	0.881
Number of heavy atoms	0.957	TPSA/Molecular weight	0.880
VABC volume descriptor	0.954	TPSA/Atomic polarizabilities	0.878
Number of bonds	0.951	TPSA/Number of heavy atoms	0.876
Number of carbons	0.950	TPSA/Total number of atoms	0.873
Molecular weight	0.946	TPSA/Bond polarizabilities	0.848
Total number of atoms	0.941	TPSA/Number of bonds	0.842
Zagreb index	0.923	TPSA/Zagreb index	0.801

r—correlation coefficient, TSA—total surface area, PSA—polar surface area, TPSA—topological polar surface area, VABC—sum of atomic and bond contributions volume (modified after [12])

The two implementations of the model based on logD at pH 7.4 as estimated by the ACD/Percepta or ChemAxon tools are presented in Eqs. 13 and 14, respectively:

$$\log Pe = -2.20(\pm 0.21) + 0.49(\pm 0.04) \log D - 10.14(\pm 0.74) \text{ TPSA/MW}$$
$$n = 251, \ R^2 = 0.75, \ SEE = 1.10, \ F = 371.3,$$
$$\text{LOO } q_{cv}^2 = 0.74, \text{ external validation } q_{pr}^2 = 0.79 \ (200/51) \tag{13}$$

$$\log Pe = -2.11(\pm 0.22) + 0.47(\pm 0.05) \log D - 10.71(\pm 0.78) \text{ TPSA/MW}$$
$$n = 248, \ R^2 = 0.74, \ SEE = 1.11, \ F = 345.1,$$
$$\text{LOO } q_{cv}^2 = 0.73, \text{ external validation } q_{pr}^2 = 0.77 \ (198/50) \tag{14}$$

The very close values of R^2 and LOO q_{cv}^2 demonstrate that the models are stable. The external predictivity coefficients vary in a narrow range (0.69–0.79) and are similar to those using PSA/TSA (0.68 and 0.79 for ACD/Percepta and ChemAxon tools calculated logD-based models, respectively). The plot of calculated versus experimental PAMPA logPe values is presented on Fig. 2.

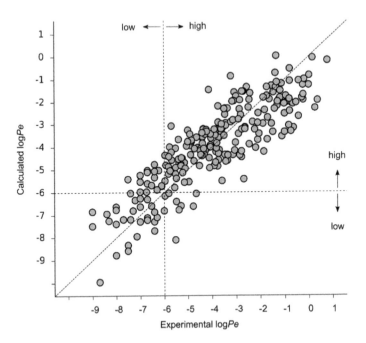

Fig. 2 Plot of calculated versus experimental logPe values of the compounds used to derive the PAMPA QSAR model; the dashed line represents the border between low-to-medium and high permeability (modified after [12])

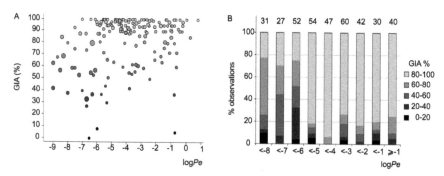

Fig. 3 Correspondence between logPe and GIA (%) for 167 compounds present in both PAMPA Pe and GIA datasets: **A** mean GIA values versus PAMPA log Pe; size of the circles corresponds to the number of averaged GIA values for the compound. **B** Distribution of GIA classes among PAMPA Pe classes, numbers on top of the columns correspond to the number of distinct GIA values in each PAMPA Pe class (from [12])

The models' applicability to GIA prediction was assessed on an external dataset (https://biomed.bas.bg/qsarmm) of 783 compounds with 1227 distinct values of GIA reported in the literature. One hundred sixty seven of them with 383 distinct GIA values had Double-Sink PAMPA Pe in the training set of the model. The collected data does not allow to discriminate between low and medium GIA, due to the low percentage of compounds with low and medium GIA in the dataset (Fig. 3a). However, an abrupt decrease in the percentage of observations in the highest GIA class (>80%) is observed for compounds with PAMPA logPe lower than –6 (Fig. 3b). The model classified the remaining 616 compounds into high or low-to-medium GIA classes with accuracy of classification ~77%, sensitivity of ~84%, and specificity of ~59%.

The developed QSAR models use descriptors calculated by open-source software tools or obtainable from free online resources that makes it appropriate for a broader application. They have already been applied for estimation of the membrane permeability of natural compounds derived from the plant *Silybum marianum* (L.) Gaertn and their derivatives [12].

3.3 Combined Pharmacophore, Docking, Virtual Screening and 3D QSAR Study of PPARγ Full Agonists

The nuclear peroxisome proliferator-activated receptor gamma (PPARγ) has intensively been studied in the last decade, because of its key role for regulation of glucose and lipid homeostasis. It appears as an important pharmacological target for treating metabolic diseases, including type 2 diabetes. Thus, the elucidation of the mode of action of its ligands at a molecular level is permanently in the focus of our research studies.

A complex *in silico* study of PPARγ full agonists was carried out applying various ligand- and structure-based approaches. As a first step structural and biological data for PPARγ agonists were harvested and curated that resulted in the most complete and extensive publicly available virtual library of PPARγ agonists: 439 structures represented by a number of structural identificators (IUPAC name, trivial name, where presented in the source, SMILES code, InChI key, etc.) as well as biological data where available (binding affinity, IC_{50}; potency, EC_{50}; maximal relative efficacy, %max) (https://biomed.bas.bg/qsarmm/).

As a second step the available X-ray structures of PPARγ in the Protein Data Bank (PDB) (https://www.rcsb.org/) were extracted. The initial X-ray structures of PPARγ were modified to reflect the physiologically relevant conditions (temperature: 310 K; pH = 7.4; ion concentration: 0.152 mol/L) in MOE software. The PPARγ-ligand complexes were further analysed to identify important interactions (HBs, salt bridges, hydrophobic interactions, cation-π, sulphur-lone pair, halogen bonds and solvent exposure) between the ligand and the receptor. According to the analysis Ser289, His323, His449 and Tyr473 are represented in the HB interactions with the most active agonists in the PPARγ binding site.

The collected and optimised structural information was implemented in the developed multistage virtual screening protocol to predict PPARγ full agonists (Fig. 4) [2].

In the first stage the optimised ligands' structures were docked and scored by London dG scoring function (MOE software) into the binding site of the prepared PPARγ structure (X-ray structure of PPARγ with rosiglitazone, PDB ID 1FM6). The highly scored poses of each ligand were selected and used in the further development of the virtual screening protocol. In order to refine the predictions, a

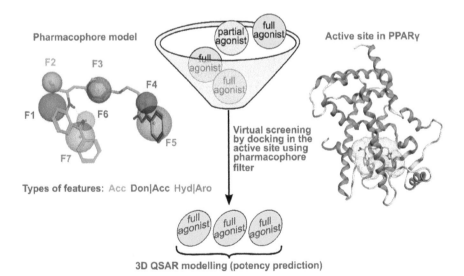

Fig. 4 Two-step *in silico* approach for virtual screening of PPARγ full agonists

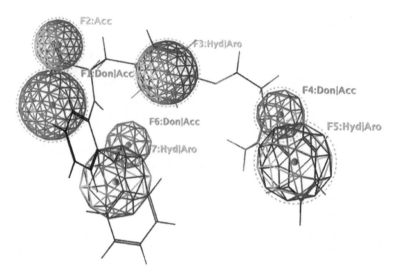

Fig. 5 Pharmacophore model of PPARγ full agonists: the essential pharmacophoric features are outlined with dashed line (from [1])

pharmacophore model of PPARγ full agonists (Fig. 5) was built and integrated in the docking procedure as a filter thus only poses that satisfy the pharmacophore were selected [46]. In the development of the pharmacophore model the three most active PPARγ full agonists were used. The model consists of 7 important pharmacophore features: four polar atoms and functional groups capable of performing HB and ionic interactions (F1, F2, F4 and F6) and three hydrophobic and aromatic structural elements (F3, F5 and F7).

In the second stage a 3D QSAR modelling based on the CoMSIA approach (implemented in SYBYL software (Tripos International, Inc., https://www.certara.com/) was performed to predict the transactivation activity of PPARγ full agonists. In order to build the CoMSIA model of PPARγ full agonists, first the 170 structures in the modelling set were spatially aligned. In the alignment procedure for the ligands extracted from the PDB complexes the experimental bioactive conformations were used. For the rest of the structures the alignment was based on their docking poses in the PPARγ ligand binding domain obtained by the virtual screening procedure. Visual inspection against the template structure and consideration of the docking score (the smallest negative scores preferred) were the criteria driving the final conformer selection for each ligand. For the aligned structures the electrostatic, steric, HB donor, HB acceptor, and hydrophobic CoMSIA fields were calculated.

The PLS method was used for modelling and the LOO cross-validation procedure was applied to select the model's optimal number of principle components (p) and to evaluate the model's statistical performance. The best model selection was based on the following statistical characteristics: q_{cv}^2, p, and SEP. The final (non-cross-validated) model (characterised by R^2, SEE, and F-value) obtained for the best cross-validated model with p, was used in external validation by prediction of the pEC$_{50}$ (log(1/EC$_{50}$)) values of a predefined test set of PPARγ full agonists and calculation of the predictive q^2_{pr}. After the preliminary modelling two categories of compounds were excluded from the initial set of 170 agonists: (i) applicability domain outliers, identified with the "extent of extrapolation" approach [37, 45] as implemented in Enalos domain leverage node [33] in the KNIME analytics platform [9]; (ii) response outliers, identified in the residual analysis. The remaining dataset was split into training (n = 83) and test (n = 39) sets in a way that ensured large structural representativeness in both sets. A number of CoMSIA models were developed. The best model included three fields (electrostatic, HB-acceptor and hydrophobic) and had the following statistical parameters: $q^2_{cv} = 0.610$, p = 7, SEP = 0.505, $q^2_{pr} = 0.552$.

The team conducted also numerous molecular dynamics simulations to elucidate the behaviour of the receptor in its antagonistic form. It was found that the antagonists introduced significant changes in both the PPARγ ligand and co-activator binding pockets affecting the activation helix H12 conformation which appears a main reason for their PPARγ antagonism [17]. The results of the PPARγ studies are important for the development of new drugs and for the safety assessment of chemical compounds.

3.4 In silico *Studies of the Human Estrogen Receptor Alpha (ERα)*

The nuclear estrogen receptor alpha (ERα) is involved in many physiologically and pharmacologically significant phenomena and is permanently in the focus of research by various *in silico* methods [47]. Our recent studies on this receptor aimed at further elucidation of the pharmacophore of its ligands for potential use in virtual screening protocols and QSAR modelling.

The analysis of the ERα X-ray complexes in the PDB outlined the ligand ETZ as a very potent ERα agonist with affinity of a picomolar range (PDB ID 2P15) [36]. The complex has been used together with two other complexes of strong ERα agonists (estradiol and diethylstilbestrol with PDB complexes ID 1QKU and 3ERD, respectively) to generate the pharmacophore model (Fig. 6). A pharmacophore core consisting of F1, F2, F3 and F4 features is shared by all three agonists in accordance with the generally recognized pharmacophore model of ERα agonists [42]. An additional hydrophobic/aromatic feature (F5: Hyd|Aro) was identified and added to the pharmacophore model derived from the EZT bioactive conformation. F5 is located deep inside the hydrophobic pocket opposite to the activation helix 12 and stabilises the position of the ligand in the active site (confirmed by molecular dynamics simulation [22]).

Using this model a virtual ligand screening was performed on databases available in the Enhanced version of the Directory of Useful Decoys (DUD-E): (i) 1315 ERα binders with reported activity; (ii) 20,685 decoys (compounds that physically resemble ERα binders but are topologically dissimilar

Fig. 6 Five-feature
pharmacophore model of
ERα agonists including:
F1—HB donor (Don)
groups; F2—HB acceptor
(Acc) groups; F3, F4 and
F5—hydrophobic (Hyd)
and/or aromatic (Aro)
substructures

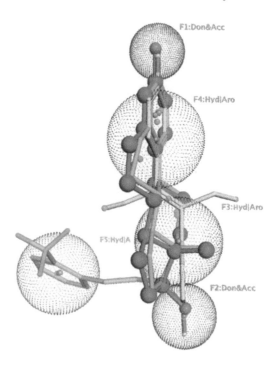

to them) [35]. Prior to screening reasonable 3D conformations of the compounds in these databases were generated from their SMILES codes by molecular mechanics energy minimisation (MOE software).

Two types of virtual screening were applied using the pharmacophore model as a filter: a ligand-based utilising the 3D structures of the ligands, and a structure-based (docking) utilising the 3D structure of the receptor and the ligands. The newly suggested 5-feature and the currently accepted 4-feature pharmacophore models were employed. The results demonstrate that the 5-feature pharmacophore is more restrictive (Table 4A) compared to the 4-feature one suggesting that the new hydrophobic/aromatic pharmacophoric feature might be helpful in identification of potential ERα binders with high agonistic activity. Both models are suitable for recognition of ERα non-binders (Table 4B). The structure-based and ligand-based pharmacophore screening yielded similar results thus confirming their validity.

Table 4 Results of structure- and ligand-based virtual screening using the ERα pharmacophore model of agonists

A. Binders with known activity (values below 1 μM)

Range, $\log(1/EC_{50})$	Total number	Pharmacophore model			
		5-features (F1–F5)		4-features (F1–F4)	
		Number of scored compounds/hits	%	Number of scored compounds/hits	%
Structure-based					
6.0–9.0	190	75	39	123	65
9.0–12.0	15	5	33	13	87
6.0–12.0	205	80	39	136	66
Ligand-based					
6.0–9.0	190	65	34	118	62
9.0–12.0	15	4	27	13	87
6.0–12.0	205	69	34	131	64

B. Decoys

In silico protocol	Pharmacophore model			
	5-features (F1–F5)		4-features (F1–F4)	
	Number of scored compounds	%	Number of hits	%
Structure-based	22	0.12	27	0.13
Ligand-based	15	0.07	20	0.10

A QSAR model was further proposed that utilises shape and electronic (polarity) molecular descriptors (MOE software used for calculation) and allows differentiation of compounds with potential pro- and anti-estrogen activity. Figure 7 illustrates the ability of the model to discriminate ERα agonists from antagonists on the example of ligands extracted from PDB ERα complexes. The results direct to the combined use of *in silico* ligand- and structure-based models for effective screening of ERα ligands.

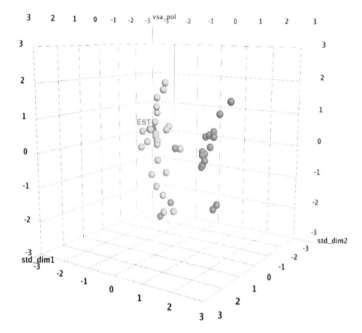

Fig. 7 Clusters of 45 ERα PDB complexes (26 agonists, in yellow; 19 antagonists, in blue) according to the QSAR model utilising the following descriptors: std_dim1, std_dim2—the two most deviating atomic coordinates along the 3D axes of the ligands; vsa_pol—van der Waals surface area of polar atoms in the ligand structures; EST—the endogenic agonist Estradiol, in magenta. PDB ID of agonists: 1GWQ, 1L2I, 1X7E, 1X7R, 2B1Z, 2FAI, 2G50, 2I0J, 2P15, 2POG, 2Q70, 2QA6, 2QE4, 2QGT, 2QGW, 2QR9, 2QSE, 2QXM, 2YJA, 3ERD, 3HM1, 3Q95, 3Q97, 3Q97, 3UU7, 4DMA; PDB ID of antagonists: 1ERR, 1R5K, 1SJ0, 1UOM, 1XP1, 1XP6, 1XP9, 1XPC, 1XQC, 1YIM, 1YIN, 2AYR, 2IOG, 2IOK, 2OUZ, 2R6W, 2R6Y, 3DT3, 3ERT. Three antagonists are related to the agonists' cluster 1R5K, 2IOK, 3DT3

In line with the growing interest in bioactive compounds of natural origin the research of the Department has recently been directed toward assessment of potential toxic effects and metabolic transformations of flavonolignans from milk thistle (*Silybum marianum* L.) using Meteor Nexus and Derek Nexus knowledge-based expert systems for metabolism and toxicity predictions [31]. Our studies revealed several plausible toxic effects in mammals, including ERα modulation. Docking of the stereoisomeric forms A and B of the main flavonolognan silybin demonstrated that the enantiomers could occupy the active site in the ligand-binding domain of ERα (Fig. 8) by showing stereospecific poses and, respectively, interactions with the receptor [11]. The results are in agreement with the experimentally observed effects of these stereoisomers toward ERα and explain at a molecular level the differences in the observed interactions of the compounds and the toxic effects potentially related to them.

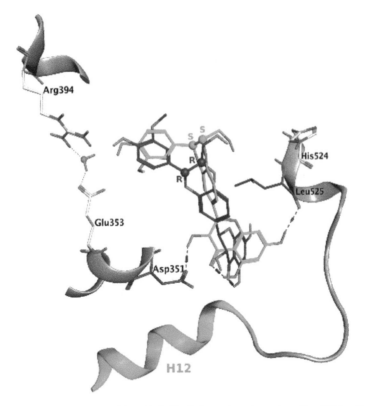

Fig. 8 Poses of silybin A (purple) and silybin B (green) at the active site of ERα. Hydrogen bonds between silybin B and the receptor are shown. R and S stereoatoms of the flavonolignans are represented as spheres; the oxygen atom of the water molecule located in the active site is represented as a red sphere (from [11])

3.5 AMMOS2 Software Platform for Post-Docking Optimisation in Virtual Screening of Bioactive Compounds

An interactive web server AMMOS2 has been developed for efficient computational refinement of protein-small organic molecules complexes for the purposes of a post-docking optimisation step in virtual screening projects [28]. The web server is based on the previously developed software platform AMMOS (Automated Molecular Mechanization optimization tool for *in silico* Screening) [40, 41] upgrading the AMMOS_ProtLig module. Another software package of AMMOS software platform, namely DG-AMMOS (Distance Geometry and Automatic Molecular Mechanics optimization for *in silico* Screening) [29], is a software package that performs an automated procedure for generating 3D conformations of small molecules and for optimising them by applying the distance geometry methods and molecular mechanics. DG-AMMOS is an open access web server, available at https://mobyle.rpbs.univ-paris-diderot.fr/cgi-bin/portal.py#forms::DG-AMMOS.

AMMOS2 provides tools for minimisation of the energies of a large number of experimental or modelled protein–ligand complexes that can be generated via a user-chosen docking program or via virtual screening web servers. AMMOS2 employs the physics-based force field of AMMP (all-featured molecular mechanics program, embedded in the platform) and allows for optimisation of protein–ligand interactions via five levels of flexibility of the protein, while ligand is always flexible:

- *Case 1*—entire protein is flexible;
- *Case 2*—only protein sidechains are flexible;
- *Case 3*—all protein atoms within a radius around the ligands are flexible (radius is of a user-choice);
- *Case 4*—protein sidechains within a radius around the ligands are flexible (radius is of a user-choice);
- *Case 5*—the entire protein is rigid.

The new version of AMMOS platform embedded in the AMMOS2 web server allows explicit water molecules and individual metal ions to be considered in the optimisation of protein–ligand complexes. AMMOS2 can be used for refinement of protein–ligand complexes containing water molecules that fill the whole protein or the binding site, as well as for the detection of key water molecules serving as bridges between the protein and ligand. In addition, AMMOS2 can take into account the presence of metal ions that can be added by several online services. Metal ions and explicit water molecules are allowed to move when they are located in the flexible part of the protein receptor.

The overall workflow of the AMMOS2 web server is illustrated in Fig. 9 representing required inputs, preparation, minimisation and ranking procedures. AMMOS2 is user-friendly and suitable for non-advanced users.

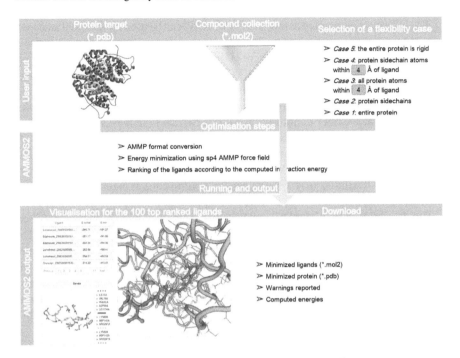

Fig. 9 Workflow of AMMOS2 web server showing inputs, execution, and outputs

As input, AMMOS2 requires the protein in *pdb* format (properly protonated and containing a maximum of 1000 residues) and bound ligands, docked or co-crystallized small organic molecules (with a maximum of 300 atoms), in *mol2* format. The AMMOS2 web server can minimise different number of small molecules depending on the chosen level of protein flexibility—up to 1000 ligands for *Cases 1* and *2* and up to 5000 ligands in *Cases 3* to *5*.

As output, AMMOS2 returns the coordinates of the geometrically optimised ligands, the coordinates of the optimised protein–ligand complex and the interaction energies thus allowing the users to perform additional analysis for drug discovery projects. Additionally, AMMOS2 provides an interactive page displaying the 100 top ligands ranked by the protein–ligand binding energies, as well as protein–ligand complex structures before and after minimisation visualised in 3D.

During the validation of the new version AMMOS2, embedded in the web server, the minimisation of 21 protein–ligand complexes showed a sustained improvement towards the initial complex structures in terms of binding energies as well as optimised water molecules positions in the protein–ligand complexes.

The software platform AMMOS2 offers valuable solutions to assist docking and structure-based virtual screening experiments keeping in mind that water molecules mediating protein–ligand interactions are of key importance to design high-affinity hit molecules. AMMOS2 is an open access web server, available at https://drugmod. rpbs.univ-paris-diderot.fr/ammosHome.php.

The information returned as output of AMMOS2 can be used for different analyses related to drug discovery projects, including generation of the enrichment curves. Figure 10 demonstrates the enrichment curves obtained for all five cases of protein atom flexibility implemented in AMMOS2 for the protein–ligand complex of coagulation factor X (FX) (https://www.rcsb.org, PDB ID 1F0R). As seen from the figure, 90% of the actives are retrieved in the top 1% (0.07% in *Case 1*) of the proceeded database after minimisation, while after docking they appear in the first 15%. All AMMOS cases that consider levels of receptor flexibility (*Cases 1 to 4*) achieved excellent enrichment results, while *Case 5* with a rigid receptor did not show improvement compared to docking. This result reflects the reality that even small local receptor flexibility may be sufficient to optimise the protein–ligand complex interaction.

Results obtained by AMMOS2 were used as input data in the recently developed approach to decision making, known as InterCriteria Analysis (ICrA) [6] described in detail in another chapter of this book. ICrA was applied to reveal the possible relations between different cases of protein flexibility. Based on the AMMOS2 results, an input matrix was constructed containing the calculated energies in the protein–ligand complexes (considered as observations in terms of ICrA) at different cases of protein flexibility (considered as criteria in terms of ICrA). Figure 11 represents the results from the implementation of ICrA approach using ICrAData software [21].

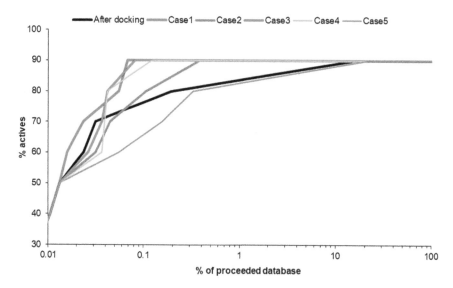

Fig. 10 Enrichment graphs for coagulation factor X (FX) inhibitors after post-docking optimisation with AMMOS2. The y-axis is the % of retrieved actives versus the percentage of the database screened (x-axis): enrichment results after flexible docking (black); enrichment results after rescoring with AMMOS2 minimisation in *Case 1* (red), *Case 2* (grey), *Case 3* (green), *Case 4* (dark yellow) and *Case 5* (magenta)

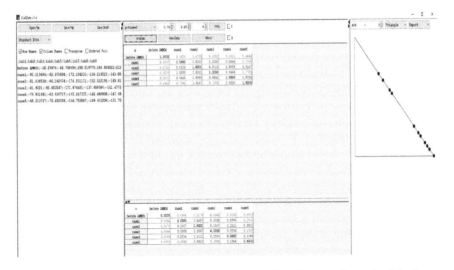

Fig. 11 ICrA application to the results obtained by AMMOS2 web-server: the matrix of the degrees of agreements μ (upper matrix in the middle panel); the matrix of the degrees of disagreements v (lower matrix in the middle panel); graphical representation by the intuitionistic fuzzy triangle (right panel); pairs of criteria in dissonance (in magenta); pairs of criteria in positive consonance (in green); the threshold of 0.75 set for the degrees of agreements μ

If one considers the relations between different cases of protein flexibility, the following intercriteria relations can be outlined:

- *Case 1* and *Case 3* are in strong positive consonance ($\mu = 1.00$). This dependence was expected since both mentioned cases of flexibility include all protein atoms—in *Case 1* all atoms of the protein, while in *Case 3*—all protein atoms in the sphere around the ligand.

- four pairs of criteria are in positive consonance ($\mu \in (0.85, 0.95]$): *Case 1—Case 4*, *Case 2—Case 4*, *Case 2—Case 5*, and *Case 3—Case 4*. Reported results were also expected, since *Cases 2* and *4* have the common "nature", namely flexible sidechains; *Cases 3* and *4* treat atoms only in the sphere around the ligand. The relatively high dependence between *Case 1—Case 4* and *Case 2—Case 5* are very promising due to the fact that *Cases 4* and *5* are significantly faster than *Cases 1* and *2* and it gives the user a choice to use much faster *Cases 4* and *5* without worry of losing the impact of *Cases 1* and *2*.

- three pairs of criteria are in weak positive consonance ($\mu \in (0.75, 0.85]$): *Case 1—Case 2*, *Case 2—Case 3*, and *Case 3—Case 4*. This fact was also expected, since all these dependencies are between cases of flexibility of different "nature"—such with all moving atoms (*Case 1*) versus one with moving atoms only from sidechain (*Case 2*), etc.

- two pairs of criteria are in weak dissonance ($\mu \in (0.67, 0.75]$), but at the edge of weak positive consonance ($\mu \in (0.75, 0.85]$): *Case 1—Case 5*, and *Case 3—Case 5*. Even at the edge of the weak positive consonance, these results show the distance between *Case 5* (rigid protein) and *Case 1* (fully flexible protein) and *Case 3* (fully flexible protein in the sphere around the ligand).

As seen from Fig. 11, there is no agreement between all considered cases of protein flexibility (*Cases 1* to *4*) and the results before the AMMOS2 application, which are, in fact, results after docking. Usually, the input files for AMMOS2 are generated via a user-chosen docking program or via virtual screening web servers that use different scoring function. This fact explains the disagreement between the results before the AMMOS2 application and all considered cases of protein flexibility (*Cases 1* to *4*). It is worth to be noted, that all AMMOS2 energies included in the input matrix (data not shown) were much lower than the energies before AMMOS2 application. This fact is in accordance with the expectation for improved interaction energies after the post-docking optimisation with AMMOS2.

The implementation of ICrA approach demonstrates once again the need of post-docking optimisation at different levels of the protein flexibility and confirms the effectiveness of the proposed post-docking procedure.

4   Conclusions

The case studies described in this chapter aimed at demonstrating the capabilities of the *in silico* methods to investigate biologically active molecules of different type— from small drugs and drug-like molecules to biomacromolecules. This research is a part of the significant modelling advances in the recent years that have afforded a variety of tools helpful to optimise the complex decision-making process in the drug discovery, namely elucidation of binding pockets of prospective targets, identification of key interactions between ligands and their receptors, prediction of binding affinity, and ADMET properties of candidate compounds, etc. [8]. The methods are equally applicable and increasingly encouraged in both pharmaceutical and safety assessment area and can help in optimisation of the research process and increased efficiency of the solutions. In general, they are considered as alternatives to animal testing thus saving laboratory animals, reducing time and cost. However, the complexity of the biological mechanisms and the numerous factors involved in the drug action represent a challenge and set further new and higher requirements to the theoretical predictive approaches. The application of *in silico* approaches requires careful consideration of the specific features of the studied biomolecules, availability of appropriate high quality biological and structural data, selection of relevant modelling methods, implementation of robust validation procedures, precise determination of the applicability domain of the models applied and critical estimation of the usefulness and reasonability of the obtained results.

Acknowledgements The authors acknowledge the financial support of the National Scientific Fund of Bulgaria, grant DN 17/6 "A New Approach, Based on an Intercriteria Data Analysis, to Support Decision Making in *in silico* Studies of Complex Biomolecular Systems" and the Bulgarian Ministry of Education and Science under the National Research Programme "Healthy Foods for a Strong Bio-Economy and Quality of Life" approved by DCM # 577/17.08.2018.

The authors would like to acknowledge also the contributions to the above described achievements of the colleagues from a number of European research institutions (Faculty of Pharmacy, University Paris Descartes (France); Institute of Pharmacy, University of Bonn (Germany); Institute for Health and Consumer Protection, Joint Research Center of the European Commission Ispra (Italy), John Moores University, Liverpool (United Kingdom); Italian Institute of Technology, Rome (Italy); Institute of Microbiology, Czech Academy of Sciences, Prague (Czech Republic)) with whom the QSAR and Molecular Modelling Department is intensively collaborating in the recent years.

References

1. Al Sharif, M., Tsakovska, I., Alov, P., Vitcheva, V., Diukendjieva, A., Pajeva, I.: Molecular modeling approach to study the PPARγ–ligand interactions. In: Badr, M.Z. (ed.) Nuclear Receptors: Methods and Experimental Protocols, pp. 261–289. Springer New York, New York, NY (2019)
2. Al Sharif, M., Tsakovska, I., Pajeva, I., Alov, P., Fioravanzo, E., Bassan, A., Kovarich, S., Yang, C., Mostrag-Szlichtyng, A., Vitcheva, V., Worth, A.P., Richarz, A.-N., Cronin, M.T.D.: The application of molecular modelling in the safety assessment of chemicals: a case study on ligand-dependent PPARγ dysregulation. Toxicology **392**, 140–154 (2017)
3. Alov, P., Tsakovska, I., Pajeva, I.: Computational studies of free radical-scavenging properties of phenolic compounds. CTMC. **15**(2), 85–104 (2015)
4. Alov, P., Tsakovska, I., Pajeva, I.: Hybrid classification/regression QSAR modelling of polyphenol antioxidant activity measured in stoichiometric assays. Presented at the ICNPU, Bansko, Bulgaria (2013)
5. Alov, P., Tsakovska, I., Pajeva, I.: Quantitative structure-property relationship modelling of antiradical properties of natural polyphenols using EVA vector descriptor approach. Comptes rendus de l' Académie bulgare des Sciences **69**(9), 1145–1152 (2016)
6. Atanassov, K., Mavrov, D., Atanassova, V.: Intercriteria decision making: a new approach for multicriteria decision making, based on index matrices and intuitionistic fuzzy sets. Issues Intuition. Fuzzy Sets General. Nets **11**, 1–8 (2014)
7. Avdeef, A.: Absorption and drug development: solubility, permeability, and charge state. Wiley, Hoboken, NJ (2012)
8. Ban, F., Dalal, K., Li, H., LeBlanc, E., Rennie, P.S., Cherkasov, A.: Best practices of computer-aided drug discovery: lessons learned from the development of a preclinical candidate for prostate cancer with a new mechanism of action. J. Chem. Inf. Model. **57**(5), 1018–1028 (2017)
9. Berthold, M.R., Cebron, N., Dill, F., Gabriel, T.R., Meinl, T., Ohl, P., Sieb, C., Thiel, K.: KNIME: the Konstanz information miner. In: Preisach, C., Burkhardt, H., Schmidt-Thieme, L., and Decker, R. (eds.) Studies in Classification, Data Analysis, and Knowledge Organization (GfKL 2007), pp. 319–326. Springer (2007)
10. Cramer, R.D., III., DePriest, S., Patterson, D., Hecht, P.: The developing practice of comparative molecular field analysis. In: Kubinyi, H. (ed.) 3D QSAR in Drug Design: Volume 1: Theory Methods and Applications, pp. 443–486. Springer, Netherlands (1993)
11. Diukendjieva, A., Al Sharif, M., Alov, P., Pencheva, T., Tsakovska, I., Pajeva, I.: ADME/Tox properties and biochemical interactions of silybin congeners: *In silico* study. Nat. Prod. Commun. **12**(2), 175–178 (2017)
12. Diukendjieva, A., Alov, P., Tsakovska, I., Pencheva, T., Richarz, A., Kren, V., Cronin, M.T.D., Pajeva, I.: *In vitro* and *in silico* studies of the membrane permeability of natural flavonoids from Silybum marianum (L.) Gaertn. and their derivatives. Phytomedicine **53**, 79–85 (2019)

13. Ertl, P., Rohde, B., Selzer, P.: Fast calculation of molecular polar surface area as a sum of fragment-based contributions and its application to the prediction of drug transport properties. J. Med. Chem. **43**(20), 3714–3717 (2000)

14. European Food Safety Authority: EFSA Strategy 2020 Trusted Science for Safe Food (2016)

15. Ferguson, A.M., Heritage, T., Jonathon, P., Pack, S.E., Phillips, L., Rogan, J., Snaith, P.J.: EVA: a new theoretically based molecular descriptor for use in QSAR/QSPR analysis **11**, 143–152 (1997)

16. Finan, C., Gaulton, A., Kruger, F.A., Lumbers, R.T., Shah, T., Engmann, J., Galver, L., Kelley, R., Karlsson, A., Santos, R., Overington, J.P., Hingorani, A.D., Casas, J.P.: The druggable genome and support for target identification and validation in drug development. Sci. Trans. Medi. **9**(383), eaag1166 (2017)

17. Fratev, F., Tsakovska, I., Al Sharif, M., Mihaylova, E., Pajeva, I.: Structural and dynamical insight into PPARγ antagonism: *In silico* study of the ligand-receptor interactions of non-covalent antagonists. Int. J. Mole. Sci. **16**(7), 15405–15424 (2015)

18. Good, A.C., So, S.S., Richards, W.G.: Structure-activity relationships from molecular similarity matrices. J. Med. Chem. **36**(4), 433–438 (1993)

19. Hansch, C.: Quantitative approach to biochemical structure-activity relationships. Acc. Chem. Res. **2**(8), 232–239 (1969)

20. Hansch, C., Maloney, P.P., Fujita, T., Muir, R.M.: Correlation of biological activity of phenoxy-acetic acids with Hammett substituent constants and partition coefficients. Nature **194**(4824), 178–180 (1962)

21. Ikonomov, N., Vassilev, P., Roeva, O.: ICrAData—software for InterCriteria analysis. Int. J. Bioautomat. **22**(1), 1–10 (2018)

22. Jereva, D., Fratev, F., Tsakovska, I., Alov, P., Pencheva, T., Pajeva, I.: Molecular dynamics simulation of the human estrogen receptor alpha: contribution to the pharmacophore of the agonists. Math. Comput. Simul. **133**, 124–134 (2017)

23. Kansy, M., Fischer, H., Kratzat, K., Senner, F., Wagner, B., Parrilla, I.: High-throughput artificial membrane permeability studies in early lead discovery and development. In: Testa, B., van de Waterbeemd, H., Folkers, G., Guy, R. (eds.) Pharmacokinetic Optimization in Drug Research, pp. 447–464. Verlag Helvetica Chimica Acta, Zürich (2001)

24. Klebe, G.: Comparative molecular similarity indices analysis: CoMSIA. Perspect. Drug Discovery Des. **12**, 87–104 (1998)

25. Klebe, G.: Drug Design: Methodology, Concepts, and Mode-of-Action. Springer, Berlin Heidelberg (2013)

26. Kubinyi, H.: Lock and key in the real world: concluding remarks. Pharm. Acta Helv. **69**(4), 259–269 (1995)

27. Kubinyi, H. (ed.): QSAR: Hansch analysis and related approaches. New York, VCH, Weinheim (1993)

28. Labbé, C.M., Pencheva, T., Jereva, D., Desvillechabrol, D., Becot, J., Villoutreix, B.O., Pajeva, I., Miteva, M.A.: AMMOS2: a web server for protein–ligand–water complexes refinement via molecular mechanics. Nucleic Acids Res. **45**(W1), W350–W355 (2017)

29. Lagorce, D., Pencheva, T., Villoutreix, B.O., Miteva, M.A.: DG-AMMOS: a New tool to generate 3D conformation of small molecules using distance geometry and automated molecular mechanics optimization for *in silico* screening. BMC Chem. Biol. **9**, 6 (2009)

30. Lauri, G., Bartlett, P.A.: CAVEAT: a program to facilitate the design of organic molecules. J. Comput.-Aided Mol. Des. **8**(1), 51–66 (1994)

31. Marchant, C.A., Briggs, K.A., Long, A.: *In silico* tools for sharing data and knowledge on toxicity and metabolism: Derek for Windows, Meteor, and Vitic. Toxicol. Mech. Methods **18**(2–3), 177–187 (2008)

32. Martin, Y.C., Abagyan, R., Ferenczy, G.G., Gillet, V.J., Oprea, T.I., Ulander, J., Winkler, D., Zefirov, N.S.: Glossary of terms used in computational drug design, part II (IUPAC Recommendations 2015). Pure Appl. Chem. **88**(3), 239–264 (2016)

33. Melagraki, G., Afantitis, A., Sarimveis, H., Koutentis, P.A., Kollias, G., Igglessi-Markopoulou, O.: Predictive QSAR workflow for the *in silico* identification and screening of novel HDAC inhibitors. Mol. Divers. **13**(3), 301–311 (2009)

34. Moore, T.J., Zhang, H., Anderson, G., Alexander, G.C.: Estimated costs of pivotal trials for novel therapeutic agents approved by the US Food and Drug Administration, 2015–2016. JAMA Int. Med. **178**(11), 1451–1457 (2018)
35. Mysinger, M.M., Carchia, M., Irwin, J.J., Shoichet, B.K.: Directory of useful decoys, enhanced (DUD-E): better ligands and decoys for better benchmarking. J. Med. Chem. **55**(14), 6582–6594 (2012)
36. Nettles, K.W., Bruning, J.B., Gil, G., O'Neill, E.E., Nowak, J., Hughs, A., Kim, Y., DeSombre, E.R., Dilis, R., Hanson, R.N., Joachimiak, A., Greene, G.L.: Structural plasticity in the oestrogen receptor ligand-binding domain. EMBO Rep. **8**(6), 563–568 (2007)
37. Netzeva, T.I., Worth, A.P., Aldenberg, T., Benigni, R., Cronin, M.T.D., Gramatica, P., Jaworska, J.S., Kahn, S., Klopman, G., Marchant, C.A., Myatt, G., Nikolova-Jeliazkova, N., Patlewicz, G.Y., Perkins, R., Roberts, D.W., Schultz, T.W., Stanton, D.T., van de Sandt, J.J.M., Tong, W., Veith, G., Yang, C.: Current status of methods for defining the applicability domain of (quantitative) structure-activity relationships: the report and recommendations of ECVAM workshop 52. Altern. Lab. Anim. **33**(2), 155–173 (2005)
38. Pearlman, R.S.: 3D Molecular structures: generation and use in 3D searching. In: Kubinyi, H. (ed.) 3D QSAR in Drug Design: Volume 1: Theory Methods and Applications, pp. 41–79. Springer, Netherlands (1993)
39. Pencheva, T., Lagorce, D., Pajeva, I., B. O, V., Miteva, M.: AMMOS: a software platform to assist *in silico* screening. Int. J. Bioautomat. **13**(4), 143–150 (2009)
40. Pencheva, T., Lagorce, D., Pajeva, I., Villoutreix, B.O., Miteva, M.A.: AMMOS: automated molecular mechanics optimization tool for *in silico* screening. BMC Bioinf. **9**, 438 (2008)
41. Pencheva, T., Lagorce, D., Pajeva, I., Villoutreix, B.O., Miteva, M.A.: AMMOS software: method and application. In: Baron, R. (ed.) Computational Drug Discovery and Design, pp. 127–141. Springer New York, New York, NY (2012)
42. Pike, A.C.W.: Lessons learnt from structural studies of the oestrogen receptor. Best Pract. Res. Clin. Endocrinol. Metab. **20**(1), 1–14 (2006)
43. Schulz-Gasch, T., Stahl, M.: Scoring functions for protein–ligand interactions: a critical perspective. Drug Discov. Today Technol. **1**(3), 231–239 (2004)
44. Stewart, J.J.P.: Optimization of parameters for semiempirical methods V: modification of NDDO approximations and application to 70 elements. J. Mol. Model. **13**(12), 1173–1213 (2007)
45. Tropsha, A., Gramatica, P., Gombar, V.K.: The importance of being earnest: validation is the absolute essential for successful application and interpretation of QSPR models. QSAR Comb. Sci. **22**(1), 69–77 (2003)
46. Tsakovska, I., Al Sharif, M., Alov, P., Diukendjieva, A., Fioravanzo, E., Cronin, M.T.D., Pajeva, I.: Molecular modelling study of the PPARγ receptor in relation to the mode of action/adverse outcome pathway framework for liver steatosis. Int. J. Mole. Sci. **15**(5), 7651–7666 (2014)
47. Tsakovska, I., Pajeva, I., Alov, P., Worth, A.: Recent advances in the molecular modeling of estrogen receptor-mediated toxicity. In: Christov, C. (ed.) Advances in Protein Chemistry and Structural Biology, pp. 217–251. Academic Press (2011)
48. Wold, S.: PLS for multivariate linear modelling. In: van de Waterbeemd, H. (ed.) Chemometric Methods in Molecular Design, pp. 195–218. VCH Verlagsgesellschaft mbH (1995)

Survey on Theory and Applications of InterCriteria Analysis Approach

Elena Chorukova, Pencho Marinov, and Ivo Umlenski

Abstract Short notes on the concept of InterCriteria Analysis (ICrA), a new method for estimating similarity with regards to estimations of criteria and objectives, are described. The main results of the theory development since ICrA's inception are considered. Several software implementations of ICrA are presented. Various applications of the approach to a wide variety of problems are discussed.

Keywords InterCriteria Analysis · Index matrix · Intuitionistic fuzzy set · ICrAData

1 Introduction

The InterCriteria Analysis (ICrA) problem has been formulated in September 2013 as a solution to a specific production problem. It was reported outside the program of the next one International Workshop on Intuitionistic Fuzzy Sets and Generalized Nets in Warsaw and was published in [17]. In the coming years, interest in this concept increased significantly. It has become the subject of both theoretical studies and applications in various fields. In Sect. 2, we will outline the basic elements of ICrA theory, and in Sect. 3, its applications, developed mainly by scientists from the Bulgarian Academy of Sciences. In conclusion, we will formulate open issues that we plan to work on in the future.

E. Chorukova (✉) · I. Umlenski
Institute of Biophysics and Biomedical Engineering, Bulgarian Academy of Sciences, Acad G. Bonchev Str., Bl.21, 1113 Sofia, Bulgaria
e-mail: elena@microbio.bas.bg

E. Chorukova
The Stephan Angeloff Institute of Microbiology, Bulgarian Academy of Sciences, Acad G. Bonchev Str., Bl.26, 1113 Sofia, Bulgaria

P. Marinov
Institute of Information and Communication Technologies, Bulgarian Academy of Sciences, Acad G. Bonchev Str., Bl.25A, 1113 Sofia, Bulgaria

© The Author(s), under exclusive license to Springer Nature Switzerland AG 2021
K. T. Atanassov (ed.), *Research in Computer Science in the Bulgarian Academy of Sciences*, Studies in Computational Intelligence 934,
https://doi.org/10.1007/978-3-030-72284-5_20

453

2 Basic Theoretical Research over ICrA

Following [10, 17], we will give a brief account of ICrA theory, but in the beginning we will begin with brief definitions of the terms Intuitionistic fuzzy set, intuitionistic fuzzy pair and index matrix.

Intuitionistic fuzzy sets (IFSs, see [8, 9]) introduced by Atanassov, are an extension of the concept of fuzzy sets, defined by Zadeh. Let E be a universe. The IFS A^* is defined by

$$A^* = \{ \langle x, \mu_A(x), \nu_A(x) \rangle \mid x \in E \},$$

where $\mu_A(x)$, $\nu_A(x) : E \rightarrow [0, 1]$, represent the functions that give the degrees of membership and non-membership for each element $x \in E$ to a fixed subset A of E, respectively, such that $0 \le \mu_A(x) + \nu_A(x) \le 1$.

The Intuitionistic Fuzzy Pairs (IFPs, see [11]) is an object in the form of an ordered pair $\langle a, b \rangle$, where $a, b \in [0, 1]$ and $a + b \le 1$. It is used as for the evaluation of objects or processes. Its components (a and b) are interpreted respectively as degrees of membership (validity, etc.) and non-membership (non-validity, etc.). For two IFPs $x = \langle a, b \rangle$ and $y = \langle c, d \rangle$ the following relations are defined:

$$x < y \quad \text{iff } a < c \text{ and } b > d$$
$$x \le y \quad \text{iff } a \le c \text{ and } b \ge d$$
$$x = y \quad \text{iff } a = c \text{ and } b = d$$
$$x \ge y \quad \text{iff } a \ge c \text{ and } b \le d$$
$$x > y \quad \text{iff } a > c \text{ and } b < d$$

In [7, 10], Atanassov introduced and developed a theory of the concept of an Index Matrix (IM) and studied its basic properties. Let I be a fixed set of indices and $\mathfrak{R}$ be the set of the real numbers. Let K and $L(K, L \subset I)$, be sets of indices. The IM is an object with the following form:

$$K, L, \left[a_{k_i, l_j} \right] \equiv$$

	l_1	$\cdots$	l_n
k_1	a_{k_1, l_1}	$\cdots$	a_{k_1, l_n}
$\vdots$	$\vdots$	$\ddots$	$\vdots$
k_m	a_{k_m, l_1}	$\cdots$	a_{k_m, l_n}

where $K = \{k_1, k_2, \ldots, k_m\}$, $L = \{l_1, l_2, \ldots, l_n\}$ and for $1 \le i \le m$, and $1 \le j \le n : a_{k_i, l_j} \in \mathfrak{R}$.

An extension of the IM is the Intuitionistic Fuzzy IM (IFIM, see [10]) that has the form:

$$\left[K, L, \left\{\left\langle \mu_{k_i,l_j}, \nu_{k_i,l_j}\right\rangle\right\}\right] \equiv$$

	l_1	$\cdots$	l_n
k_1	$\langle \mu_{k_1,l_1}, \nu_{k_1,l_1}\rangle$	$\cdots$	$\langle \mu_{k_1,l_n}, \nu_{k_1,l_n}\rangle$
$\vdots$	$\vdots$	$\ddots$	$\vdots$
k_m	$\langle \mu_{k_m,l_1}, \nu_{k_m,l_1}\rangle$	$\cdots$	$\langle \mu_{k_m,l_n}, \nu_{k_m,l_n}\rangle$

where $1 \le i \le m$, $1 \le j \le n$, $0 \le \mu_{k_i,l_j}, \nu_{k_i,l_j}, \mu_{k_i,l_j} + \nu_{k_i,l_j} \le 1$, i.e. $\langle \mu_{k_i,l_j}, \nu_{k_i,l_j}\rangle$ is an IFPs.

Let us have an IM:

$A \equiv$	O_1	$\cdots$	O_n
C_1	a_{C_1,O_1}	$\cdots$	a_{C_1,O_n}
$\vdots$	$\vdots$	$\ddots$	$\vdots$
C_m	a_{C_m,O_1}	$\cdots$	a_{C_m,O_n}

where for every p, q ($1 \le p \le m$, $1 \le q \le n$):

- C_p is a criterion, taking part in the evaluation;
- O_q is an object, being evaluated;
- a_{C_p,O_q} is a real number or another object, that is comparable to relation R with another a—object, so that for each i, j, k : $R\left(a_{C_k,O_i}, a_{C_k O_j}\right)$ is defined. Let $\bar{R}$ be the dual relation of R in the sense that if R is satisfied, then $\bar{R}$ is not satisfied and vice versa. For example, if 'R' is the relation '$>$', then $\bar{R}$ is the relation '$<$', and vice versa.

Let $S_{k,l}^{\mu}$ be the number of cases in which $R\left(a_{C_k,O_i}, a_{C_k O_j}\right)$ and $R\left(a_{C_l,O_i}, a_{C_l O_j}\right)$ are simultaneously satisfied. Let $S_{k,l}^{\nu}$ be the number of cases in which $R\left(a_{C_k,O_i}, a_{C_k O_j}\right)$ and $\bar{R}\left(a_{C_l,O_i}, a_{C_l O_j}\right)$ are simultaneously satisfied. Obviously,

$$S_{k,l}^{\mu} + S_{k,l}^{\nu} \le \frac{n(n-1)}{2}.$$

Now, for every k, l such that $1 \le k < l \le m$, and for $n \ge 2$, it can be defined:

$$\mu_{C_k,C_l} = \frac{2S_{k,l}^{\mu}}{n(n-1)}, \quad \text{and} \quad \nu_{C_k,C_l} = \frac{2S_{k,l}^{\nu}}{n(n-1)}.$$

Therefore, $\langle \mu_{C_k,C_l}, \nu_{C_k,C_l}\rangle$ is an IFP. Now, we can construct the IFIM:

$B \equiv$	C_1	$\cdots$	C_m
C_1	$\langle \mu_{C_1,C_1}, \nu_{C_1,C_1}\rangle$	$\cdots$	$\langle \mu_{C_1,C_m}, \nu_{C_1,C_m}\rangle$
$\vdots$	$\vdots$	$\ddots$	$\vdots$
C_m	$\langle \mu_{C_m,C_1}, \nu_{C_m,C_1}\rangle$	$\cdots$	$\langle \mu_{C_m,C_m}, \nu_{C_m,C_m}\rangle$

that determine the degrees of correspondence between criteria $C_1, \ldots, C_m$.

Let $\alpha, \beta \in [0, 1]$ be given, so that $\alpha + \beta \le 1$. We say that criteria C_k and C_l are in:

- (α, β)—*Positive consonance*, if $\left(\mu_{C_k,C_l} > \alpha\right)$ and $\left(\nu_{C_k,C_l} < \beta\right)$;
- (α, β)—*Negative consonance*, if $\left(\mu_{C_k,C_l} < \beta\right)$ and $\left(\nu_{C_k,C_l} > \alpha\right)$;
- (α, β)—*Dissonance, otherwise.*

After analyzing various types of data—medical, chemical, economic, environmental and others, it was concluded that it is necessary to refine the algorithms for determining the values of similarity and difference. Here, we propose the following rules when ICrA is applied to objects evaluated by criteria with real numbers, where π is the degree of uncertainty.

>>	>=	><	=>	==	=<	<>	<=	<<
$\mu+$	$\pi+$	$\nu+$	$\pi+$	$\pi+$	$\pi+$	$\nu+$	$\pi+$	$\mu+$

The motive for determining how to increase the three functions in some cases is clear, but the case $==$ needs comment. When we have in two rows equality between the corresponding elements of two columns, for example 2.1 and 2.1, according to the algorithm we have equalities. But the first of the numbers 2.1, if it had to be more than one decimal place, could be, for example, 2.11, and the second number—2.12, i.e. then there should be no sign of equality between them. On the other hand, if the two numbers are different, then as much as we look at them, they will continue to be different.

It is important to note that the scale described in this way guarantees a balance between the positive and negative consonants, which should really be expected to have a symmetrical appearance.

Then, when the equality of two pairs of values of objects by a pair of criteria is determined by the specificity of a specific task (for example, when comparing solutions to problems of integer, optimization and other types), then it is convenient to use the following table

>>	>=	><	=>	==	=<	<>	<=	<<
$\mu+$	$\pi+$	$\nu+$	$\pi+$	$\mu+$ or half the values are to $\mu+$, and the other half – to $\nu+$	$\pi+$	$\nu+$	$\pi+$	$\mu+$

The scale must be completely different when working with integers. It will be convenient to have the appearance

>>	>=	><	=>	==	=<	<>	<=	<<
$\mu+$	$\nu+$	$\nu+$	$\nu+$	$\mu+$	$\nu+$	$\nu+$	$\nu+$	$\mu+$

It is now apparent that symmetry is breaking. However, the presumption here is that we can recognize whether the two endpoints are the same or not and classify the relation between them correctly. Now, if we have all the data in the IM, we will get a fuzzy, not an IF score. The IF nature of the assessment will manifest itself in cases where there is insufficient data.

When we have to work with data that we know to be true (+), false (−) or missing (~), then it is convenient to work with the following table

+ +	+ −	+ ~	− +	− −	− ~	~ +	~ −	~ ~
$\mu +$	$\nu +$	$\pi +$	$\nu +$	$\nu +$	$\pi +$	$\pi +$	$\pi +$	$\pi +$

It is seen that, although it has greatly reduced cases in which a component will grow, it is symmetrical about its individual components. Such a table could be of real use, for example, in the study of logic schemes by axioms.

The described algorithm is modified when working with a-elements of IM A, which are IFPs (see [39]). It describes the procedure for detecting correlations by ICrA on intuitionist fuzzy inputs. The article presents the application of the ICrA over truth values for some logical formulas, some of which are axioms of intuitionist fuzzy logic.

In [20] ICrA that uses intuitionistic fuzzy implications from a special type are described.

In [18, 28] geometric interpretation of the results of the ICrA based on the IF-interpretation triangle is proposed (see [8, 9, 11]).

For example, when $\alpha = 0.75$, $\beta = 0.25$, we obtain as a region of a positive consonance the region marked by PC in Fig. 1; when $\alpha = 0.25$, $\beta = 0.75$, we obtain as a region of a negative consonance the region marked by NC in the same figure, while region marked by D corresponds to the zone of dissonance.

Ways to determine values for α and β are discussed in [27, 29, 44]. In [34], the case of the choice of three similar criteria is considered. The idea of a-elements normalizing before applying ICrA is described in [14].

In [23] a new operator N_γ of type level operator was defined. It assumes a threshold value γ, corresponding to the ratio of the values of the membership function μ to the non-membership function ν and the set of elements whose ratio of the values of the two functions is greater than or equal to the value γ. This is a fundamentally new way of building a level operator in the theory of IFSs, and one that allows decisions to be made even at otherwise high nominal uncertainty values due to the high ratio μ/ν in favor of the membership function. Four statements about the properties of the operator are proved.

In [104] a new ordinance was proposed between intuitionist blurry couples, which set a more restrictive order than the classical one. The results obtained in the article were subsequently further developed to create new operators.

In [103] the notion of structurally similar intuitionist fuzzy sets have been introduced—i.e. sets for which the "fuzzy" and "intuitionist fuzzy" parts on the universal set coincide. Thanks to it, it is possible to consider operators that act not on the whole universal set but on some of its restriction—in this way it is naturally possible to describe a formally different set of operators. The proposed scheme is also applicable to the results obtained from intercritical analysis.

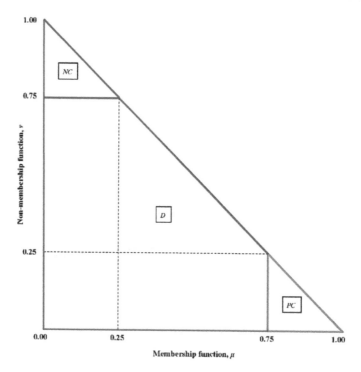

Fig. 1 Zones of *positive* consonance, *negative* consonance and *dissonance* in the IFSs interpretation triangle

In [20, 21, 105, 106], the set of objects and rules to which ICrA applies has been expanded. Extension of the theoretical ICrA is a prerequisite for improving the applicability of the method to decision-making procedures under uncertainty, in the absence of missing, incomplete or uncertain data, such as medical and patient data, such as history and functional studies.

In [16] an *n*-type IFSs study is conducted and the idea of how intuitionist fuzzy pairs of this type could be used in intercritical analysis is discussed.

In [86, 93] four different algorithms performing ICrA, namely *μ-biased, balanced, ν-biased* and *unbiased*, are proposed. The work is based on the [18] where several rules for defining the ways of estimating the degrees of "agreement" and "disagreement", with respect to the type of data, are defined.

3 ICrA Software

Some of the works use the authors' Matlab realization of the ICrA algorithm [68, 86, 89], but most of them utilize the specialized software developed by Mavrov for application of ICrA and presented in [64, 65]. The software of Mavrov "takes two matrices of input data and outputs the intuitionistic fuzzy pairs that describe the intercriterial relationship as two tables. The application can work with Microsoft Excel workbooks or text files and provides ways to transfer the output data to other programs. It can also include functionality to display graphics of the output data [64]. In [65] additional graphical interpretation of the results of ICrA in the intuitionistic fuzzy interpretational triangle, has been implemented. While the software presented in [64, 65, 102] has a user friendly interface and is easy to use, it suffers from several limitations: the software only works under windows; the copied data has to be compatible with Microsoft Excel, and in this regard imposes restriction on the number of objects (placed in columns).

A new cross-platform software implementing ICrA, called ICrAData, is proposed in [54]. The software is written in Java programming language. ICrAData previews the results in two panels—matrix and graphical views. The results can be exported in different formats: arrays, vectors and graphs. In the matrix view, the column data can be sorted in ascending or descending order. The graphical view has options for resizing the intuitionist blurred interpretation triangle, for displaying a grid and for setting different colors of dots. In addition, the selected point in the graph is highlighted in the matrix view and vice versa. One of the additional advantages of ICrAData is that it also enables the analysis of data presented through intuitionist blurry pairs. ICrAData has no limitation on the number of objects and has much additional functionality compared to the software of Mavrov. Moreover, five different algorithms (μ-biased, unbiased, ν-biased, balanced and weighted) for intercriteria relations calculation are implemented in ICrAData, in accordance with the current development of the ICrA theory [18, 86].

4 ICrA Applications so Far

ICrA has found applications in various fields: biomedicine and quality of life, petrochemical industries, economics and transport, evaluation of university ratings in different countries, ecology and meteorology, artificial intelligence, metaheuristic algorithms, neural networks, etc.

In [19] is described an approach to a constructive simplification of multiagent multicriteria decision making problems via intercriteria analysis. This approach proposes that decision makers be provided with a list of criteria that they can work on and that they can expand with new criteria. Any decision maker can use all or part of the criteria and evaluate all or part of the set objects. At the end of the decision-making process, when not only subjective but objective assessments already exist, the ICrA performs the data above in order to find pairs of similar criteria. The one of two similar criteria, which is more laborious, more expensive, or which takes longer, may be dropped, so that in the next procedure the same or other decision makers will work with the new list of criteria. The article demonstrates that multiagents (such as a summary involving experts, automated decision-making systems and other tools) can make decisions based on their criteria, as well as criteria proposed by them (especially in the case of experts) that the multiagents can arrange on their own at the vertices of oriented tree columns. Upon completion of the procedures, based on the evaluation of the sites by the criteria used, an ICrA is conducted in order to remove the excess criteria, so that in the next procedure the following multiagents are offered the criteria remaining after the removal of the excess ones.

In [62], a detailed comparison was made between the results obtained by ICrA on the one hand and the correlation analyzes of Spearman, Pearson, Kendall and ICrA on the other.

In [100], the analysis was prepared for four separate predefined subsets according to the calorimetric profiles of patients with colorectal cancer and it was found that the specific intercritical patterns differ between the different profiles and conclusions were made about the enthalpy of the thermal profiles, the thermal capacity of the main plasma proteins, albumin and immunoglobulins and other parameters.

For the first time the ICrA approach is applied to data connected with the centers for Transfusion Haematology in Bulgaria [1].

Investigations performed in [107–109] present an application of ICrA to medical data, aimed at the discovery of correlations between important healthy indicators, based on available medical data of patients with Behterev's Disease.

The authors of [75, 76] propose the application of the approach of ICrA to data of intellectual and physical parameters obtained from university students practicing sports activities in order to evaluate the appropriateness of the used tests.

A possible application of ICrA in search of correlations between the body composition measurements and obesity among college students is presented in [77].

In [61, 102], ICrA is applied to explore the performance of various docking scoring functions available in molecular modelling software packages for the purposes of in silico studies of bioactive compounds.

In [97] a method for constructing a smaller neural network with weight coefficient matrices is proposed. The idea is illustrated by applying the ICrA method to 140 crude oil samples, evaluated according to 8 physicochemical criteria and submitted to the input of a neural network. The results obtained show that the use of ICrA can reduce some of the irrelevant data used in training neural networks, reduce the number of inputs and reduce the number of weights and offsets. As a result of these optimizations, the memory required for the learning process, the conversion time and the number of iterations are reduced.

The ICrA approach was applied for the first time for the purpose of parametric identification of cultivation processes and assessment of the quality of execution of genetic algorithms (GA). In this relation, ICrA allows an establishment of relationships and dependencies between some of the basic parameters of GA, namely number of individuals in the population, number of iterations of algorithm execution, generation gap, probabilities of crossing and mutation, on the one hand, and algorithm convergence time, model parameter estimates and accuracy, on the other hand. The obtained results assist in clarification of the relations between the parameters of cultivation process models of yeast and bacteria and reveal some possibilities for the improvement of the quality of presentation of GA.

The efficacy of ICrA for assessment of the quality of GA performance has been demonstrated by its application to both assessment of different kinds of GA (SGA—simple, single-population GA and MpGA—multi-population GA), as well as to evaluate the impact of main operators of GA, namely selection, crossing and mutation on the performance of the different kinds of GA. The implementation of ICrA has assisted in the assessment of the performance of:

- six different kinds of SGA for the purposes of parameter identification of yeast fed-batch cultivation process model [69];
- six different kinds of MpGA for the purposes of parameter identification of yeast fed-batch cultivation process model [3];
- SGA at 51 different values of the generation gap for the purposes of parameter identification of yeast fed-batch cultivation process model [68];
- SGA at different values of number of individuals in the population and number of iterations of algorithm execution for the purposes of parameter identification of yeast and bacteria fed-batch cultivation process models [5, 67];
- SGA at different values of probabilities for crossover and mutation (crossover and mutation rates), in pairs and triples, for the purposes of parameter identification of yeast and bacteria fed-batch cultivation process models [6, 87];
- different kinds of SGA for the purposes of parameter identification of yeast and bacteria fed-batch cultivation process models [85];
- six different kinds of SGA and six different kinds of MpGA for the purposes of parameter identification of yeast fed-batch cultivation process model [4];
- analysis of mutation rate influence [82];
- generation gap influence on genetic algorithms performance [81];
- model parameters identification using genetic algorithm [84].

ICrA has found application also in the fields of structural and parametric identification in the cultivations of *E. coli* [56], *L-lysine* fed-batch processes [60] and selection of growth rate model for *Kluyveromyces marxianus var. lactis* MC 5 batch processes [59, 73, 74].

Four algorithms for calculation of the degrees of agreement and disagreement in cases of equality between two criteria, aforementioned as implemented in the ICrAData, have been applied for the purpose of parameter identification of fed-batch cultivation process model of bacteria [70, 86] and of yeast [86]. μ-biased algorithm is shown to be the most reliable one, assigning the pair of criteria to the degree of agreement. Further, another algorithm for calculation of the degrees of agreement and disagreement has been introduced, *weighted* one, in which weights are assigned to the degrees of agreement and disagreement. Altogether five algorithms have been implemented for both parameter identification of fed-batch cultivation process models of bacteria and yeast [93].

Some further analyses of ICrA approach are also published. Discovering knowledge from predominantly repetitive data by ICrA is presented in [95]. ICrA results based on different number of objects are discussed in [110]. Results about the computational complexity and influence of numerical precision on the results of intercriteria analysis in the decision making process are presented in [22]. An application of topological operators over data from ICrA is proposed in [91].

Following the main idea of the assessment of the quality of GA performance for the purposes of parameter identification of yeast fed-batch cultivation process model, ICrA has been applied to assess the performance of another well-known and widely used metaheuristic technique of artificial bee colony (ABC) [2]. In [83] ICrA is applied over ordered pairs applied to ABC optimization results.

Further, ICrA is applied for investigation of the Ant Colony Optimization (ACO) algorithm performance. A series of papers use ICrA to analyse variants of ACO algorithm for wireless sensor network positioning [47, 80]; different hybrid ACO algorithms for workforce planning [52, 94]; comparison of different ACO start strategies based on ICrA [50, 92]; ACO application to GPS surveying problems [46, 49, 51]; ICrA of relations between model parameters estimates and ACO performance is investigated in [78]; multi-objective ACO algorithm for Wireless Sensor Networks (WSN) layout [48].

A comparison of different metaheuristic algorithms based on ICrA is presented in [79]. ICrA of ACO and GA hybrid algorithms is proposed in [90]. ICrA of Bat Algorithm with parameter adaptation using Type-1 and Interval Type-2 fuzzy systems is done in [88].

ICrA is also applied in order to analyze the universities rankings system in the Slovak Republic [38] and rankings of Indian universities [66].

ICrA is applied for the forest fire analysis. Some results are presented in [112] where ICrA is used as a complementary approach to the Lubenov's methodology in order to refine and improve the classification of the regions in Bulgaria into forest fires risk groups. ICrA is able to identify regions which are classified in a particular risk group with ambiguity. Further, the obtained results in [111] show that the ICrA could be used as an additional approach in order to refine the classification of the regions in the specific risk groups. In such cases, ICrA will confirm the regions which are unconditionally in the particular group with high value of the degree of agreement (positive consonance). ICrA can distinguish those regions which are classified in a particular group with a degree of uncertainty. Thus, the result of ICrA can prevent the neglect of certain areas identified as out-of-risk, which might be actually in risk. In [96] ICrA is applied for acoustic monitoring of forest for early detection fires.

In [55, 57, 58, 71, 72] the ICrA method is used for analysis and assessment of pollution indices of river ecosystems, regarding the Struma River in the Bulgarian section at ten stations along the river. In this study the selected criteria are ten stations along the river. These criteria (stations) are analysed by thirteen indices (objects) for pollution. ICrA analysis allows excluding those stations in which we have a positive agreement in the modelling of pollution.

ICrA approach is applied also to evaluate the quality of water treatment processes in a typical wastewater treatment plant (WWTP) [53]. Based on real data, the interdependencies between various parameters characterizing the quality of processes in

WWTP, such as water quantity, pH, petroleum products, mechanical additives, phosphates, nitrates, etc., were analyzed. The application of ICrA led to additional knowledge about the processes in WWTP and showed some new relationships between the considered criteria, especially with regard to seasonal dependencies.

In [99] the authors are using the approach of ICrA order to select potentially beneficial new crudes for processing in a refinery.

ICrA found interesting applications in neural network based procedures [98], in medicine [40] and in the economy [24–26, 30–33, 35–37, 41–45].

5 Conclusion

ICrA has appeared seven years ago and their theory is still under development. Detailed algorithms for working with intercritical analysis on three types of parameters are to be developed, for example criteria, objects, time points. Then one will have to work with 3-dimensional IM (see [10]), as the first step is already done in [101].

In [12] an ICrA algorithm is described when the a-elements of IM A are images, but further development of this idea and its implementation are forthcoming. Also in the beginning is the idea set out in [15], the estimates should not be based on the IFSs apparatus, but on interval-valued IFSs (see [13]). A detailed comparison of ICrA with other types of correlation analyzes is forthcoming, the first step is being done in [63].

Those wishing to implement the ICA face other important problems awaiting their solution in the near future. After them are the following:

1. Specify the minimum number of criteria under which the algorithms will be correct.
2. Specify the minimum number of evaluated objects for which the algorithms will be correct.
3. Specify for which types of tasks, which scale (or new) is appropriate to use.

Solving each of these problems will contribute to the development of ICrA theory.

References

1. Andreev, N., Sotirova, E., Ribagin, S.: Intercriteria analysis of data from the centers for transfusion haematology in Bulgaria. Comptesrendus de l'Académiebulgare des Sciences (2017)
2. Angelova, M.: InterCriteria analysis of control parameters relations in artificial bee colony algorithm. WSEAS Trans. Math. **18**, 123–128 (2019)
3. Angelova, M., Pencheva, T.: Intercriteria analysis of multi-population genetic algorithms performance. Ann. Comput. Sci. Inf. Syst. **13**, 77–82 (2017)

4. Angelova, M., Pencheva, T.: Intercriteria analysis approach for comparison of simple and multi-population genetic algorithms performance. In: Fidanova, S. (ed.) Recent Advances in Computational Optimization. Studies in Computational Intelligence, vol. 795, pp. 117–130 (2019)

5. Angelova, M., Roeva, O., Pencheva, T.: Intercriteria analysis of a cultivation process model based on the genetic algorithm population size influence. Notes Intuit. Fuzzy Sets 21(4), 90–103 (2015)

6. Angelova, M., Roeva, O., Pencheva, T.: Intercriteria analysis of crossover and mutation rates relations in simple genetic algorithm. Ann. Comput. Sci. Inf. Syst. 5, 419–424 (2015)

7. Atanassov, K.: Generalized index matrices. Comptes rendus de l'Academie Bulgare des Sciences 40(11), 15–18 (1987)

8. Atanassov, K.: Intuitionistic Fuzzy Sets. Springer, Heidelberg (1999)

9. Atanassov, K.: On Intuitionistic Fuzzy Sets Theory. Springer, Berlin (2012)

10. Atanassov, K.: Index Matrices: Towards an Augmented Matrix Calculus. Springer, Cham (2014)

11. Atanassov, K.: Intuitionistic Fuzzy Logics. Springer, Cham (2017)

12. Atanassov, K.: Intercriteria analysis over patterns. In: Sgurev, V., Piuri, V., Jotsov, V. (eds.) Learning Systems: From Theory to Practice. Studies in Computational Intelligence, vol. 756. Springer (2018)

13. Atanassov, K.: Interval-Valued Intuitionistic Fuzzy Sets. Springer, Cham (2020)

14. Atanassov, K., Atanassova, V., Chountas, P., Mitkova, M., Sotirova, E. Sotirov, S., Stratiev, D.: Intercriteria analysis over normalized data. In: Proceedings of the 8th IEEE Conference "Intelligent Systems", Sofia, 4–6 September, pp. 136–138 (2016)

15. Atanassov, K., Marinov, P., Atanassova, V.: InterCriteria Analysis with Interval-Valued Intuitionistic Fuzzy Evaluations. Lecture Notes in Computer Science, vol. 11529, pp. 329–338. Springer (2019)

16. Atanassov, K., Vassilev, P.: On the intuitionistic fuzzy sets of n-th type. In: Advances in Data Analysis with Computational Intelligence Methods. Studies in Computational Intelligence, vol. 738, pp. 265–274 (2018)

17. Atanassov, K., Mavrov, D., Atanassova, V.: Intercriteria decision making: a new approach for multicriteria decision making, based on index matrices and intuitionistic fuzzy sets. Iss. Intuit. Fuzzy Sets Gen. Nets 11, 1–8 (2014)

18. Atanassov, K., Atanassova, V., Gluhchev, G.: InterCriteria analysis: ideas and problems. Notes Intuit. Fuzzy Sets 21(1), 81–88 (2015)

19. Atanassov, K., Szmidt, E., Kacprzyk, J., Atanassova, V.: An approach to a constructive simplification of multiagent multicriteria decision making problems via intercriteria analysis. Comptes rendus de l'Academie bulgare des Sciences 70(8), 1147–1156 (2017)

20. Atanassov, K., Ribagin, S., Sotirova, E., Bureva, V., Atanassova, V., Angelova, N.: Intercriteria analysis using special type of intuitionistic fuzzy implications. NIFS 23(5), 61–65 (2017)

21. Atanassov, K., Ribagin, S., Doukovska, L., Atanassova, V.: Implication → 190. NIFS 23(4), 79–83 (2017)

22. Atanassova, V., Roeva, O.: Computational complexity and influence of numerical precision on the results of intercriteria analysis in the decision making process. Notes Intuit. Fuzzy Sets 24(3), 53–63 (2018)

23. Atanassova, V.: New modified level operator Nγ over intuitionistic fuzzy sets. In: Christiansen, H., Jaudoin, H., Chountas, P., Andreasen, T., Larsen, H.L. (eds.) Proceedings of 12th International Conference on Flexible Query Answering Systems (FQAS 2017), London, UK, June 21–22, 2017. LNAI, vol. 10333, pp. 209–214. Springer (2017)

24. Atanassova, V., Doukovska, L., Atanassov, K., Mavrov, D.: InterCriteria decision making approach to EU member states competitive analysis. In: Proceedings of 4th International Symposium on Business Modeling and Software Design, Luxembourg, Grand Duchy of Luxembourg, 24–26 June 2014, pp. 289–294 (2014)

25. Atanassova, V., Doukovska, L., Atanassov, K., Mavrov, D.: Intercriteria decision making approach to EU member states competitiveness analysis. In: Proceedings of 4th International Symposium on Business Modelling and Software Design, Luxembourg, June, 24–26 (2014)

26. Atanassova, V., Doukovska, L.: Business dynamism and innovation capability in the European Union Member States in 2018 through the Prism of InterCriteria analysis. In: International Conference on Flexible Query Answering Systems, pp. 339–349. Springer, Cham (2019)

27. Atanassova, V., Vardeva, I., Sotirova, E., Doukovska, L.: Traversing and ranking of elements of an intuitionistic fuzzy set in the intuitionistic fuzzy interpretation triangle. In: Novel Developments in Uncertainty Representation and Processing, pp. 161–174. Springer, Cham (2016)

28. Atanassova, V.: Interpretation in the intuitionistic fuzzy triangle of the results, obtained by the intercriteria analysis. In: 2015 Conference of the International Fuzzy Systems Association and the European Society for Fuzzy Logic and Technology (IFSA-EUSFLAT-15). Atlantis Press (2015)

29. Atanassova, V., Vardeva, I.: Sum-and average-based approach to criteria shortlisting in the InterCriteria analysis. Notes Intuit. Fuzzy Sets **20**(4), 41–46 (2014)

30. Atanassova, V., Mavrov, D., Doukovska, L., Atanassov, K.: Discussion on the threshold values in the InterCriteria decision making approach. Notes Intuit. Fuzzy Sets **20**(2), 94–99 (2014)

31. Atanassova, V., Doukovska, L., Karastoyanov, D., Čapkovič, F.: InterCriteria decision making approach to EU member states competitiveness analysis: trend analysis. In: Intelligent Systems' 2014, pp. 107–115. Springer, Cham (2015)

32. Atanassova, V., Doukovska, L., Mavrov, D., Atanassov, K.: InterCriteria decision making approach to eu member states competitiveness analysis: temporal and threshold analysis. In: Intelligent Systems' 2014, pp. 95–106. Springer, Cham (2015)

33. Atanassova, V., Doukovska, L., Mavrov, D., Atanassov, K.: InterCriteria decision making approach to EU member states competitiveness analysis: temporal and threshold analysis. In: Angelov, P., Atanassov, K., Doukovska, L., Hadjiski, M., Jotsov, V., Kacprzyk, J., Kasabov, N., Sotirov, S., Szmidt, E., Zadrozny, S. (eds.) Intelligent Systems 2014, pp. 95–106. Springer, Cham (2015)

34. Atanassova, V., Doukovska, L., Michalikova, A., Radeva, I.: Intercriteria analysis: from pairs to triples. Notes Intuit. Fuzzy Sets **22**(5), 98–110 (2016)

35. Atanassova, V., Doukovska, L., De Tré, G., Radeva, I.: Intercriteria analysis and comparison of innovation-driven and efficiency-to-innovation driven economies in the European Union. Notes Intuit. Fuzzy Sets **23**(3), 54–68 (2017)

36. Atanassova, V., Doukovska, L., Sotirova, A., Kacprzyk, E., Radeva, I., Vassilev, P.: Inter-Criteria analysis of the global competitiveness reports: from efficiency-to innovation-driven economies. Multiple-Valued Logic Soft Comput. **31**(5–6), 469–494 (2018)

37. Atanassova, V., Doukovska, L., Krawczak, M.: Intercriteria analysis of countries in transition from factor-driven to efficiency-driven economy. Notes Intuit. Fuzzy Sets **24**(2), 84–96 (2018)

38. Bureva, V., Michalíková, A., Sotirova, E., Popov, S., Riečan, B., Roeva, O.: Application of the InterCriteria analysis to the universities rankings system in the Slovak Republic. In: 21st International Conference on Intuitionistic Fuzzy Sets, 22–23 May 2017, Burgas, Bulgaria, Notes on IFS, vol. 23, No. 2, pp. 128–140 (2017)

39. Bureva, V., Sotirova, E., Atanassova, V., Angelova, N., Atanassov, K.: Intercriteria Analysis over Intuitionistic Fuzzy Data. Lecture Notes in Computer Science, vol. 10665, pp. 333–340. Springer International Publishing AG (2018)

40. Bureva, V., Atanassov, K., Andreev, N.: Intercriteria analysis applied to Healthcare Rankings. In: Annual of Assen Zlatarov University, Burgas, Bulgaria, vol. XLVIII, No. 1, pp. 81–85 (2019)

41. Doukovska, L., Atanassova, V.: InterCriteria analysis of the most problematic factors for doing business in the European Union, 2017–2018. In: International Conference on Flexible Query Answering Systems, June 2019, pp. 353–360. Springer, Cham (2019)

42. Doukovska, L., Atanassova, V., Shahpazov, G., Capkovic, F.: InterCriteria analysis applied to various EU enterprises. In: Proceedings of the Fifth International Symposium on Business Modeling and Software Design, Milan, Italy, July 2015, pp. 284–291 (2015)

43. Doukovska, L., Atanassova, V., Mavrov, D., Radeva, I.: InterCriteria analysis of EU competitiveness using the level operator N γ. In: Advances in Fuzzy Logic and Technology 2017, pp. 631–647. Springer, Cham (2017)

44. Doukovska, L., Atanassova, V., Sotirova, E., Vardeva, I., Radeva, I.: Defining consonance thresholds in intercriteria analysis: an overview. In: Intuitionistic Fuzziness and Other Intelligent Theories and Their Applications, pp. 161–179. Springer, Cham (2019)
45. Doukovska, L., Shahpazov, G., Atanassova, V.: Intercriteria analysis of the creditworthiness of SMEs. A case study. Notes Intuit. Fuzzy Sets 22(2), 108–118 (2016)
46. Fidanova, S., Roeva, O.: InterCriteria analysis of ant colony optimization application to GPS surveying problems. Iss. Intuit. Fuzzy Sets Gen. Nets 12, 20–38 (2016)
47. Fidanova, S., Roeva, O.: InterCriteria analysis of different variants of ACO algorithm for wireless sensor network positioning. In: Dimova, S. (ed.) Numerical Methods and Applications. Lecture Notes in Computer Science, vol. 11189, pp. 88–96. Springer (2019)
48. Fidanova, S., Roeva, O.: Multi-objective ACO Algorithm for WSN Layout: InterCriteria Analysis. Lecture Notes in Computer Science (including subseries Lecture Notes in Artificial Intelligence and Lecture Notes in Bioinformatics), vol. 11958, LNCS, pp. 501–509 (2020)
49. Fidanova, S., Roeva, O., Mucherino, A., Kapanova, K.: InterCriteria analysis of ant algorithm with environment change for GPS surveying problem. In: Dichev, C., Agre, G. (eds.) AIMSA 2016. LNAI, vol. 9883, pp. 271–278 (2016)
50. Fidanova, S., Roeva, O., Paprzycki, M., Gepner, P.: InterCriteria analysis of ACO start strategies. In: Proceedings of the Federated Conference on Computer Science and Information Systems, 9-th Workshop on Computational Optimization—WCO'2016, Gdansk, Poland, September 11–14, 2016. Ann. Comput. Sci. Inf. Syst. 8, 547–550 (2016)
51. Fidanova, S., Atanassova, V., Roeva, O.: Ant colony optimization application to GPS surveying problems: InterCriteria analysis, uncertainty and imprecision in decision making and decision support: cross fertilization, new models and applications. Adv. Intell. Syst. Comput. (Springer) 559, 251–264 (2018)
52. Fidanova, S., Roeva, O., Luque, G., Paprzycki, M.: InterCriteria analysis of different hybrid ant colony optimization algorithms for workforce planning. In: Fidanova, S. (ed.) Recent Advances in Computational Optimization. Studies in Computational Intelligence, vol. 838, pp. 61–81. Springer, Cham (2020)
53. Georgieva, V., Angelova, N., Roeva, O., Pencheva, T.: Intercriteria analysis of wastewater treatment quality. J. Int. Sci. Publ.: Ecol. Saf. 10, 365–376 (2016)
54. Ikonomov, N., Vassilev, P., Roeva, O.: ICrAData—software for intercriteria analysis. Int. J. Bioautom. 22(1), 1–10 (2018)
55. Ilkova, T., Petrov, M.: Application of intercriteria analysis to the Mesta River pollution modelling. Notes Intuit. Fuzzy Sets 21(2), 118–125 (2015)
56. Ilkova, T., Petrov, M.: Intercriteria analysis for identification of Escherichia coli fed-batch mathematical model. J. Int. Sci. Publ.: Mater. Methods Technol. 9, 598–608 (2015)
57. Ilkova, T., Petrov, M.: Intercriteria analysis for evaluation of pollution of the Struma river in the Bulgarian section. Notes Intuit. Fuzzy Sets 22(3), 120–130 (2016)
58. Ilkova, T., Petrov, M.: Using intercriteria analysis for assessment of the pollution indexes of the Struma River. Adv. Intell. Syst. Comput. 401, 351–364 (2016)
59. Ilkova, T., Petrov, M.: Intercriteria analysis for modelling of process for the unicellular protein production for training people. J. Int. Sci. Publ.: Mater. Methods Technol. 10, 455–467 (2016)
60. Ilkova, T., Roeva, O., Vassilev, P., Petrov, M.: InterCriteria analysis in structural and parameter identification of L-lysine production model. Iss. Intuit. Fuzzy Sets Gen. Nets 12, 39–52 (2015)
61. Jereva, D., Pencheva, T., Tsakovska, I., Alov, P., Pajeva, I.: Exploring applicability of intercriteria analysis to evaluate the performance of MOE and GOLD scoring functions. Stud. Comput. Intell. (2020)
62. Krumova, S., Todinova, S., Mavrov, D., Marinov, P., Atanassova, V., Atanassov, K., Taneva, S.: Intercriteria analysis of calorimetric data of blood serum proteome. Biochim. Biophys. Acta Gen. Subj. 1861, 409–417 (2017)
63. Marinov, P., Fidanova, S.: Intercriteria and correlation analyses: similarities, differences and simultaneous use. Ann. Inform. Sect. 8, 45–53 (2015–2016)
64. Mavrov, D.: Software for InterCriteria analysis: implementation of the main algorithm. Notes Intuit. Fuzzy Sets 21(2), 77–86 (2015)

65. Mavrov, D., Radeva, I., Capkovic, F., Doukovska, L.: Software for InterCriteria analysis: graphic interpretation within the intuitionistic fuzzy triangle. In: Proceedings of 5th International Symposium on Business Modeling and Software Design, pp. 279–283 (2015)
66. Parvathi, R., Atanassova, V., Doukovska, L., Yuvapriya, C., Indhurekha, K.: InterCriteria analysis of rankings of Indian universities. Notes Intuit. Fuzzy Sets 24(1), 99–109 (2018)
67. Pencheva, T., Angelova, M., Atanassova, V., Roeva, O.: Intercriteria analysis of genetic algorithm parameters in parameter identification. Notes Intuit. Fuzzy Sets 21(2), 99–110 (2015)
68. Pencheva, T., Angelova, M., Vassilev, P., Roeva, O.: Intercriteria analysis approach to parameter identification of a fermentation process model. In: Atanassov, K.T., Castillo, O., Kacprzyk, J., Krawczak, M., Melin, P., Sotirov, S., Sotirova, E., Szmidt, E., De Tré, G., Zadrożny, S. (eds.) Novel Developments in Uncertainty Representation and Processing. Advances in Intelligent Systems and Computing, vol. 401, pp. 385–397 (2016)
69. Pencheva, T., Angelova, M.: Intercriteria analysis of simple genetic algorithms performance. In: Georgiev, K., Todorov, M., Georgiev, I. (eds.) Advanced Computing in Industrial Mathematics. Studies in Computational Intelligence, vol. 681, pp. 147–159 (2017)
70. Pencheva, T., Roeva, O., Angelova, M.: Investigation of genetic algorithm performance based on different algorithms for intercriteria relations calculation. In: Lirkov, I., Margenov, S. (eds.) Large-Scale Scientific Computing. Lecture Notes in Computer Science, vol. 10665, pp. 390–398 (2018)
71. Petrov, M.: An approach to analysing and assessment pollution index for the Bulgarian section of the Struma river. In: International Conference Automatics and Informatics'18, 4–6 October 2018, Sofia, Bulgaria, pp. 147–150 (2018)
72. Petrov, M.: Intercriteria analysis and assessment of the pollution index for the Bulgarian Section of the Struma River. Inf. Technol. Control 3, 25–31 (2018)
73. Petrov, M., Ilkova, T.: Application of the intercriteria analysis for selection of growth rate models for cultivation of strain Kluyveromyces marxianus var. lactis MC 5. Notes Intuit. Fuzzy Sets 21(5), 49–60 (2015)
74. Petrov, M., Ilkova, T.: Intercriteria decision analysis for choice of growth rate models of batch cultivation by strain Kluyveromyces marxianus var. lactis MC 5. J. Int. Sci. Publ.: Mater. Methods Technol. 10, 468–486 (2016)
75. Ribagin, S., Stavrev, S.: InterCriteria Analysis of data from intellectual and physical evaluation tests of students practicing sports activities. NIFS 25(4), 83–89 (2019)
76. Ribagin, S., Stavrev, S.: InterCriteria Analysis of Data Obtained from University Students Practicing Sports Activities. Advances in Intelligent Systems and Computing, Springer (2020)
77. Ribagin, S., Grozeva, A., Popova, G., Stoyanova, Z.: InterCriteria analysis of body composition measurements data, associated with obesity among college students. NIFS 25(4), 78–82 (2019)
78. Roeva, O., Fidanova, S.: InterCriteria analysis of relations between model parameters estimates and ACO performance. In: Advanced Computing in Industrial Mathematics, Part of the Studies in Computational Intelligence book series, vol. 681, pp. 175–186 (2017)
79. Roeva, O., Fidanova, S.: Comparison of different metaheuristic algorithms based on InterCriteria analysis. J. Comput. Appl. Math. 340, 615–628 (2018)
80. Roeva, O., Fidanova, S.: Different InterCriteria analysis of variants of ACO algorithm for wireless sensor network positioning. In: Fidanova, S. (eds.) Recent Advances in Computational Optimization. Studies in Computational Intelligence, vol. 838, pp. 83–103. Springer, Cham (2020)
81. Roeva, O., Vassilev, P.: InterCriteria analysis of generation gap influence on genetic algorithms performance. In: Atanassov, K.T., Castillo, O., Kacprzyk, J., Krawczak, M., Melin, P., Sotirov, S. Sotirova, E., Szmidt, E., De Tré, G., Zadrożny, S. (eds.) Novel Developments in Uncertainty Representation and Processing, Part V. Advances in Intelligent Systems and Computing, vol. 401, pp. 301–313 (2016)
82. Roeva, O., Zoteva, D.: Knowledge discovery from data: InterCriteria analysis of mutation rate influence. Notes Intuit. Fuzzy Sets 24(1), 120–130 (2018)

83. Roeva, O., Zoteva, D.: ICrA over Ordered Pairs Applied to ABC Optimization Results. Studies in Computational Intelligence. Springer (2020)
84. Roeva, O., Fidanova, S., Vassilev, P., Gepner, P.: InterCriteria analysis of a model parameters identification using genetic algorithm. In: Proceedings of the Federated Conference on Computer Science and Information Systems. Ann. Comput. Sci. Inf. Syst. **5**, 501–506 (2015)
85. Roeva, O., Vassilev, P., Angelova, M., Pencheva, T.: Intercriteria analysis of parameters relations in fermentation processes models. In: Nunes M., Nguyen, N., Camacho, D., Trawinski, B. (eds.) Computational Collective Intelligence. Lecture Notes in Artificial Intelligence, vol. 9330, pp. 171–181 (2015)
86. Roeva, O., Vassilev, P., Angelova, M., Su, J., Pencheva, T.: Comparison of different algorithms for intercriteria relations calculation. In: IEEE 8th International Conference on Intelligent Systems, Sofia, Bulgaria, September 4–6, pp. 567–572 (2016)
87. Roeva, O., Pencheva, T., Angelova, M., Vassilev, P.: Intercriteria analysis by pairs and triples of genetic algorithms application for models identification. In: Fidanova, S. (ed.) Recent Advances in Computational Optimization. Studies in Computational Intelligence, vol. 655, pp. 193–218 (2016)
88. Roeva, O., Perez, J., Valdez, F., Castillo, O.: InterCriteria analysis of bat algorithm with parameter adaptation using type-1 and interval type-2 fuzzy systems. Notes Intuit. Fuzzy Sets **22**(3), 91–105 (2016)
89. Roeva, O., Vassilev, P., Fidanova, S., Paprzycki, M.: InterCriteria analysis of genetic algorithms performance. Stud. Comput. Intell. **655**, 235–260 (2016)
90. Roeva, O., Fidanova, S., Paprzycki, M.: InterCriteria analysis of ACO and GA hybrid algorithms. Stud. Comput. Intell. (Springer) **610**, 107–126 (2016)
91. Roeva, O., Vassilev, P., Chountas, P.: Application of topological operators over data from InterCriteria analysis. In: Christiansen, H. et al. (eds.) FQAS 2017. LNAI, vol. 10333, pp. 215–225 (2017)
92. Roeva, O., Fidanova, S., Paprzycki, M.: Comparison of different ACO start strategies based on InterCriteria analysis. Stud. Comput. Intell. **717**, 53–72 (2018)
93. Roeva, O., Vassilev, P., Ikonomov, N., Angelova, M., Su, J., Pencheva, T.: On different algorithms for intercriteria relations calculation. In: Hadjiski, M., Atanassov, K.T. (eds.) Intuitionistic Fuzziness and Other Intelligent Theories and Their Applications. Studies in Computational Intelligence, vol. 757, pp. 143–160 (2019)
94. Roeva, O., Fidanova, S., Luque, G., Paprzycki, M.: Intercriteria analysis of ACO performance for workforce planning problem. In: Recent Advances in Computational Optimization. Studies in Computational Intelligence, vol. 795, pp. 47–67. Springer, Cham (2019)
95. Roeva, O., Ikonomov, N., Vassilev, P.: Discovering knowledge from predominantly repetitive data by InterCriteria analysis. In: Fidanova, S. (ed.) Recent Advances in Computational Optimization. Studies in Computational Intelligence, vol. 795, pp. 213–233. Springer, Cham (2019)
96. Sonkin, M.A., Khamukhin, A.A., Pogrebnoy, A.V., Marinov, P., Atanassova, V., Roeva, O., Atanassov, K., Alexandrov, A.: Intercriteria analysis as tool for acoustic monitoring of forest for early detection fires. In: Atanassov, K.T., Kacprzyk, J., Kaluszko, A., Krawczak, M., Owsiński, J.W., Sotirov, S., Sotirova, E., Szmidt, E., Zadrożny, S. (eds.) Uncertainty and Imprecision in Decision Making and Decision Support: New Challenges, Solutions and Perspectives. Springer (2020)
97. Sotirov, S., Atanassova, V., Sotirova, E., Doukovska, L., Bureva, V., Mavrov, D., Tomov, J.: Application of the intuitionistic fuzzy InterCriteria analysis method with triples to a neural network preprocessing procedure. Comput. Intell. Neurosci. Hindawi **2017**, Article ID 2157852, 9 pp. (2017)
98. Sotirov, S., Sotirova, E., Atanassova, V., Atanassov, K., Castillo, O., Melin, P., Petkov, T. Surchev, S.: A hybrid approach for modular neural network design using intercriteria analysis and intuitionistic fuzzy logic. Complexity **2018**, Article ID 3927951, 11 pp. (2018)
99. Stratiev, D., Sotirov, S., Shishkova, I., Nedelchev, A., Sharafutdinov, I., Vely, A., Mitkova, M., Yordanov, D., Sotirova, E., Atanassova, V., Atanassov, K., Stratiev, D.D., Rudnev, N.,

Ribagin, S.: Investigation of relationships between bulk properties and fraction properties of crude oils by application of the intercriteria analysis. Pet. Sci. Technol. **34**(13), 1113–1120 (2016)

100. Todinova, S., Mavrov, D., Krumova, S., Marinov, P., Atanassova, V., Atanassov, K., Taneva, S.: Blood plasma thermograms dataset analysis by means of InterCriteria and correlation analyses for the case of colorectal cancer. Int. J. Bioautom. **20**(1), 115–124 (2016)

101. Traneva, V., Tranev, S., Szmidt, E., Atanassov, K.: Three dimensional intercriteria analysis over intuitionistic fuzzy data. In: Advances in Fuzzy Logic and Technology' 2017, vol. 3, no. 1, pp. 442–449. Springer, Cham (2018)

102. Tsakovska, I., Alov, P., Ikonomov, N., Atanassova, V., Vassilev, P., Roeva, O., Jereva, D., Atanassov, K., Pajeva, I., Pencheva, T.: Intercriteria analysis implementation for exploration of the performance of various docking scoring functions. Stud. Comput. Intell. (2020)

103. Vassilev, P.: On similarly structured intuitionistic fuzzy sets. Notes Intuit. Fuzzy Sets **23**(2), 13–16 (2017)

104. Vassilev, P.: A note on new partial ordering over intuitionistic fuzzy sets. Ann. Inform. Sect. (Union of Scientists in Bulgaria) **8**, 18–22 (2015–2016)

105. Vassilev, P., Ribagin, S.: A note on intuitionistic fuzzy modal-like operators generated by power mean. In: Advances in Intelligent Systems and Computing, vol. 643. Springer (2017)

106. Vassilev, P., Ribagin, S., Todorova, L.: On an aggregation of expert value assignments using index matrices. NIFS **23**(4), 75–78 (2017)

107. Zaharieva, B., Doukovska, L., Ribagin, S., Mihalicova, A., Radeva, I.: Intercriteria Analysis of Behterev's Kinesi therapy Program. Notes on IFS, vol. 23, No. 3 (2017)

108. Zaharieva, B., Doukovska, L., Ribagin, S., Radeva, I.: InterCriteria approach to Behterev's disease analysis. In: Proceedings of the 21st ICIFS, 22–23 May, Burgas, Bulgaria (2017)

109. Zaharieva, B., Doukovska, L., Ribagin, S., Radeva, I.: InterCriteria analysis of data obtained from patients with Behterev's disease. Int. J. Bioautom. **24**(1), 5–14 (2020)

110. Zoteva, D., Roeva, O.: InterCriteria analysis results based on different number of objects. Notes Intuit. Fuzzy Sets **24**(1), 110–119 (2018)

111. Zoteva, D., Roeva, O., Delkov, A., Tsakov, H.: Intercriteria analysis of forest fire risk. In: Proceedings of the 4th International Conference on Numerical and Symbolic Computation—Developments and Applications (SYMCOMP 2019), Porto, 11–12 April 2019, ©ECCOMAS, Portugal, pp. 215–229 (2019)

112. Zoteva, D., Roeva, O., Tsakov, H.: Forest fire analysis based on InterCriteria analysis. In: Advances in Intelligent Systems and Computing. Springer (2020)

Academic ICT Research for Defence and Security

Todor Tagarev

Abstract This chapter presents research activities in the interest of defence and security, conducted in the Bulgarian Academy of Sciences, and their main results. The uniquely formed capacity is applied in largely demand-driven research to support defence planning and policy-making, crisis management and critical infrastructure protection, cybersecurity and countering hybrid threats. The research methods are multidisciplinary and draw, in addition to technology-related subjects, from the fields of security studies, operations research, foresight, systems analysis, risk management, complex systems, multi-criteria decision making, and governance studies. The concluding section presents obstacles to and drivers for the future development of the academic ICT defence and security-related research and the outlines of a strategy to make this development sustainable in the long run.

Keywords Information security · Cybersecurity · Defence planning · Capability · Architecture · Operations research · C4ISR · Crisis management · Hybrid threats · Foresight

1 Introduction

Nowadays, the importance of information and communications technologies (ICT) for the economic development, the functioning of the public administration, provision of security, social interaction and entertainment is well understood, and their innovative implementation provides various competitive advantages. This has not always been so. The emergence of the electromechanical calculators in the 1930s, and later—that of the computers and the early days of the Internet—was the subject of interest of a relatively small number of researchers and practitioners. This is one of the likely explanations why during the Cold War Bulgaria—a relatively small country in the former Soviet-led bloc—managed to position itself as the primary

T. Tagarev (✉)
Institute of Information and Communication Technologies, Bulgarian Academy of Sciences, "Acad. G. Bonchev" Str., Bl. 2, Sofia 1113, Bulgaria
e-mail: tagarev@bas.bg

© The Author(s), under exclusive license to Springer Nature Switzerland AG 2021
K. T. Atanassov (ed.), *Research in Computer Science in the Bulgarian Academy of Sciences*, Studies in Computational Intelligence 934,
https://doi.org/10.1007/978-3-030-72284-5_21

producer of computers and software in the Council for Mutual Economic Assistance (Comecon—the format for economic integration of the Soviet Union and the communist countries in Eastern Europe [1]). Among the explanations are also the high level of education in science and engineering and the initiative of academic researchers reaching a perceptive political leadership [2].

The development of the first computers in Bulgaria, analogues of IBM 360, started in the 1960s, and the first exports were realised in the early 1970s. In the 1970s and 1980s researchers and designers developed for the Comecon market analogues of PDP 8, VAX 11, Apple IIe and IBM PC/XT, and Bulgaria strengthened its niche position in that market with a particularly robust capacity in software development and the production of computer memories [2].

In the process of democratisation and transition to a market economy, and the dissolution of the Soviet Union and Comecon, however, the guaranteed Eastern European market for Bulgarian made computers and software disappeared. The leadership of the sector was not able to put a realistic strategy or a transition plan in place, and the computer design and production sector quickly declined and went into oblivion.

With a few exceptions, the processes of adaptation of related sectors, e.g. the design and production of sensors and communications equipment, had a similar fate. Likewise, the state was unable to maintain sizeable governmental research institutes conducting security and defence research. As a result, in the 1990s, many researchers moved to universities or the Bulgarian Academy of Sciences (BAS). Among them were researchers in the fields of information and communications technologies, automation of organisational processes, e.g. command and control, computer-assisted exercises, and operations research.

Coincidentally, towards the end of the 1990s, some of these researchers with experience in defence were called to plan and implement much-delayed reforms of the Bulgarian armed forces and, towards that purpose, were appointed at senior governmental positions. After the period of profound and comprehensive reforms of 1997–2001, and depending on the appetite for reforms of the government in power, some of these researchers moved back and forth between the academy of sciences and the government, taking positions of increased responsibility (two of them served in their expert capacity as defence ministers in Bulgaria's caretaker governments, respectively in the Spring of 2013 and the Fall of 2014).

Reforms and transformation of the armed forces and the security sector and achieving interoperability with NATO allies require reliable methodologies and tools supporting both planning and implementation. Inevitably, this requirement had a profound impact on shaping the research agenda in the Bulgarian Academy of Sciences in the field of ICT, defined broadly to include operations research, organisational design, business process improvement, serious gaming, and other research disciplines. The second driving factor was the integration of research teams into the European and Euro-Atlantic research community.

The next section of this chapter provides an overview of the transition process that shaped the academic ICT research in the following two decades. The next four sections present major research activities and results in providing support to defence policy-making, crisis management, cybersecurity, understanding and countering hybrid threats. The final section concludes the chapter by outlining the foreseen development of the ICT security and defence-related research in the Bulgarian Academy of Sciences.

2   The End of the Military-Industrial Complex—Reorientation and Integration with the Western Research Community

2.1   The Transition

At the end of the Cold war Bulgaria, a nation in South-Eastern Europe with a population of just under nine million at the time, maintained one of the largest military per capita and a military-industrial complex involving directly some 115 000 people and placing the country on the top ten list of arms exporters, with more than USD 1 billion of annual exports at its peak [3].

The Ministry of Defence had three institutes—a military-technical institute, a logistics institute, and an IT and automation institute directly subordinated to the General Staff of the armed forces—and several smaller research centres employing approximately 1 200 researchers and support staff in total. During the 1990s, this defence research infrastructure became unsustainable. After several rounds of transformation, the defence ministry maintains two institutes at present:

- "Prof. Tsvetan Lazarov" Defence Institute, conducting technology-oriented research and supporting the defence acquisition process; and
- Defence Advanced Research Institute, part of the "G.S. Rakovski" National Defence College, conducting studies in strategy, doctrine, and military history.

with a combined staff of under 150. In the transition process, a number of the remaining researchers decided to join the Bulgarian Academy of Sciences, thus strengthening the backbone of the academic defence and security research.

On the industrial end, during the Cold War Bulgaria's military-industrial complex specialised in small arms and light weapons, ammunition, overhaul and repair of a variety of land, sea and aerial platforms and engines. Further, it produced more sophisticated weapon systems such as sea surveillance radars, surface-to-air missiles and communications equipment, mostly under Soviet license [3]. It also invested in indigenous technologies, e.g. in ammunitions, optics, laser technologies, and electrical batteries, involving in some cases research units from the academy of sciences.

With the end of the Cold war, and against the background of lacking guidance from the state and decimated defence procurement budgets during the 1990s, state-owned companies were not able to maintain a sufficient Research and Development (R&D) capacity in-house, nor to sub-contract the necessary R&D in universities and academic research institutes. Nevertheless, entrepreneurship allowed to maintain Bulgaria's defence industrial specialisation in small arms and light weapons till this day. It remains exclusively export-oriented, with significant volumes of export in periods of high demand.

The fate of the companies in more research-intensive sectors varied widely. Some companies, primarily in the field of optical sensors and small radar systems, managed to adapt to the new political and market environment combining products developed in-house and functioning as sub-contractors. Others moved to lower tiers of the supply chain, producing components based on foreign technology and know-how or focused on the maintenance of more sophisticated systems. Yet others had to convert to the civilian industry or close down.

Important for the discussion in this chapter is the growing understanding that a company producing sensors, communications equipment and ICT systems cannot compete internationally, nor domestically, unless it invests extensively in R&D, and the academia is a good source of research capacity and knowledge.

In fact, that opportunity was recognised already at the end of the 1980s. In 1988, a research unit supporting the development of sea and river radars was moved from the defence industry to the Central Laboratory for Parallel Processing (later Institute for Parallel Processing) of the Bulgarian Academy of Sciences; it was organised as "Multiple Sensors Data Fusion" (MSDF) department, also known as Laboratory "Signal" [4]. In the early 1990s, other R&D units were transferred to the Institute of Control and Systems Research and the Institute of Metal Science, while some recognised researchers joined other institutes individually.

To coordinate the activities of the dispersed research units and researchers and streamline the interaction with the Ministry of Defence and other security agencies, in 2002 the Bulgarian Academy of Sciences established the Centre for National Security and Defence Research. Since its establishment, this centre became the focal point for coordinating the contribution of the academy of sciences to the research programme of the Ministry of Defence and, later, research in the interest of the Ministry of Emergency Situations.

One of the most significant steps in regard to the ICT research in the academy of sciences was the establishment in 2004 of the C4ISR[1] department at the Institute of Parallel Processing (currently part of the Institute of Information and Communication Technologies under the name "IT for Security" department). This newly established department brought together researchers from the MSDF department and defence organisations. The author of this chapter led the "IT for Security" department from 2010 till 2016 and, hence, most of the research activities described below were conducted by this department, as a rule in cooperation with other units of the academy of sciences (see Table 1) and internationally.

---

[1]C4ISR is the abbreviation for Command, Control, Communications, Computers, Intelligence, Surveillance and Reconnaissance, popular in the military parlance.

Table 1 Main ICT-related research areas of BAS units with application in security and defence

Organisation	Main research areas
Information and Communication Technologies (IICT), https://iict.bas.bg/	Cybersecurity, data and information fusion, grid technologies, artificial intelligence, language processing, decision support systems, simulations, foresight
Institute of Mathematics and Informatics, https://math.bas.bg/	Cryptography, coding, operations research, software technologies and information systems
Space Research and Technology, https://space.bas.bg/	Remote sensing, geographic information systems
Institute of Robotics, https://ir.bas.bg	Sensors, robotics, quantum communications, risk management
Institute of Mechanics, https://www.imbm.bas.bg/	Autonomous platforms, human–machine interfaces
National Laboratory for Computer Virology, www.nlcv.bas.bg	Computer virology; computer, communications and information security
Centre for National Security and Defence Research, www.cnsdr.bas.bg	Coordination of security and defence research in the Bulgarian Academy of Sciences, technology transfer

2.2 International Cooperation Directs Westwards

The need to cooperate with the research community of NATO and the European Union, as well as individual Wester countries, was recognised already in the early 1990s. The MSDF department was one of the first units specialised in defence and security-related research to seek actively cooperation opportunities, building primarily of the available knowledge and know-how in multiple-target tracking with applications of recursive pseudo-Bayesian algorithms, parallel algorithms, and parallel and distributed computer architectures (e.g. [5]). That quickly raised the interest of scientists from the United Kingdom and the United States and led initially to small joint benchmarking studies [6]. The cooperation was later expanded to other countries. A special mention deserves the long-standing and mutually beneficial cooperation with the French aeronautics, space and defence research laboratory ONERA that led to profound new theoretical results in Bayesian reasoning with belief functions [7, 8] and applicable defence technologies [9, 10].

The opportunities for international cooperation in defence research increased after the decision of Bulgaria to join NATO. As director for defence planning from 1999 to 2001, the author of this chapter managed the research programme of the Ministry of Defence. In that capacity, he became the first national representative to NATO's Research and Technology Board and extended an invitation to host activities of NATO's Research and Technology Organization (RTO) in Bulgaria. The Board responded positively, and in the following years RTO conducted several lecture series and other activities with massive participation of researchers from the Bulgarian Academy of Sciences.

In addition to getting advanced knowledge in specific areas, this provided opportunities to exchange information, establish working contacts in multi-lateral and bi-lateral formats, and quickly launch development and demonstration projects. One example is the project with The Netherlands Aerospace Centre (NLR), implemented in 2000 with the Space Research Institute of the Bulgarian Academy of Sciences as the leading local partner, demonstrating the capabilities of a mobile ground station for real-time acquisition of satellite data and its further processing for defence and disaster management purposes.

Once Bulgaria became a member of NATO, academic researchers, primarily from the C4ISR Department, became involved in RTO study groups, with the main interest in the activities of:

- RTO System Analysis and Studies panel, contributing to SAS-063 "Benchmarking Studies and Capability Costing" and other exploratory teams, research groups, specialist meetings and symposia; and
- The RTO Modelling and Simulation Group, contributing to MSG-049 "Modelling and Simulation System for Emergency Response Planning and Training.".

This cooperation had a significant impact on the research agenda, presented in the following two sections.

NATO's Science for Peace programme provided additional opportunities to develop the Bulgarian capacity to provide scientific support to the defence and security sector. Most important in that regard was the four-year project "Operations Research Support to Force and Operations Planning in the New Security Environment" (SfP 981149) that brought together researchers from the C4ISR department, the Operations Research department of the Institute of Mathematics and Informatics, and the Defence Management department of the "G.S. Rakovski" National Defence College with German and Dutch counterparts [11].

At that stage, personal contacts, knowledge, and shared experience allowed for the successful cooperation in the framework of the security research programme, launched by the European Union, as well as joint studies for the European Defence Agency.

3 Supporting Defence Policy-Making

One of the major challenges to the post-Cold war adaptation of the Bulgarian defence to the new security, political and economic realities was the lack of indigenous capacity for long-term planning (in comparison with the operations planning capacity) and defence-policy making—a direct consequence of the 45 years of centralised planning in the Warsaw Pact [12]. The challenge was even more prominent since, and notwithstanding the continuous conflicts in the Western Balkans, there was no credible direct military threat to the country. The methodological void was filled in by defence official and civil servants with both military and academic background, who laid the foundations of a defence planning system [13]. That included the adaptation of the Defense Resource Management Model (DRMM), developed by the U.S. Institute for Defense Analysis, to account for the weapon systems and equipment of the Bulgarian Armed Forces and local cost factors.

DRMM was then used to calculate the combat potential of a unit, or a set of force units, depending on the level of manning, level of individual and unit training, maintenance expenditures, and investment budget. Varying these parameters, a team of military and civilians designed various force structures and used DRMM to estimate their combat potential. The planning task was formulated as an optimisation problem, i.e. to find a force size and structure that has the maximum combat potential within anticipated financial constraints (at that time—3% of the foreseen Gross Domestic Product) [12, 14]. Experienced military officers designed the force structures. The force structure with the highest combat potential was that with a peacetime force size of 45 000—nearly 2.5 times lower than the size of the then-existing force structure. Military planners tested it in a notional scenario of aggression against the country—already agreed in the draft military doctrine—and proved to be sufficient to deter the aggression and regain the territorial integrity of the country with consequences, considered as acceptable [12].

In April 1999, the Parliament approved the Military Doctrine of the Republic of Republic of Bulgaria and, thus, the notional scenario and the force structure of 45 000, and tasked the defence ministry to prepare a transition plan for achieving the target force structure by 2004. One way to support the transition was through design and implementation of a program-based defence resource management system, allowing to relate in a transparent manner long-term goals with short-term plans and activities, as well as expenditures to results [15].

The success of this joint effort of military and civilians, planners and academics set a precedent, used in the organisation of later studies to facilitate defence and security policy-making and implementation.

The research methods and techniques used to support policy-making from 1998 to 2001 are presented in the second column of Table 2. The next stage of research

Table 2 Main ICT-related research areas of BAS units with application in security and defence planning

Representation of uncertainty	Notional scenarios [12]	Plausible scenarios [17]	Alternative futures [17, 22, 24, 25]
Focus	Combat potential	Defence capabilities	Capabilities + capacity to adapt
Methodological approach	Optimisation [14]	Scenario-based planning [17, 19, 21] Optimisation [18] Serious gaming [20, 23]	Foresight [22–25] Proactive risk management [25, 26]
Support for implementation	Reengineering [13] Process design and process improvement [15, 27, 31] Operational, system and technical architectures [28, 29] Multi-criteria decision making [30]		

support to the introduction of "capabilities-based planning" is presented in column 2 of the same table.

The focus on combat potential was instrumental for 'rightsizing' the armed forces, allowing them to adapt to the changing security environment. In itself, however, this is not a sufficient condition. With the accession to NATO and the increasing demand for units deployed in out-of-area operations, it became painfully clear that a force structure with maximum combat potential is not necessarily adequate to the diverse requirements of operations different than a symmetric force-on-force conflict. By the beginning of the new century, an approach known as "capabilities-based planning" was getting traction in NATO and partner countries [16].

In 2007, the Ministry of Defence turned to the Institute for Parallel Processing with a request, and its C4ISR department took the lead in organising and performing a study to develop a methodology and scenarios for "scenario-based, concept-led and capabilities-oriented" defence planning. The resulting compendium, published by the ministry's publishing house [17], set out the conceptual basis, proposed a planning methodology, a methodology for development and selection of plausible scenarios, and described 31 selected scenarios.

Jointly with researchers from the Operations Research department of the Institute of Mathematics and Informatics, the task to determine a 'best' set of defence capabilities was formulated as an optimisation problem [18].

In parallel, and working with faculty members of the Bulgarian Naval Academy, the planning methodology was tailored to the needs of planning of capabilities for maritime sovereignty, with the elaboration of respective organisational roles and scenarios [19]. Further, teaming with faculty members from the "G.S. Rakovski" National Defence College, academic researchers designed and implemented multiple times a role game, simulating decision-making on defence resource allocation [20]. While not a planning tool per se, this game, typically played with senior officers and defence civilians, allows to explore the decision space, reveal caveats and focus the attention of planners on the most promising options [20].

In a follow-on study on the assignment from the Ministry of Defence, C4ISR researchers developed further methodological tools, including a capability taxonomy and a library of capability building blocks, and delivered a comprehensive case study, defining levels of ambition, projecting the force structure ten years ahead, and proposing advanced defence management arrangements [21].

In 2010, the Institute for Parallel Processing merged with two other institutes, jointly forming the Institute of Information and Communication Technologies. The C4ISR Department changed its name to "Information Technologies for Security" (IT4Sec, www.IT4Sec.org). Although the change took away some of the name's military flavour, the department continued to support through its research defence policy-making and planning.

The next advancement of the planning process aimed to reflect on the increasing uncertainty of the security environment (the last column of Table 2). The research team developed further the concept of 'alternative futures,' designated in its 2007 study as 'context scenarios' [17, 22]. First, in the EU Seventh Framework Programme-funded project FOCUS (Foresight Security Scenarios: Mapping

Research to a Comprehensive Approach to Exogenous EU Roles), the team adapted a scenario analysis tool suite supporting morphological analysis and clustering, and applied it in a two-day game with 26 mid- and upper-level military officers and defence civilians [23]. In an interactive cycle, preparing the game and analysing its results, we teamed with researchers from the Munich-based Centre for European Security Strategies and developed alternative futures and roles for the European Union as a global security actor [24].

Practically at the same time, the Bulgarian government decided to update the state plan for maintaining the national wartime reserves and stocks. Given the lack of a credible military threat to the country (already member of NATO), the Ministry of Defence turned to the IT4Sec department for methodological support. The research team developed a set of alternative futures and planning scenarios, plausible in one or more of the alternative futures, a taxonomy of the wartime capabilities of the country and a model of their generation in a situation of imminent military crisis [25]. Further, and building on its earlier work [26], the team developed a planning methodology allowing to balance in a transparent manner the investments in:

- the information-analytical capacity to monitor the security environment and issue warnings when significant changes are foreseen;
- the adaptiveness of the reserves' management system, i.e. the ability to react and generate quickly additional resources and capabilities, and
- the level of reserves and stocks maintained in peacetime,

thus providing opportunities for significant reduction of the level of wartime reserves and the respective costs to the budget and the economy [25].

While planning in uncertainty is a significant intellectual challenge, implementation is not easy too, in particular when it involves complex acquisition projects and development of C4ISR systems. To support these processes, academic researchers have conducted several dedicated studies, including:

- streamlining the process of programme-based defence resource management and enhancing its information support [27];
- designing an architectural model of the defence acquisition system to identify opportunities for its improvement [28];
- designing the architecture of a naval sovereignty operations centre, intended to support the evolution of the respective C4ISR system [29];
- elaborating a framework for the multi-criteria selection of the future fighter plane for the Bulgarian Air Force [30];
- streamlining the process of delivering capabilities funded by NATO's common budget [31].

Some of these methodological approaches are applicable to the domain of disaster management and critical infrastructure protection and are outlined in the following section.

4 Crisis Management

The academic research on crisis management can be structured around four topics: organisational arrangements, technical solutions, computer-assisted exercises, and critical infrastructure protection, and each of these is presented here in a sub-section.

Starting in 2003, the then-new Ministry of Emergency Situations[2] (MoE, closed down in 2009) decided to follow the example the Ministry of Defence and—through the Permanent Committee for Protection of the Population from Disasters, Accidents, and Catastrophes to the Council of Ministers, and in particular its budget for prevention—set the grounds for cooperation with the Bulgarian Academy of Science. That allowed the MoE to tap into the academic capacity for methodological support in organisational development and innovation, development of tailored technical solutions, and training and exercises.

The other major platform for crisis management research is the international projects, including several small regional projects and the large FP7 project DRIVER+ (Driving Innovation for Crisis Management for European Resilience, https://www.driver-project.eu, 2014–2020).

4.1 Crisis Management Organisation and Innovation

Since the early 1990s, the system for civil protection underwent several reorganisations – from a fully militarised "Civil Defence" force in the Ministry of Defence as the core, to a semi-civilian agency in the same ministry, to a civilian State Agency "Civil Protection" under the Council of Ministers and, then, to becoming the operational backbone of the newly established Ministry of Emergencies. Even so, the MoE did not have the capacity to perform the civil protection function by itself and, given the disrupted ties with the military and organisational rivalries with other ministries and agencies, it needed a sound mechanism for organising the national capacity for prevention, disaster response and recovery. For that purpose, it turned to the Bulgarian Academy of Sciences.

An academic research team, organised by the Centre for National Security and Defence Research, analysed the evolving security environment, existing norms, identified best practices from the experience of other countries and designed six alternative architectural models for Bulgaria's system for civil security, with focus on the protection of the population and critical infrastructures. Then it solicited the opinion of subject matter experts, applying Saaty's Analytical Hierarchy Process methodology to rank the alternatives. The alternative, envisioning maximum interdepartmental coordination of both operations and capability development topped the list (thus contradicting the very idea of a separate Ministry of Emergencies). Finally, the study outlined the main steps in the transformation of the civil security system.

---

[2]Existing for a short period of time as "Ministry for State Policy on Disasters and Accidents," and then renamed to "Ministry of Emergency Situations".

The results of the study were summarised and presented in English in a journal article [32].

Of key importance in designing such type of studies—considering the challenges of the practical application of study results—is their framing, since 'civil protection' is never a stand-alone area of activities but interacts with many other fields of civil, or 'societal' security [33]. The IT4Sec research team believes that 'essential services' is a more adequate framing concept and, to that effect, within the FOCUS project, developed a taxonomy of essential services [34].

Another foundational study, this time performed on the request of the Ministry of Transport and Communications, described and recommended the application of an architectural approach to the development of emergency management communications and information systems [35].

The participation in the DRIVER+ project provided access to a wealth of know-how from across the European Union. Instead of the traditional top-down approach based on a rigorous definition of needs and requirements, and then developing and implementing respective technological and organisational solutions, DRIVER+ relies on bottom-up initiative and collateral ties between practitioners, policy-makers, researchers and solution providers. It creates a methodological and technical infrastructure allowing to identify crisis management gaps and potential solutions (even at medium TRLs) and then conduct trials to assess their utility in realistic settings. This in itself is a significant innovation in developing the European crisis management and disaster risk reduction capacity.

The IT4Sec team contributes in a number of ways, including by designing a taxonomy of crisis management functions as the primary tool to navigate the information space and match gaps and solutions. Currently, the team explores the potential for using the taxonomy as a framework for planning disaster risk reduction measures [36] and standardising the assessment of the societal impact of crisis management solutions.

4.2 Technical Solutions for Crisis Management

In the ICT field, the academy of sciences searches for talent in a highly competitive labour market. Therefore, as a rule, it does not aim to develop final products. Instead, in various cooperation formats, research teams contribute primarily by defining requirements to technical solutions, their testing, demonstration, and evaluation, all with a good understanding of operational context and policy ramifications.

One such example is the development and the demonstration of the capabilities of mobile IT modules for field command and control posts that can be used both in emergency management operations and by military units deployed in coalition operations abroad. In this example, the C4ISR department teamed with a small private company [37].

Another example proves the utility of international cooperation. In this case, academic researches worked with a small US company in the development of a tool named "Computer Aided Dispatch." The tool facilitates the cooperation of military and civilians in multinational and interagency operations in emergencies of various intensity. Given the complexity of such operations, the tool was tested in a computer environment following NATO's Concept Development and Experimentation process [38].

4.3 Computer-Assisted Exercises

New technical solutions may be tested and demonstrated in designated trials, as in the DRIVER+ project, or in exercises. As a norm, these exercises are 'computer-assisted' and may pursue a variety of additional goals, including:

- train command post and field personnel;
- identify capability gaps;
- assess organisational arrangements and procedures;
- evaluate training capacity and gaps.

The C4ISR department demonstrated its capacity to organise a complex exercise pursuing all of the listed goals in the TACOM 2006 project – "European Union Terrorist Act Consequences Management in South-East Europe 2006" [39].

TACOM 2006 was a distributed exercise, combining field activities and simulations, and bringing in international participants. In fact, the academic institute hosted the "National Crisis Management Centre" with the personal participation of Bulgaria's Vice Prime minister and MoE minister.

Given that the organisation of such large-scale and complex exercises is very labour- and time-intensive, the C4ISR department developed and tested a simpler and more cost-efficient environment for organising and conducting computer-assisted exercises, known as "Basic Environment for Simulation and Training" (BEST) [40].

4.4 Critical Infrastructure Protection

With an account of the growing societal reliance on infrastructures, that are becoming increasingly interdependent, a crisis may be triggered by the failure of or an attack on one or more assets in critical infrastructures. The related policy discussion in the European Union became particularly active in 2005, with the publication by the Commission of the so-called 'Green Paper' [41].

At that point, the Bulgarian government turned to the Bulgarian Academy of Sciences, and the author of this chapter led the first comprehensive study in the country, coordinated by the Centre for National Security and Defence Research. More than 30 researchers from 10 institutes of the academy contributed to the elaboration of a methodology for identification and designation of critical infrastructures, analysis of interdependencies, evaluation of the consequences with an account of potential cascading effects, and the planning of strategies and measures for protection. The study results are summarised and presented in English in a journal article [42]. Afterwards, the methodology was adapted for practical application at the municipal level [43].

After the adoption of the European Directive 2008/114 on critical infrastructure protection, IT4Sec researchers were invited by and worked with the Ministry of the Economy and Energy to transpose the Directive in the national legislation.

The accumulated know-how in critical infrastructure protection facilitated the academic research in the fields of cybersecurity and hybrid threats, presented in the following two sections.

5   Cybersecurity and Resilience

The subject of critical infrastructures also applies to information, or 'digital,' infrastructures and has been a subject of study by the National Laboratory for Computer Virology (NLCV) and its late director, Prof. Eugene Nikolov [44]. NLCV was the first academic unit created to address information security and is well-established as the national laboratory for protection of computers, information systems and communications from malicious attacks. It conducts studies of all types of malware and its spreading mechanisms (e.g. [45]) and develops methods and procedures for real-time assessment of the information security of networks [46]. Further, NLCV develops and tests cyber forensics methods and tools [47].

The Institute of Mathematics and Informatics is another academic unit researching aspects of information security. Primarily, that includes the development of cryptographic methods and techniques [48] and coding methods (e.g. [49]).

With the establishment of the Institute of Information and Communication Technologies in 2010, the study of information and cyber security became one of the main tasks of its IT4Sec department. Researchers from the department contributed to the EU-funded SysSec project, aiming to establish "a European network of excellence in managing threats and vulnerabilities in the future internet," primarily by tracking cyber threats and vulnerabilities and applying foresight methodologies [50].

Building on the experience on the broader security problematics, in its internal research programme the department studies models and processes for developing cybersecurity capabilities, relevant organisational architectures [51] and other management aspects (e.g. [52]), as well as the role of the human factor in cybersecurity and the resilience to cyber threats [53]. In that strand of work, an IT4Sec researcher was formally included in the Interagency Working Group tasked to develop Bulgaria's cybersecurity strategy. That is a possible explanation for the focus of the national strategy on dealing with uncertainty and building resilience to threats from cyberspace [54].

Still, in most of its studies, the IT4Sec department cooperates both nationally and internationally. In the national research programme "Information and Communication Technologies for Science, Education, and Security," involving several Bulgarian universities and academic institutes, senior researchers from the IT4Sec department lead main tasks in the 'security' component of the programme:

- monitoring, prevention and reaction to information security incidents, with focus on the analysis of cyber threats [55] and development of organisational information security policies [56];
- developing and auditing of secure software;
- developing information security curricula for the training of the public administration.

On the request of the Bulgarian Institute of Public Administration, the department teamed with the NLCV, the Bulgarian Defence Institute and the European Software Institute-Centre Eastern Europe to develop and rank alternatives for the organisation for cybersecurity in the public administration and analyse the opportunities for implementation of innovative technologies in its work. The study team focused on blockchain technologies and artificial intelligence and discussed the results [57] in both interagency and international settings.

The biggest current project of the department is ECHO (European Network of Cybersecurity Centres and Competence Hub for Innovation and Operations), funded by the European Union's Horizon 2020 programme. ECHO is a four-year project, involving 30 organisations from 14 European countries, including three research organisations and one company from Bulgaria.

In ECHO, the IT4Sec department leads the work package dedicated to the design and implementation of a business and governance model of a future European network of cybersecurity competence centres. Towards that goal, international research teams already performed a comprehensive study to identify and prioritise needs, objectives, and requirements to the governance of collaborative networked organisations [58], identified and analysed business and governance models of such organisations [59], and developed a methodology and tools to evaluate and rank governance alternatives [60]. Currently, the team solicits expert opinions to rank the governance alternatives and then will detail and support the implementation of the selected alternative.

In addition to the work on cybersecurity governance, the IT4Sec team provides major contributions to the project research on:

- framing the problem of cyber risk management to account for cross-sectoral dependencies in cyber-physical systems and transversal effects [61], and contributing to the development of the ECHO Multi-sector Assessment Framework (E-MAF);
- development of cyber-attack scenarios with sectoral and cross-sectoral effects;
- definition of computationally intensive functions in the provision of cybersecurity [62] and exploration of the opportunities to include the 'Avitohol' supercomputer, operated by the academy, in the ECHO Federated Cyber Range;
- development of the ECHO technology roadmap and selection of prototype tools to be tested and demonstrated within the project;

- design of demonstration cases and analysis of the results of new cybersecurity solutions;
- analysis of the demand for main cybersecurity products and services developed within the project;
- development of exploitation strategy and sustainability plans.

The ECHO project provides excellent opportunities to integrate the academic cybersecurity-related research capacity in the European research and technology landscape, contributing at the same time to meeting the increasing needs of public and private organisations in Bulgaria effectively.

6 Hybrid Threats

The final research area addressed in this chapter aims to provide for a better understanding of the hybrid influence of foreign interests and develop effective countermeasures. There is no universal definition for a "hybrid threat"; yet, it is commonly understood as a threat of using a variety of tools, e.g. military and civilian, in a coordinated way to achieve political goals.

With the background of research and policy-making experience in defence, crisis management, critical infrastructure protection and cybersecurity, the IT4Sec department is ideally positioned to study the subject and provide practical advice to policy-makers and practitioners.

The understanding of hybrid threats requires adequate frameworks, suitable to reflect the actual interdependencies and complexity of modern society, its vulnerabilities, and thus to estimate the impact of the combined application of a set of tools, e.g. propaganda, disinformation, cyberattacks, energy coercion, political influence, corruption, and others [63]. The creation of such frameworks is only one part of the newly emerging research topics [64].

To this point, and in addition to theoretical studies, the research of the IT4Sec Department addresses three application areas:

- Implications of recent examples of hybrid warfare for defence policies, with a focus on the need for organisational agility and adaptiveness [65];
- Countering radicalisation and transnational terrorism [66], with focus on attempts at social engineering in cyberspace [67, 68];
- Application of proven foresight methods and techniques to identify and foresee trends in specific contexts [69].

The hybrid influence is a new area of research, requiring multidisciplinary competencies. Yet, the demand for scientific support to the formulation of policies and development of measures to counter hybrid influence will grow. The author already participates in an advisory capacity in the Horizon 2020 project "Empowering a Pan-European Network to Counter Hybrid Threats" (EU-HYBNET), launched in May 2020.

In sum, combining in-house research and networking, the academy will be able to meet demands for support to policy-making by public institutions.

7   The Future of Defence- and Security-Related ICT Research

Since the early 1990s, in the process of transition to a market economy, the Bulgarian Academy of Sciences lost some of its attractiveness to younger people and was gradually reduced in size. In contrast to other fields, however, its capacity to conduct research in the interest of defence and security grew for two main reasons. First, it was able to attract reputable researchers from both the defence industry and the former institutes of the Ministry of Defence. Secondly, with the Westward orientation of the country, both NATO and the European Union opened their research programmes to Bulgaria, and individual researchers and teams succeeded in finding adequate cooperation opportunities and funding.

One very specific, possibly unique feature is that several senior researchers in the field combine policy-making experience (and attitude) and useful contacts—nationally and internationally. That had an evident impact on the defence and security research agenda:

- it is demand-driven, and public agencies often rely on receiving quickly methodological support and advice of high quality and in line with international best practice; and
- as a rule, this research requires inter- and multidisciplinary competencies delivered through a flexible network with academic, other national, and international partners.

This chapter provided a partial overview of the conducted research and its results, focusing on the "IT for Security" (formerly, C4ISR) department of the Institute of Information and Communication Technologies. It did not touch on related activities such as teaching, consultancy, participation in national and international expert bodies, educating the public through popular publications and media interviews, and related activities, all contributing to enhancing the impact of academic research.

In the humble opinion of the author, this impact is positive and noticeable, thus bringing value to the state, the academy, and the Bulgarian society. Hence, the vital question is whether the current model is sustainable. And the answer is not entirely clear.

First, on the negative side:

- The academic ICT research units look to find, educate, and retain young people in a highly competitive labour market, in particular given the mobility of researchers, nationally and internationally;

- The unique case of researchers with practical and even policy-making experience moving from the defence sector to the academia, that occurred in the 1990s and the early 2000s, is unlikely to repeat itself;
- Usually, ministries and other governmental agencies seek support only when in need and do not have long-term research programmes to cultivate the necessary capacity and have guaranteed access to it.

There are positive signs as well:

- The funding for the EU security research programme is expected to increase in the next programming period, and academic research units may build on established international contacts and trust to expand their participation in Horizon Europe and the following research programmes of the European Union;
- The EU will launch a full-scale Defence research programme in the forthcoming programming period;
- Bulgarian companies will try to enter the EU defence capabilities development programme, financed through the multi-billion European Defence Fund, and might look for contributions from and continuous cooperation with the academia;
- The Ministry of Defence started several main rearmament projects, looking for Western suppliers, and the offset clauses of each contract will likely be open to cooperation with research institutes and transfer of know-how;
- The Ministry of Education and Science announced plans to launch a national research programme "Security and Defence";
- There is an informal but well-functioning collaboration between institutes from the Bulgarian Academy of Sciences and other Bulgarian research organisations.

This latter advantage is visible from the list of references to this chapter, where "Information & Security: An International Journal," https://isij.eu, acts as a platform for the exchange of ideas, studies on future concepts, needs and requirements in the defence and security field, on the one hand, and results in technology development on the other.

The series of annual international scientific conferences "Digital Transformation, Cyber Security and Resilience" (DIGILIENCE), https://digilience.org/, launched in 2019 as a cooperative endeavour of several Bulgarian organisations, provides another venue for national and international networking, exchange of knowledge, and discussion on needs and opportunities for cooperation.

This combination of factors provides both threats and opportunities, and success, in this case, will be measured primarily by the ability of the Bulgarian Academy of Sciences to attract and retain qualified researchers.

Curiously, the current competitive advantages of academic ICT research units, working in the field of security and defence, are not in creating specific technologies, nor in some niche competence, but in maintaining interdisciplinary and multidisciplinary competencies.

On that basis, the author considers viable in long-term a three-prong strategy: (1) specialise in niche capabilities, e.g. governance of cybersecurity organisations, the security of digital infrastructures or resilience to cyber threats; (2) collaborate with business partners to develop advanced security and defence solutions; and (3) maintain multidisciplinary competencies through the national network of research organisations. All three approaches should be open to international collaboration.

The author prefers the third one, but common sense dictates a broader portfolio of activities. The leadership of the Bulgarian Academy of Sciences should be open to provide arrangements that are sufficiently flexible to allow the implementation of all three approaches.

Acknowledgements The preparation of this paper is supported by the National Scientific Program "Information and Communication Technologies for a Single Digital Market in Science, Education and Security (ICT in SES)," financed by the Ministry of Education and Science of the Republic of Bulgaria.

References

1. Comecon [online]. https://en.wikipedia.org/wiki/Comecon. Accessed 25 Feb 2020
2. Boyanov, K.: Notes on the Development of Computing in Bulgaria. Prof. Marin Drinov Academic Publishing House, Sofia (2010)
3. Coward, C.M., Bialos, J.P.: The Bulgarian Defense Industry: Strategic Options for Transformation, Reorientation & NATO Integration. The Atlantic Council of the United States, Washington, DC (2001)
4. Semerdjiev, E., Semerdjiev, Tz.: Bulgarian MSDF R&D national program and activities. Inf. Secur. Int. J. **2**, 71–81 (1999). https://doi.org/10.11610/isij.0206
5. Angelova, D.S., Semerdjiev, Tz.A., Jilkov, V.P., Semerdjiev, E.A.: Application of a Monte Carlo method for tracking maneuvering target in clutter. Math. Comput. Simul. **55**(1–3), 15–23 (2001). https://doi.org/10.1016/S0378-4754(00)00242-1
6. Jilkov, V.P., Mihaylova, L.S., Rong-Li, X.: An alternative IMM filter for benchmark tracking problem. In: 1st International Conference on Multisource-Multisensor Information Fusion, FUSION' 98, Las Vegas, Nevada, July 1998, vol. 2, pp. 924–929 (1998)
7. Dezert, J., Tchamova, A., Han, D.: Total belief theorem and generalized Bayes' theorem. In: 21st International Conference on Information Fusion (FUSION), Cambridge, 10–13 July 2018, pp. 1040–1047. https://doi.org/10.23919/ICIF.2018.8455351
8. Dezert, J., Tchamova, A., Han, D. Wickramarathne, T.: A simplified formulation of generalized Bayes' theorem. In: 22nd International Conference on Information Fusion (FUSION), Ottawa, ON, Canada, 2–5 July 2019, pp. 1–8 (2019)
9. Dezert, J., Tchamova, A., Konstantinova, P., Blasch, E.: A comparative analysis of QADA-KF with JPDAF for multitarget tracking in clutter. In: 20th International Conference on Information Fusion (FUSION), Xi'an, China, 10–13 July 2017, pp. 1–8 (2017). https://doi.org/10.23919/ICIF.2017.8009736
10. Dezert, J., Tchamova, A., Konstantinova, P.: Multitarget tracking performance based on the quality assessment of data association. In: 19th International Conference on Information Fusion (FUSION), Heidelberg, Germany, 5–8 July 2016, pp. 201–208 (2016)
11. Niemeyer, K., Shalamanov, V., Tagarev, T., Tsachev, Ts., Rademaker, J.G.M.: Operations Research Support to Force and Operations Planning in the New Security Environment. Artgraf, Sofia (2008)
12. Tagarev, T.: Shaping the security environment in South-Eastern Europe: Bulgarian armed forces and national security policy. In: Krupnick, C., Lanham, Md. (eds.) Almost NATO: Partners and Players in Central and Eastern European Security, pp. 119–155. Rowman & Littlefield

13. Shalamanov, V., Tagarev, T.: Reengineering the Defense Planning in Bulgaria. Research Report # 9. Institute for Security and International Studies, Sofia (1998)

14. Totev, D., Roussanov, P.: Implementing the defense resource management model (DRMM) in the development of the new Bulgarian military doctrine and the plan for organizational structure and development of the MOD by the year 2004. In: Proceedings "Applications of Operations Analysis Techniques to Defense Issues", George-Marshall Center for European Security Studies, Garmish-Partenkirchen, Germany, 14–17 Mar 2000

15. Tagarev, T.: Introduction to program-based defense resource management. Connect. Q. J. **5**(1), 55–69. https://doi.org/10.11610/Connections.05.1.05

16. Young, T.-D.: Capabilities-based defense planning: techniques applicable to NATO and partnership for peace countries. Connect. Q. J. **5**(1), 35–54 (2006). https://doi.org/10.11610/Connections.05.1.04

17. Ratchev, V., et al.: Methodology and Scenarios for Defence Planning. Military Publishing House, Sofia (2007).(in Bulgarian)

18. Tagarev, T., Tsachev, Ts., Zhivkov, N.: Formalizing the optimization problem in long term capability planning. Inf. Secur. Int. J. **23**(1), 99–114 (2009). https://doi.org/10.11610/isij.2309

19. Tagarev, T., Mednikarov, B.: Planning of security sector capabilities for protection of maritime sovereignty. In: Kounchev, O., Willems, R., Shalamanov, V., Tsachev, Ts. (eds.) Scientific Support for the Decision Making in the Security Sector, pp. 72–86. IOS Press, Amsterdam (2007)

20. Tagarev, T., Stankov, G.: A gaming approach to enhancing defense resource allocation. Connect. Q. J. **8**(2), 7–15 (2009). https://doi.org/10.11610/Connections.08.2.02

21. Tagarev, T., Ratchev, V.: Bulgaria's Defence Policy and Force Development 2018. Military Publishing House, Sofia (2008).(in Bulgarian)

22. Ratchev, V.: Context scenarios in long-term defense planning. Inf. Secur. Int. J. **23**(1), 62–72 (2009). https://doi.org/10.11610/isij.2306

23. Tagarev, T., Ivanova, P.: Analytical support to foresighting EU roles as a global security actor. Inf. Secur. Int. J. **29**(1), 21–33 (2013). https://doi.org/10.11610/isij.2902

24. Ratchev, V., Nerlich, U., Tagarev, T.: Context scenarios and alternative future EU roles as a global security actor. IT4Sec Reports 100 (2014). https://doi.org/10.11610/it4sec.0100

25. Tagarev, T., Ratchev, V., Georgiev, V., Ivanova, P., Bizov, L.: Methodology for Planning Wartime Defence Capabilities. Centre for Security and Defence Management, Sofia (2012).(in Bulgarian)

26. Tagarev, T., Ivanova, P.: Classic, modern, and post-modern approaches to making security strategy. Int. J. Inf. Technol. Secur. **2**(1), 58–67 (2010)

27. Tagarev, T., et al.: Information system ISURO for programme-based planning of defence resources. Technical Report. Centre for National Security and Defence Research, Sofia (2002)

28. Dimov, A., Stankov, G., Tagarev, T.: Using architectural models to identify opportunities for improvement of acquisition management. Inf. Secur. Int. J. **23**(2), 188–203 (2009). https://doi.org/10.11610/isij.2315

29. Tagarev, T., Ivanova, P.: Developing an architecture for naval sovereignty operations center. Inf. Secur. Int. J. **16**, 29–38 (2005). https://doi.org/10.11610/isij.1604

30. Tagarev, T., Stankov, G., Bizov, L., Natchev, A.: A framework methodology to support the selection of a multipurpose fighter. Inf. Secur. Int. J. **21**, 82–91 (2007). https://doi.org/10.11610/isij.2107

31. Borsboom, M.J.M., et al.: NATO Governance and Delivery of Commonly Funded Capabilities: Improving Support to NATO Commanders. NATO HQ, Brussels [Online]. https://www.nato.int/nato_static_fl2014/assets/pdf/pdf_2017_04/20170518_170418-gse-report.pdf. Accessed 28 Feb 2020

32. Shalamanov, V., Hadjitodorov, S., et al.: Civil security: architectural approach in emergency management transformation. Inf. Secur. Int. J. **17**, 75–101 (2005). https://doi.org/10.11610/isij.1706

33. Sundelius, B.: A brief on embedded societal security. Inf. Secur. Int. J. **17**, 23–37 (2005). https://doi.org/10.11610/isij.1702

34. Tagarev, T., Georgiev, V., Ratchev, V.: A taxonomy of essential services. Radioelectron. Comput. Syst. **6**(58), 191–196 (2012)
35. Shalamanov, V., et al.: Architectural approach for planning communication and information systems for emergency management. Technical Report. Centre for National Security and Defence Research, Sofia (2004)
36. Tagarev, T., Papadopolous, G., Hagenlocher, M., Sliuzas, R., Ishiwatari, M., Gallego, E.: Integrating the disaster risk management cycle. In: Science for DRM 2020: Acting Today, Protecting Tomorrow. Luxembourg: Publications Office of the European Union (2020).
37. Shalamanov, V., Avramov, S.: Ruggedized commercial IT modules for field C2 posts in coalition operations and emergencies. Inf. Secur. Int. J. **16**, 93–103 (2005). https://doi.org/10.11610/isij.1608
38. Shalamanov, V.: Computer aided dispatch—a tool for effect-based multinational and inter-agency operations. Inf. Secur. Int. J. **22**, 123–137 (2007). https://doi.org/10.11610/isij.2211
39. Shalamanov, V., et al. (2008). *Security Research and Change Management in the Security Sector. The Bulgarian Example in the Period 1999–2008*. Sofia: Demetra (in Bulgarian).
40. Shalamanov, V., Minchev, Z.: Scenario generation and assessment framework solution in support of the comprehensive approach. In: RTO-SAS-081 Symposium "Analytical Support to Defence Transformation," Sofia, 26–28 April 2010, paper # 22 (2010)
41. Commission of the European Communities: Green paper on a European programme for critical infrastructure protection. Brussels (2005)
42. Tagarev, T., Pavlov, N.: Planning measures and capabilities for protection of critical infrastructures. Inf. Secur. Int. J. **22**, 38–48 (2007). https://doi.org/10.11610/isij.2205
43. Hadjitodorov, S., et al.: Methodology for assessing critical infrastructures at municipal level. In: 2nd National Emergency Management and Civil Protection Conference, Sofia, 9 Nov 2007, pp. 44–53 (2007) (in Bulgarian)
44. Nickolov, E.: Critical information infrastructure protection. Inf. Secur. Int. J. **17**, 105–119 (2005). https://doi.org/10.11610/isij.1707
45. Bontchev, V., Yosifova, V.: Analysis of the global attack landscape using data from a telnet honeypot. Inf. Secur. Int. J. **43**(2), 264–282 (2019). https://doi.org/10.11610/isij.4320
46. Polimirova, D., Nickolov, E.: Real-time system for assessing the information security of computer networks. In: Camenisch, J., Kisimov, V., Dubovitskaya, M. (eds.) Open Research Problems in Network Security, pp. 123–133. Springer, Berlin (2010). https://doi.org/10.1007/978-3-642-19228-9_11
47. Polimirova, D., Nickolov, E.: Analysis of malicious attacks accomplished in real and virtual environment. In: New Trends in Intelligent Technologies, pp. 53–60. Institute of Mathematics and Informatics, Sofia (2009)
48. Preneel, B., Dodunekov, S., Rijmen, V., Nikova, S.I.: Enhancing Cryptographic Primitives with Techniques from Error Correcting Codes. IOS Press, Amsterdam (2009)
49. Boyvalenkov, P., Landjev, I.: The Mathematical Aspects of Some Problems from Coding Theory. In the current volume (2021)
50. Balzarotti, D., Markatos, E., Minchev, Z., et al.: The Red Book: A Roadmap for Systems Security Research. SysSec Consortium (2013). https://www.red-book.eu/. Accessed 8 May 2020
51. Gaydarski, I., Minchev, Z.: Conceptual modeling of an information security system and its validation through DLP systems. In: 9th International Conference on Business Information Security (BISEC-2017), 18 Oct 2017, Belgrade (2017)
52. Georgiev, V.: Cybersecurity and Management. Avangard Prima, Sofia (2018).(in Bulgarian)
53. Minchev, Z.: Human factor role for cyber threats resilience. In: Multigenerational Online Behavior and Media Use: Concepts, Methodologies, Tools, and Applications, pp. 215–241. IGI Global, Hershey, PA (2019). https://doi.org/10.4018/978-1-5225-7909-0.ch013
54. Sharkov, G.: From cybersecurity to collaborative resiliency. In: 2016 ACM Workshop on Automated Decision Making for Active Cyber Defense (SafeConfig '16), Oct 2016, pp. 3–9. https://doi.org/10.1145/2994475.2994484

55. Minchev, Z., Gaydarski, I.: Cyber risks, threats & security measures associated with COVID-19. CSDM Views **37** (2020). https://doi.org/10.11610/views.0037
56. Tagarev, T., Polimirova, D.: Main considerations in elaborating organizational information security policies. In: Vassilev, Tz., Smrikarov, A. (eds.) ACM International Conference on Computer Systems and Technologies (CompSysTech'19), Ruse, Bulgaria, 21–22 June 2019, pp. 68–73 (2019). https://doi.org/10.1145/3345252.3345302
57. Polimirova, D., Shalamanov, V., Stoianov, N. (eds.): Cybersecurity and Opportunities for Implementation of Innovative Technologies in the Work of the State Administration in Bulgaria. Institute of Public Administration, Sofia (2018).(in Bulgarian)
58. Tagarev, T.: Towards the design of a collaborative cybersecurity networked organisation: identification and prioritisation of governance needs and objectives. Future Internet **12**(4), 62 (2020). https://doi.org/10.3390/fi12040062
59. Tagarev, T., Yanakiev, Y.: Business models of collaborative networked organisations: implications for cybersecurity collaboration. In: 11th IEEE International Conference on Dependable Systems, Services and Technologies, (DESSERT'2020), 14–18 May 2020, Kyiv, Ukraine (2020)
60. Shalamanov, V., Penchev, G.: Methodology for organizational design of cyber research networks. In: Tagarev, T., Atanassov, K., Kharchenko, V., Kacprzyk, J. (eds.) Digital Transformation, Cyber Security and Resilience of Moern Societies. Cham: Springer (2021), pp. 25–47
61. Tagarev, T., Stoianov, N., Sharkov, G.: Integrative approach to understand vulnerabilities and enhance the security of cyber-bio-cognitive-physical systems. In: Cruz, T., Simoes, P. (eds.) Proceedings of the 18th European Conference on Cyberwarfare and Security, (ECCWS19), Coimbra, Portugal, 4–5 July 2019, pp. 492–500
62. Tagarev, T., Sharkov, G.: Computationally intensive functions in designing and operating distributed cyber secure and resilient systems. In: Vassilev, Tz., Smrikarov, A. (eds.) International Conference on Computer Systems and Technologies (CompSysTech'19), Ruse, Bulgaria, 21–22 June 2019, pp. 8–17 (2019). https://doi.org/10.1145/3345252.3345255
63. Tagarev, T.: Understanding hybrid influence: emerging analysis frameworks. In: Tagarev, T., Atanassov, K., Kharchenko, V., Kacprzyk, J. (eds.) Digital Transformation, Cyber Security and Resilience of Modern Societies. Cham: Springer (2021), pp. 449–463.https://doi.org/10.1007/978-3-030-65722-2_29
64. Tagarev, T.: Hybrid warfare: emerging research topics. Inf. Secur. Int. J. **39**(3), 289–300 (2018). https://doi.org/10.11610/isij.3924
65. Tagarev, T.: Reflecting developments in hybrid warfare into defence policy. In: Iancu, N., Fortuna, A., Barna, C., Teodor, M. (eds.) Countering Hybrid Threats: Lessons Learned from Ukraine, pp. 27–33. IOS Press, Amsterdam. https://doi.org/10.3233/978-1-61499-651-4-27
66. Wither, J.K., Mullins, S. (eds.): Combating Transnational Terrorism. Procon, Sofia (2016)
67. Minchev, Z.: Human factor dual role in modern cyberspace social engineering. In: Ogun, M.N. (ed.) Terrorist Use of Cyberspace and Cyber Terrorism: New Challenges and Responses, pp. 116–128. IOS Press, Amsterdam (2015)
68. Minchev, Z.: Hybrid challenges to human factor in cyberspace. In: Minchev, Z., Bogdanoski, M. (eds.) Countering Terrorist Activities in Cyberspace, pp. 32–43. IOS Press, Amsterdam (2018)
69. Minchev, Z.: Foreseeing of hybrid security trends in the eastern partnership wider context. In: Ionescu, M. (ed.) Eastern Partnership: A Civilian Security Perspective, pp. 212–225. Military Publishing House, Bucharest (2016)

Informatics Approaches for Forest Fire Spread Prediction

Emilia Velizarova and Alexander Alexandrov

Abstract The informatics approaches for the forest fire spread prediction are developed based on the relative importance of the weather and/or the fuel accumulation, developing the "weather hypothesis" and the "fuel hypothesis". The purpose of the study was to analyse development of the informatics approaches for forest fire detection and spread prediction using intelligent techniques in Bulgaria. Three main models were developed and tested: Modelling of the forest fire spread, based on hexagonal cells with intuitionistic fuzzy estimations; Game Method for Modelling (GMM) of forest fire not and influenced by wind and 3-Dimensional Game Method for Modelling. The results show, that the hexagonal lattice presents spread of the fire in more realistic way. The so described 3-dimensional case of the GMM) will obtain application for modelling of different processes and in particularly applications of the GMM for modelling of forest dynamic and of forest fire. The intuitionistic fuzzy estimations for the areal of the fire will be given.

Keywords Forest fire · Game method for modelling

1 Introduction

Changes in climate have the potential to affect wildfire frequency, size and intensity, while more severe fire effects are expected due to longer fire season [11, 12]. Therefore, these possible future changes will increase the risk of wildfire-related hazards as economic losses well as and human casualties. Fire regimes in the Mediterranean region are influenced not only by climatic conditions, but also by human activities and land use changes [10]. In 2018 (last reported year), for territory of Bulgaria these

E. Velizarova (✉)
Ministry of Environment and Water, 22 Mariya Luiza Blvd, Sofia, Bulgaria
e-mail: velizars@abv.bg

A. Alexandrov
Department of Agricultural and Forestry Sciences, Forest Research Institute, Bulgarian Academy of Sciences, Kliment Ohridski Boul. 132, 1756 Sofia, Bulgaria
e-mail: forestin@bas.bg

cases represent 72%. The European Forest Fire Information System [10] also reports that about 1500 ha have been affected by forest fires in Bulgaria in 2018. For period 2009–2018 according to the EFFIS dataset, the total burned area in Bulgaria were 49 thousand ha. Historically, fire occurrence, expressed as the annual number of fires and total burned area, was found to correlate with the absolute air temperatures maximum [13, 22]. In recent years researches have been studying detection systems and fire prediction systems. The informatics approaches are developed based on the relative importance of weather and fuel on fire behaviour, creating the "weather hypothesis" and the "fuel hypothesis" [3, 7]. In this paper, the informatics approaches for forest fire detection and spread prediction using intelligent techniques in Bulgaria is reviewed.

2   Theoretical Conception of the Informatic Approaches

The Game Method for Modelling (GMM) was used for modelled development of the fire front. One of the modifications of the John Horton Conway's Game of Life is called – Game Method for Moddelling" (GMM) [4]. Conway's Game of Life was devised by John Horton Conway in 1970, and already 40 years it is an object of research, software implementations and modifications. The GMM notations are described by Atanassov [4].

The standard Conway's Game of Life (CGL, see, e.g., [8, 23]) has a "universe" which is an infinite two-dimensional orthogonal grid of square cells, each of which is in one of two possible states, alive or dead, or (as an equivalent definition) in the square there is an asterisk, or there is not. The first situation corresponds to the case when the cell is alive and the second corresponds to the case when the cell is dead.

In case of consideration of a set of symbols S and an n-dimensional simplex comprising of n-dimensional cubes (when $n = 2$, a two-dimensional grid of squares).

The main assumption of the approach is that material points (referred in brief as objects) can be found in some of the centres of the n-dimensional cells. The GMM-grid can be either finite or infinite. In the first case, for ith dimension of the grid there is natural number g_i that corresponds to the number of the sequential cells of the grid in the present dimension. Therefore, when the n-dimensional GMM-grid is finite, there is a vector $\langle g_1, g_2, \ldots.. g_n \rangle$ of the lengths of its sides. In the case of the finite grids, a set of rules A are as follows:

1. rules for the motion of the objects along the vertices of the simplex;
2. rules for the interactions among the objects, e.g., when they are collected in one cell.

Let the rules from the ith type be denoted as i-rules, where $i = 1, 2$.

When S $= \{*\}$, we obtain the standard CGL.

We can call an initial configuration every set of (ordered) $(n + 2)$-tuples with an initial component being the number of the object; the second, third, etc. until the $(n + 1)$-st its coordinates, and the $(n + 2)$-nd—its corresponding symbol from S.

We can call a final configuration the ordered set of $(n+2)$-tuples having the above form and being a result of the modifications that occurred during a certain number of applications of the rules from A over a (fixed) initial configuration.

The single application of a rule from A over a given configuration K is called an *elementary step* in the transformation of the model and is denoted by $A_1(K)$.

When we have some initial configuration, we obtain new configurations in a stepwise manner. We must determine some constructive criteria for stopping the process. For example, such a condition may be the following.

1. The rules of the GMM are applied over the initial configuration and its derivates for exactly n iterations.
2. A predefined configuration is obtained on the GMM-grid. For example, if we model a process of interaction between some objects, the process should stop when the grid contains only one of these objects.
3. A previous configuration is obtained on the GMM-grid, i.e., the process oscillates. This criterion is applicable for deterministic processes.

3 Application of the Informatic Approaches—Case Studies

1. Modelling of the forest fire spread based on hexagonal cells with intuitionistic fuzzy estimations

The existing models are based on square or hexagonal lattice. The model, represented from Sotirova et al. [21] is extended with intuitionistic fuzzy estimations for the areal of the fire. The theoretic description of the intuitionistic fuzziness is discussed by Atanassov [2, 5].

This type of models is based on a finite grid, having the form of a hexagonal lattice, with size 11×11 in which the development of forest fire processes is checked. The existing of the river in the field is marked by letter R, while territory of stones are marked by S and on the rest part of the field there is a homogeneous forest. The digits correspond to the wood mass per one square. These digits are specific for different types of forests, but here the forest is homogeneous and initially, the digits are only "9". After the beginning of the fire, the digits will decrease by specific rules. In the model, the fire will develop in concentric circles (Fig. 1).

The rules for the GMM are the following.

1. $R \to R$;
2. $S \to S$;
3. $0 \to 0$;
4. In the initial time-step, the fire starts from a fixed cell containing digit "9" that represents the density of the trees in that cell of the forest. On the second time-step, for the same cell $9 \to 8$. On the third time-step, for the same cell $8 \to 7$. In the same moment, all neighbouring cells of the cell with the fire change their digits with the previous digit.

Fig. 1 Model for fire spread
based on a finite grid, having
the form of a hexagonal
lattice

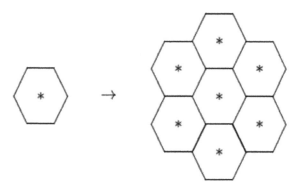

5.   On the next time-steps, for the cells with fire

$$n \rightarrow \begin{cases} 0, & if \; n = 0 \\ n - 1, & in \; n > 1 \end{cases}$$

6.   The process continues until all cells in the region contain only digit 0. In the
     opposite case, go to 5.

The propagation of a forest fire is shown within iterations on Fig. 2.
The intuitionistic fuzzy estimations for the areal of the fire were determined by
the following formulas:

$$\mu(i) = \frac{\text{number of cells with symbol } "0"}{\text{number of all cells}}$$

$$v(i) = \frac{\text{number of cells in which there is not fire}}{\text{number of all cells}}$$

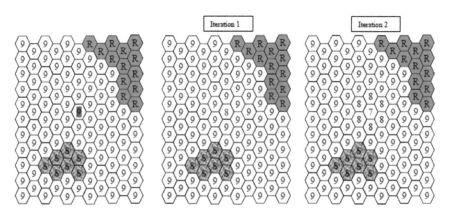

Fig. 2 Possible iterations for forest fire propagation modelling

$$\pi(i) = \frac{\text{number of cells in which there is fire}}{\text{number of all cells}}$$

The i value, the number of the iteration and the number of all cells could be determined for each case. The values of functions $\mu(i)$; $v(i)$ and $\pi(i)$ also could be calculated based on previous parameters of the model.

2.   GMM-model of forest fire, influenced by wind

Wind existence during forest fire events is a common weather component, which influence the fire spread direction and fires and burn severity.

For modelling purpose, the direction of wind was assumed to be parallel to the horizontal lines (Fig. 3A), with slope of 45° (Fig. 3B) and vertical direction (Fig. 3C).

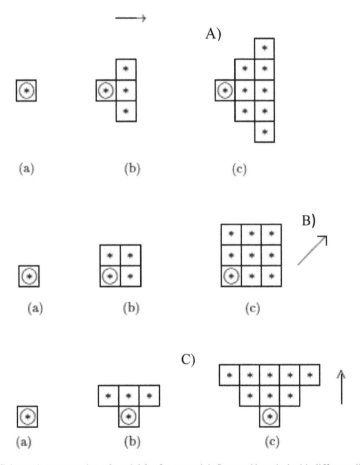

Fig. 3 Schematic presentation of model for fire spread, influenced by wind with different directions

At low wind speed (low angle terrain and gentle wind speed) the back and flank rates decrease and the elliptical shape of fire spread is transformed to the more rounded shape. For better adaption of the model a 3-dimensional grid was suggested.

3.   3-Dimensional Game Method for Modelling

The GMM is described in general, i.e., in n-dimensional case, but in practice, all real models [1, 3, 9, 14–18, 21] are related to the case, when n = 2, i.e., to the 2-dimensional (plane) case. However, following Atanassov [2, 5] and possibility of the cells with different forms to be used, a 3-dimensional case of model was developed [6].

In case of 3-dimensional models, a set of symbols **S** and **n**-dimensional cubes (when n = 3, a 3-dimensional grid of cubes) or other types of simplexes, which could be called *cells* have been chosen.

The assumption that material points (referred in brief as objects) can be found in some of the centres of the n-dimensional cells. The GMM-grid can be either finite or infinite. In the first case, for **i**-th dimension of the grid there is natural number of g_i that corresponds to the number of the sequential cells of the grid in the present dimension. Therefore, when the n-dimensional GMM-grid is finite, there is a vector $\{g_1, g_2,....g_n\}$ of the lengths of its sides. In general, finite grids are used. A set of rules **A** are as follows:

(1)   rules for the motion of the objects along the vertices of the simplex;
(2)   rules for the interactions among the objects, e.g., when they are collected in one cell.

In case of the initial configuration type $(n + 2)$-tuples with an initial component being the number of the object, the new configuration will be obtained in a stepwise manner. For stopping the model, constructive criteria should be developed, as follows for example.

1.   The rules of the GMM are applied over the initial configuration and its derivates for exactly s iterations, where the natural number s is fixed in advance.
2.   A preferred configuration is obtained on the GMM-grid. The process should stop when the grid contains only one of these material points (referred as objects).
3.   This is criterion is applicable also for deterministic processes.

In the 3-dimensional case of the GMM, the cells can have the forms from Fig. 4. They are differed from that in the 2-dimensional case of the GMM, where, the cells can have forms of an equilateral triangle, square or an equilateral hexagon.

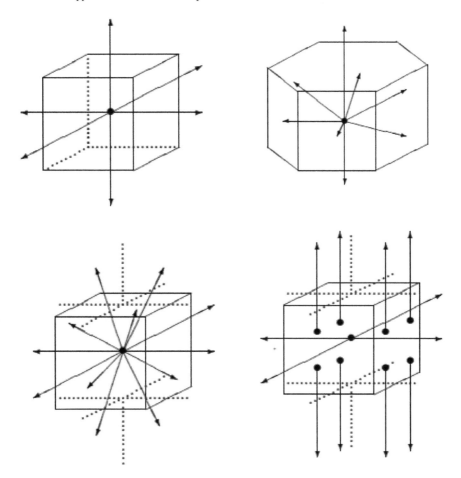

Fig. 4 Forms of the 3-dimensional GMM cells

4 Conclusion

The hexagonal lattice presents spread of the fire in more realistic way. The intuitionistic fuzzy estimations for the areal of the fire will be given. The so described 3-dimensional case of GMM will obtain application for modelling of different processes and in particularly applications of the GMM for modelling of forest dynamic and of forest fire.

References

1. Agee, J.K.: The severe weather wildfire – too hot to handle? Northwest Sci. **71**, 153–156 (1997)

2. Atanassov, K.: On Intuitionistic Fuzzy Sets Theory. Springer, Berlin (2012)
3. Atanassov, K.: On a combinatorial game-method for modelling. Adv. Model. Anal. (AMSE Press) **19**(2), 41–47 (1994)
4. Atanassov, K.: Game Method for Modelling, "Prof. M. Drinov". Academic Publishing House, Sofia (2011)
5. Atanassov, K.: On game-method for modelling. Adv. Stud. Contemp. Math. **22**(2), 189–207 (2012)
6. Atanassova, L., Atanassov, K., Angelova, N.: Short remark on 3-dimensional game method for modelling. In: Uncertainty and Imprecision in Decision Making and Decision Support: New Challenges, Solutions and Perspectives—Selected Papers from BOS-2018, held on September 24–26, 2018, and IWIFSGN-2018, held on September 27–28, 2018 in Warsaw, Poland. Springer (in press) (2020)
7. Bessie, W.C., Johnson, E.A.: The relative importance of fuels and weather on fire behavior in subalpine forests. Ecology **76**, 747–762 (1995)
8. Deutsch, A., Dormann, S.: Cellular Automaton Modeling Biological Pattern Formation. Birkhauser, Boston (2005)
9. Dobrinkova, N., Fidanova, S., Atanassov, K.: Game-method model for field fires. Lect. Notes Comput. Sci. **5910**(2010), 173–179 (2010)
10. EFFIS: Forest Fires in Europe, Middle East and North Africa 2018 (2018). https://effis.jrc. ec.europa.eu/media/cms_page_media/40/Annual_Report_2018_final_pdf_06.11.2019_n208 KFB.pdf
11. Intergovernmental Panel on Climate Change (IPCC): Climate change 2007: impacts, adaptation and vulnerability. In: Solomon, S., Qin, D., Manning, M., Chen, Z., Marquis, M., Averyt, K.B., Tignor, M., Miller, H.L. (eds.) Contribution of Working Group II to the Intergovernmental Panel on Climate Change, Fourth Assessment Report, pp. 1–22. Cambridge University Press, Cambridge, UK; New York, NY, USA (2007)
12. Kalabokidis, K., Palaiologou, P., Gerasopoulos, E.,, Giannakopoulos, Ch., Kostopoulou, E., Zerefos, Ch.: Forests **6**(6), 2214–2240 (2015). https://doi.org/10.3390/f6062214
13. Koutsias, N., Xanthopoulos, G., Founda, D., Xystrakis, F., Nioti, F., Pleniou, M., Mallinis, G., Arianoutsou, M.: On the relationships between forest fires and weather conditions in Greece from long-term national observations (1894–2010). Int. J. Wildland Fire **2012**(22), 493–507 (2012)
14. Mahdipour, E., Dadkhah, C.: Automatic fire detection based on soft computing techniques: review from 2000 to 2010. Artif. Intell. Rev. **42**(4), 895–934 (2014)
15. Sotirova, E., Atanassov, K., Fidanova, S., Velizarova, E., Vassilev, P., Shannon, A.: Application of the game method for modelling the forest fire perimeter expansion. Part 1: A model fire intensity without effect of wind. In: IFAC Workshop on Dynamics and Control in Agriculture and Food Processing, Plovdiv, 13–16 June 2012, pp. 159–163 (2012)
16. Sotirova, E., Atanassov, K., Fidanova, S., Velizarova, E., Vassilev, P., Shannon, A.: Application of the game method for modelling the forest fire perimeter expansion. Part 2: A model fire intensity with effect of wind. In: IFAC Workshop on Dynamics and Control in Agriculture and Food Processing, Plovdiv, 13–16 June 2012, pp. 165–169 (2012)
17. Sotirova, E., Atanassov, K., Fidanova, S., Velizarova, E., Vassilev, P., Shannon, A.: Application of the game method for modelling the forest fire perimeter expansion. Part 3: A model of the forest fire speed propagation in different homogenous vegetation types. In: IFAC Workshop on Dynamics and Control in Agriculture and Food Processing, Plovdiv, 13–16 June 2012, pp. 171–174 (2012)
18. Sotirova, E., Dimitrov, D., Atanassov, K.: On some application of game-method for modelling. Part. 1: Forest dynamics. In: Proceedings of the Jangjeon Mathematical Society, vol. 15, No. 2, pp. 115–123 (2012)
19. Sotirova, E.N., Bureva, V., Velizarova, E., Fidanova, S., Marinov, P., Shannon, A., Atanassov, K.: Modelling forest fire spread through a game method for modelling based on hexagonal cells with intuitionistic fuzzy estimations. Notes Intuit. Fuzzy Sets **19**(3), 73–80 (2013)

20. Sotirova, E., Velizarova, E., Fidanova, S., Atanassov, K.: Modeling forest fire spread through a game method for modeling based on hexagonal cells. In: Large Scale Scientific Computing. Lecture Notes in Computer Science, vol. 8353, pp. 321–328. Springer (2014)
21. Sotirova, E., Sotirov, S., Atanassova, L., Atanassov, K., Bureva, V., Doukovska, L.: Game method for modelling with intuitionistic fuzzy rules. In: Atanassov, K., Kacprzyk, J., Kałuszko, A., Krawczak, M., Owsiński, J., Sotirov, S., Sotirova, E., Zadrożny, S. (eds.) Uncertainty and Imprecision in Decision Making and Decision Support: Cross-Fertilization, New Models, and Applications. Advances in Intelligent Systems and Computing, vol. 559, pp. 153–168. Springer, Cham (2018). https://doi.org/10.1007/978-3-319-65545-1_15
22. Velizarova, E., Nedkov, R., Avetisyan, D., Radeva, K., Stoyanov, A., Georgiev, N., Gigova, I.: Application of remote sensing techniques for monitoring of the climatic parameters in forest fire vulnerable regions in Bulgaria. In: Seventh International Conference on Remote Sensing and Geoinformation of the Environment (RSCy2019) (2019). https://doi.org/10.1117/12.253 3656
23. Velten, K.: Mathematical Modeling and Simulation. Wiley-VCH Verlag GmbH & Co. KgaA, Weinheim (2009)